SEED TREATMENT: PROGRESS AND PROSPECTS

BCPC Monograph No 57

SESSION 6:
BIOLOGICAL SEED TREATMENTS

Mechanisms of protection of seed and seedlings by biological seed
treatments: implications for practical disease control
G E HARMAN and E B NELSON 283

Bacterization to protect seed and rhizosphere against disease
B J J LUGTENBERG, L A DE WEGER and B SCHIPPERS 293

Biological seed treatments - the development process
D J RHODES and K A POWELL 303

The seed industry's view on biological seed treatments
R J SCHEFFER 311

Rhizosphere constraints affecting biocontrol organisms
applied to seeds
J W DEACON 315

POSTERS:

Incorporation of microencapsulated beneficial organisms
into environmentally acceptable seed coatings to
enhance crop performance in soyabeans
J S BURRIS and Y CHEN 327

Tests for biological control of seed and seedling damping-off
diseases of peas and beans using *Bacillus* species
R WALKER, A A POWELL and B SEDDON 333

Spore movement of *Mucor hiemalis* in the rhizosphere
of groundnut in natural field conditions
U KRAUSS 339

Biocontrol of seed-borne *Botrytis allii* using
an antagonistic bacterium
L PEACH, R B MAUDE and G M PETCH 345

SESSION 7:
APPLICATION OF SEED TREATMENTS, COATINGS, PELLETING AND OTHER TECHNIQUES

Large-scale seed priming techniques and their integration
with crop protection treatments
D GRAY 353

The development of quality seed treatments in commercial
practice - objectives and achievements
P HALMER 363

The influence of seed treatment uniformity on the biological
performance of chlorfenvinphos and chlorpyrifos
A A JUKES, D L SUETT, K PHELPS, S SIME and S J COOKE 375

The life cycle of a seed treatment formulation development
L R FRANK 381

Temperature effects on osmotic priming of leek seeds
W BUJALSKI, A W NIENOW, R B MAUDE and D GRAY 385

POSTERS:

The physiological advancement of sugar-beet seeds
T H THOMAS, K W JAGGARD, M J DURRANT and S J MASH ... 391

The ethirimol content of commercially treated cereal seeds
D L SUETT, P J HEWETT, A A JUKES, L MORGAN and V FOX ... 397

The effect of pellet weight on the distribution of
imidacloprid applied to sugar-beet pellets
F WESTWOOD, K M BEAN and A M DEWAR ... 403

The coating of carrots with chlorfenvinphos
H WEBER and O FUSS ... 409

Interactions between hymexazol, furathiocarb and some
clay materials used for seed treatment
L PUSSEMIER, Ph DEBONGNIE and Y VAN ELSEN ... 413

A small scale laboratory fluidized bed seed-coating apparatus
J S BURRIS, L M PRIJIC and Y CHEN ... 419

SESSION 8:
REGULATORY REQUIREMENTS

Seeds standards in legislation: assessing seed quality
and effect of seed treatment
J H B TONKIN ... 427

Efficacy and physical/mechanical data requirements for
approval of seed treatments in the UK
D D SLAWSON and M J GILLESPIE ... 441

Research based improvements in the regulation of hazards to
wildlife from pesticide seed treatments
A D M HART and M A CLOOK ... 449

Operator exposure during seed treatment
H CHAMBERS ... 455

Regulatory controls on fungicidal cereal seed treatments - past
experience and future prospects
H EHLE ... 461

POSTERS:

The number of exposed dressed seeds in the field;
an outline for field research
W L M TAMIS, M GORREE, J DE LEEUW,
G R DE SNOO and R LUTTIK ... 471

Environmental hazard assessment of the use of pesticides
for seed treatment: the Dutch concept
R LUTTIK and G R DE SNOO ... 477

Preface

The need to protect seeds and seedlings was one of the earliest imperatives known to agrarian man. Palladius, a Roman writer on agricultural matters, records that 'to obtain vigorous seedlings they should first be nurtured in a small plot surrounded by a hyena's skin'; another ancient recommendation was to treat all seed with the juice of the house leek. Even in the Middle Ages there was little or no comprehension of the causes of crop damage and it was not until the 17th and 18th centuries that phytopathology, especially for seed-borne and soil-borne plant diseases, was placed on a more sound scientific basis by such scientists as Tillet, Prévost, de Bary and Hiltner. At that time, of course, the search for suitable seed treatment compounds could only be made by testing existing chemicals. Among the most effective were the copper salts and formalin. With most of these early chemical methods seed rates had to be increased by up to 20% to compensate for phytotoxic effects. In the early part of this century the remarkable effectiveness of organo-mercurial compounds was recognised and mercury remained the mainstay of cereal protection (also used in other crops) for many years, and in the UK until 1992.

In January 1988 a BCPC Symposium entitled *'Application to Seeds and Soil'* was held at the University of Surrey. This was largely concerned with seed treatments. Six years will have elapsed and the BCPC and SCI considered that the seed treatment topic deserved another forum. Research and development has continued strongly during the six year period, the chemists have made more progress in synthesis and the withdrawal of mercurial products in the UK and elsewhere increased the search momentum.

Seed treatments are generally viewed favourably by those concerned with the environmental impact of modern farming techniques. For similar reasons there has been renewed interest in integrated crop protection programmes and considerable research effort is being expended into the control of pests and pathogens by treating seeds with other 'beneficial' organisms. The various parameters used to measure the quality of application including the quantity of seed treatment retained by seeds, the magnitude of between seed variation, the effects of formulation and the influence of coating and pelleting on some of these factors have also received considerable attention.

There is little doubt that there have been remarkable advances in recent times: notably the discovery of systemic fungicides and their development as seed treatments and the more sophisticated machinery used for accurate dosing of seeds. Whilst the chemists produce more refined crop protection agents and the biologists discover useful biological seed treatments, the Regulatory Authorities need to keep a watchful eye on the diverse registration criteria and data requirements to be supplied by applicants before allowing a product to be sold and recommended in commercial practice.

The interest in the whole science and technology of seed treatments, their application and use, is very evident in this monograph, and the depth and breadth of the seed treatment theme emphasise the relevance of a dedicated symposium.

T J MARTIN

Acknowledgements

The organisation of a Symposium of this nature requires much effort from many individuals. Thanks are due to those who attended the first advisory meeting and for their constructive assistance in formulating the structure of the programme. Special credit must, of course, be given to the Programme Committee, all members of which have done battle with deadlines and in the case of Session Organisers also risked unpopularity with authors. Derek Soper was particularly helpful in administrative matters and as Secretary to the Committee.

I am grateful to the BCPC Programme Policy Committee for choosing the theme and promoting the idea of the Symposium and to the SCI for encouragement and support. The AAB kindly helped to publicise the event.

Financial assistance with speakers' travel expenses was given by the following organisations:

> Bayer plc
> British Sugar plc
> Ciba Agriculture
> Elsoms Seeds Limited
> Germain's (UK) Limited
> Uniroyal Chemical Limited
> Zeneca Crop Protection

Ciba-Geigy AG, Crop Protection Division, Basle sponsored the drinks reception. Bayer plc provided the symposium brief-cases.

Finally I am grateful for the help and support of Sherrie Simpson and Lisa Davies of Conference Associates.

T J MARTIN

Symposium Programme Committee

Chairman:	T J Martin	Bayer plc, Crop Protection Business Group, Eastern Way, Bury St Edmunds, Suffolk.
Deputy Chairman & Secretary	D W G Soper	Consultant, Grooms Cottage, Great Easton, Dunmow, Essex.

Session Organisers:

Session 1.	T J Martin	Bayer plc, Crop Protection Business Group, Eastern Way, Bury St Edmunds, Suffolk.
Session 2.	W J Rennie	Official Seed Testing Station for Scotland, SASA, Edinburgh.
Session 3.	Dr R A Noon	Zeneca Crop Protection, Fernhurst, Haslemere, Surrey.
	J N Oakley	ADAS Reading, Coley Park, Reading, Berkshire.
Session 4.	Dr M J C Asher Dr A M Dewar	Institute of Arable Crops Research, Broom's Barn Experimental Station, Bury St Edmunds, Suffolk.
Session 5.	Dr G A Hide	Institute of Arable Crops Research, Rothamsted Experimental Station, Harpenden, Hertfordshire.
Session 6.	Dr J W Henfling	DLO-Research Institute for Plant Protection, Wageningen, The Netherlands.
	Dr R B Maude	Horticulture Research International, Wellesbourne, Warwickshire.
Session 7.	D L Suett	Horticulture Research International, Wellesbourne, Warwickshire.
Session 8.	Dr P J Brain	Bayer plc, Crop Protection Business Group, Eastern Way, Bury St Edmunds, Suffolk.
Exhibition Manager:	G Beaumont	Linden House, Old Stowmarket Road, Woolpit, Bury St Edmunds, Suffolk.
Posters Organiser:	G C Scott	Schering Agriculture, Chesterford Park, Saffron Walden, Essex.
Public Relations Officer:	R Pierce	Chiltern View Farmhouse, Kingston Stert, Chinnor, Oxford.

Chairmen of Sessions

Session 1.	T J Martin	Bayer plc, Crop Protection Business Group, Eastern Way, Bury St Edmunds, Suffolk.
Session 2.	R P Davis	MAFF Pesticides Safety Directorate, Rothamsted Experimental Station, Harpenden, Hertfordshire.
Session 3.	Dr R A Noon	Zeneca Crop Protection, Fernhurst, Haslemere, Surrey.
Session 4.	Dr A M Dewar	Institute of Arable Crops Research, Broom's Barn Experimental Station, Bury St Edmunds, Suffolk.
Session 5.	Dr G A Hide	Institute of Arable Crops Research, Rothamsted Experimental Station, Harpenden, Hertfordshire.
Session 6.	Dr R B Maude	Horticulture Research International, Wellesbourne, Warwickshire.
Session 7.	D L Suett	Horticulture Research International, Wellesbourne, Warwickshire.
Session 8.	Dr P J Brain	Bayer plc, Crop Protection Business Group, Bury St Edmunds, Suffolk.

Exhibitors

BAYER PLC
Crop Protection Business Group, Eastern Way, Bury St Edmunds, Suffolk.
On-planter applicators for powder and liquid treatments with pencycuron formulations in potatoes (Monceren DS & Monceren flowable).

DOWELANCO LTD
Latchmore Court, Brand Street, Hitchin, Hertfordshire.
Latest seed treatment developments and information on product range; also details of seed treatment machines.

ELSOMS SEEDS LTD
Pinchbeck Road, Spalding, Lincolnshire.
"Drum Priming - Elsoms Seeds Limited", a technique developed by HRI and licensed by the BTG.

GERMAIN'S (UK) LTD
Hansa Road, Hardwick Industrial Estate, King's Lynn, Norfolk.
"Seed enhancement technology - Germain's UK Limited". Pelleting, coating and efficient application of seed treatments.

PHOSYN PLC
Manor Place, Wellington Road, The Airfield, Pocklington, York.
Provision of essential micro-nutrients as seed coatings. A technical break through which provides plant nutrients at crucial stages of crop development.

RHÔNE-POULENC AGRICULTURE LIMITED
Fyfield Road, Ongar, Essex
"Helping seed succeed". Application of cereal and potato seed treatments in the UK. Includes equipment and data on a range of available products including Panoctine, Panoctine Plus, Rovral and Fungaryl.

RHÔNE-POULENC CERES
14 rue de la Pierre-Follège, Z.1. - 91669 Mereville, France.
Pelleting and film-coating processes, featuring Ceremar 91, Ceretop, and Cerpro. Also coating formulation - Peridiam.

SEPPIC
76, Quai D'Orsay, 75321 Paris, Cedex 07
Manufactures and supplies seed coating products and adjuvants for agricultural formulations.

SILSOE RESEARCH INSTITUTE
Wrest Park, Silsoe, Bedford.
Development of seed treatment equipment for on-farm use in Africa. Equipment on display together with results from project work done to support the design study.

UNIROYAL CHEMICAL LTD
Brooklands Farm, Cheltenham Road, Evesham, Worcestershire.
Seed treatment and service from Uniroyal Chemical and Gustafson.

WISEPRESS LTD
The Old Church Hall, 89A, Quicks Road, London SW19 1EX
Display of a range of books and journals relevant to the symposium theme.

ZENECA CROP PROTECTION
Fernhurst, Haslemere, Surrey
"A world of seed treatments". Comprehensive product range for arable crops; also application equipment for powered and non-powered use; returnable containers for better storage and handling of seed treatments.

Abbreviations

acid equivalent	a.e.		nuclear magnetic resonance	nmr
active ingredient	AI		number average diameter	n.a.d.
boiling point	b.p.		number median diameter	n.m.d.
British Standards Institution	BSI		organic matter	o.m.
centimetre(s)	cm		page	p.
concentration x time product	ct		pages	pp.
concentration required to kill			parts per million by volume	mg/l
50% of test organisms	LC50		parts per million by weight	mg/kg
correlation coefficient	r		pascal	Pa
cultivar	cv.		percentage	%
cultivars	cvs.		post-emergence	post-em.
day(s)	d		power take off	p.t.o.
days after treatment	DAT		pre-emergence	pre-em.
degrees Celsius (centigrade)	°C		probability (statistical)	P
dose required to kill 50% of			relative humidity	r.h.
test organisums	LD50		revolutions per minute	rev./min
dry matter	d.m.		second (time unit)	s
Edition	Edn		standard error	SE
Editor	Ed		standard error of means	SEM
Editors	Eds		soluble powder	SP
emulsifiable concentrate	EC		species (singular)	sp.
freezing point	f.p.		species (plural)	spp.
gas chromatography-mass			square metre	m^2
spectrometry	gcms		subspecies	ssp.
gas-liquid chromatography	glc		surface mean diameter	s.m.d.
gram(s)	g		suspension concentrate	SC
growth stage	GS		temperature	temp.
hectare(s)	ha		thin-layer chromatography	tlc
high performance (or pressure)			tonne(s)	t
liquid chromatography	hplc		ultraviolet	u.v.
hour	h		vapour pressure	v.p.
infrared	i.r.		variety (wild plant use)	var.
International Standardisation			volume	V
Organisation	ISO		weight	W
Kelvin	K		weight by volume	W/V
kilogram(s)	kg		(mass by volume is more correct)	(m/V)
least significant difference	LSD		weight by weight	W/W
litre(s)	Litre		(mass by mass is more correct)	(m/m)
litres per hectare	l/ha		wettable powder	WP
mass	m			
mass per mass	m/m		approximately	c.
mass per volume	m/V		less than	<
mass spectrometry	m.s.		more than	>
maximum	max.		not less than	<
melting point	m.p.		not more than	>
metre(s)	m		Multiplying symbols-	Prefixes
milligram(s)	mg		mega	(x 10^6)	M
millilitre(s)	ml		kilo	(x 10^3)	k
millimetre(s)	mm		milli	(x 10^{-3})	m
minimum	min.		micro	(x 10^{-6})	μ
minute (time unit)	min		nano	(x 10^{-9})	n
molar concentration	M		pico	(x 10^{-12})	p

Session 1
Symposium Opening

Chairman and
Session Organiser   T J MARTIN

SEED TREATMENT - A PANACEA FOR PLANT PROTECTION?

F.J. SCHWINN

Institute of Botany, University, Basle, Switzerland

ABSTRACT

Milestones in seed dressing development, main products and uses, characteristics of the market (by products, crops, regions) and untapped potentials/chances are described.

INTRODUCTION

Before discussing the question raised in the title a brief overview will be given highlighting the past development, present status and some market aspects of seed treatment.   At the same time, this general introduction is intended to set the scene for the following papers dealing with concrete contributions to this topic.

Worldwide, around 90% of all food crops are propagated by seed.   The seed, besides being the carrier of the genetic potential of the plant developing from it, is an involuntary carrier of propagation units of pathogens and pests and an excellent food basis for the structures emerging from them.   The seedling/plantlet is the most vulnerable stage of plant development, susceptible to adverse growing conditions in general and specifically to parasitic organisms.   Thus, seed treatment, that is, the application of bioactive chemicals, or antagonistic or symbiotic micro-organisms to the seed prior to sowing may aim at:

- the protection against seed- and soil-borne pathogens and animal pests (Hewett and Griffiths, 1986)
- the protection of the seedling from damage by herbicides by safening agents (Hatzious and Hoagland, 1989)
- physical treatments (heat) to control deep-seated seed-borne pathogens (Appel and Gassner, 1907) including viruses
- the stimulation of germination and/or enhancement of growth during the seedling stage by application of nutrients
- the speedy establishment of beneficial micro-organisms on the roots, for fixing nitrogen, enhancing nutrient uptake (VAM), or stimulating growth of seedlings (bacteria)
- the facilitation of sowing by standardising the shape of seeds by pelleting techniques.

The many-fold advantages and benefits of seed treatment as a major crop protection measure in comparison to soil application or foliar sprays have been described by several authors (Graham-Bryce, 1973; Powell and Matthews, 1988; Suett, 1988; Taylor and Harman, 1990), therefore, they will not be repeated here. However, by briefly describing the history and main characteristics of seed treatment they will become evident.

SEED TREATMENT CHRONICLE

The history of chemical seed treatment, the oldest practice in plant protection, has been documented in a range of reviews (Keitt, 1959; Frohberger, 1969; Neergard, 1977; Ainsworth, 1981; Jeffs, 1986). Major milestones in this development are summarised in Table 1.

TABLE 1. Selected milestones in the development of chemical seed treatments (compiled from Ainsworth, 1981; Frohberger, 1969; Jeffs, 1986; Neergard, 1977).

Since 1755	First proof of contagious nature of wheat bunt based on field trials (Tillet, 1755) and of activity of copper sulfate against it (Schulthess, 1761; Tesier, 1779).
Since 1800	Use of various inorganic salts against bunt and smuts.
Since 1807	Increasing use of copper sulphate against cereal smuts/ bunts, based on scientific work by Prévost, 1807, and Kühn, 1873.
1895-1897	Formalin against smuts/bunt (Geuther, 1895; Bolley, 1897).
Since 1914	Organic mercury compounds against bunt, snow mould etc. become dominating based on pioneering work of Wesenberg (1938) and Riehm (1914).
1942	Thiram (Tisdale and Flenner), broad spectrum fungicide.
1945	γ-hexachlorocyclohexane (γ-HCH) (Slade, 1945), first organic insecticidal seed treatment.
1945	Introduction of hexachlorobenzene (Yersin *et al*) against cereal bunt.
1952	Captan as a broad spectrum fungicide (Kittleson).
1962	Methiocarb (Unterstenhöfer), insecticide/bird repellent.
1965	Carbofuran (McEwen and Davis), broad spectrum systemic insecticide.
1966	Carboxin, first systemic fungicide against cereal smuts (von Schmeling and Kulka).
1968	Guazatine, broad spectrum fungicide (Jackson *et al*, 1973).
1969	Ethirimol, first systemic fungicide against powdery mildew by seed treatment (Bebbington *et al*).
1973	Imazalil against seed-borne *Helminthosporium* spp (Bartlett and Ballard, 1973).
1977	Metalaxyl, first systemic fungicide against downy mildews (Schwinn *et al*).
1978	Triadimenol, broad spectrum systemic fungicide, including control of powdery mildew by seed treatment (Frohberger).
1981	Furathiocarb, systemic insecticide for maize, sugar beet, oilseed rape, vegetables (Bachmann and Drabek).
1981	Pencycuron, residual fungicide against *Rhizoctonia* on potatoes (Frohberger and Grossmann).
1982-1992	Ban of mercury-based products in EC countries.
1986	Tefluthrin, insecticide for sugar beet and maize (Jutsum *et al*).
1988	Fenpiclonil, broad spectrum residual fungicide (Koch and Leadbeater, 1992).
1990	Imidacloprid, broad spectrum insecticide (Elbert *et al*).

From the beginning, it focused on small grain cereals as the most important staple crop in Europe which had to be protected from attack by fungal pathogens. Without covering the centuries since Roman times during which cereal seed was treated with products of questionable performance against unspecified problems, Table 1 shows the stepwise development from treatments with inorganic salts against bunt since the 19th century, to formalin as the first organic compound and the first to control smut (replacing hot water treatment), to the organo-mercuries, which dominated the scene from the 1920s to the 1980s, the introduction of Gamma-BHC as the first insecticidal treatment in the late 1940s and finally to the introduction of novel residual and apoplastically systemic fungicides (the latter against deep-seated pathogens) and to a lesser extent, insecticides, since the 1960s. The systemic fungicides (for literature see Davidse and de Waard, 1984) opened the door to a novel approach to disease control, that is, the control of foliar and systemic pathogens by seed treatment. This is a highly attractive approach particularly from the viewpoint of efficacy, targeted application and environmental aspects. Ethirimol and triadimenol against cereal powdery mildew and metalaxyl against systemic downy mildews were the pioneers in this field. However, there are three limiting factors to this approach:

1.   The risk of resistance development. We know, from broad experience, that the duration of exposure of a pathogen population to selection pressure by a fungicide, is a major risk factor for resistance-prone chemicals such as the single-site systemic fungicides. Despite cases of resistance to residual seed-applied fungicides, starting with hexachlorobenzene/bunt (Kuiper, 1965) and leading on to mercury compounds/*Pyrenophora graminicola* (Jones *et al*, 1989) and benzimidazoles *Gerlachia nivalis* (Hartke and Buchenauer, 1981; Locke *et al*, 1987), respectively, the risk has become much more serious with the introduction of systemic compounds. Fortunately, in contrast to foliar diseases, resistance in seed- and soil-borne pathogens still is comparatively rare. In order to stay on the safe side, industry and advisory services have to find the balance between the farmer's interest in long-lasting control and their responsibility for avoiding resistance and offering a practicable compromise to the user. With regard to the control of foliar pathogens by seed treatment, the solution could mean a use restriction to spring cereals and/or winter cereals in mild climates without an extended dormant phase of the crop.

2.   The persistence of the fungicide, which on the one hand should be limited in order to avoid residues in the edible parts of the crop; on the other hand, longevity of the molecule expands the period of protection, making the product more attractive to the farmer.

3.   Crop tolerance, that is, inhibition/retardation of germination and/or seedling growth which may be an inherent feature of such compounds.

ANALYSIS OF SEED TREATMENT USE

Seed treatment started (Table 1) in small grain cereals in temperate zones for their protection against fungal pathogens, firstly bunt and smut, later snow mould and other seed- and soil-borne pathogens (for literature see Bateman *et al*, 1986).

Since then and till now they have held the leading position in the fungicide seed treatment market (Table 2).   However, considering the vast variety of biological problems and the marginal role of seed treatment in many parts of the world, there are untapped market potentials in bulk crops like rice and cotton and many speciality crops.   In contrast to fungicides, for insecticidal seed treatments maize, followed by sugar beets, cotton and rice, are the dominating crops.

TABLE 2.   The seed-dressing market by crops (figures represent percentage of total).   1991 figures.

Crop	Total	Fungicides		Insecticides	
Small-grain cereals	54	64	(1)	3	(5)
Potatoes	11	15	(2)	-	
Maize	7	2	(6)	44	(1)
Rice	6	7	(3)	14	(4)
Cotton	6	5	(4)	16	(3)
Oilseed rape	3	?		?	
Sugar beet	2	-		17	(2)
Legumes	2	3	(5)	1	(6)
Soybeans	2	2	(6)	-	
Vegetables	?	1	(7)	3	(5)
Total	93	99		98	

( ) = ranking within fungicides/insecticides.   Source: Ciba-Geigy.

Fungicides clearly lead the seed treatment market, as shown in Table 3.   The various chemical classes were compiled by Martin and Woodcock, (1983), in comparison, insecticide treatments play a minor role, they are used in combination with fungicides as much as they are used alone.   Safeners and PGR's are only used on a very limited scale, although they may play a significant role in certain crops under specific growing conditions.   There is ample room for much broader use.

The market shares of the different chemicals are shown in Table 4.   It reveals a strong position of the top two products in all three categories.   In addition, it shows in the case of fungicides the enormous number of active ingredients, and the present insignificance of mercury-based treatments after more than seventy years of dominance.   It also illustrates the limited importance of the 51 chemicals not specified here, despite their significant role in certain niche crops and uses.   They range from old products like thiram, maneb/mancozeb, quintozene and guazatine to novel types like bitertanol, metalaxyl, oxadixyl, and prochloraz.   It can be assumed that these will increase in importance with the decrease and eventual removal of the old products from the market.

TABLE 3.   The seed treatment market by product segments in % of total .   1991 figures.

Product group	% market share
Fungicides (F)	68
Insecticides (I)	11
Dual treatments (F + I)	11
Safeners	1
Plant growth regulators	1
Various	7
Total	99

Source: Ciba-Geigy

TABLE 4.   Seed treatment market 1991.   Ranking of products by market share.

Fungicides (F)		Insecticides (I)		Mixtures (F + I)	
Carboxin + mixtures	) ~ 40%	Carbofuran Methiocarb	) ~ 70%	γ- BHC + fungicides	) ~ 90%
Triadimenol + mixtures	)	Thiodicarb Acephate Disulfoton	) ~ 25%	Tefluthrin + Oxine-Cu	)
Oxine-Cu + mixtures, Pencycuron, Captan, Benomyl, TBZ, Mercury.	) ~ 30%	Chlorpyrifos Rest (= 14 products) ~ 5%	)	Rest (= 6 products) ~ 5%	

Rest (= 51 products) ~ 30%

Source: Ciba-Geigy

Just two insecticidal compounds, namely carbofuran and methiocarb, cover 70% of the market, both introduced some 30 years ago.   Likewise the remaining insecticides in Table 4 were introduced 20 to 30 years ago.   Thus, in contrast to fungicides, no novel chemical groups have entered the market or reached a sizeable position in the market.   Newcomers like furathiocarb, imidacloprid or fipronil may change this picture within the next few years.

Even more surprising is the situation in the segment of dual mixtures. The 10% share of the total seed treatment market they hold is dominated by Gamma-BHC plus a wide range of commodity fungicide mixture partners, followed by tefluthrin plus oxine-Cu, both holding some 90% of this segment.

A glimpse at the geographical pattern of the seed treatment market shows the strong position of Western Europe (Table 5) with France and Germany as the leading countries. This is not surprising in view of the high market value of cereal crops in this part of the world, and, correspondingly, the high inputs the farmer grants them. To what extent this will change in connection with the new agricultural policy (CAP) of the EC remains to be seen. What can be said already, apart from the decrease in cultivated acreage, is that the cost pressure both on the farmer and the agrochemical industry is increasing, a trend which can only make seed treatments even more attractive, particularly if they manage to control early season stem and foliar diseases. Table 5 also underlines the importance of the CIC and the USA, with markets in the order of the leading Western European countries. If the productivity of cereal growing in these countries improves, there would be tremendous potential for more sophisticated seed treatments.

TABLE 5. World seed treatment market by regions (1991 figures) and ranking of countries within them.

Region/Countries	% market share
Western Europe (France, Germany, UK, rest)	40
Rest of the world (CIC [= ex USSR], USA, Canada, Brazil Central/Eastern Europe, Japan, rest)	60

Source: Ciba-Geigy

When considering the percentage of treated seed within crop species in the highest developed market (the EC), sugar beet and oilseed rape rank highest with more than 90% (Table 6). Over 80% of all other crops are treated with fungicides, with the exception of potatoes, but less than 20% receive insecticidal seed treatments alone.

Looking briefly at the development of application technology, a line can be drawn from powder formulations for dry seed treatment via water dispersible powders for slurry treatment to ready-to-use liquid formulations such as emulsions, flowables or solutions and micro-encapsulated active ingredients. In speciality crops such as sugar beet and vegetables, seed pelleting is common practice providing uniform seed shape for precision sowing and offering room for multiple chemical treatments, leading to lower quantities of seed/surface units. Film coating with special binders added to the chemical formulations provides good product adherence during handling, and protection from mechanical damage to the seed. Novel approaches in seed technology, like seed priming and precision sowing in crops other than sugar beet and vegetables will have an impact on seed treatments (for literature see Clayton, 1988; Taylor and Harman, 1990).   This aspect will be dealt with in Session 7 of this meeting.

TABLE 6.   Ranking of crops with regard to quantities and percentage treated in the EC.

By quantities (in t)	Percentage of seed treated with			
	Fungicides		Insecticides	
wheat	sugar beet)		sugar beet)	
barley	oilseed rape)	> 90%	oilseed rape)	> 90%
potatoes	maize)		wheat)	
maize	barley)		maize)	
peas	wheat)	> 80%	barley)	< 20%
oilseed rape	peas)		peas)	
sugar beet	potatoes	- 40%	potatoes)	

Source: Ciba-Geigy

Depending on the seed species, treatments are carried out by breeders, seed propagators, seed processors or farmers.   In the case of sugar beet, maize and vegetables the end user buys the treated seed, whereas in small grain cereals there is a trend towards on-farm treatments, a trend which may have implications with regard to formulation and the safety profile of the product.   On-farm treatment is the method of choice in potatoes, tropical millet, sorghum and maize against the systemic downy mildews in developing countries where home saved seed predominates.   In this segment, industry is selling special sachets and containers for mixing small seed quantities assuring simple and safe product use.

The development of seed treatment machinery would be a topic of its own right with many interesting aspects including those pointing at opportunities for further improvement with regard to uniformity of seed loading, economy of product use etc. In addition it would show the urgent need for improvement in many countries outside Western Europe which holds a leading position in this field.   Some aspects of seed treatment techniques will be covered in Session 7.

In summarising, it can be seen that the history of seed treatment began with the development of non-defined agents, then inorganic molecules followed by non-selective organic biocides, mainly mercury-based molecules, then organic selective residual and lately systemic chemicals.   New molecules are under development in industry such as phenylpyrroles, methoxyacrylates (ICIA 5504 and BAS 490 F), triticonazole, MON 24000 and triazoxide.

As a new element, biological products are about to enter the scene. They are preparations of living bacteria or fungi, antagonistic to seed- and soil-borne pathogens. After a long lead time, the first products of this type are now being introduced into the market, such as Kodiak, a *Bacillus subtilis* preparation or Blue Circle, a *Ps. cepacia* preparation against damping-off and nematodes respectively. There is a lot of potential for such products (for literature see Becker and Schwinn, 1993) and we will see more of them coming along provided the active principles are competitive with regard to biological performance, applicability, storage lifetime and cost. An interesting approach in this context is their combination with chemicals. Session 6 will deal with this topic in more detail.

Cereals as a crop and fungicides as treatments clearly dominate the scene both from the diversity of pathogens (for literature see Hewett and Griffiths, 1986), and variety of chemical solutions available (Tables 1, 2, 3). They also have a high potential outside the developed countries, and in other crops, such as, tropical graminaceous species, soybeans, and cotton.

Insecticidal seed treatments are only 11% of the fungicide market in size probably due to the fact that development of novel molecules was insignificant until recently. Turning this argument around, it means that their importance can increase with the development of more potent chemicals. This would hold particularly true if such novel compounds could replace the granular soil-applied insecticides. Considering the superiority of seed versus soil-application in terms of targeting, required dose and environmental side effects this development would be highly desirable.

Seed treatment is a small part of both the seed and the plant protection industry (Table 7). In addition, it is a highly diverse segment of their markets in terms of crops and products, scattered over a wide range of regions and price levels. Till now, the main markets are in Western Europe, the USA and Canada. In terms of application technology and price level Western Europe leads the seed treatment stakes. There is large potential outside these countries and in many more crops.

TABLE 7. Share of seed treatments in the plant protection and seed industry market (1991 figures) worldwide.

Market segment	Value in billion US $	% of total
Plant protection	25	100
of which seed treatments	0.88	3.5
Fungicides	5.2	100
of which seed treatments	0.5	10
Seed industry	10	100
of which seed protection	0.5	5

Source: Ciba-Geigy

Whereas small grain cereals, potatoes and maize are bulk crops (Table 7) which are still seed treated with relatively simple machinery in many countries, seed treatment of sugar beet and vegetables has reached a much higher level of sophistication.

As long as the cheap and biologically highly active, broad range organic mercury compounds dominated the cereal seed treatment market, this segment had low priority in the plant protection industry's research and development activities. It is interesting to reflect that the development of more selective, toxicologically and environmentally safer products was blocked for decades by the low price and excellent cost/benefit ratio of the existing products, despite their high acute toxicity. They required considerable safety measures at the application sites and created occasional problems in developing countries, when treated surplus seed was consumed by livestock or humans.   The ban of mercury-based products has resulted in significant changes;   the immediate need for toxicologically and environmentally safer, biologically sound replacements, at the expense of a significant increase in treatment costs, or, conversely, a more attractive price level for industry.   All in all, these changes strongly stimulated industry's interest in this market segment.   The development of alternative products also increased the trend away from professional seed treaters to on-farm application and towards new developments in formulation and machinery.   Thus, the ban of mercury-based products, imposed in Germany as the first European country in 1982, in the UK only recently, marked the transition of seed treatment from a commodity to a speciality market.   Here, the systemic fungicides have opened up a new dimension which has yet to be fully explored.

CONCLUDING COMMENTS

So can we expect seed treatment to become a panacea for crop protection, as the title of this paper suggests or does the question mark reflect uncertainty? What do we mean by "panacea"?   Webster's Dictionary defines it as "a remedy for all ills or difficulties" or "a universal remedy".   In this broad, general sense seed treatment does not live up to this definition yet.

However, the level of protection of seeds from pathogens and insect pests achieved with the contemporary products and technology, clearly underlines the advantages and benefits of this approach.   The stimulating effect of the ban of mercury-based products on the development of superior alternatives has been illustrated above.   As the value of quality seeds is more appreciated by the farmer, and if environmental restrictions on the use of plant protection chemicals continue to increase, seed treatment will become more and more attractive to industry and the whole profession.   Already the traditional fungicidal and insecticidal seed treatments offer opportunities for considerable improvement, such as products against stem and early season foliar diseases or better chemicals against soil-borne pests.   Agriculture still awaits seed treatments to control seed-borne bacteria and viruses; to protect against nematodes, slugs and rodents; chemicals inducing systemic plant resistance; novel bird repellents; seed-applied volatile herbicides; safeners against non-selective herbicides such as glyphosate and sulfonylureas; plant growth regulators and micronutrients; symbiotic mycorrhizal fungi; and the exploitation of film coating technology to provide product combinations with slow release properties.

Finally, if the huge markets outside the industrial countries and the traditional crops can be developed, seed treatment will attract even more interest from industry. So, with some imagination seed treatment has the potential of becoming a panacea for crop protection. On the other hand, life has taught us that "remedies for all ills or difficulties" are rare. Therefore, it is wiser to have less demanding expectations. But even a partial remedy or one which overcomes a major difficulty would mean an optimistic outlook for seed treatment, embracing stimulating challenges and attractive potentials (Graham-Bryce, 1988). During this meeting we hope to hear about progress in many aspects of seed treatment which will uphold my optimistic outlook.

REFERENCES

Ainsworth, G.C. (1981) *Introduction to the history of plant pathology.* Cambridge University Press, Cambridge, 315 pp.

Appel, O.; Gassner, G. (1907) Der derzeitige Stand unserer Kenntnisse von den Flugbrandarten des Getreides und ein neuer Apparat zur einfachen Durchführung der Heisswasserbeize. *Mitteilungen der kaiserlichen Biologischen Anstalt für Land-und Forstwirtschaft, 3,* 1-20.

Bachmann, F.; Drabek, J. (1981) CGA 73102, a new soil-applied systemic carbamate insecticide. *Brighton Crop Protection Conference - Pests and Diseases, 1,* 51-58.

Bartlett, D.H.; Ballard, N.E. (1973) The effectiveness of guazatine and imazalil as seed treatment fungicides in barley. *Proceedings of the 8th British Insecticide and Fungicide Conference, 1,* 205-211.

Bateman, G.L.; Ehle, H.; Wallace, H.A.H. (1986) Fungicidal treatments of cereal seeds. In: *Seed treatment,* K.A. Jeffs (Ed.), British Crop Protection Council Publications, Thornton Heath, pp. 83-112.

Bebbington, R.M.; Brooks, P.H.; Geoghegan, M.; Snell, B.K. (1969) Ethirimol: a new systemic fungicide for the control of cereal powdery mildew. *Chemistry and Industry,* 1512.

Becker, J.O.; Schwinn, F.J. (1993) Control of soil-borne pathogens with living bacteria and fungi: status and outlook. *Pesticide Science, 37,* 355-363.

Bolley, H.L. (1897) New work upon the smuts of wheat, oats and barley and a resumé of treatment experiments for the last three years. *North Dakota Agricultural Experiment Station, Bulletin 77,* 109-164.

Clayton, P.B. (1988) The implications for industry of technological advances in seed treatment. *Brighton Crop Protection Conference - Pests and Diseases, 3,* 833-843.

Davidse, L.C.; de Waard, M.A. (1984) Systemic fungicides. In: *Advances in Plant Pathology,* D.S. Ingram, P.H. Williams (Eds.). Academic Press, London, pp. 191-257.

Elbert, A.; Overbeck, H.; Iwaya, K.; Tsuboi, S. (1990) Imidacloprid, a novel systemic nitromethylene analogue insecticide for crop protection. *Brighton Crop Protection Conference - Pests and Diseases, 1,* 21-28.

Frohberger, P.E. (1969) Über díe Entwicklung von Mitteln zur Bekämpfung samen- und bodenbürtiger pilzlicher Pflanzenkrankheiten bei den Farbenfabriken Bayer AG. *Pflanzenschutz-Nachrichten Bayer, 22*, 23-48.

Frohberger, P.E. (1978) Baytan, ein neues systemisches Breitband - Fungizid mit besonderer Eignung fúr díe Getreidebeizung. *Pflanzenschutz-Nachrichten Bayer, 31*, 11-24.

Frohberger, P.E.; Grossmann, F.K. (1981) Monceren, ein neues Kartoffelbehandlungsmittel gegen Auflaufschäden durch *Rhizoctonia solani. Mitteilungen aus der Biologischen Bundesanstalt fúr Land-und Forstwirtschaft, 203*, 230-231.

Geuther, T. (1895) Über díe Einwirkung von Formaldehydlösungen auf Getreidebrand. *Berichte der Pharmazeutischen Gesellschaft, 5*, 325-329.

Graham-Bryce, I.J. (1973) Cereal seed treatment: problems and progress. *Proceedings of the British Insecticide and Fungicide Conference,* 921-932.

Graham-Bryce, I.J. (1988) Pesticide application to seeds and soil: unrealised potential? In: *Application to Seeds and Soil,* T.J. Martin (Ed.), *BCPC Monograph No. 39,* Thornton Heath: BCPC Publications, pp. 3-14.

Hartke, Sabine; Buchenauer, H. (1981) Untersuchungen zur Resistenz von *Gerlachia nivalis* gegenúber Wirkstoffen in Hg-freien Saatgutbehandlungsmitteln. *Mitteilungen der Biologischen Bundesanstalt für Land-und Forstwirtschaft, 203*, 240-241.

Hatzios, K.K.; Hoagland, R.E. (1989) *Crop safeners for herbicides.* Academic Press, New York, London, 400 p.

Hewett, P.D.; Griffiths, D.C. (1986) Biology of seed treatment. In: *Seed Treatment,* K.A. Jeffs (Ed.), British Crop Protection Council Publications, Thornton Heath, pp. 7-14.

Jeffs, K.A. (1986) A brief history of seed treatment. In: *Seed Treatment,* K.A. Jeffs (Ed.), British Crop Protection Council Publications, Thornton Heath, pp. 1-5.

Jones, D.R.; Slade, M.D.; Birks, Kathryn A. (1989) Resistance to organomercury in *Pyrenophora graminea. Plant Pathology, 38*, 509-513.

Jutsum, A.R.; Gordon, R.F.S.; Ruscoe, C.N.E. (1986) Tefluthrin, a novel pyrethroid soil insecticide. *British Crop Protection Conference - Pests and Diseases, 1*, 97-106.

Keitt, G.W. (1959) History of Plant Pathology. In: *Plant Pathology. An advanced treatise,* J.G. Horsfall and A.E. Dimond (Eds.). Academic Press, New York and London, *1*, pp. 61-97.

Kittleson, A.R.A. (1952) A new class of organic fungicides. *Science, 115*, 84-86.

Kühn, J. (1873) Die Anwendung des Kupfervitriols als Schutzmittel gegen den Steinbrand des Weizens. *Botanische Zeitung, 31*, 502-505.

Koch, E.; Leadbeater, A.J. (1992) Phenylpyrroles - a new class of fungicides for seed treatment. *Brighton Crop Protection Conference - Pests and Diseases, 3*, 1137-1146.

Kuiper, J. (1965) Failure of hexachlorobenzene to control common bunt of wheat. *Nature, 206*, 1219-1220.

Jackson, D.; Roscoe, R.J.; Ballard, N.E. (1973) The effectiveness of bis (8-guanidine-octyl) amine as a seed dressing, alone or in mixture with other fungicides, against diseases of wheat. *Proceedings of the British Insecticide and Fungicide Conference,* 143-150.

Locke, T.; Moon, L.M.; Evans, J. (1987) Survey of benomyl resistance in *Fusarium* species on winter wheat in England and Wales in 1986. *Plant Pathology*, <u>36</u>, 589-593.

Martin, H.; Woodcock, D. (1983) *The scientific principles of crop protection*, 7th edition. Edward Arnold, Baltimore, 486 pp.

McEwen, F.L.; Davis, A.C. (1965) Tests with insecticides for seed-corn maggot control in Lima beans. *Journal of Economic Entomology*, <u>58</u>, 369-371.

Neergard, P. (1977) *Seed Pathology*, Vol. 1 and 2. MacMillan, London, 1187 pp.

Powell, A.A.; Matthews, S. (1988) Seed Treatments: developments and prospects. *Outlook in Agriculture,* <u>17</u>, 97-103.

Prévost, B. (1807) Mémoire sur la cause immédiate de la carie ou charbon des blés, et de plusieurs autres maladies des plantes, et sur les préservatifs de la carie. Paris, Bernard.

Riehm, E. (1914) Prüfung einiger neuer Beizmittel. *Mitteilungen der kaiserlichen Biologischen Anstalt fúr Land-und Forstwirtschaft,* <u>15</u>, 7-8.

Schmeling, B. von; Kulka, M. (1966) Systemic fungicidal activity of 1,4-oxathiin derivatives. *Science*, <u>152</u>, 659-660.

Schulthess, H. (1761) Vorschlag einiger durch die Erfahrung bewãhrter Hilfsmittel gegen den Brand in Korn. *Abhandlungen der Züricher Naturforschenden Gesellschaft,* <u>1</u>, 498-506.

Schwinn, F.J.; Staub, T.; Urech, P.A. (1977) Die Bekämpfung falscher Mehltaukrankheiten mit einem Wirkstoff aus der Gruppe der Acylalanine. *Mitteilungen der Biologischen Bundesanstalt fúr Land-und Forstwirtschaft,* <u>178</u>, 145-146.

Slade, R.E. (1945) The γ-Isomer of Hexachlorohexane (Grammoxane). An insecticide with outstanding properties. *Chemical Industry,* <u>40</u>, 314-319.

Suett, D.L. (1988) Application to seeds and soil: recent developments, future prospects and potential limitations. *Brighton Crop Protection Conference - Pests and Diseases, 2,* 823-832.

Taylor, A.G.; Harman, G.E. (1990) Concepts and technologies of selected seed treatments. *Annual Review of Phytopathology,* <u>28</u>, 321-339.

Tessier, A.H. (1779) *Traité des maladies des grains.* La Veuve Herissant, Paris. 349 pp.

Tillet, M. (1755) Dissertation sur la cause qui corrompt et noircit les grains de bled dans les épis; et sur les moyens de prevenir ces accidents. P. Bruin, Bordeaux.

Tisdale, W.H.; Flenner, A.L. (1942) Derivatives of dithiocarbamic acid as pesticides. *Industrial and engineering chemistry,* <u>34</u>, 501-502.

Unterstenhöfer, G. (1962) Mesurol, ein polyvalentes Insektizid und Akarizid. *Pflanzenschutz-Nachrichten, Bayer,* <u>15</u>, 181-194.

Wesenberg, G. (1938) Wíe das Uspulun entstand. *Nachrichten über Schädlingsbekâmpfung (Bayer),* <u>13</u>, 103-111.

Yersin, H.; Chomette, H.; Baumann, G. (1945) L'Hexachlorobenzene produit organique de synthèse utilisé dans la lutte contre la carie du ble. *Comptes rendues des Séances de l'Académié Agricole de France,* <u>31</u>, 24-27.

Session 2
Cereal Seed Treatment Strategies

Chairman	R P DAVIS
Session Organiser	W J RENNIE

STRATEGIES FOR CEREAL SEED TREATMENT

W J RENNIE AND VALERIE COCKERELL

Official Seed Testing Station, Scottish Agricultural Science Agency, East Craigs, Edinburgh, EH12 8NJ

ABSTRACT

Following the withdrawal of organo-mercury fungicides in 1992 cereal growers have a choice of seed treatments or can consider sowing seed untreated. The paper reports on cereal seed treatment practice in a number of countries and discusses seed treatment strategies appropriate to the UK.

INTRODUCTION

For more than 50 years seed-borne pathogens of cereals were effectively controlled in the United Kingdom by the extensive and routine use of organo-mercury fungicides. Because they were effective against a number of potentially damaging diseases, were relatively inexpensive and easily applied, organo-mercury treatments were rapidly and almost universally accepted by cereal growers. In 1977 95% of all UK cereal seed was treated with organo-mercury fungicides (Steed <u>et al</u>., 1979). It has been suggested (Yarham and Jones, 1992) that organo-mercury treatments were so successful that, for many years, the diseases which they had been introduced to control were rare and almost unknown to UK farmers. Nevertheless, largely because of their low cost, they continued to be used as routine treatments by the great majority of cereal growers.

Organo-mercury treatments were ineffective against loose smut (<u>Ustilago nuda</u>) since inoculum occurs within the embryos of infected seeds and it required the development of the systemic fungicide carboxin to facilitate effective routine chemical control. The introduction of seed treatment fungicides for the control of seedling foliar pathogens offered growers additional benefits and eventually led to competition with organo-mercury for a share of the UK cereal seed treatment market.

Richardson (1986) made comparisons of spring barley and winter wheat crops grown from untreated and organo-mercury treated seed. He concluded that seed treatment, to protect against seed-borne pathogens, was not necessary for certified seed used to produce a non-seed crop. However, the low cost of organo-mercury fungicides meant that, at that time, there was no economic incentive to sow untreated seed.

In response to concerns over the toxicity and persistence of mercury in the environment, EC Council Directive 79/117/EC (Anon., 1979) prohibited the use of mercury in agriculture, although a derogation permitted its use in the UK until March 1992. With the withdrawal of organo-mercury growers were faced with a range of significantly more expensive treatments, some of which had a different spectrum of activity compared with the mercury-based fungicides. This led growers and their advisers to seek detailed information on the risks posed by seed-borne cereal pathogens and question the need for continued routine treatment.

According to Brodal (1993) only a proportion of cereal seed is treated in Norway and Sweden and seed treatment decisions are based on the results of tests for seed-borne pathogens. In many countries organo-mercury was withdrawn some years before 1992 and it was considered useful to seek information on seed treatment practice in some of these countries.

METHODS

A questionnaire was sent to seed certification authorities in Canada, Denmark, Finland, France, Ireland, New Zealand, Norway and Sweden during 1993. Respondents were asked to provide detailed answers to a number of questions and were encouraged to confirm their information, where appropriate, with extension and trade colleagues. Separate forms were issued for barley and wheat. Among the questions asked were:-

a.   What proportion of your national crop is grown from (a) certified, (b) farm-saved seed?

b.   What proportion of (a) certified, (b) farm-saved seed is sown untreated?

c.   Are there standards for seed-borne pathogens in your seeds regulations/certification schemes?

d.   What proportion of seed is tested for seed-borne pathogens?

e.   Are decisions on seed treatment made as a result of seed testing information?

f.   What do you consider to be the most important seed-borne pathogens?

g.   What are the most commonly applied seed treatments?

h.   Are different treatments applied to different generations of certified seed?

In 1992 the Pesticide Usage Survey Group at The Scottish Agricultural Science Agency (SASA) was asked to collect detailed information on cereal seed treatment usage on autumn sown cereals from Scottish growers during an Arable Survey of Pesticide Usage and in 1993 6 major seed processing companies and 5 operators of mobile seed dressers in Scotland were asked to provide information on autumn cereal seed treatment usage.

RESULTS

Table 1 shows the proportion of certified and farm-saved seed sown in each of the countries that responded to the questionnaire and indicates the proportion of certified seed sown untreated. Information on the proportion of farm-saved seed sown untreated was usually not available.

Table 1   Proportion of cereal seed sown as certified and farm-saved seed and percentage of certified seed sown untreated

Country	Certified seed %	Farm-saved seed %	Percentage of certified seed sown untreated	
			Wheat	Barley
Canada	40	60	Nil	2
Denmark	90	10	10	10
Finland	35	65	30	30
France	50	50	nil	nil
Ireland	94	6	nil	nil
New Zealand	55	45	nil	nil
Norway	65	35	10	38
Sweden	55	45	5	5-35

Certified seed, or commercial seed of equivalent quality, accounted for a higher proportion of the barley and wheat area than farm-saved seed in most cases, but there was considerable variation between countries. In Denmark and Ireland more than 90% of the area of these crops was sown with certified seed whereas in Canada, Finland, France and Sweden half the cereal area was sown with farm-saved seed.

In all countries except Ireland a small percentage of farm-saved cereal seed was sown untreated. Generally, relatively little certified seed was sown untreated, but in Finland 30% of certified cereal seed was claimed to be sown untreated and in Norway and Sweden up to 40% of certified spring barley seed was sown untreated. There was a general tendency to sow a higher proportion of spring varieties untreated.

The most important seed-borne pathogens on barley were considered to be _Ustilago nuda_ and _Pyrenophora graminea_ (Table 2). There was general agreement that _Fusarium_ seedling blight was of little importance on barley. _Cochliobolus sativus_ was considered to be important on susceptible barley varieties. Bunt (_Tilletia_ spp) and seedling diseases caused by _Fusarium nivale_ and _Septoria nodorum_ were considered important on wheat.

Few countries have specific standards for seed-borne pathogens in seeds regulations (Table 2). Several quoted the EC Council Directive (Anon., 1966) which requires that "Harmful organisms which reduce the usefulness of the seed shall be at the lowest possible level" but this

Table 2 Seed-borne pathogens considered important (Y) in seed production

Country	Pathogen						Standards applied through certification or in seeds regulations
	Ustilago nuda	Fusarium spp	Drechslera spp	Septoria nodorum	Tilletia spp	Other spp	
Canada (Barley only)	Y	–	Y	–	–	Y*	U.nuda (seed standard)
Denmark	Y	Y	Y	Y	Y	–	None
Finland	Y	–	Y	Y	Y	–	None
France	Y	Y	Y	Y	–	–	None
Ireland	Y	Y	Y	–	–	–	None
New Zealand	–	–	–	–	–	–	U.nuda (field standard)
Norway	Y	Y	Y	Y	–	Y♦	U.nuda (field standard) Voluntary standards for seed-borne pathogens of barley to facilitate seed treatment decisions, because of pesticide reduction policies
Sweden	Y	Y	Y	Y	Y	YØ	U.nuda, Fusarium spp S. nodorum, Tilletia spp., Drechslera spp Cochliobolus sativus

*Ustilago hordei
 Ustilago nigra
 Barley stripe mosaic virus

♦Rhynchosporium secalis

ØCochliobolus sativus

vague requirement is open to wide interpretation and is difficult to enforce. Sweden has quite specific standards for cereal seed health in its certification scheme and these are tied to requirements for the treatment of seed that fails to meet the standards. Seed that has relatively low levels of seed-borne pathogens is often sown untreated. Norway has no officially enforced standards but has recently introduced a voluntary scheme in which all spring cereal seed is tested and fungicide treatment is required only if threshold levels for seed-borne pathogens are exceeded. Fungicide seed treatment is positively discouraged if disease thresholds are not reached.

Although cereal seed treatment usage was reported to be high in all countries there is no general requirement for seed treatment in national regulations or in certification schemes except in Sweden where seed must be treated if the threshold levels for seed-borne pathogens are exceeded. A number of countries have specific standards for loose smut and effective treatment is required if the threshold is exceeded, either in laboratory tests (Canada) or in control plots of multiplication grades (Norway; UK).

Table 3   Proportion of barley and wheat tested for seed-borne pathogens

Country	Percentage seed tested Wheat	Barley	Seed treatment decisions made as a result of seed testing
Canada	-	<25	No
Denmark	Some farm-saved seed	Most winter barley for loose smut	Yes, particularly to control loose smut
Finland	50	60	Yes
France	0.5	Nil	No
Ireland	45	20	Yes
New Zealand	Nil	Nil	No
Norway	30	100	Yes (barley)
Sweden	80	80	Yes

There is considerable variation in the amount of non-statutory testing being done for seed-borne pathogens and the extent to which seed treatment decisions are made on the basis of advisory laboratory test results (Table 3). In France and New Zealand there is almost no advisory testing and treatments are applied on a routine basis irrespective of seed health. In Canada, Denmark, Ireland and the UK tests are occasionally made for loose smut infection in barley and seed treatment decisions are made on the basis of these results. In Finland, Norway and Sweden a high proportion of seed (30-100%) is tested for a range of seed-borne pathogens and treatment decisions are influenced by seed-borne inoculum. The highest proportion of untreated seed is sown in these countries.

Tables 4 and 5 list the active ingredients most often used on wheat and barley in the countries that responded to the questionnaire. Tables 6 and 7 give an indication of the relative proportions of fungicides applied to winter wheat and winter barley sown in Scotland in 1992 and 1993. The data are rough estimates, from relatively small samples and are included to show the range of fungicides used.

Table 4   Fungicide usage on wheat seed

Country	Active ingredients	Estimated percentage of treated seed
Denmark	bitertanol + fuberidazole	90
	guazatine	< 10
Finland	carboxin + imazalil	70
	guazatine + imazalil	15
	triadimenol + imazalil	5
France	oxyquinolate + anthraquinone	40
	oxyquinolate + anthraquinone + lindane + endosulfan	20
New Zealand	triadimenol + imazalil	93
	flutriafol + imazalil	5
Norway	guazatine	80
	guazatine + imazalil	20
Sweden	guazatine	90
	bitertanol + fuberidazole	10

There was little evidence that seed treatment fungicides were regularly applied below recommended rates, except in the UK and Ireland where triadimenol + fuberidazole was sometimes applied at reduced rates, especially on wheat.

Only in Ireland and the UK did respondents indicate that different active ingredients were applied to different generations during seed multiplication. In both cases a much higher proportion of systemic fungicides effective against U.nuda were applied to multiplication grades of seed where the aim was to produce final generation seed that met loose smut standards in seed regulations. Respondents were asked to indicate whether their country had a policy to reduce seed treatment usage. There are general moves to reduce pesticide usage in a number of countries, notably in Scandinavia, but only in Norway is there a specific policy to reduce cereal seed treatment usage.

Table 5   Fungicide usage on barley seed

Country	Active ingredients	Estimated percentage of treated seed
Canada	carboxin	70
Denmark	imazalil + carboxin (or thiabendazole or fuberidazole)	90
Finland	carboxin + imazalil	70
	guazatine + imazalil	5
	triadimenol + imazalil	10
France	oxyquinolate + anthraquinone + flutriafol + ethirimol	80
Ireland	tebuconazole + lindane	65
	triadimenol + fuberidazole	20
	guazatine	15
New Zealand	triadimenol + imazalil	88
	flutriafol + imazalil	8
	carboxin + thiram	4
Norway	guazatine + imazalil	45
	imazalil	25
	thiabendazole + imazalil	20
	carboxin + imazalil	5
	guazatine	5
Sweden	guazatine + imazalil	95
	fenfuram or carboxin	5

DISCUSSION

Until the withdrawal of organo-mercury seed treatments in 1992, the choice for UK cereal growers was relatively simple. An inexpensive single purpose organo-mercury seed treatment could be used as an insurance, or a systemic seed treatment could be applied to both control seed-borne pathogens and give protection against early infection by foliar pathogens. Seed for further multiplication was usually treated with a systemic fungicide effective against U.nuda. UK cereal growers now have the choice of a range of treatments, all significantly more expensive than organo-mercury and with different spectra of disease control. With increasing economic pressures on cereal growers and the need to look critically at inputs, some will consider using the least expensive treatments, some will apply reduced rates of treatment, especially on farm-saved seed and a number may consider sowing some of their cereal area with untreated seed. Advice and information on seed treatments and the pathogens they control may come from different sources and will inevitably reflect the interests of the adviser. Seed merchants are generally reluctant to sell certified seed untreated, partly because fungicide seed

Table 6   Fungicide usage on Scottish winter wheat sown in 1992 and 1993

Fungicide	Percentage of treated seed			
	Certified seed		Farm-saved seed	
	1992[1]	1993[2]	1992[1]	1993[3]
untreated	< 1	< 1	< 1	< 1
guazatine	20	55	30	60
carboxin + thiabendazole	25	10	35	15
triadimenol + fuberidazole (full rate)	45	25	20	15
triadimenol + fuberidazole (reduced rate)	10	< 5	15	5
fenpiclonil	-	< 5	-	< 1

Table 7   Fungicide usage on Scottish winter barley seed sown in 1992 and 1993

Fungicide	Percentage of treated seed			
	Certified seed		Farm-saved seed	
	1992[1]	1993[2]	1992[1]	1993[3]
untreated	nil	< 1	< 2	< 2
guazatine + imazalil	25	40	10	35
carboxin + thiabendazole + imazalil	30	15	15	15
triadimenol + fuberidazole (full rate)	10	20	10	5
triadimenol + fuberidazole (reduced rate)	< 5	< 1	5	5
ethirimol + flutriafol + thiabendazole (full rate)	25	25	25	20
ethirimol + flutriafol + thiabendazole (reduced rate)	5	nil	35	20

[1] Data collected by PUS, SASA
[2] Information from Scottish seed merchants
[3] Information from Scottish operaters of mobile seed-cleaning equipment

treatments generate an important element of added value and also because of concerns over crop performance.

Information from other countries suggests no uniform approach to cereal seed treatment usage. In New Zealand almost all cereal seed is routinely treated with relatively expensive broad spectrum systemic fungicides. The seed is not tested prior to treatment and seed-borne diseases are said to be of no importance in cereal production. With modern seed treatments this is a simple and effective means of controlling seed-borne pathogens, but there is a significant cost to the grower. On the other hand a relatively high proportion of untreated cereal seed is sown in Finland, Norway and Sweden. Seed is tested prior to treatment and treatment decisions depend on the incidence of seed-borne pathogens. In Norway and Sweden treatment is discouraged if seed-borne pathogens are present only at low levels.

In the UK, Denmark and Ireland a small but significant proportion of cereal seed is farm-saved and some of this is tested before treatment. Seed treatment decisions are based on seed testing results, for example a guazatine or fenpiclonil treatment may be favoured if wheat seed is heavily infected with F.nivale; carboxin or flutriafol may be used if U.nuda is present. In the UK and Denmark the seed may occasionally be sown untreated if only low levels of seed-borne pathogens are found to be present.

A similar approach could be extended to certified seed and UK growers could save £30-£80 for each tonne of seed sown untreated. If seed is effectively treated during multiplication the final generation should be relatively free from bunt, covered smut (Ustilago hordei), loose smut and barley leaf stripe - the diseases which are mainly seed-borne and which can "build-up" rapidly in successive generations of untreated seed. However, seed treatment cannot guarantee freedom from infection in the subsequent generation since U.nuda and P.graminea can infect developing seed from neighbouring diseased crops and T.caries and U.hordei can contaminate healthy seed from cleaning equipment. Seed would therefore have to be tested to ensure relative freedom from these pathogens. Pathogens which cause seedling blights, particularly F.nivale and S.nodorum, are not specifically seed-borne and do not multiply only in the absence of seed treatment. Their development on seed is encouraged by rainfall soon after flowering and disease incidence on harvested seed is probably not influenced by seed treatment in the previous generation. It would be necessary to test untreated seed to ensure that these pathogens were unlikely to affect germination and emergence. The cost of testing would reduce savings made through not using seed treatment but, with large lots of certified seed, testing costs would be low.

More significant factors would be the delay in processing winter cereal seed until a test result was available and the cost to the merchant of rejecting seed stocks found to be infected. Spring barley seed would offer the best opportunity to reduce seed treatment usage, or target fungicides for specific disease control, since a high proportion of UK seed is likely to meet seed-borne disease thresholds and there would be adequate time for testing. In Scandinavia it is mainly spring sown seed that is untreated.

In the short term, interest in sowing untreated seed is likely to be limited mainly to farm-saved seed. Pressure to sell certified seed untreated will come only if the relative costs of farm-saved and certified seed change, for example by applying royalties to farm-saved seed, or through environmental pressure to reduce pesticide usage.

There are risks in extending the area of untreated cereal seed. Seed-borne pathogens can multiply rapidly between years, loose smut and leaf stripe can infect neighbouring seed crops and bunt can contaminate clean seed during processing. To avoid the multiplication of seed-borne pathogens seed should never be sown untreated for 2 or more generations and untreated seed should always be tested before sowing. However, growers are free to ignore such guidance and may, by so doing, increase inoculum levels in cereal crops, thereby increasing the risks to other growers.

Growers now have a wider choice of cereal seed treatments than ever before and they have opportunities to choose seed treatments most suited to their needs or make savings by sowing untreated seed. Decisions on seed treatment should be made in the light of professional advice and with knowledge of the health status of the seed.

ACKNOWLEDGEMENTS

We are grateful to our seed testing and certification colleagues who took time to complete the questionnaires.

REFERENCES

Steed, J M; Sly, J M A; Tucker, G G; Cutler, J R (1979) _Pesticide Usage Survey Report 18_, _Arable Farm Crops 1977_, MAFF/DAFS, London, p 155.

Yarham, D J; James, D R (1992) The forgotten diseases: why we should remember them. _Proceedings of the 1992 Brighton Crop Protection Conference, - Pests and Diseases_, 1117-1126.

Richardson, M J (1986) An assessment of the need for routine use of organomerculial cereal seed treatment fungicides. _Field Crops Research_, _13_, 3-24.

Anon. (1979) Council Directive 79/117/EEC of 12 December 1978 prohibiting the placing on the market and use of plant protection products containing certain active substances. _Official Journal of the European Communities_, _22_ L33, 8.2. 1979, 36-40.

Brodal, G (1993) Fungicide treatment of cereal seeds according to needs in the Nordic Countries. _Proceedings Crop Protection in Northern Britain 1993_, 7-16.

Anon. (1996) Council Directive 66/402/EEC of 14 June 1966 on the marketing of cereal seed. _Official Journal of the European Communities, 125_, 11 July 1966 p2306.

CEREAL SEED TREATMENT - RISKS, COSTS AND BENEFITS.

N.D. PAVELEY, J.M.Ll. DAVIES

ADAS Terrington, Terrington St Clement, Kings Lynn, Norfolk. PE34 4PW.

ABSTRACT

Fungicide seed treatments are routinely applied to over 90% of winter wheat and barley crops in the UK; primarily to control the seed borne diseases, bunt (*Tilletia caries*), loose smut (*Ustilago nuda*) and leaf stripe (*Pyrenophora graminea*), and seed and soil-borne fusarium seedling blight. A 15 site field experiment on winter wheat and barley quantified the risk from soil-borne seedling blight, and the benefits of seed treatment derived foliar disease and take-all control. Single year results: i) support previous studies in suggesting that where healthy seed is sown there is on average little or no yield benefit from seed treatment, and ii) conclude, tentatively, that the risk from soil-borne seedling blight is low and that the benefits of broad spectrum seed treatment against foliar pathogens and take-all are likely to be small in all but a few specific circumstances. Data are presented on the costs, risks and benefits of seed treatment, to assess the consequences of moving to a strategy of disease control where treatments are applied in response to risks quantified by seed testing.

INTRODUCTION

Concluding a major review of the need for routine cereal seed treatment, Richardson (1986) stated that: "[from the results of over 220 treated vs. untreated comparisons] there were no significant differences in overall mean yield associated with the absence of treatment for either wheat or barley". He estimated that nearly two thirds of UK cereal crops could be left untreated. In contrast Yarham and Jones (1992), reviewing the use of cereal seed treatment, concluded: "Any general advocacy of non-treatment of seed would be irresponsible in the extreme".

Risk / cost / benefit analysis for the use of cereal seed treatment is complicated by: i) the wide range of target pathogens (seed-borne, soil-borne and foliar), ii) fear of achieving short term savings from non-treatment, at the long term expense of a population increase in currently minor pathogens, iii) savings from non-treatment by an individual leading to communal losses via wind-borne spores (Yarham, 1993), iv) the risk of systemic conazole seed treatments increasing the probability of fungicide resistance developing in 'non-target' or 'marginal target' foliar diseases, v) the absence of data or techniques to quantify the risk from soil-borne pathogens, vi) the short period available in the autumn for seed health to be assessed and for appropriate treatment decisions to be implemented, and vii) the risk of phytotoxic effects from treatment (Skou, 1989).

It is this diversity of considerations that leads to the apparently contradictory statements in the first paragraph.

One area of uncertainty arises because the risk of seedling loss due to soil-borne *Fusarium* (principally *F. nivale*) has not been quantified. Hence, even when seed has been tested for pathogens and found to be within tolerances, the risk of soil-borne fusarium seedling blight has made it: "...advisable to treat where there is a risk of winter crops being sown late, or where seed may be sown in unusually cold or wet seedbeds that may delay seedling emergence" (Rennie & Cockerell, 1993). As seedbed conditions cannot be predicted at the time of the treatment decision, most cereal growers err on the side of caution. This is a contributory factor to 97% and 93% of winter wheat and barley crops in England and Wales being grown from treated seed (Polley and Slough, 1992a;1992b).

With cereal producers re-evaluating the use of seed treatments following the withdrawal of organomercury - and with immunological and nucleic acid techniques offering the prospect of rapid seed health assessment (Ball & Reeves, 1992) - consideration of the costs, risks and benefits of seed treatment seems timely. This paper presents data to address these issues and reports initial results from a multi-site experiment to quantify the risk of soil-borne fusarium seedling blight on winter wheat and barley.

COSTS AND RISKS OF SEED TREATMENT

Economic cost

The cost to the grower of cereal seed treatment ranges from approximately £35 to £80 tonne^{-1} at current prices, depending on the product used.

Fungicide resistance risk

The use of some systemic conazole materials as seed treatments exposes foliar pathogen populations to fungicide during the autumn, winter and early spring - a period when selection pressure for the development of fungicide resistance would not otherwise occur. It is now widely accepted that the period of exposure is an important determinant of the risk of pesticide resistance (Staub, 1991), although the increase in risk is difficult to quantify.

Phytotoxicity

In the absence of a compensating positive response from disease control, any negative effect of seed treatment on crop yield is an additional cost to be borne by the grower. Deleterious effects of treatments on crop establishment and yield, whilst noted from practical experience when severe, have seldom been quantified in the literature (Skou, 1989). As few field crops have the benefit of an untreated control for comparison, such effects as may occur are not widely appreciated.

Environmental

Whilst seed treatments are a relatively benign form of pesticide use, they nevertheless carry some environmental cost. It is now widely accepted that routine use of pesticides should be replaced by a strategy where applications are made in response to quantified risks.

BENEFITS OF TREATMENT

Growers should at least be able to expect a yield or quality gain from seed treatment, meaned across sites and seasons, that would compensate for the treatment costs and risks defined above.

Conversely, any scheme that encouraged the use of untreated seed in some controlled circumstances should ensure that the grower would not be exposed to even a small probability of a catastrophic loss, which the farm business might not survive.

Protection from bunt, loose smut, covered smut and leaf stripe

Without an effective seed treatment strategy, bunt (*Tilletia caries*), loose smut (*Ustilago nuda*), leaf stripe (*Pyrenophora graminea*) and covered smut (*Ustilago hordei*) have the combined potential to render UK cereal production uncompetitive. In recent years routine treatment, predominantly with cheap organomercurial materials, has kept these diseases in check; although problems of fungicide resistance (to carboxin and organomercury) have allowed a limited upsurgence of loose smut and leaf stripe in barley.

If the speed, logistics and economics of seed testing allowed it, all of these diseases (even those, such as leaf stripe, capable of contaminating seed from neighbouring crops) are completely vunerable to a strategy involving routine testing of seed, followed by treatment or sale for ware of batches found to be contaminated.

The only exception to the rule is bunt, which has occasionally been shown to persist as soil contamination (Yarham, 1993). Control of this soil-borne phase would continue to rely on vigilance on the part of wheat producers, and the use of triadimenol + fuberidazole treatment where soil contamination was suspected.

<u>Protection from foliar diseases</u>

Use of 'broad spectrum' seed treatments is often encouraged by claims of useful control of foliar diseases such as mildew (*Erysiphe graminis*), yellow rust (*Puccinia striiformis*) and in some cases *Septoria tritici*. This is a separate issue from the control of those foliar diseases which can have a significant seed borne phase, such as net blotch (*Pyrenophora teres*) and *Septoria nodorum*.

To assess the benefit of foliar disease control it is necessary to quantify the level and consistency of the control achieved, and the probability of this control either producing a yield benefit or replacing a foliar fungicide application. The data presented later in this paper help to quantify the former. A series of winter barley experiments in 1989 (ADAS unpublished) produced substantial mildew epidemics during the winter and early spring and provide some guide to the value of seed treatment derived foliar disease control (Table 1). In these experiments, the flutriafol + ethirimol + thiabendazole seed treatment generally provided good control of mildew during the winter and early spring.

Table 1. Effect of seed treatments and foliar sprays on yield of winter barley cv. Magie (mean of three sites where mildew affected 6-40% of leaf 2 or 3 during the autumn, winter or early spring in phenyl mercuric acetate treated plots).

| | Grain yield (tonnes ha^{-1} at 85% dry matter) | | | |
| | Number of sprays in foliar fungicide programme | | | |
Seed treatment	None	One	Two	Three
Phenyl mercuric acetate	5.11	6.12	6.63	6.92
Flutriafol + ethirimol + thiabendazole	5.11	6.33	6.77	6.94

There was no significant (P<0.05) yield benefit from the mildew active seed treatment compared to the organomercury product, or indication of mildew activity allowing a reduced spray programme.

Practical experience suggests that foliar disease control by seed treatment can be valuable where a yellow rust susceptible variety is being grown in a yellow rust prone area such as coastal Norfolk or Lincolnshire. In seasons with few frosts during the winter, crops grown from seed with no treatment active against yellow rust can require foliar sprays as early as February, wheras crops grown from triadimenol + fuberidazole treated seed seldom require further yellow rust control until late April.

<u>Protection from seed and soil-borne fusarium seedling blight</u>

Although soil-borne *Fusarium* can act as a source of seedling blight (Bateman, 1977), the importance of natural soil-borne inoculum has not previously been quantified in isolation from the seed-borne inoculum. Experiments where seed lots carrying different levels of seed-borne *Fusarium nivale* were grown untreated have shown that: i) the relationship between percentage seed infection and seedling loss is almost linear (W J Rennie, pers.comm.), and ii) more than a few percent seedling loss from seed lots carrying nil or low levels of disease is rare. Seed-borne seedling blight is clearly a significant risk, but one which can be quantified by seed testing. The data below were obtained from an experiment conducted at a wide range of locations, where the risk of soil-borne seedling blight might be expected to represent that experienced by winter cereals in England and Wales.

MATERIALS AND METHODS

Field experiments were conducted at 15 sites; nine sites on winter wheat cv. Riband and six on winter barley cv. Puffin. Sites were selected and managed to give a range of soil types, weather conditions, rotational positions, seedbed conditions and sowing dates representative of cereal growing in England and Wales (Table 2).

Certified seed was obtained centrally, randomly sub-divided and treated with the seed treatments listed in Table 3. Treatments were applied to the sub-samples either by a Rotostat (ICI) mobile seed treatment apparatus or by the manufacturer. After treatment the sub-samples were further divided and despatched to the sites. Additional coded product treatments were also included in the experiments at all sites. Data from these treatments were included in the analyses, but are not presented here.

Table 2. Site location, rotational position, soil type and drilling date

Site No.	Location	Rotation *	Soil texture	Drilling date
Winter wheat				
1.	ADAS Boxworth Cambs.	4th	Clay loam	10 October
2.	ADAS Bridgets, Hants.	1st	Silty clay loam	22 October
3.	Owstwick, N. Humberside	1st	Silty clay loam	10 October
4.	Fishtoft, Lincs.	1st	Silty clay loam	30 October
5.	Kneesall, Notts.	2nd	Clay loam	11 October
6.	Pancross, S. Glamorgan	1st	Silty loam	2 October
7.	Preston Deanery, Northants.	1st	Silt loam	7 October
8.	ADAS Arthur Rickwood, Cambs.	1st	Sandy peat	21 October
9.	ADAS Rosemaund, Hereford	1st	Silty clay loam	29 October
Winter barley				
10.	Attlebridge, Norfolk	3rd	Sandy loam	2 October
11.	ADAS Bridgets, Hants.	1st	Silty loam	27 September
12.	Pancross, S. Glamorgan	3rd	Silty clay	2 October
13.	Chipping Warden, Northants.	2nd	Sandy clay loam	2 October
14.	Nocton, Lincs.	4th	Sandy loam	27 September
15.	ADAS Rosemaund, Hereford	2nd	Silty clay loam	2 October

* - position of the experimental crop in the run of cereal crops since the last non-cereal 'break'.

Seed tests showed that both the wheat and barley samples were of high health status (Table 4). Occasional plants with loose smut were found at some barley sites. No bunt was detected at wheat sites.

Table 3. Seed treatments.

Treatment	Active ingredient/s	Dose (c.p. tonne^{-1})
Winter wheat		
1. Untreated	-	-
2. 'Baytan'	Triadimenol + fuberidazole (Tr.+Fu)	2.0 litres
3. 'Panoctine'	Guazatine (G)	2.0 litres
4. 'Cerevax'	Carboxin+thiabendazole (C+Th)	2.5 litres
5. 'Panogen M'	Methoxyethyl mercuric acetate (MEMA)	1.0 litres
6. 'Beret'	Fenpiclonil (Fe)	4.0 litres
7. 'Ceresol'	Phenyl mercuric acetate (PMA)	1.0 litres
Winter barley		
1. Untreated	-	-
2. 'Cerevax extra'	Carboxin+thiabendazole+imazalil (C+Th+I)	2.0 litres
3. 'Ferrax IM'	Flutriafol+ethirimol+thiabendazole+imazalil (Fl+Et+Th+Im)	5.0 litres
4. 'Panoctine Plus'	Guazatine+imazalil (G+Im)	2.2 litres
5. 'Beret Extra'	Fenpiclonil+imazalil (Fe+Im)	4.0 litres
6. 'Ceresol'	Phenyl mercuric acetate (PMA)	1.1 litres
7. 'Panogen M'	Methoxyethyl mercuric acetate (MEMA)	1.1 litres

Table 4. Seed testing results

Cultivar	Germination (%)	*Fusarium nivale* (%)	*Drechslera graminea* (%)
Riband	98	1	-
Puffin	97	-	1

All the sites were sown by plot drill as a randomised block design with four replicates. Each replicate contained one plot of seed treated with each of the treatments listed in Table 2. Plots were a minimum of 24 m^2, and were typically 2m by 18m. All the experiments were oversprayed with foliar fungicide as appropriate to control foliar diseases. No foliar fungicides were applied earlier than growth stage 31 (Tottman, 1987). Plant establishment was measured as the number of plants along 10 randomly selected 0.5m drill row lengths within each plot, when the most advanced plants were at GS 13. These data were converted to number of seedlings m^{-2}. Foliar diseases were assessed on ten main tillers per plot at GS 31 (prior to any foliar fungicides being applied), using methods and keys described in the MAFF Manual of Plant Growth Stages and Disease Assessment Keys (Anon., 1976). Leaf numbers quoted in the results section are counted down the tiller from the youngest fully expanded leaf (leaf 1). Lodging was assessed, as percentage of plot area lodged, immediately prior to harvest. Experiments were harvested by plot combine and grain yields calculated as tonnes ha^{-1} at 85% dry matter. Statistical analysis of all data was by analysis of variance. Treatment means were separated by calculation of least significant difference (LSD) P=0.05.

RESULTS

Establishment

At the majority of sites none of the treatments significantly affected the number of seedlings established. Data from sites which produced significant treatment effects, and the cross-site analysis from all sites, are shown in Table 5. Vigour scores (data not presented here) showed that where seedling establishment was reduced by seed treatment the effect on crop ground cover and plant height tended to persist into the spring.

Table 5. Effect of treatment on seedling establishment

	Seedlings m^{-2} at GS 13				
Winter wheat	Site 2	Site 5	Site 8	Site 9	Cross-site
1. Untreated	274	210	387	288	293
2. Tr+Fu	273	158	383	260	279
3. G	296	204	353	232	287
4. C+Th	318	.188	334	256	287
5. MEMA	283	199	402	250	286
6. Fe	314	204	352	252	285
7. PMA	313	175	372	253	278
SED 27df (cross-site 243df)	15.1	19.8	16.4	16.4	6.1
LSD	31.1	40.5	33.4	33.4	12.4

Winter barley	Site 10	Cross-site
1. Untreated	303	286
2. C+Th+Im	315	290
3. F1+Et+Th+Im	251	273
4. G+Im	268	284
5. Fe+Im	305	283
6. PMA	308	299
7. MEMA	298	300
SED 27df (cross-site 243df)	15.4	7.2
LSD P< 0.05	31.7	14.4

Foliar diseases

Foliar diseases were present at significant levels in untreated plots at winter wheat site 6 (12.5% *S. tritici* on leaf 2), site 1 (22.2% *S. tritici* on leaf 3) and site 5 (11.6% *S. tritici* on leaf 3); and at winter barley site 13 (21.6 % mildew on leaf 3), site 12 (1.6% mildew and 2.2% *Rhynchosporium secalis* on leaf 3), site 11 (1.2% mildew on leaf 2) and site 10 (3.7% mildew on leaf 4). Although in some cases there appeared to be marginal suppression of foliar disease, no significant (P<0.05) control was obtained from any of the treatments.

Lodging

Lodging or brackling (lodging of tillers where the 'break' occurs part way up the stem, usually at a node) occured at four sites, but significant treatment differences were found at only two sites - both of which had untreated plot area affected scores of less than 10%. Significant control was achieved at these sites by Baytan and Ferrax.

<u>Grain yield</u>

At the majority of sites grain yield was not affected by any of the treatments. Data from sites where significant (P<0.05) effects were recorded and data from a cross-site analysis are presented in Table 6.

Table 6. Effect of treatments on grain yield.

Grain yield (t ha^{-1} at 85% dry matter)

Winter wheat	Site 1	Site 2	Site 3	Site 5	Site 7	Cross-site
1. Untreated	6.98	8.40	6.03	4.62	8.03	7.31
2. Tr+Fu	6.88	8.56	6.53	4.39	8.13	7.32
3. G	6.98	8.56	6.22	4.46	8.35	7.37
4. C+Th	6.81	8.70	6.61	4.45	7.87	7.36
5. MEMA	6.63	8.47	6.25	4.52	8.21	7.33
6. Fe	7.12	8.55	6.42	4.57	8.36	7.45
7. PMA	6.84	8.64	6.10	4.44	8.32	7.29
SED 27df (x-site 243df)	0.129	0.129	0.232	0.120	0.160	NS
LSD	0.264	0.264	0.473	0.244	0.326	

Winter barley	Cross-site
1. Untreated	7.39
2. C+Th+Im	7.52
3. Fl+Et+Th+Im	7.48
4. G+Im	7.46
5. Fe+Im	7.49
6. PMA	7.49
7. MEMA	7.50
SED (cross-site 144df)	NS

<u>Seed bed conditions and treatment responses</u>

Mean establishment response and yield response data for all wheat sites were regressed individually for each treatment against two measures of seedbed conditions; soil temperature at drilling (mean of 10cm soil temperatures during the 10 days following drilling) and soil moisture (by cumulative rainfall during the 10 days before, and 10 days after, drilling). These data were derived from the nearest meteorological recording site to each experiment location. The spread of 10 day mean temperatures and cumulative rainfall between sites was 7.2°C to 12.3°C and 1mm to 103mm respectively. No significant (P<0.05) relationships to establishment or yield were observed. As no significant treatment yield effects were detected at the winter barley sites, and establishment effects were small, regression of responses against seed bed conditions was considered inappropriate.

<u>Rotational position and treatment responses</u>

There were no rotational position effects that implied a benefit from control of take-all; generally the 1st wheat sites were the more responsive. This is perhaps not surprising as most of the treatments tested do not have take-all activity, and responses to take-all control by triadimenol + fuberidazole treatment have previously been associated with very early sown 2nd to 4th wheat crops where severe disease developed (Hornby and Bateman, 1991).

DISCUSSION

Although overall there was no significant effect from any of the treatments on seedling establishment or yield, this result hides considerable site to site differences, which were not explained by foliar disease, lodging or take-all control. At some sites, positive yield responses were obtained from treatment and these were occasionally associated with improved establishment. At other sites deleterious effects of treatments on seedling establishment, combined with adverse seedbed conditions, reduced the plant population to a level where the crop was unable to compensate and yield was reduced.

Practical experience suggests that fusarium seedling blight is most damaging in cold, dry seedbeds, but regression analysis did not suggest that positive responses to treatment were associated with low seedbed temperature or moisture. Possibly the range of temperatures experienced was not sufficiently extreme to expose such an effect. The spread of soil moistures between sites was probably as wide as would be encountered in practice.

Clearly there are a wide range of variables that can interact to determine the response of cereals to seed treatment. Some of these factors may be more to do with the physiological effects of the treatment on the plant, than responses to disease control. The data gathered so far reinforce the findings of Richardson (1986), in suggesting that where healthy seed is intended for ware production, a consistent positive yield response to seed treatment is unlikely. The results, albeit from a single year's work, suggest that the risk from soil-borne fusarium seedling blight is low.

Seed-borne *Fusarium* remains a significant, but quantifiable, problem. In seasons when conditions during flowering are conducive to fusarium ear blight development, levels of *Fusarium* on the resulting grain may require extensive use of treatments - due to a shortage of grain below the treatment threshold - even if seed treatment was entirely determined by the results of seed health tests. This is illustrated by results from wheat seed testing at NIAB, quoted by Lovelidge (1993): in 1992/3 the proportion of samples failing to meet the advisory limit of <5% of seeds infected with *Fusarium nivale* was 77% (compared to 14% in the previous season).

The Scandinavian countries are already moving towards a strategy where cereal seed treatments are applied in response to results from seed health tests (Brodal, 1993). The data reviewed here suggest no scientific reason why such an approach should not be successful under UK conditions.

REFERENCES

Anon. (1976) Manual of Plant Growth Stages and Disease Assessment Keys. Ministry of Agriculture, Fisheries and Food, London.

Ball, S.; Reeves, J. (1992) Application of rapid techniques to seed health testing - prospects and potential. IN: Techniques for the Rapid Detection of Plant Pathogens. Ed. Duncan, J.M.; Torrance, L. Blackwell Scientific Publications, Oxford.

Bateman, G.L. (1977) Effect of organomercury seed treatment of wheat seed on *Fusarium* seedling disease in inoculated soil. *Annals of Applied Biology*, **85**, 195-201.

Brodal, G. (1993) Fungicide treatment of cereal seeds according to need in the Nordic countries. *Crop Protection in Northern Britain - Proceedings*, 7-16.

Hornby, D.; Bateman, G.L. (1991) Take-all disease of cereals. HGCA Research Review No. 20. Home-Grown Cereals Authority, London.

Lovelidge, B. (1993) Seed-borne disease trends are cause for concern. *Arable Farming*, **20** No. 8, 34-37.

Polley, R.W.; Slough, J.E. (1992a) Survey of winter wheat diseases in England and Wales - 1992. Internal Report, CSL Harpenden and ADAS.

Polley, R.W.; Slough, J.E. (1992b) Survey of winter barley diseases in England and Wales - 1992. Internal Report, CSL Harpenden and ADAS.

Rennie, W.J.; Cockerell, V. (1993) A review of seed-borne pathogens of cereals in the post-mercury period. *Crop Protection in Northern Britain - Proceedings*, 17-24.

Richardson, M.J. (1986) An assessment of the need for routine use of organomercurial cereal seed treatment fungicides. *Field Crops Research*, **13**, 3-24.

Skou, J.P. (1989) Phytotoxic effect of seed-dressing chemicals. *Nordic Plant Protection Conference* 1989, Helsingor, 47-56.

Staub, T. (1991) Fungicide resistance: practical experience with antiresistance strategies and the role of integrated use. *Annual Review of Phytopathology*, **29**, 421-442.

Tottman, D.R. (1987) The decimal code for the growth stages of cereals, with illustrations. *Annals of Applied Biology*, **110**, 441-454.

Yarham, D. (1993) Soil-borne spores as a source of inoculum for wheat bunt (*Tilletia caries*). *Plant Pathology*, **42**, 654-656.

Yarham, D.J.; Jones, D.R. (1992) The forgotten diseases: why we should remember them. *Brighton Crop Protection Conference: Pests and Diseases - Proceedings*, **3**, 1117-1126.

ACKNOWLEDGEMENTS

Funding by ADAS is gratefully acknowledged. Thanks are due to the staff at ADAS Research Centres and Morley Research Centre who conducted these experiments, and to Mandy Jones for statistical analysis. Seed analysis by Valerie Cockerell at SASA East Craigs and provision of seed treatment material by various companies is also acknowledged.

SEED TESTING, SEED CERTIFICATION AND SEED TREATMENT IN THE CONTROL OF CEREAL SEED-BORNE DISEASE.

J. C. REEVES, M. W. WRAY

National Institute of Agricultural Botany, Cambridge, UK.

ABSTRACT

Results are presented from a certification and advisory seed testing programme for the 1993 harvest illustrating, by comparison with results from previous seasons, how disease occurrence has varied. Percentage infections of loose smut (*Ustilago nuda*) in barley and *Fusarium nivale* in wheat have increased. Results from field examination of certification plots for the incidence of loose smut in the period 1990 to 1993 are presented and related to the usage of seed treatments. In addition the effect of various seed treatments on germination rates is presented.

INTRODUCTION

Seed health testing

Field and other environmental conditions during recent growing seasons in England and Wales appear to have been particularly favourable for the development of a number of plant pathogens. The National Institute of Agricultural Botany (NIAB) has tested a range of crops, either for seed certification purposes or as an advisory service to growers. Seed certification schemes are designed to ensure the adequate supply of seed meeting prescribed standards of quality in relation to germination, purity and seed health. Advisory testing is carried out for farmers and seed merchants for their own purposes. Summary results are presented from the seed testing programme for the 1993 harvest illustrating infection levels pooled across both certification (where appropriate) and advisory test categories. These results are presented in conjunction with those of previous seasons, where available, in order to demonstrate the seasonal variation in occurrence for a range of pathogens.

Seed certification and treatment

In addition to seed testing results, data from NIAB's Seed Production Department on the incidence of loose smut (*Ustilago nuda*) in the cereal certification growing-on plots are presented and the effects of a number of seed treatments on disease incidence are shown. The effect of seed treatments on germination is also given with this effect ascribed, at least in part, to the presence of *Fusarium* infection on the seed.

Loose smut is a potentially serious seed-borne disease which can affect most commonly grown cereal species and at present its incidence in the UK is currently lower than when the disease was at its peak. This reduction has been achieved by a combination of applying infection standards in seed certification schemes, the introduction of effective varietal resistance in wheat and the availability of seed treatments with extremely good activity against the *Ustilaginae*. A measure of this success is that in the last decade there has been no reported incidence of loose smut in seed certification plots for oats, rye or triticale. Loose smut is still common in seed of barley although certification failures are infrequent.

Loose smut has been readily controlled in oats by mercury seed treatments, because the spores are carried outside the seed coat. The low cost of mercury lead to very widespread usage until it was banned in 1992 and hence an extremely low incidence of this disease in oats. The first fully effective seed treatment for wheat and barley was carboxin, which was introduced in 1969. Until this time control methods were based on hot water treatment (effective but inconvenient and so not widely used) and control through seed certification. Annual reports of the NIAB (Anon,1968-1973) indicate that the incidence of loose smut in seed stocks fell significantly between 1969 and 1973. In 1969, 34% of wheat lots and 16% of barley lots failed the certification standard of 0.2% for the "Field Approved" category, whereas in 1973 the failure rates were around 4% and 2% respectively. Carboxin provided good, although not complete, protection against loose smut in both wheat and barley until the mid-1980's, when tolerant strains were first found in the winter barley cultivar Panda. This coincided with a peak in loose smut incidence in winter barley and 13% of seed stocks being multiplied for the 1984 harvest failed a seed certification standard. The equivalent figure for 1993 was 0.4%. This reduction in loose smut frequency can be attributed to the efficacy of triazole seed treatments, which largely replaced carboxin on seed for further multiplication.

Shortly after the introduction of carboxin there was a shift from very susceptible winter wheat varieties to those with greater resistance against the prevalent C4 race of loose smut. In 1969, susceptible varieties (NIAB rating of 3 or less) accounted for about 80% of certified seed. This situation had changed by 1973, when it had dropped to around 55%, mainly owing to the success of Maris Huntsman. Varietal resistance and seed treatment were both exploited in the certification scheme at the time; treatment was obligatory for all Basic and Certified seed of susceptible varieties of winter wheat. In barley there have been varieties with some resistance, although it has been morphological rather than physiological as in wheat (Wray and Pickett, 1985).

Since 1976 all cereal seed marketed in the United Kingdom must have been certified in the 'United Kingdom Seed Certification Scheme for Cereals'. This is a statutory scheme with procedures and seed standards applied through Seeds Regulations. The Ministry of Agriculture Fisheries and Food (MAFF) is the Certifying Authority for England and Wales, but much of the technical supervision is carried out by the NIAB. Seed is multiplied in a controlled way, with the number of multiplications being restricted by a so-called generation system (Anon,1985). Most of the seed bought by farmers is Certified Seed 2nd Generation, (C2). This may be certified at either of two standards; 'HVS' (higher voluntary standard) or 'Minimum'. The maximum level of loose smut infection allowed in certified seed is 0.2% for C2 HVS and 0.5% for C2 Minimum seed. The procedure for applying this standard is described below.

<u>Germination and seed treatment</u>

Loose smut and ergot (*Claviceps purpurea*) are the only seed-borne diseases of cereals for which the Cereal Seed Regulations (Anon,1985) specify standards in certified seed. However a number of other pathogens are of importance and if any of these are present at an economically significant level, purchasers may have a legitimate claim against the supplier. One such disease is *Fusarium nivale* which, amongst other effects, reduces germination and field emergence in wheat. Since one of the requirements of certified seed is that the minimum germination should be at least 85%, the presence of this disease can have an effect on certification.

In most years *F. nivale* has little effect on seed production, but when conditions favour its development in the time approaching harvest, the effects can be severe. A range of seed treatments is available which give adequate control of *F. nivale* in most circumstances, but seed can become too highly infected or rendered in an otherwise poor condition so that the seed should be discarded. In the last decade conditions have favoured severe *Fusarium* infections in English grown wheat three times; in 1982, 1987 and 1992. The incidence and severity of *Fusarium* can only be accurately measured by specifically testing for this pathogen but because of its known effects some correlation with germination can be expected although a number of other factors will have an affect.

METHOD

<u>Seed health testing</u>

Seed testing methods followed are as set out by the International Seed Testing Association (ISTA) unless otherwise stated (Anon., 1987). Data from 1993/4 seed testing year (1993 harvest) are incomplete at the time of writing and testing is still continuing. This arises from the delayed harvest in this season.

Ustilago nuda (loose smut) in barley
Seed samples received for testing originated from all areas of England with no apparent bias towards any region; however, the majority were from winter barley varieties. The total number of samples received in 1991/2 was 257; in 1992/3 the number was 240, and in1993 97 samples have been received so far. Almost all of these tests were advisory.

Pyrenophora graminea (leaf stripe) and *P. teres* (net blotch) in barley
The regional distribution of seed samples was broadly similar to that for loose smut tests. All tests were advisory, mostly on seed of winter barley varieties. In 1991/2, 287 samples were tested; in 1992/3 220 samples were tested and in 1993 78 tests have been carried out to date.

Tilletia caries (bunt) in wheat
Regional distribution of seed samples was biased to parts of central and southern England. All seed testing was advisory and was mostly on winter varieties of wheat. The number of samples received was 105 in 1991/2, 115 in 1992/3, and 49 in 1993.

Fusarium nivale (Microdochium nivale) in wheat

Seeds samples, the majority of which were from winter wheat varieties, came from all regions and were tested on an advisory basis only.

Septoria nodorum in wheat

The majority of samples received were from central and eastern England, and were tested on an advisory basis.

Seed certification and treatment

A representative sample of all seed stocks being grown for further multiplication in the certification scheme in England and Wales is grown-on in plots by NIAB. The suitability of seed stocks for further seed production is assessed by recording the level of loose smut and varietal impurities, as well as verifying the varietal identity of the sample. Thus it is the level of infection in the multiplication seed which is being measured, rather than routine testing of all C2 seed lots produced from it. Nevertheless seed producers have a legal obligation to ensure that all the seed which they market complies with the certification standards

The plots measure 1.2m by 8m and mean plant and ear populations are about 2000 and 6000 per plot respectively. Counts of infected ears in each plot are made, starting from the ear emergence growth stage (GS 51-55) and thereafter plots are recorded at regular intervals over the next three to four weeks. At each visit the infected ears are counted and removed from the plot. An ear population of the plot(s) is determined from ear counts in five, one-metre row sections, taken at random. The standards are not applied as a strict percentage but instead 'reject values' are used. These apply a formula, taking plot population into account, to ensure that the risk of rejecting a sample which just meets the standard is only 1%.

In the event that a seed stock has excessive infection, all seed crops sown with it are rejected at the category/level where the standard has not been met. However seed can be 'retrieved' from rejection in either of two ways:-

    i) by treating each lot of seed produced with a seed treatment which the Certifying Authority has accepted as being sufficiently effective, or

    ii) by having an embryo test caried out on a sample from each lot of seed produced, with a result which shows the seed to be within standard.

Germination and seed treatment

In some instances certified seed is sampled both before and after treatment and the data below summarise relevant results from germination tests carried out on winter wheat seed from the harvests of 1990, 1991 and 1992. The data presented come from C1 seed lots tested according to ISTA methods.

RESULTS

<u>Seed health testing</u>

Results from the 1991/2 and 1992/3 seed testing seasons are presented below for the diseases previously mentioned. Included is the certification standard(s) if applicable to a disease test, i.e. the level of a particular disease which, when exceeded, renders the seed lot unacceptable. Where these standards are available they have also been applied to any advisory testing carried out for the same pathogen. For those tests where no certification requirement exists, advisory standards have nevertheless been applied. These standards are also described but essentially they represent a level of infection above which advice on a suitable seed treatment should be sought. In other instances a nil standard has been applied, i.e. the presence of any infection is reported. This is done either because it is considered important that a seed sample is free of a particular pathogen or because insufficient data are available to devise a suitable advisory standard.

The results show that in general the incidence of seed borne cereal diseases has steadily increased over the previous four years although the data for 1993 harvest are as yet incomplete and should be interpreted with care. Only the results for *Septoria nodorum,* for which only one test on seed from the 1993 harvest has yet been done, show a decline. This general increase in incidence has been accompanied by an increase in the number of samples failing to meet "standards" for net blotch in barley, loose smut in wheat, for which only a small number of tests have been done, and for *Fusarium nivale* in wheat.

<u>Seed certification and treatment</u>

During the period 1990 to 1993 the number of seed lots with excessive infection by loose smut in the control plots has remained at a fairly low level, with a mean failure rate of around 0.5%. However 5.3% of the 5926 samples tested showed some infection. Around 32% of the samples were treated with a product known to give good control of loose smut and in these the incidence of infection was about 1%. In samples not treated with such products the incidence was 7%.

The following examples of seed pedigrees with loose smut infection data were taken from certification records for 1990 to 1993. They show how certification controls are applied and also how patterns of infection build-up can vary.

Results are expressed as percentage of infected ears per plot.

<u>Example 1</u>   <u>Sown seed 1990</u>

Cultivar	Camargue
Treatment:	Ethirimol + flutriafol + thiabendazole
Infection:	Nil

<u>Progeny 1991</u>

	Lot 1	Lot 2	Lot 3	Lots 4	Lots 5-9	Lot 10
Treatment:	None	None	Mercury	Mercury	Ethirimol + flutriafol + thiabendazole	Carboxin + imazalil+ thiabendazole
Infection:	0.42%	0.47%	0.82%	0.42%	NIL	0.06%

The results in example 1 demonstrate that there is always a risk of crops becoming infected with loose smut. Despite the use of a seed treatment, which on the evidence of the control plot was very effective, a serious infection was found in all the untreated 1991 progeny.

<u>Example 2</u>   <u>Sown seed 1991</u>

Cultivar:	Treatment:	Infection:
Bronze	None	0.06%

<u>Progeny 1992</u>

	Lot 1	Lot 2	Lot 3	Lots 4	Lots 5 & 6	Lot 7
Treatment:	None	None	None	None	Ethirimol + flutriafol + thiabendazole	Carboxin + imazalil+ thiabendazole
Infection:	0.38%	0.21%	0.16%	0.11%	NIL	NIL

In example 2 there was a range of infection level and if the samples were representative of the bulk it indicates a reinfection rates of between two-fold and six-fold in the same crop.

<u>Germination and seed treatment</u>

Table 1 clearly shows the value of seed treatments in effecting the recovery of seed samples with sub-minimal levels of germination. Treatment with fungicides has markedly improved the germinability of the seed by 15 percentage points in 1991 and by 25 percentage points in 1992. Levels of germination in 1991 were generally somewhat higher than those in 1992.

TABLE 1.   Mean germination (in %) of winter wheat samples (below 85% germination) before and after treatment 1990 -1992.

Year	1990	1991	1992
Untreated	No examples	78.5 (4 samples)	69 (15 samples)
Treated	"	93.5 (")	94 (")

Figure 3. Bunt infection in wheat

No certification standards apply. Advisory limit 20 spores/g of seed.

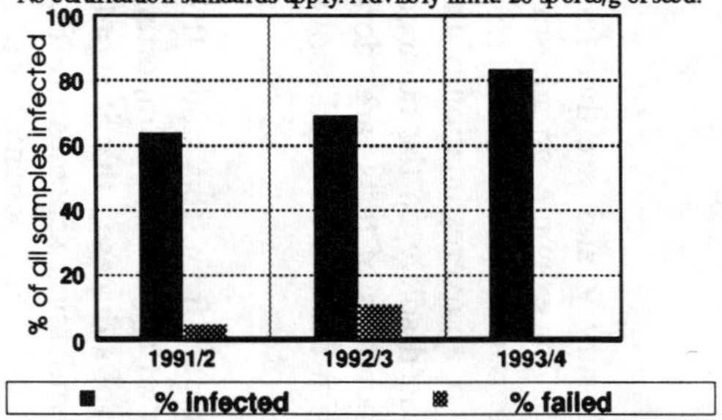

Figure 4. Fusarium nivale infection in wheat

No certification standard. Advisory limit 5%.

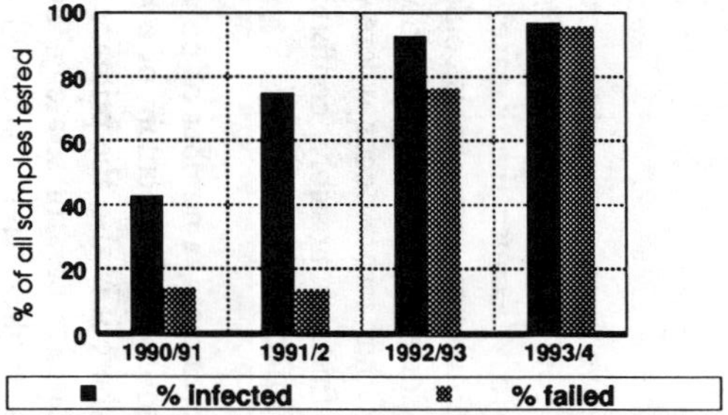

Figure 5. Septoria nodorum infection in wheat.

No certification standards apply.

Figure 1. Loose smut infection in barley

Certification standards: basic=0.1% in 1000; certified=0.2% in 1000. (At the Higher Voluntary Standard)

Figure 2. Leaf stripe and net blotch infection in barley.

No distinction was made between these 2 species in 1990/1 or 1991/2. A marketing standard of 2% infection with leaf stripe is currently applied to seed tests.

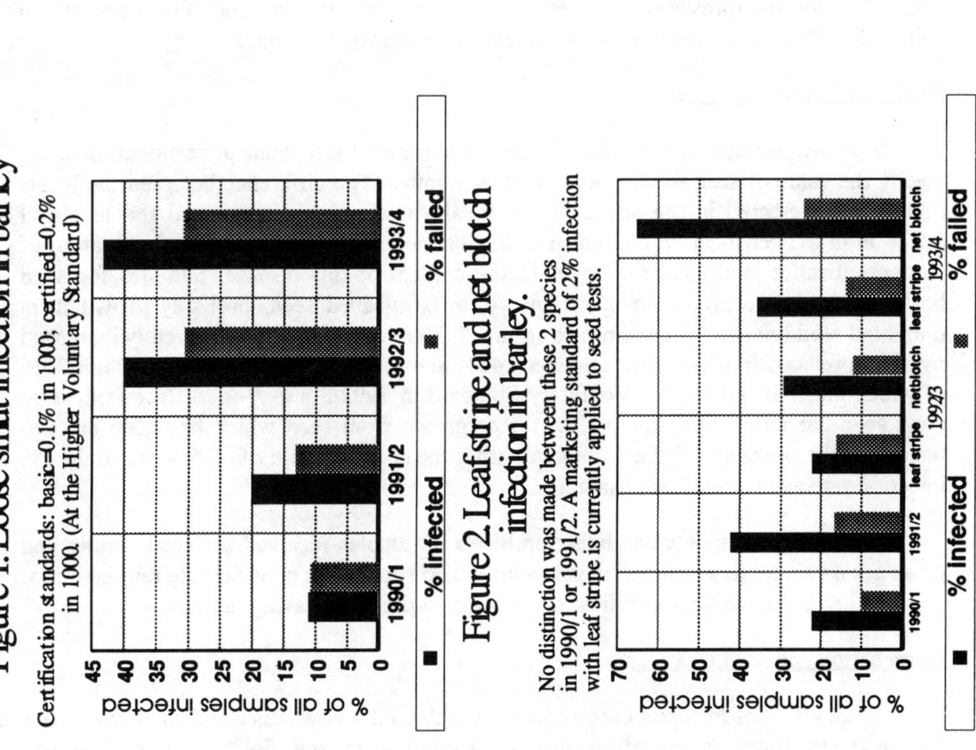

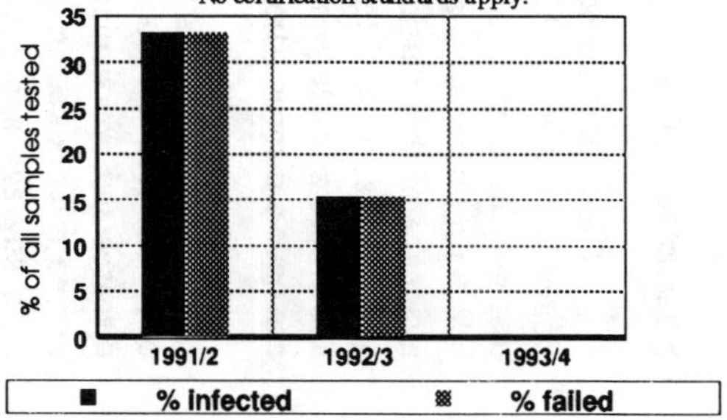

DISCUSSION

Seed health testing

These data do not derive from a properly constituted survey and are therefore potentially subject to bias errors from a number of sources. For example, an advisory test may have been requested because the disease symptoms were seen in the crop which produced the seed. Motivation such as this could lead to an overestimation of the incidence of that particular disease if the estimate were based solely on the numbers and results of seed tests. Other errors occur where small numbers of tests are done and where a single positive finding, possibly confounded with the bias outlined above, can have a disproportionate effect.

Notwithstanding these caveats, a number of points emerge from this study. It is clear that seed is a major source of inoculum of a number of agriculturally important diseases and, particularly where it is also the primary vector of a disease, that it should not be overlooked in any disease control strategy. Seed certification schemes are one such strategy and previous studies have confirmed that certification schemes are of significant value in reducing seed borne disease.

For a number of diseases no certification standards exist. This often arises because insufficient epidemiological data are available to allow standards to be devised which minimise the risk of significant disease levels developing from seed. Clearly more research is required to redress this and should be related to the efficacy of seed treatments and the threshold of infection at which they are required. This information will also affect the design and improvement of certification schemes.

Seed certification and treatment

The two examples presented of the incidence of loose smut in certification show clearly the value of seed treatments in disease control. The difference between the levels of infection detected in the advisory seed health testing programme and the levels of infection in the certification plots also estimates the effectiveness of the combination of seed certification and seed treatment in the control of this disease. Samples for seed health testing were predominantly advisory, ie farm-saved seed, probably grown from untreated seed once or twice removed from C2. Here infection has consistently increased over the previous four years with between 40% and 45% of samples infected, about 30% of which failed at the Higher Voluntary Standard in 1992/3 and 1993/4. In certification plots seed has either been treated (32%) or grown from seed which has been embryo tested for the presence of the pathogen with a mean failure rate of 0.5% and only 7% infection rate even in untreated samples.

The comparison between infection levels in samples received for seed testing and those grown on in the certification procedures also shows that considerable reinfection of the seed has occurred after certification and during the farm-saving activity.

Germination and seed treatment

High levels of *Fusarium* are associated with cool wet weather and such conditions also lead to lower germination due to sprouting in the field prior to harvest.

Additionally, if grain has too high a moisture content, damage may be caused by over rapid drying. However these effects can be separated by comparing the germination results for treated and untreated seed. The use of a treatment will have a beneficial effect on infected seed but can be expected to make no difference or may further reduce germination where seed is damaged in some other way. Table 1 gave the effect of fungicidal seed treatments on levels of germination and it is clear that germination was improved indicating that fungal infection was implicated in the initial depression of germination. *Fusarium* spp. are known to have an effect on germination in wheat seed and it is probable that it is through the control of this pathogen that the seed treatment effect is mediated.

The data indicate some variation between years in the extent to which seed treatments were beneficial in achieving the minimum germination standard for certification. In 1990 weather conditions leading up to harvest were warm and dry and wheat germination levels were relatively high. In 1991 the South West experienced wet, humid weather whilst the rest of England and Wales stayed dry and in 1992 wet weather was general throughout the country. The incidence of adverse weather therefore parallels the number of instances where seed treatment has resulted in increased germination. Mean germination over this period shows a similar pattern (Table 2).

TABLE 2. Germination levels in winter wheat 1990 -1992
% of lots with germination 95% or more

	1990	1991	1992
North	85	71	64
East	84	79	50
South West	81	30	36
% of lots with germination < 85%			
North	<0.1	0.2	0.8
East	0.2	<0.1	1.5
South West	0.2	3.8	1.6

Table 2 indicates a much lower proportion of seed lots from the South West achieving a germination of 95% or more in 1991 than in the North or East, whilst in all regions it was lower in 1992 than 1990. Similarly the number of lots with a low germination (<85%) was higher in the South West than elsewhere in 1991 and higher in 1992 generally than in 1990.

These germination data correlate closely with the incidence of *Fusarium nivale* (Figure 4). In 1992 more seed lots were infected with this pathogen than in previous years but more importantly the number of highly infected seed lots, ie. which failed to meet the advisory standard, rose from about 15% in 1991/2 to 75% in 1992/3. It is probable that this high level of infection is responsible to some extent for the poor germination in 1992. No regional data for *Fusarium* infections are available so it is possible only to speculate that the poor weather in the South West in 1991 was responsible for encouraging *Fusarium* development to the point where germination in this region was affected so markedly.

CONCLUSIONS

Seed can be an important vector of disease and the incidence of some seed borne diseases in the UK is currently at a high level, having increased over recent years. Seed health testing can be used to monitor the emergence of particular disease problems and to indicate the need for application of seed treatments. Control measures involving a combination of seed certification and seed treatment are effective in maintaining seed quality by reducing the proliferation and effects of seed borne diseases.

ACKNOWLEDGEMENTS

We are grateful to Sean Simpkins, Julia Stanton and Gary Stevens for their assistance in preparing the data presented here.

REFERENCES

Anon. (1969) British Cereal Seed Scheme Report. In: *National Institute of Agricultural Botany Fiftieth Report and Accounts 1969*, 52-54.

Anon. (1970) British Cereal Seed Scheme Report. In: *National Institute of Agricultural Botany Fifty-first Report and Accounts 1970*, 58-59.

Anon. (1971) British Cereal Seed Scheme Report. In: *National Institute of Agricultural Botany Fifty-second Report and Accounts 1971*, 51-53.

Anon. (1972) British Cereal Seed Scheme Report. In: *National Institute of Agricultural Botany Fifty-third Report and Accounts 1972*, 51-54.

Anon. (1973) British Cereal Seed Scheme Report. In: *National Institute of Agricultural Botany Fifty-fourth Report and Accounts 1973*, 58-60.

Anon. (1985) The Cereal Seeds Regulations 1985 *S.I. No 976 H.M.S.O*

Anon. (1987) ISTA Handbook on Seed Health Testing, Section 2: Working Sheets, ISTA, Zurich, Switzerland.

Taylor, J. D. (1970) The quantitative estimation of the infection of bean seed with *Pseudomonas phaseolicola* (Burkh.) Dowson. Annals of Applied Biology, 66, 29-36.

Taylor, J. D. (1984) In Report on the 1st International Work shop on Seed Bacteriology, 1982, Angers, 9-14. ISTA, Zurich.

Wray, M.W. and Pickett, A.A. (1985) Trends in loose smut (Ustilago nuda) infections in certified seed of barley in England and Wales 1976 to 1983. *Journal of the National Institute of Agricultural Botany,* Vol. XVII, No 1, 31-40.

CONTROL OF COMMON BUNT (*TILLETIA CARIES (DC) TULL.*) IN DENMARK

BENT J. NIELSEN, LISE NISTRUP JØRGENSEN

Danish Institute of Plant and Soil Science, Department of Plant Pathology and Pest Management
Lottenborgvej 2, DK - 2800 Lyngby.

ABSTRACT

Common bunt, which is normally a rare disease in Denmark, has occurred quite frequently in recent years. First in the mid 1970s and again from 1989. In both cases, unsatisfactory fungicide treatment has been a contributory cause. Trials with new seed treatments show that it is possible to obtain 98-100% control and, in order to stop the propagation of the disease, it has been decided only to give biological approval to products which have a very high degree of efficacy. In the trials, fungicide treatment with e.g. bitertanol/fuberidazole, fenpiclonil, fenpiclonil/imazalil and fenpiclonil/difenoconazole has given 99-100% control, and only these seed treatments have been approved for control of common bunt on winter wheat in Denmark. Only bitertanol/fuberidazole is presently registered and marketed in Denmark as a liquid formulation for control of common bunt. The older powder formulation carbendazim/maneb still holds an approval, but it is not used very often. In Denmark, farmers primarily use fungicide treated certified seed and disease control should be good. In 1993, soil infestation of common bunt was demonstrated in a few localities in Denmark. Soil infestation can make disease control more difficult, in particular if dry summers predominate.

INTRODUCTION

Common bunt (*Tilletia caries*) is normally a very rare disease in Denmark and has only been of minor importances since the use of fungicide treated seed became general. However, in the 1970s rather frequent occurrences of common bunt were observed after several years without attacks. The disease was controlled by means of efficient seed treatments, however, since 1989 there have again been more frequent reports on common bunt (Figure 1).

In Denmark, there is a long tradition of using certified seed, and it is estimated that most of the seed used is certified and fungicide treated. However, in the late 1960s the use of fungicide treated seed declined, and this led to the multiplication of common bunt (Figure 1). From 1976, prebasic, basic and certified seed of the 1st generation was again treated with a full dose of mercury-based products, whereas 2nd generation seed was treated with a full dose of other non-mercury products. In this way the rather frequent occurrences of common bunt in Denmark disappeared for some time (Jørgensen & Nielsen, 1990).

In the 1980s, mercury was gradually replaced by other products. In 1983/84, 22% of the seed grain was treated with mercury which is equivalent to the first three generations of seed grain (normally 20-25%). The use of mercury dropped over a 10-year period and stopped completely in 1989/90 (Table 1).

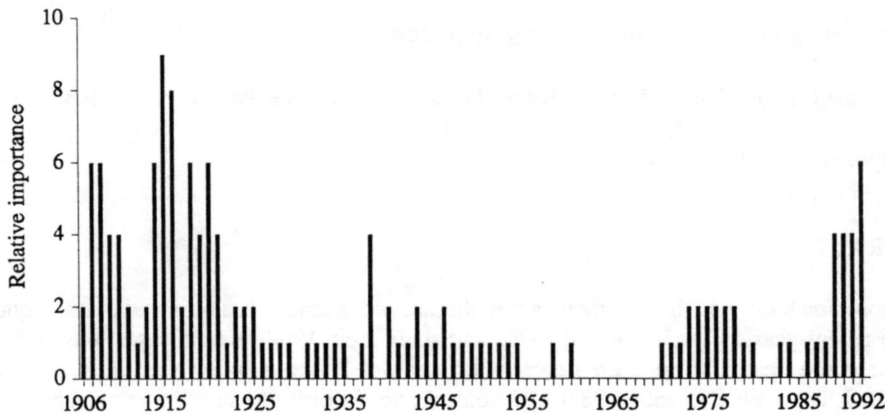

Figure 1. Attacks of common bunt in Denmark, estimated from disease surveys made by the Danish Institute of Plant and Soil Science (Stapel, Jørgensen & Hermansen, 1976 and Stapel & Nielsen, 1992).

Table 1. Use of Hg-seed treatment for cereals in Denmark 1982/83-1989/90. (Jørgensen & Nielsen, 1990).

Year	Cereal seed lots treated with Hg Per cent
1982/83	22.2
1983/84	21.8
1984/85	12.5
1985/86	13.7
1986/87	9.4
1987/88	1.7
1988/89	1.1
1989/90	0

The products, which since then have been used for fungicide treatment of wheat, are Na-N-dimethyldithiocarbamate/fuberidazole ('Neo-Voronit' which was on the market until 1991), carbendazim/maneb ('Derosal M bejdse'), guazatine ('Panoctine 30') and bitertanol/fuberidazole ('Sibutol LS 280' which came on the market in 1989, Table 2).

In 1989, after some years with sporadic occurrences, more common attacks of bunt were reported (Figure 1). Again, the propagation of common bunt was expected to be a consequence of unsatisfactory fungicide treatment. However, the role of soil borne inoculum which in the same period has been reported from the U.K. (Yarham & Jones, 1992) cannot be excluded.

In the light of this experience it was decided to intensify the demands on the efficacy of seed treatments and only approve the most effective products. In addition, it was impressed on the processors of seed grain that the quality of the fungicide treatment should be high.

The data below are from a number of the trials that have been established in order to examine the effect of new seed treatments on common bunt. The trials form part of the basis for the official approval of seed treatments (Nielsen & Jørgensen, 1986, 1989). In Denmark, the biological approval carried out at the Danish Institute of Plant and Soil Science is voluntary and is independent of the registration made by the National Agency of Environmental Protection. In practice, however, only seed treatments that have obtained a biological approval for the area in question are recommended and used.

MATERIALS AND METHODS

Trials to examine the effect of seed treatments were established as small plot trials with rows of 400 plants and with 4 replicates. In addition, yield trials with bigger plots (1.5 x 10 m) were established to examine the yield effect and possible phytotoxic effects of the seed treatments. In the trials, healthy seed, which prior to fungicide treatment had been inoculated with 0.25 g and 5.0 g bunt spores per kg, respectively, was used. Fungicide treatment was carried out in glass cylinders, which were rotated for 5 minutes after treatment. The standard dosages used in the official testing experiments are now 1/2, 3/4, 1/1 and 2/1 for observation trials and 1/2, 3/4 and 1/1 dose for yield trials.

Table 2. Seed treatment products and active ingredients.

Product	Dose/100 kg	Appr.	Active ingredients (a.i.)	g a.i. per l/kg
Beret FS 050	400 ml	(w)	fenpiclonil	50
Beret FS 060	400 ml	(w)	fenpiclonil/imazalil	50/10
Beret Combi	200 ml	(w)	fenpiclonil/difenoconazole	50/50
Derosal M bejdse	150 g	w	carbendazim/maneb	150/600
DLG Manebbejdse	200 g	sw	maneb	700
Fungazil C	200 ml		carboxin/imazalil	400/25
Neo-Voronit	250 ml		Na-N-dimethyldithiocarbamate/-fuberidazole	300/5
Panoctine 30	200 ml		guazatine	300
Sibutol LS 280	100 ml	w	bitertanol/fuberidazole	250/18
Vitavax 200 FF	250 ml		carboxin/thiram	200/200

w : Products with a biological approval against common bunt in winter and spring wheat in Denmark.
sw : Spring wheat only. (w) : Products which have an approval in wheat, but are not registered for use yet.

RESULTS

Table 3 shows the results of trials with the seed treatments that were used in the late 1980s. Only bitertanol/fuberidazole has given full control, whereas the other products have had weaker effects and show a rapidly decreasing effect when treated with half dose.

Table 3. Control of common bunt, 6 trials on winter wheat, 1987-90.

	Standard dose (1/1) g a.i./100 kg seed	Percentage control		
		Dose		
		1/2	1/1	2/1
Na-N-dimet.dt.carb./fuberidazole	75/1.3	68[a]	87[c]	97[d]
Guazatine	60	81[b]	88[c]	96[d]
Bitertanol/fuberidazole	50/3.6	100[e]	100[e]	100[e]
Percentage attacked plants in untreated control			42,0	

Values with the same letter do not differ significantly (p ≤ 0.05)

Table 4 shows the results of trials carried out in recent years with two disease levels. Even at the low disease level, guazatine has a relatively weak effect compared to bitertanol/fuberidazole (which in this trial has been tested with 100 ml product).

Table 4. Control of common bunt, 3 trials on winter wheat, 1991-93 using different doses at two disease levels. Artificial inoculation with 0.25 and 5.0 g bunt spores per kg wheat.

	Standard dose (1/1) g a.i./ 100kg seed	Percentage control 0.25 g spores/kg wheat				Percentage control 5 g spores/kg wheat			
		Dose				Dose			
		1/2	3/4	1/1	2/1	1/2	3/4	1/1	2/1
Carb./maneb[1)]	22.5/90	100[d]	100[d]	100[d]	100[d]	96[c]	98[d]	98[d]	99[de]
Guazatine	60	80[a]	93[b]	97[c]	95[b]	89[a]	90[b]	95[c]	96[c]
Bitert./fub.	25/1.8	100[d]	100[d]	100[d]	100[d]	100[e]	100[e]	100[e]	100[e]
Percentage attacked plants in untreated control				6.6				22.5	

Values with the same letter at each dicease level do not differ significantly (p ≤ 0.05)
[1)] Carb./maneb = Carbendazim/maneb, Bitert./fub. = bitertanol/fuberidazole

Common bunt can lead to severe yield losses and the close correlation between percentage diseased plants and yield loss is shown in Table 5.

A summary of the trials carried out in recent years with seed treatments on wheat is shown in Figure 2. The results of the individual products originate from different trials and the effects can therefore not be compared directly. However, the figures provides a good picture of the potential effects of the various seed treatments on common bunt, and it appears that there are several seed treatments which are very efficient.

Table 5. Control of common bunt, 7 trials on winter wheat, 1991-93.

	Rate g a.i./100 kg seed	Percentage control of common bunt	Yield t per ha
Guazatine	60	89[a]	6.48[b]
Bitertanol/fuberidazole	12.5/0.9	98[d]	6.57[ab]
Bitertanol/fuberidazole	18.8/1.4	99[e]	6.74[a]
Bitertanol/fuberidazole	25.0/1.8	99[e]	6.72[a]
Carboxin/thiram	30/30	95[b]	6.62[ab]
Carboxin/thiram	40/40	97[c]	6.67[ab]
Carboxin/thiram	50/50	98[d]	6.69[ab]
Untreated control		26.6% [1]	5.13

Values with the same letter in each column do not differ significantly (p $\leq$ 0.05)
[1] Percentage attacked plants in untreated control.

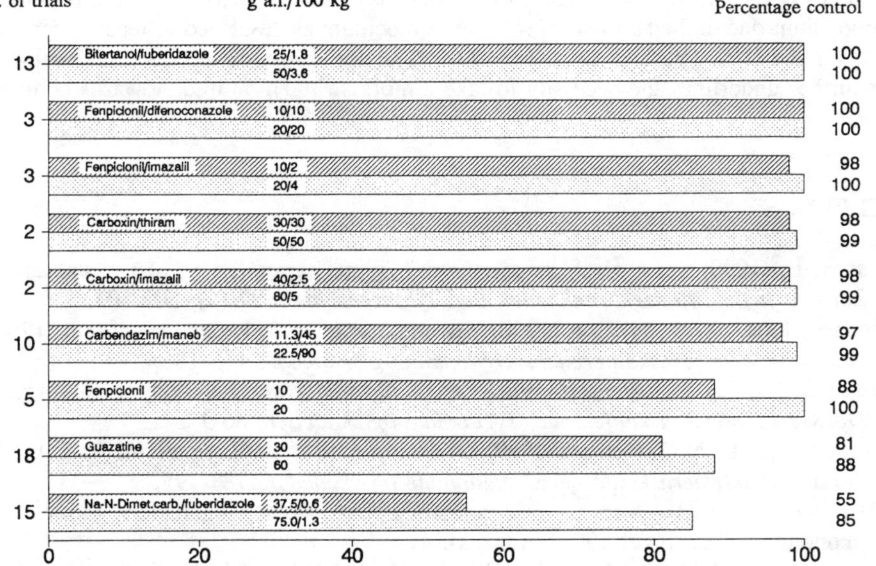

Figure 2. Efficacy of different products against common bunt in winter wheat in Denmark.

DISCUSSION

Trials with fungicide treatment against seed borne common bunt shows that it is possible to obtain almost 100% control. In all trials, bitertanol/fuberidazole has shown 99-100% control, even at low doses. The product was approved at a rate of 200 ml/hkg in 1987, which based on trials results was changed to 150 ml in 1990 and finally to 100 ml/hkg in 1991. The efficacy against Fusarium spp. and Septoria nodorum using this dose is considered satisfactory under Danish conditions. It was shown that the older seed treatment Na-N-dimethyldithiocarbamate/-

fuberidazole and guazatine had a lower efficacy as regards common bunt, which was also seen in Swedish trials (Olofsson & Johnson, 1985). Seed treatments with fenpiclonil, which proved to be effective against bunt, especially in combination with e.g. difenoconazole, are still under evaluation for registration in Denmark.

In order to stop the development of common bunt, it was decided in 1991 only to approve the most effective seed treatments for control of common bunt. The biological approvals are shown in Table 2 and among these products only bitertanol/fuberidazole (100 ml) is now on the market and can be used for control of common bunt in wheat. Guazatine now only holds an approval for control of <u>Fusarium spp.</u> and <u>Septoria nodorum</u> and is recommended for wheat only when the seed is free of common bunt.

In 1993, signs of soil infestation with common bunt were found in Denmark. The soil infestation was observed in an area where clean, untreated wheat had been sown in September 1992, following a crop with 10% bunt in summer 1992. Soil infestation has also been demonstrated in other Danish trials (Nielsen, 1993). Earlier it had been reported in Swedish trials that spores of <u>T. caries</u> could survive for 10 years in the soil (Johnson, 1990), but soil-borne spores are only regarded as having epidemiological significance in winter wheat regions with arid summers (Wiese, 1987). The recent dry years may have contributed to the development of common bunt due to the survival of soil borne inoculum as described in recent years in certain regions in the U.K. (Yarham & Jones, 1992). The possibility of soil infestation with common bunt further underlines the necessity to take a more stringent attitude towards controlling the disease.

REFERENCES

Johnsson, L. (1990) Survival of common bunt (<u>Tilletia caries</u> (DC) Tul.) in soil and manure. *Zeitschrift für Pflanzenkrankheiten und Pflanzenschutz*, **97** (5), 502-507.

Jørgensen, Johs. ; B. J. Nielsen (1990) Bekæmpelse af udsædsbårne sygdomme i hvede. *7. Danske Planteværnskonference, sygdomme og skadedyr*, 169-186.

Nielsen, B. J. ; L. N. Jørgensen (1986) Bejdsning mod svampesygdomme på korn, 1985. *3. Danske Planteværnskonference, sygdomme og skadedyr*, 86-97.

Nielsen, B. J. ; L. N. Jørgensen (1989) Bejdsning mod svampesygdomme på korn, 1988. *6. Danske Planteværnskonference, sygdomme og skadedyr*, 183-195.

Nielsen, G. C. (1993) Oversigt over landsforsøgene. Forsøg og undersøgelser i de Landøkonomiske foreninger 1993 (in press).

Olofsson, B.; L. Johnson (1985) Försök rörande kvicksilverfria betningsmedel för stråsäd. *Växtskyddsrapporter*, 67 pp.

Stapel, Chr.; Johs. Jørgensen ; J. E. Hermansen (1976) Sædekornets sygdomme i Danmark, deres udbredelse, betydning og bekæmpelse ved afsvampning, især i perioden 1906-1975. *Tidsskrift for Landøkonomi*, **163**, 185-283.

Stapel, Chr. ; G. C. Nielsen (1992) Unpublished survey report, Danish Institute of Plant and Soil Science.

Wiese, M. V. (1987) Common bunt (stinking smut). In : *Compendium of Wheat Diseases, APS Press*, 19-20.

Yarham, D. J. ; D. R. Jones (1992) The forgotten diseases : Why we should remember them. *Brighton Crop Protection Conference - Pest and Diseases*, 1117-1126.

EVALUATION OF BROAD SPECTRUM SEED TREATMENTS FOR THE CONTROL OF CEREAL
FOLIAR DISEASES IN SCOTLAND

K.G. Sutherland, S.J. Wale
Scottish Agricultural College, 581 King Street, Aberdeen, AB9 1UD

S.J.P. Oxley
Scottish Agricultural College, West Mains Road, Edinburgh, EH9 3JG

ABSTRACT

The use of broad spectrum cereal seed treatments for control of
seed-borne diseases has increased over the past two seasons with
the withdrawal of organomercury products. The advantage of
these seed treatments is their activity against foliar diseases
as well as seed-borne diseases. This paper reviews their
efficacy at controlling foliar diseases and increasing yields in
cereals grown under Scottish conditions.

INTRODUCTION

Cereal fungicide seed treatments, in particular the organomercury
fungicides, have been used for over 50 years for the control of seed-
borne diseases such as barley leaf stripe (*Pyrenophora graminea*),
seedling blight (*Monographella nivalis = Fusarium nivale*) and wheat
covered smut or bunt (*Tilletia tritici = Tilletia caries*). These early
seed treatments had no activity against foliar diseases. The development
of ethirimol in the late 1960s (Brooks, 1970) provided the first systemic
seed treatment specifically for the control of a foliar disease, powdery
mildew (*Erysiphe graminis*) in barley. In 1977, 93% of the cereal seed
sown in Scotland received an organomercury seed treatment for control of
seed-borne diseases (Steed *et al.*, 1979), most being applied to the
spring barley crop which comprised 85% of the cereals grown at this time
(Table 1). The use of ethirimol in combination with organomercury
occurred on 43% of spring barley seed.

Subsequent to 1977 the area of winter barley and winter wheat grown
in Scotland began to rise, with a corresponding reduction in the spring
barley area (Table 1). The rise in winter barley acreage posed a
potential threat to the spring barley crop, providing a 'green bridge'
for carry over of powdery mildew (Yarham *et al.*, 1971). Likewise, spring
barley posed a threat to early sown winter barley crops.

The development of several broad spectrum seed treatments in the
1970s and 80s, including fuberidazole + triadimenol (Baytan - Bayer UK
Ltd) and ethirimol + flutriafol + thiabendazole (Ferrax - Zeneca) which
could give control of both seed-borne and foliar diseases provided a
single alternative to the organomercury/ethirimol two treatment option.
Unlike ethirimol these new broad spectrum seed treatments were also
effective against wheat mildew, yellow rust (*Puccinia striiformis*) and
other foliar diseases. However, their use was limited in comparison to
that of the cheap organomercury products until the withdrawal of
organomercury from the UK market in 1992 (Table 1). Growers must now
make a choice between a broad spectrum seed treatment or one of the new
alternatives to mercury specifically for seed-borne disease control.

Table 1. Summary of fungicide seed treatment used on cereals in Scotland.

Crop	Active ingredient	% Crops Treated/Year					
		1974	1977	1982	1988	1990	1992
Spring barley	organomercury	92	95		79	82	48
	ethirimol	-	43		0.5	0.5	-
	fuberidazole+ triadimenol	-	-		8	5	11
	ethirimol + flutriafol + thiabendazole	-	-		8	7	22
(% of total cereal crop)		78	85	78	61	57	55)
Winter barley	organomercury	-	-		86	69	39
	fuberidazole + triadimenol	-	-		20	7	17
	ethirimol + flutriafol + thiabendazole	-	-		15	18	25
(% of total cereal crop		0	+	8	13	14	13)
Winter wheat	organomercury	43	89		86	69	39
	fuberidazole + triadimenol	-	-		15	23	48
(% of total cereal crop		7	4	8	19	23	26)
Spring oats	organomercury	90	83		79	69	66
	fuberidazole + triadimenol	-	-		0	1	0
	ethirimol + flutriafol + thiabendazole	-	-		1	0	2
(% of total cereal crop		16	11	6	7	6	5)

Chapman *et al.*, 1977; Steed *et al.*, 1979; Snowden *et al.*, 1991a, 1991b; Bowen *et al.*, 1993.

This paper reviews work carried out by the Scottish Agricultural College (SAC) over the past eleven years to determine the effectiveness of systemic broad spectrum seed treatments for control of foliar diseases in the major cereal crops grown in Scotland. Unless stated, in

the trials presented comparing broad spectrum seed treatments with
organomercury seed treatment foliar fungicide oversprays have also been
applied; but for any one trial the overspray(s) has been the same
fungicide applied at the same timing.

SPRING BARLEY

The main disease of spring barley in Scotland is powdery mildew but
Rhynchosporium (*Rhynchosporium secalis*) can occur in wet seasons. Net
blotch (*Pyrenophora teres*) and brown rust (*Puccinia hordei*) are of less
importance. Trials carried out in north-east Scotland during the 1970s
showed the use of a broad spectrum seed treatment of fuberidazole +
triadimenol gave good foliar disease control and significant yield
increases in spring barley (Wale & Shipton, 1981). These results preceded
the development of insensitivity by *E. graminis* to 'triazole' fungicides.
Prior to this time the use of ethirimol had not been generally
recommended in Scotland. For example in the south of Scotland Channon &
Boyd (1973) found no yield advantage from the use of ethirimol, except
where mildew infections occurred early or were known to occur in local
areas when its use was advised (Channon & Clark, 1980).

When the effect of broad spectrum seed treatments alone (without
subsequent foliar sprays) was tested, the yield responses when disease
pressure was high, as when fuberidazole + triadimenol was evaluated,
generally covered the cost of the seed treatment (Table 2). In years
when mildew levels were low, when ethirimol + flutriafol + thiabendazole
was evaluated, the cost of the seed treatment was not covered and
resulted in an overall loss of £14.21/ha. These trials were done on the
cultivar Golden Promise which was fully susceptible to powdery mildew
(Anon., 1987). Most modern varieties show better resistance to powdery
mildew (Anon., 1992). Mildew susceptibility must be considered when seed
treatment decisions are taken.

**Table 2. Effect of broad spectrum seed treatments on yield of spring
barley, margin over cost and mildew infection compared to an
organomercury control, 1982-92; cv. Golden Promise.**

Seed treatment	Yield (t/ha) at 15% MC)		Margin/cost (£/ha)	mildew - mean % LAI	GS 45-65
organomercury	4.85	4.97		25.4	tr
fuberidazole+ triadimenol	5.02	-	2.93	23.3	-
ethirimol+ flutriafol+ thiabendazole	-	4.98	-14.21	-	tr
no. of trials	3	10			

Margin/cost calculated as (yield broad spectrum - yield organomercury x
£94) - (cost of broad spectrum - cost mercury). Based on prices at time
of publishing. No fungicide oversprays.

In trials where broad spectrum seed treatments with subsequent foliar sprays were compared to one or two foliar sprays, they resulted in relatively good control of mildew but were not always cost effective. In one series of trials (Table 3, Series A), mildew control from a fuberidazole + triadimenol seed treatment with an over-spray was equivalent to a two spray programme. The yields, however, were no greater than a single well timed spray.

In a second series of trials, an ethirimol + flutriafol + thiabendazole seed treatment with a single 'early' overspray gave a lower yield and was less cost effective than a two spray programme 'early' and 'late' (Table 3, Series B).

Table 3. Effect of broad spectrum seed treatment plus foliar spray(s) at different timings on yield of spring barley, fungicide cost and mildew infection, cv. Golden Promise.

SERIES A (3 trials) 1981-82

Seed treatment	Spray timing			Yield (t/ha) at 15% MC	Cost of fungicide (£/ha)	mildew % LAI GS 71-80
	early	mid	late			
organomercury	x			5.52	23.60	38.8
organomercury		x		5.67	23.60	23.8
organomercury	x		x	5.88	37.41	11.8
fuberidazole+ triadimenol			x	5.60	36.65	10.5

SERIES B (1987-89)

Seed treatment	Spray timing		No. trials	Yield (t/ha) at 15% MC		mildew - % LAI at GS 65-81	
	early	late		G.P.	Golf	G.P.	Golf
organomercury *			3	4.01	4.83	46.1	30.8
organomercury	x	x	3	5.03	5.66	7.9	4.5
fuberidazole+ triadimenol	x		2	5.00	5.35	36.8	22.0
ethirimol+ flutriafol+ thiabendazole	x		3	4.77	5.22	23.6	11.7

* - organomercury seed treatment with no foliar fungicide

One of the main reasons for use of a seed treatment with action against foliar diseases in Scotland is 'peace of mind'. Their use in spring barley gives growers the chance to delay the timing of the foliar spray without affecting yield. In practice, the delay may only be a few days, but in an area where mixed arable farms are common and silage

making is an important feature of the farm calender, one week delay in
spraying spring barley crops can give the grower breathing space.

Where a suceptible variety such as Blenheim is grown, on farms where
mildew is a problem or where spring barley is grown in close proximity to
winter barley an early foliar spray may have to be followed up by a
second spray. At full dose, two foliar sprays are usually more expensive
than a seed treatment plus one spray. The use of broad spectrum seed
treatments in these situations is more favourable.

Rhynchosporium is an important disease of spring barley in wet
seasons, when grown in damp coastal areas and in areas known to show high
levels of infection, especially where a suceptible variety such as
Prisma, Blenheim or Derkado is grown.

The use of fuberidazole + triadimenol gave only a small yield
benefit in eight field trials where Rhynchosporium infections was absent
or at low levels (Table 4). The yield response just covered the cost of
treatment. In four trials where Rhynchosporium levels were higher, the
average yield response was 0.78 t/ha, making this a very profitable
treatment. However, unlike mildew, Rhynchosporium is more sporadic in
occurence and there is less justification for routine treatment.

**Table 4. Effect of broad spectrum seed treatment on yield of spring
barley, margin over cost and control of Rhynchosporium compared to an
organomercury control, cv. Maris Mink 1978-83.**

Seed treatment	Yield (t/ha) at 15% MC	
	Rhynchosporium infection:	
	nil - slight	Mod-sev
organomercury	5.43	2.88
fuberidazole+ triadimenol	5.58	3.67
Margin/cost (£/ha)	1.05	61.21
no. of trials	8	4

Yield equivalent to cost of fuberidazole + triadimenol = 0.14 t/ha.
No fungicide oversprays.

WINTER BARLEY

On average, treatment with fuberidazole + triadimenol gave no yield
advantage over an organomercury control in 17 trials carried out between
1982 and 1990, resulting in an overall economic loss to the grower (Table
5). Ethirimol + flutriafol + thiabendazole consistently gave slightly
higher yields than fuberidazole + triadimenol. Where the later was used,
brairding was consistently delayed for 3 to 5 days compared to that of
the organomercury control. In some trials the seed treatment reduced
autumn mildew or early spring Rhynchosporium infection but in others
there was no effect. In all trials levels of both powdery mildew and
Rhynchosporium at the end of the season, after foliar sprays had been

applied, were similar irrespective of whether a broad spectrum seed
treatment had been used or not. The value of a broad spectrum seed
treatment was much greater the earlier trials were sown (Wale, 1987).

**Table 5. Effect of broad spectrum seed treatment on yield of winter
barley, margin over cost and disease control compared to an organomercury
control, 1982-1990.**

Seed treatment	No. trials	Yield (t/ha) at 15% MC	Margin/cost (£/ha)	Disease assessments % LAI at GS 45-71	
				mildew	Rhyncho
organomercury	19	6.80		8.8	3.5 [11]
fuberidazole + triadimenol	17 (3)	6.73	-19.63	7.7	6.0 [9]
ethirimol + flutriafol + thiabendazole	13 (0)	7.11	13.99	5.1	6.0 [7]

() no. trials where yield increase significant at p=0.05.
[] no. trials where diseases assessments carried out

WINTER WHEAT

The area of winter wheat grown in Scotland has gradually increased
from 30,000 ha (6.6% of the cereal crop) in 1974 to 120,000 ha (26.4% of
the cereal crop) in 1992 (Table 1), making it the second most important
cereal crop in Scotland. Until 1988 much of the crop was still treated
with organomercury seed treatment for control of seed-borne diseases but
this rapidly decreased until 1992 when only 39% of seed was organomercury
treated. There was a corresponding increase in broad spectrum seed
treatments, culminating in almost half the seed sown in 1992 being
treated.

The most important foliar diseases of winter wheat are powdery
mildew, *Septoria tritici* and yellow rust. In a series of trials carried
out by SAC between 1984 and 1992 in east and north-east Scotland, the use
of a broad spectrum seed treatment consistently increased the yield of
winter wheat compared to the organomercury treated control (Table 6).
One in five of the trials gave a significant yield increase. Increases
in yields were accompanied by a corresponding (but not significant)
decrease in foliar disease infections: *Septoria tritici* was the main
disease. In general, the cost of the seed treatment was covered by yield
benefits.

Of the trials where the use of a fuberidazole + triadimenol seed
treatment gave significant yield benefit over the organomercury control,
the increase in yields could not be associated with large reductions in
disease levels (Table 7). Economic benefits were large and seed
treatment use was cost effective.

Table 6. Effect of broad spectrum seed treatment on foliar diseases control and yield of winter wheat compared to an organomercury control, 1984-92, various cultivars.

Seed treatment	Yield (t/ha) at 15% MC	Disease assessments-% LAI at GS 31-83		
		Powdery mildew	Septoria tritici	Yellow rust
organomercury*	6.60	3.2	11.2	13.1
organomercury	8.34	1.7	6.1	2.1
fuberidazole+ triadimenol	8.71	1.4	4.4	0.6
Margin/cost (£/ha)	21.64			
total no. trials	20	14	15	10
no. where increased	12 (4)	4 (0)	4 (0)	0 (0)
no. where decrease	7 (0)	7 (0)	8 (0)	7 (1)

() no. trials where increase or decrease significant at p = 0.05
* organomercury seed treatment with no foliar fungicides.
Yield equivalent of cost of fuberidazole + triadimenol = 0.14 t/ha.

Table 7. Relationship between yield increase and disease reduction in seed treated winter wheat compared to organomercury control where yield response was significant

Seed treatment	Yield (t/ha) at 15% MC	Disease assessments-% LAI at GS 59-83		
		Powdery mildew	Septoria tritici	Yellow rust
organomercury*	5.42	2.7	4.4	9.4
organomercury	7.52	0.8	3.0	2.0
fuberidazole + triadimenol	8.62	0.8	1.4	1.1
Margin/cost (£/ha)	93.18			

* organomercury seed treatment with no foliar fungicides applied

Broad spectrum seed treatments gave reductions in foliar disease
levels at the time of the first foliar fungicide spray (GS 31) compared
to the organomercury control (Table 8), which were maintained through to
later growth stages.   Subsequent foliar fungicide oversprays gave equal
control of disease in both the mercury and fuberidazole + triadimenol
seed treatments: there were no interactions. Foliar sprays, at GS 31, 39
and 59, usually give large yield responses, much larger than that of the
broad spectrum seed treatment (Sutherland *et al.*, 1993).

Sowing dates of winter wheat in Scotland vary from September through
to November. The later the crop is sown, the colder the seed-bed and
hence the more prone the crop is to delayed emergence, slower growth and
increased risk from seedling blight. Field observations have shown the
use of a fuberidazole + triadimenol seed treatment can delay emergence
from a few days up to a few weeks depending on sowing date. There was a
trend towards earlier sown crops giving a positive yield benefit from the
use of a broad spectrum seed treatment (Table 9) but there was no
corresponding reduction in disease levels.

**Table 8. Persistance of seed treatment for control of foliar diseases on
winter wheat, compared to an organomercury control.**

Seed treatment	Disease Assessments			-	% LAI	
	GS 31-32 (before fungicide oversprays)			GS 59-83 (after fungicide oversprays)		
	Pm	St	Yr	Pm	St	Yr
organomercury*	0.9	2.9	0.2	1.9	13.6	9.1
organomercury	0.9	3.8	0.3	1.2	4.6	1.3
fuberidazole+ triadimenol	0.4	2.6	0.1	1.3	3.5	0.6

total no. trials = 10   no. of trials where decrease sig. = 0

Pm = powdery mildew; St = Septoria tritici;   Yr = Yellow rust
*organomercury seed treatment with no foliar fungicides applied

CONCLUSIONS

Although early seed treatment trials on spring barley showed good
disease control and yield responses, more recent work has shown their use
solely for the control of foliar diseases is not cost effective.
However, in situations of high disease pressure (susceptible cultivar,
locality, proximity to winter barley) or where the grower, because of
other on-farm commitments, wants to delay the time of spraying, broad
spectrum seed treatments are advisable. Similar advice can be given for
winter barley.

The use of fuberidazole + triadimenol seed treatment in winter
wheat for the control of foliar disease is not essential. Only where a
threat of early infection of yellow rust exists (early sowing,
susceptible variety) is this seed treatment easily justified since where
late attacks occur, warnings of outbreaks south of the border allow
growers in Scotland to tailor their GS 31 sprays accordingly. On average

fuberidazole + triadimenol gave a cost-effective yield response. This may
be due in part to foliar disease control but it may also be related to
other factors such as reduction in autumn take-all and stem base disease
and improved rooting which occur sometimes. These other factors are most
important in early sown crops and, because fuberidazole + triadimenol can
cause delays in emergence in later sowings, this seed treatment is most
strongly recommended for use in early sowing. The trial results broadly
support this contention.

**Table 9. Effect of sowing date on yield of winter wheat, foliar disease
control and margin over cost for a broad spectrum seed treatment compared
to an organomercury control.**

Yield (t/ha) at 15% MC

Sowing date	Seed treatment organo-mercury*	organo-mercury	fuberid+ triad	Margin/ cost (£/ha)	No. trials
30 Sept - 10 Oct	6.27	7.23	7.68	29.48	7 (3)
11 Oct - 20 Oct	7.55	9.10	9.14	-10.70	8 (0)
21 Oct - 31 Oct	7.76	8.76	8.88	- 2.86	2 (0)
1 Nov - 6 Nov	6.09	9.13	10.25	95.14	2 (1)

* organomercury seed treatment with no foliar fungicide applied
( ) - no. where yield increase significant

Disease assessments - % LAI at GS 59-83

	Seed treatment organomercury*			organomercury			fuberidazole + triadimenol		
	Pm	St	Yr	Pm	St	Yr	Pm	St	Yr
30 Sept- 10 Oct	4.9	7.8	4.4	1.4	2.4	0.2	1.0	1.3	0.1
11 Oct - 20 Oct	2.1	14.9	3.0	2.0	5.7	0.4	1.7	4.7	0.1
21 Oct - 31 Oct	1.0	4.1	3.0	0.0	0.4	0.0	0.0	0.5	0.0
1 Nov - 6 Nov	1.0	0.5	15.7	0.6	0.2	3.7	0.8	0.04	1.9

* organomercury seed treatment with no foliar fungicide applied
Pm = powdery mildew   St = Septoria tritici   Yr = Yellow rust

With the withdrawal of organo-mercury products and the introduction
of more expensive non-mercury alternatives, the price differences between
these and broad spectrum seed treatments are narrower making the use of
broad spectrum seed treatments with the added bonus of foliar disease
control more inviting to growers.

REFERENCES

Anon. (1987) Cereals recommended list 1988, Scottish Agricultural Colleges, Edinburgh, 12pp.

Anon. (1992) SAC cereal recommended list for 1993, SAC/HGCA, Edinburgh, 16pp.

Bowen, H.M.; Snowden, J.P.; Thomas, L.A. (1993) Pesticide Usage in Scotland Survey Report 117, Arable Crops 1992, SASA, Edinburgh (in press).

Brooks, D.H. (1970) Powdery mildew of barley and its control. *Outlook on Agriculture*, **6**, 122-127.

Channon, A.G.; Clark, R.M. (1980) Disease control in winter and spring barley. *West of Scotland Agricultural College Technical Note* No. **90**.

Channon, A.G.; Boyd, A.G. (1973) The effect of some fungicides on mildew of spring barley in the south-west of Scotland. *Proceedings 7th British Insecticide and Fungicide Conference 1973*, 21-28.

Chapman, P.J.; Sly, J.M.A.; Cutler, J.R. (1977) Pesticide Usage Survey Report 11, Arable Farm Crops 1974, MAFF/DAFS, London, 139pp.

Snowden, J.P.; Bowen, H.M.; Dickson, J.M. (1991a) Pesticide Usage in Scotland Survey Report 77, Arable Crops 1988, SOAFD, Edinburgh, 83pp.

Snowden, J.P.; Bowen, H.M.; Dickson, J.M. (1991b) Pesticide Usage Survey in Scotland Survey Report 87, Arable Crops 1990, SAFD, Edinburgh, 84 pp.

Steed, J.M.; Sly, J.M.A.; Tucker, G.G.; Cutler, J.R. (1979) Pesticide Usage Survey Report 18, Arable Farm Crops 1977, MAFF/DAFS, London, 155 pp.

Sutherland, K.G.; Wale, S.J.; Oxley, S.J.P. (1993) Effect of GS 31 fungicide sprays on yield benefit and disease control in winter wheat. *Proceedings Crop Protection in Northern Britain 1993*, 115-120.

Wale, S.J.; Shipton, P.J. (1981) Spring barley seed treatments for the control of foliar diseases. *Proceedings Crop Protection in Northern Britain 1981*, 27-32.

Wale, S.J. (1987) Effect of fungicide timing on yield of winter barley in northern Britain. *Proceedings Crop Protection in Norther Britain 1987*, 61-66.

Yarham, D.J.; Bacon, E.T.G.; Haywood, C.F. (1971) The effect on mildew development of the widespread use of fungicide on winter barley. *Proceedings 6th Insecticide and Fungicide Conference 1971*, 15-25.

THE USE OF GUAZATINE-BASED PRODUCTS FOR THE CONTROL OF SEED-BORNE DISEASES OF CEREALS

T.W.COX

Rhône-Poulenc Agriculture, Fyfield Road, Ongar, Essex. CM5 0HW, U.K.

G.MUSSARD

Rhône-Poulenc Secteur Agro, 14-20 Rue Pierre Baizet, BP 9163, 69263, Lyon, Cedex 09, France

ABSTRACT

Guazatine, used as a treatment for wheat seed has had a significant impact in the UK since the withdrawal of organomercury. The seed-borne diseases - *Fusarium nivale*, *Tilletia caries*, and *Leptosphaeria nodorum* have been shown to be well controlled by the fungicide.

In barley, guazatine plus imazalil has been shown to give improved control of *Pyrenophora graminea*, *Cochliobolus sativus* and a useful effect against *Ustilago nuda*.

INTRODUCTION

Fungicidal properties of the salts of guazatine were first reported by Catling <u>et al.</u>(1968). The triacetate was later developed as a cereal seed treatment by Murphy Chemical in the UK and by Kenogard in Sweden.

The first reports of activity against *Leptosphaeria nodorum, Monographella nivalis (Fusarium nivale), Tilletia caries, Pyrenophora graminea, P.avenae* and *Ustilago avenae* were confirmed by Jackson <u>et al</u>. (1973).

In an attempt to provide the same level of control as organomercury against *P.graminea*, on barley, Bartlett and Ballard (1975) were successful when a mixture of guazatine (0.6g/kg seed) and imazalil (0.04g/kg seed) was used.

Guazatine alone, and in mixture with imazalil , has been used in a number of European countries since 1973. It has been jointly developed in the UK by Rhône-Poulenc as 'Panoctine' (wheat) or 'Panoctine Plus' (barley) and by DowElanco Ltd as 'Rappor' or 'Rappor Plus'.

MATERIALS AND METHODS

The products tested are listed below:

Product name	active ingredient	dose rate → product/tonne seed
Panoctine	guazatine	2L
Panoctine Plus	guazatine + imazalil	2.2L
Baytan	tradimenol + fuberidazole	2.0L
Cerevax	carboxin + thiabendazole	2.5L
Cerevax Extra	carboxin + thiabendazole + imazalil	2.0L
Ferrax	flutriafol + thiabendazole + ethirimol	5.0L
Panogen	organo-mercury	1L

Seed was treated in the laboratory using a Hege seed treater no. II. In the case of *Fusarium* on wheat and *P.graminea* and *U. nuda* on barley, naturally contaminated seed was used, but with *T. caries* on wheat seed was infected prior to treatment with 2g spores/kg seed.

Plot sizes in the case of Scottish Agricultural College (SAC) were 2 x 5m, all other sites being 2m x 12m replicated 3 or 4 times, drilling was achieved using a Hege small plot drill.

Results are expressed as plant population /m² or percent disease incidence. An analysis of variance (LSD or SED) was carried out on all means.

RESULTS

1) *Fusarium nivale* - winter wheat.

The results of 3 independent trials carried out in 1992 using infected seed stocks of winter wheat are shown in Table 1.

TABLE 1. *Fusarium nivale* - effect on seedling emergence of winter wheat

Plant population/m^2

Treatment	Site % seed infection	Bratton, Wilts 60	Lavant, Sussex 60	Rosemaund, Herefs 48
Untreated		73.3	112.4	141.7
Guazatine		200.0	211.6	229.0
Triadimenol +		201.3	-	230.9
fuberidazole		166.2	-	199.1
Carboxin thiabendazole				
	LSD	31.6	12.4	28.0
	CV (%)	7.2	3.4	9.1

2) *Tilletia caries* - winter wheat

Trials carried out by SAC in 1990 and 1992, using high grade, artificially infected seed gave excellent bunt control with guazatine.

TABLE 2. Control of *Tilletia caries* - SAC results 1990 and 1992

Winter wheat C1 seed, artificially infected - 2g spores/kg seed

% bunted heads

Treatment	Year	1990	1992
Untreated		85.1	32.2
Guazatine		3.1	0.0
Organomercury		7.4	-
Triadimenol +		0.6	0.7
fuberidazole		-	6.7
Carboxin +			
thiabendazole			
	SED +/-	12.9	4.03

3) *Pyrenophora graminea* - Barley

Results from trials carried out at SAC during 1990-92 show good control of *P.graminea* with guazatine /imazalil (see Table 3.).

TABLE 3. Control of *Pyrenophora graminea*

SAC trials - spring barley 1990-1992 (infected seed)

% leaf stripe

Treatment	Year	1990	1991	1992
Untreated		5.54	19.4	21.3
Guazatine & imazalil		0.00	0.00	0.00
Organomercury		1.32	-	-
Flutriafol + thiabendazole + ethirimol		0.00	0.00	0.00
Triadimenol + fuberidazole		-	0.1	0.9
	SED +/-	1.99	2.51	4.46

4) *Ustilago nuda* - Barley

Although there is currently no label claim for the control of loose smut, trials data (Table 4) confirm that activity is equivalent to that given by other products.

TABLE 4. Control of *Ustilago nuda*

SAC trials 1991 and 1993 - Winter barley

Treatment	1991 no. smut/m. drill	1993 % smut
Untreated	9.21	8.38
Guazatine + imazalil	0.04	0.00
Triadimenol + fuberidazole	0.00	-
Carboxin + thiabendazole + imazalil	-	0.00
Flutriafol + thiabendazole + ethirimol	-	0.00
SED +/-	0.44	1.168

DISCUSSION

<u>Crop Tolerance</u>

Jackson <u>et al</u>. (1973) quote the proven crop safety of guazatine in both laboratory and field emergence studies in wheat , barley and oats, as equivalent to that for organo-mercury. Bartlett and Ballard (1975), further showed the safety of a guazatine/imazalil mix in barley.

In 17 winter wheat trials carried out by Rhône-Poulenc, the crop vigour in guazatine plots showed a 10% increase over the untreated control. Noon and Jackson (1992) also commented on the safety of guazatine in winter wheat. In the 20 winter barley RP sites, results were equivalent to both untreated and organomercury with guazatine/imazalil. Emergence was equivalent to, or greater than untreated control, in both spring-sown wheat and barley and winter and spring oats. Although no recommendation presently exists, tolerance was also found to be good on durum wheat, rye, triticale and naked oats.

<u>Winter wheat disease control</u>

1) *Fusarium nivale*. A major strength of guazatine lies in the control of *Fusarium nivale* seedling blight particularly as the fungicide provides a suitable replacement for strains of the fungus shown to be resistant to the benzimidazoles. Guazatine controls both surface and deep-seated infections of *Fusarium*, as a result of its penetrant properties.

F.nivale is increasing in importance, and is now probably the major seed-borne disease of wheat. According to Reeves and Simpkins (1993), 76% of seed samples submitted in 1992 to the Official Seed Testing Station for England and Wales were above the 5% advisory limit for the disease.

Jackson et al. (1973) showed guazatine to give a similar level of control to mercury in spring wheat and barley, whilst Roberti <u>et al</u>. (1992) found that this fungicide gave good control of *F culmorum* in winter wheat in Italy.

Seed treatment trials conducted by ADAS in 1986 (Jones, 1993) used seed stocks of wheat heavily infected with *F.nivale* . Guazatine was found to be the most effective treatment, significantly increasing plant emergence compared with untreated seed and organo-mercury. Referring to a 1986 survey (Locke <u>et al</u>, 1987) showing high levels of resistance of the fungus to benomyl, the author noted that guazatine was the only non-benzimidazole fungicide available for *F.nivale* control, and suggest this material to be the most suitable treatment for seed stocks with high levels of the disease.

In a trial carried out by the SAC in 1992 using winter wheat with 50% *F.nivale* infection, guazatine gave a statistically significant increase in plant emergence over other standard fungicides. Noon and Jackson (1992). showed that, although not statistically different from other standards, guazatine gave the highest plant counts and yield increases where *F. nivale* was controlled.

2) *Tilletia caries* Jackson et al.(1973) confirmed original reports that guazatine was effective against this disease, but showed a slightly lower control compared to organomercury, results confirmed in later trials by Noon and Jackson (1992). Skorda (1981) showed activity of guazatine/imazalil against hexachlorobenzene insensitive strains of *T.foetida* in Greece, whilst researchers in Canada also found levels of bunt ,caused by the same fungus, to be significantly reduced by guazatine (Atkinson, 1974).

Trials conducted by ADAS in the UK during 1990-92 (Jones D.R. - Pers. comm.), indicate good control of *T.caries* with guazatine. Treatments were applied to artificially infected graded seed, or to naturally infected uncleaned, ungraded seed. With the graded seed, guazatine was found to give similar control to mercury and other standard fungicides over the three seasons of the trial; however, where ungraded seed was used, inferior control was experienced with most products. These latter poor results may possibly be explained by reduced seed surface loading due to the presence of higher numbers of small grains.

Control of bunt has been found to be in excess of 90% in situations where normal grade seed is involved. Recent work has shown guazatine to give similar control to current standards. Evidence on bunt control collected from 60 sites across Europe (not reported here), demonstrate the superiority of guazatine over organomercury, and its consistent achievement of high levels of control.

Barley disease control

1) *Pyrenophora graminea* Jackson et al.(1973) reported that guazatine alone was not sufficiently active against leaf stripe, and Bartlett and Ballard (1975) showed superior control compared to organomercury when imazalil was added to guazatine. During the 1980's instances occurred of *P.graminea* resistance to organomercury seed treatments. Jones et al.(1989) showed that guazatine + imazalil was effective against both organomercury-sensitive and resistant strains of the fungus. Noon and Jackson (1992) showed that in 3 trials with winter barley carried out in 1991, guazatine/imazalil gave excellent control of leaf stripe, equivalent to that given by mercury. In a further 2 spring barley trials the mixture eliminated the disease, giving superior control compared to mercury, which at one site completely failed to arrest the leaf stripe.

2) *Ustilago nuda* Although control of *nuda* is not currently claimed on the label text for the product , recent results indicate that control may be obtained, although in view of the systemic nature of the disease the mechanism of control by the mixture is not understood.

ACKNOWLEDGEMENTS

The authors are grateful to Dr S.Oxley of SAC for much of the data included and to Dr D.Jones of ADAS and G.Ingram of Rhône-Poulenc for their helpful comments. We also wish to thank Elizabeth Court of Rhône-Poulenc for the literature searches.

REFERENCES

Atkinson, T.G. (1974) Evaluation of seed treatment fungicides for control of bunt in winter wheat. <u>Pesticide Research Report</u> 1974 meeting pp300-301.

Bartlett,D.H ; Ballard,N.E.(1975) The effectiveness of guazatine and imazalil as seed treatment fungicides in barley. <u>Proceedings of the 8th British Insecticide and Fungicide Conference</u> ,**1** , 205-211.

Catling,W.S. ; Cook,I.K. ; McWilliam,R.W. ; Rhodes,A.(1968) Bis(8-guanadinoctyl) amine sulphate, a new broad spectrum fungicide especially effective against seed-borne diseases of cereals. <u>First International Congress Plant Pathology 1968</u> p27 (abstr).

Jackson, D ; Roscoe,R.J. ; Ballard,N.E.(1973) The effectiveness of Bis (8-guanadino-octyl) amine as a seed dressing alone or in mixture with other fungicides, against diseases of cereals. <u>Proceedings of the 7th British Insecticide and Fungicide Conference</u> ,**1**, 143-150.

Jones, D.R; Slade, M.D. ; Birks,K.A.(1989) Resistance to organomercury in <u>Pyrenophora graminea.</u> <u>Plant Pathology</u> ,**38**, 509-513.

Jones, D.R.(1993) Evaluation of seed treatments for control of <u>Fusarium nivale</u> in winter wheat. <u>Tests of Agrochemicals and Cultivars **14** (Annals of Applied Biology 122 supplement)</u>, 60-61.

Locke,T. ; Moon,L.M. ; Evans,J.(1987) Survey of benomyl resistance in <u>Fusarium spp</u> on winter wheat in England and Wales in 1986. <u>Plant Pathology</u>, **36**, 589-593.

Noon,R.A. ; Jackson,D.(1992) Alternatives to mercury for cereal seed-borne diseases. <u>Proceedings British Crop Protection Conference - Pests and Diseases</u> ,**3**, 127-136.

Reeves,J.C., Simpkins,S.E.(1993) Plant health and the European Single market. <u>BCPC monograph</u> **54**, 333-338.

Roberti,R ; Flori,P. ; Busi,L.(1992) Evaluation of chemical seed treatment for the control of seed-borne <u>Fusarium culmorum</u> and <u>Bipolaris sorokiniana</u> on wheat. <u>Meded Fac Landbouwet Rijksuniv, Gent</u>,**57**, no2a 223-230.

Skorda,E.A.(1981) Evaluation of fungicides as seed dressings against new strains of wheat bunt in Greece. <u>Proceedings British Crop Protection Conference - pests and diseases,</u> **1**, 317-323.

Session 3
Cereal Seed Treatments

Chairman	R A NOON
Session Organisers	R A NOON
	J N OAKLEY

CGA 219417: A NOVEL FUNGICIDE FOR THE CONTROL OF *PYRENOPHORA* SPP. ON BARLEY

N.J. LEADBITTER, B. STECK, L.R. FRANK

Ciba Plant Protection, Basle, Switzerland

A.J. LEADBEATER

Ciba Agriculture, Whittlesford, Cambridge, UK

ABSTRACT

CGA 219417 is a novel fungicide of pyrimidine amine chemistry, being developed by Ciba. As a seed treatment CGA 219417 controls leaf stripe (*Pyrenophora graminea*) and seed-borne infections of *Pyrenophora teres* on barley.

Trials in Europe (1990-1993) have shown that CGA 219417 gives excellent control of leaf stripe, equivalent to imazalil, on both winter and spring barley at high and low disease pressure using a rate of 5g AI/100kg seed. Control of seed-borne *P. teres* is also given at 5g AI/100kg seed.

INTRODUCTION

Seed treatment provides an effective and efficient way of applying small quantities of fungicides where they are most needed and is the only efficient method for controlling many seed-borne diseases, for example *Ustilago* spp and *Tilletia* spp.

Leaf stripe of barley caused by *Pyrenophora graminea* S. Ito & Kuribay is an example of a seed-borne disease which can cause great damage in the major barley growing areas of the world (Mathere, 1982). *P. graminea* causes infected plants to be stunted and produces characteristic necrotic striping on most of the leaves which develop. In many infected plants the spikelet does not emerge and where it does little or no grain is produced.

P. graminea survives exclusively as seed-borne mycelium in the hull, pericarp and seed coat. At the time of heading and under conditions of high moisture, conidia are produced on infected leaves and are windblown to nearby heads. Seed may be infected at all stages of development although the early infections are the most severe. Infection of the seedling from the seed is greatly affected by soil temperature and soil moisture. The maximum number of seedlings is infected at temperatures below 12°C (Teviotdale & Hall, 1976) and infection tends to be greatest when the soil is moist rather than wet or dry (Prasad *et al*, 1976).

If left uncontrolled leaf stripe can develop rapidly via infected seed lots and therefore to protect future crops it is important to use a seed treatment to control the disease. Since the 1940s leaf stripe has been successfully controlled using mercury seed treatments, except in the rare cases where resistance of the fungus to mercury developed. Following the banning of mercury throughout Western Europe several products have been used which give incomplete control of the disease. Most recently the phenylpyrroles (fenpiclonil and fludioxonil) have been added to this armoury. For very high levels of control e.g. on seed crops, a single active ingredient, imazalil, has often been relied upon.

Net blotch (*Pyrenophora teres* Drechs.) is a common disease of barley everywhere the crop is grown. The disease is known in two forms, the "net" type, the most common form which gives net-like symptoms on the leaves and the "spot" form which produces dark brown elliptical lesions surrounded by a chlorotic area (Mathere, 1982). Seed-borne infection can be the primary source of inoculum from which lesions develop on the lower leaves and coleoptile. From these lesions conidia are then dispersed to cause further infection on newly developing foliage. Under favourable conditions large areas of the leaf tissue can be affected and hence yield reduced. Infection of the seed occurs when conditions allow the ear to be infected during seed development. Control of the seed-borne infection is mainly by products containing imazalil.

New compounds that will broaden the chemical basis of control of *Pyrenophora* spp. on barley are therefore desirable. This paper describes the activity of a novel compound CGA 219417 when used against *P. graminea* and *P. teres* on barley.

PRODUCT

CGA 219417 (proposed common name cyprodinil) is a new pyrimidine amine fungicide which is being developed by Ciba for the control of leaf stripe on barley as well as for the control of a range of foliar diseases on grapes, cereals and other crops (Heye *et al*, in press). This compound shows good activity against pathogenic fungi in the *Ascomycetes* and *Deutero-mycetes*. Biochemical studies show that CGA 219417 inhibits amino-acid synthesis (*Ibid*). As such it has a different mode of action from other products which control leaf stripe on barley.

METHODOLOGY

Replicated field trials were carried out in a number of European countries from 1990 to 1993 using naturally infected seed. The levels of infection on the seed were determined before treatment.

Treatments were usually applied diluted in a Hege 11 seed dresser. Seed loading analyses were made on selected batches following treatment to evaluate the quantity of CGA 219417 on the seed.

CGA 219417 formulated as either an FS025 or ES025 water based formulation containing 25g AI/litre was applied at a product rate of 0.2litre/100kg of seed. The following standards were used. On winter barley, triadimenol+fuberidazole+imazalil (37.5+4.5+5g AI/100kg seed; GB, 1990 & 1991, CH & D, 1991-1993), triadimenol+imazalil (11.3+4.5g AI/100kg seed; I, 1990 & 1991), fenfuram+guazatine+imazalil (30+60+4g AI/100kg; D, 1990, CH, 1991), fenpiclonil+imazalil (20+4g AI/100kg seed; GB, 1992), flutriafol+ethirimol+thia-bendazole+imazalil (15+200+5+4g AI/100kg seed; GB, 1993) and betaxate+anthraquinone (20+50g AI/100kg seed, F, 1991 & 1992). On spring barley, triadimenol+fuberida-zole+imazalil (37.5+4.5+5g AI/100kg seed; D, 1992); fenpiclonil+imazalil (20+4g AI/100kg seed; D, 1991, CH, 1990); triadimenol+fuberidazole (37.5+4.5 AI/100kg seed; GB, 1991); guazatine+imazalil (60+5 g AI/100kg; GB, 1993).

Per cent crop vigour was assessed using a visual score for each plot. This evaluation gives an overall evaluation of plant size, crop cover and crop appearance in comparison with other plots in the trial. In Germany the untreated plots were always given a 100 rating and other plots in the replicate assessed in comparison with these. In this case treated plots can have values greater than 100 if they are more vigorous than the untreated. In Great Britain the most vigorous plot in the replicate was assessed as 100 and other plots in the replicate assessed in comparison with it. In this case treatments can have values greater than the untreated if they are more vigorous but the value will never exceed 100. Assessments of plant stand were made by counting the number of plants in five randomly chosen 0.5m lengths of row in each plot. Leaf stripe was assessed by counting the number of infected tillers at ear emergence. Infection by *P. teres* from seed-borne infection was made at Zadoks GS10 - 20 by counting numbers of diseased plants.

RESULTS AND DISCUSSION

Crop tolerance

Field trials carried out during 1990 - 1993 in Germany, Italy, France, Switzerland and Great Britain showed that CGA219417 when used at 5g Al/100kg seed caused no adverse effects on plant stand, subsequent plant development or final yield (results not presented). Data for crop development (Table 1) also showed that when CGA 219417 was used at 5 and 7.5g Al/100kg seed there were no adverse effects on plant development compared with the untreated under conditions of rapid growth (Germany, Sept/Oct drilling) or where crop development was slower due to later drilling (UK, Oct/Nov drilling). In some cases, under the same conditions crop vigour had been reduced by the standard.

TABLE 1. Crop tolerance of CGA 219417 on winter barley, measured as % crop vigour* at Zadoks growth stage 10-11.

Treatment	Rate	Country			
	g Al/100kg seed	D	D	GB	GB
untreated	-	100	100	85.7	90.7
standard	*	90	100	38.8	36.3
CGA 219417	5.0	100	100	91.3	86.3
CGA 219417	7.5	100	100	91.3	86.9
Days after planting		13-26	24-26	22-35	55
Drilling date		26/9-19/10	21/9-28/9	24/10-31/10	15/11-28/11
No. of trials		5	4	2	2
Years		90	91	90	91

* see methodology

In 1993, trials using certified seed were conducted in Great Britain to assess the crop tolerance of CGA 219417 at 10g Al/100kg on eight varieties of winter barley. The data show that CGA 219417 at this rate does not adversely effect plant stand (Table 2)

TABLE 2. Crop tolerance of CGA 219417 on 8 winter barley varieties in Great Britain, 1993. Numbers of plants per 0.5m length of row.

Treatment	Rate g Al/100kg	Variety** 1	2	3	4	5	6	7	8
untreated	-	30.9	34.6	37.2	30.1	30.8	29.8	27.3	31.0
standard*	220	26.2	31.8	35.2	26.3	27.7	25.1	23.8	28.2
CGA219417	10	28.8	33.6	36.4	28.0	34.4	27.1	28.2	31.8
LSD(p=0.05)***		5.64	5.41	5.39	7.98	6.59	5.46	6.38	7.87

* flutriafol+ethirimol+thiabendazole 15+200+5g Al/100kg seed
** 1: Marinka, 2: Puffin, 3: Gypsy 4: Silk, 5: Pipkin 6: Sprite, 7: Princess, 8: Halcyon
*** Comparisons only possible within columns. LSD calculated using Tukey Test.

Activity of CGA 219417 against leaf stripe on winter barley

Field trials carried out in a number of European countries showed that control of leaf stripe on winter barley by CGA 219417 (5g Al/100kg seed) was at least equal to the standard products (Table 3). In all countries except France the standards contained imazalil (4-5g Al/100kg seed depending on the product) demonstrating that the control of leaf stripe given by CGA 219417 was at least as good as the most common material used for control of this disease.

TABLE 3: % Control of *P. graminea* on winter barley in European countries using CGA 219417.

Treatment	Country CH	D	GB	F	I
untreated	(16.1)*	(7.7)*	(8.7)*	(37.8)*	(8.4)*
standard **	98.8	97.3	98.8	75.8	91.6
CGA 219417 ***	98.5	98.2	99.3	96.9	96.7
No. of trials	11	16	8	5	6
Years	90-92	90-92	90-93	91-92	90-91

* % infection in untreated
** see Methodology
*** Applied at 5g Al/100kg seed

In these trials a wide range of infection levels occurred, demonstrating that at 5g Al/100kg seed CGA 219417 gives consistently high levels of disease control independent of disease pressure (Figure 1).

FIGURE 1. % Control of *P. graminea* on winter barley by CGA 219417   (5g Al/100kg seed) as related to final disease levels in the untreated (data from 41 trials).

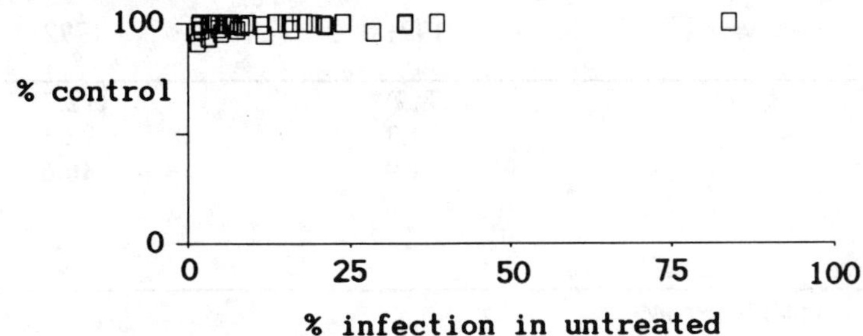

Activity of CGA 219417 against leaf stripe on spring barley

In field trials in Switzerland, Germany and Great Britain CGA 219417 (5g Al/100kg seed) also gave excellent activity against leaf stripe on spring barley, equivalent to the standards (Table 4).

TABLE 4. % Control of *P. graminea* on spring barley in European countries using CGA 219417

| Treatment | Country | | |
	CH	D	GB
untreated	(10.1)*	(12.1)*	(10.8)*
standard **	100	99.8	100
CGA 219417 **	99.6	99.5	99.6
No. of trials	1	3	3
Years	90	91-92	91 & 93

* % infection in untreated
** see Methodology
*** Applied at 5g Al/100kg seed

Activity of CGA219417 against *P. teres* on barley

In trials carried out in Denmark in 1991 and 1992 CGA 219417 gave very good control of *P. teres* generally equivalent to the standard imazalil (Table 5).

TABLE 5. % Control of *Pyrenophora teres* on spring barley in Denmark

| | Year | |
Treatment	1991	1992
untreated	(8.2)*	(12.7)*
imazalil **	79.3	92.1
CGA 219417 **	71.9	86.6
No. of trials	3	3

* No. of diseased plants per m²
** Applied at 5g Al/100kg seed

CONCLUSION

Field trials carried out over a number of years and in various countries have demonstrated the excellent and consistent activity of CGA 219417 against leaf stripe of barley. Unlike some other classes of compounds e.g. the triazoles there is no difference in the level of control on winter and spring barley. A high level of control has also been demonstrated against seed-borne infection of *P. teres*. This activity together with its excellent crop safety and low use rate make CGA 219417 a good alternative to imazalil where consistently high levels of leaf stripe control are required, thereby reducing the dependence on a limited number of active ingredients.

With both economic and environmental pressures against the use of agrochemicals continuing to increase seed treatment remains an efficient and environmentally attractive method of disease control. CGA 219417 provides another tool with which to develop future crop protection strategies.

ACKNOWLEDGEMENTS

The authors would like to thank their colleagues from Italy, Great Britain, Germany and France who carried out the field work described in this paper.

REFERENCES

Heye, U.; Speich, J.; Siegle, H; Wohlhauser, R.; Zeun, R. (1994) CGA 219417: A new broad spectrum pyrimidine amine fungicide for cereals, pome fruits, grapes and other crops. *Crop Protection* In press

Mathre, D.E. (1982) Barley stripe, In: *Compendium of barley diseases*, Mathre, D.E. (Ed), American Phytopathological Society, Minnesota.

Prasad, M.N.; Leonard, K.J.; Murphy, C.F. (1976) Effects of temperature and soil water potential on expression of barley leaf stripe incited by *Helminthosporium gramineum. Phytopathology*, **66**, 631-634.

Teviotdale, B.L.; Hall, D.H. (1976) Factors affecting inoculum development and seed transmission of *Helminthosporium gramineum. Phytopathology*, **66**, 295-301

EFFICACY OF TRITICONAZOLE SEED TREATMENT AGAINST COMMON EYE-SPOT IN WHEAT.

J.M. GAULLIARD

Rhône-Poulenc Agro France, 55 avenue René Cassin, CP 310, 69337 Lyon Cedex 09

M. CHAZALET, A. SAILLAND, J.M. GOUOT

Rhône-Poulenc Agro, 14/20 rue Pierre Baizet, BP 9163, 69263 Lyon Cedex 09

ABSTRACT

Triticonazole (RPA 400727) is a new triazole fungicide developed by Rhône-Poulenc Agro for seed treatment of cereals and other crops. Because of its excellent selectivity, seeds treated at 1200 g AI/t were protected against early attacks of cereal diseases and the programme of subsequent foliar fungicide applications could therefore be reduced. As observed in a number of trials conducted in France, triticonazole was effective against common eye-spot (*Pseudocercosporella herpotrichoides*), from emergence and during the most damaging period of the cropping season. It controlled early attacks of W (fast-growing) and R (slow-growing) strains, decreasing both the incidence and the severity of the disease.

INTRODUCTION

Rhone Poulenc Agro have discovered a new sterol demethylation inhibitor (DMI) fungicide which has shown high selectivity in cereals, especially by seed treatment, superior to most other DMI products. Some of the properties of triticonazole (RPA 400727) have been described already (Mugnier *et al.*, 1991,1993). The use of a high dose on the seeds allowed better control and a longer protection of cereal diseases such as common eye-spot caused by *Pseudocercosporella herpotrichoides* (Montfort et *al*, 1991). Work to confirm this is reported in this paper.

Physical and chemical properties of triticonazole are described below :

Chemical name (IUPAC)	(1RS)-(E)-5-(4-Chlorophenyl)methylene)-2,2-dimethyl-1-(1H-1,2,4-triazol-1-ylmethyl)-cyclopentan-1-ol

Chemical structure

$C_{17}H_{20}ClN_3O$

Molecular mass:	317.82
Melting point:	144°C
Vapour pressure:	< 0.1 x 10-7 hPa (50°C)
Solubility in water:	8.4 mg/l (20°C)
Partition coefficient: (Octanol/water)	logP = 3.29

Toxicology

Acute and subchronic toxicological studies have demonstrated that triticonazole shows a very low toxicity towards mammals (Table 1).

TABLE 1. Acute toxicity of triticonazole.

Animal	Route	Results
Rat (M + F)	Oral	LD 50 : > 2000 mg/kg
	Dermal	LD 50 : > 2000 mg/kg
	Inhalation	LC 50 : > 1.4 mg/kg
Rabbit	Eye	Irritant
	Skin	Non irritant
Guinea pig	Skin	No sensitization

Ecochemistry

In standard laboratory conditions, there was no significant hydrolysis of triticonazole in water (pH range : 5-9). The product could be unstable under strong light in water solution or in some organic solvents in the presence of chemical photoactivators but it was stable when applied on seeds stored under standard conditions.

In standard laboratory conditions, its half-life in a clay loam soil was of the order of 4 months. Triticonazole was neither toxic to birds (LD 50 > 2000 mg/kg) nor to the total microbial soil population.

Formulations

Several formulations of triticonazole have been developed. These are mainly suspension concentrates (FS), alone or in mixture with complementary fungicides such as guazatine, iprodione, or with anthraquinone as a bird repellent for specific countries and uses. In France, an FS containing 200 g triticonazole + 84 g anthraquinone / litre has received a provisional registration in 1993 and is commercialized under the trade mark REAL. The registered dose rate for this formulation is 6 l/t of seeds, for the control of the following seed-borne diseases : *Tilletia caries, Septoria nodorum, Fusarium roseum* on wheat and *Ustilago nuda* on barley . The following foliar diseases are also controlled : *Erysiphe graminis, Septoria tritici* and *S. nodorum, Puccinia striiformis* and *P. recondita* on wheat, *Rhynchosporium secalis, E. graminis* and *P. hordei* on barley.

Material and methods

Field trials have been conducted in France according to the official French methodology : CEB (Commission des Essais Biologiques) no. 64 with 4 blocks, and 20-50 m²/plot. Wheat seeds treated with triticonazole at 1200 g AI/t were sown at 160-200 kg/ha. In some trials, the inoculum was increased by broadcasting *P. herpotrichoides*-inoculated straw or grain in the field during winter. The types of strains found on diseased stems were determined by molecular biology techniques. This method allowed the specific characterization of *P. herpotrichoides* from a mixture of other fungi, even in case of very early attacks without symptoms and permitted the differentiation between the W (Wheat type, fast-growing strain) and R (Rye type, slow-growing strains) types of common eye-spot strains. It was then easier to determine the frequency of attacks caused by each strain type in treated and untreated plots. An average of 20 to 40 tillers or stems per plot were assessed.

RESULTS

Characteristics of *P. herpotrichoides* strains in France (1992 and 1993)

Analyses of the strains of the population at many locations (38 locations in 1993) were conducted and used to map their distribution in France according to the W/R type ratio. The R type strains were predominant over W strains in the North and North-West of France, while W strains were in the majority in the rest of the country. R strains were present in most of the triticonazole trials.

<u>Efficacy of triticonazole against common eye-spot</u>

The frequency of both strain types on wheat tillers was determined at growth stage Z 30, and triticonazole efficacy against early attacks of common eye-spot was calculated at this stage using the Abbott formula. As shown in Table 2, triticonazole was effective against both W and R strains which were found alone or mixed on leaf sheaths, 30 out of 38 locations showing at least 5% of tillers infected either by a W or an R strain (or by both strains together).

TABLE 2. Efficacy of triticonazole seed treatment (1200 g Al/t) against *P. herpotrichoides* on wheat at GS Z 30, as determined by molecular biology techniques (average of 30 locations in France, showing at least 5% infected stems in untreated 1992/93 trials).

	All strains	W type	R type
Efficacy (%)	52.2	58.8	47.5
% infected stems in untreated	32.9	20.9	18.7
Number of trials	30	26	25

TABLE 3. Results from trials conducted in 1992 by R.P. Agro France.

Product/dose g A.l/t or ha	% of stem section girdled and yield (t/ha) Trial reference code						Means : % control yield benefit over untreated t/ha
	trial 1	trial 2	trial n 3	trial 4	trial 5	trial 6	
triticonazole	19.5 b	21.0 b	18.6 b	7.3 bc	25.8 bc	40.9 bc	43.2
1200 g/t	-	(8.72 a)	(5.51 ab)	(6.12 a)	(6.70 ab)	(8.62a)	(0.197)
prochloraz	20.9 b	7.8 c	15.9 bc	13.2 ab	9.6 cd	38.3c	51.3
600 g/ha	-	(8.77 a)	(5.48 ab)	(5.77 ab)	(7.08 a)	(8.00 b)	(0.084)
flusilazole	19.7 b	15.6 bc	15.4 bc	10.6 abc	3.2 d	42.7 abc	50.8
300 g/ha	-	(8.84 a)	(5.33 ab)	(6.20 a)	(7.00 a)	(8.47 ab)	(0.239)
untreated	33.0 a	33.0 a	31.0 a	20.0 a	56.0 a	54.3 ab	
	-	(8.77 a)	(5.192 b)	(5.582 b)	(6.93 ab)	(8.20 ab)	
Inoculation	artificial	natural	artificial	natural	natural	artificial	
LSD %	6.25	9.37	7.95	8.88	18.3	12.59	
CV %	19.34	37.67	33.34	55.84	65.8	19.44	

Figures followed by the same letter are not statistically significantly different (P = 0.05)

Trials conducted by R.P. Agro in 1992 have demonstrated that triticonazole seed treatment at 1200 g Al/t gave a good control of eye-spot, in the range of efficacy achieved by a foliar application of standard EBIs registered against eye-spot in France, bringing an additional yield of 0.2 t/ha on average over the untreated control (Table 3). These results have been confirmed by trials conducted in 1992 by I.T.C.F. - Institut Français des Céréales et des Fourrages - (data not shown). In 1993, a large number of trials have been conducted with triticonazole at 1200 g Al/t. Early assessments have demonstrated the good efficacy of triticonazole against early attacks of eye-spot which occurred very early, from November 1992 and all through the winter and spring seasons (Table 4).

TABLE 4. Efficacy of triticonazole in 1993 R.P. Agro France trials (early assessments, average of 30 trials).

	triticonazole	prochloraz
Application	Seed treatment	Foliar treatment
Dose rate	1200 g Al/ t	600 g Al/ ha
% Efficacy at GS Z30*	44.2	Not applicable
GS Z40*	43.9	42.3
GS Z55-60**	53.0	47.7

* based on frequency of attacked stems (incidence)
** based on severity of the attack (% stem circumference girdled)

Results from most of the trials are not yet available. However, in a series of trials conducted by the RP Agro Evaluation farms, in which a W strain sensitive to EBIs and benzimidazoles has been used, triticonazole showed an acceptable efficacy versus the standard fungicides (Table 5) when observed late in the season (G.S. Z64 to Z75, 192 to 223 days after sowing). In addition, in two of these trials, lodging occurred, following heavy rainfall. In these two cases, triticonazole prevented lodging as well as the standard EBI fungicides (Table 6).

TABLE 5. Efficacy of triticonazole against *P. herpotrichoides* in 1993 R.P. Agro - Evaluation trials (assessments made at G.S. Z64 - Z75).

Trial	Growth stage (days after sowing)	Untreated	% stem circumference girdled (% efficacy)		
			triticonazole 1200 g Al / t	prochloraz 600 g Al / ha	flusilazole+ carbendazim 300 + 150 g Al / ha
1	Z64 (192 das)	70 a	45 b (37.5)	27 bc (61.4)	14.5 c (79.2)
2	Z64 (226 das)	19a	5.3 bc (72)	3.8 bc (80)	1.3 c (93.1)
3	Z75 (180 das)	26.1 a	17.4 b (33.3)	3.6 c (86.2)	1.5 c (94.2)
4	Z75 (213 das)	17.7 a	1.3 b (92.5)	1.7 b (90.2)	1.3 b (92.5)
5	Z69	28.9 a	21.4 ab (26.0)	6.0 b (79.2)	22.3 ab (22.8)

Figures followed by the same letter are not statistically significantly different (P = 0.05)

TABLE 6. Effect of triticonazole and other fungicides on winter wheat crop lodging.

% of crop lodging	Untreated	triticonazole 1200 g AI / t	prochloraz 600 g AI / ha	flusilazole + carbendazim 300+150 g AI /ha
Trial 1 (239 das)	37.5 a	3.8 b	0 b	2 b
Trial 2 (245 das)	10 a	0 b	0 b	0 b

Figures followed by the same letter are not statistically significantly different (P = 0.05)

CONCLUSION

Triticonazole seed treatment at 1200 g AI/t, i.e. of the order of 200 g AI/ha when applied on seed drilled at 0.17 t/ha, provided good protection of cereal crops against early attacks of common eye-spot occurring from autumn up to the spring season, controlling both W and R types of strains. These early attacks generally caused most of the yield, losses. justifying the use of triticonazole as a preventive treatment during this period.

REFERENCES

Montfort, F. ; Denis, S. ; Cavelier, N. (1991) Effect of two triazole fungicide seed treatments on eye-spot of cereals caused by two strains of *Pseudocercosporella herpotrichoides* (Fron) Deighton. *Troisième Conférence Internationale sur les maladies des plantes ANPP. Bordeaux,* **III** , 1065 - 1072.

Mugnier, J. ; Chazalet, M. ; Gatineau, F. (1991) RPA 400727: un nouveau fongicide pour le traitement des semences de céréales. *Troisième Conférence Internationale sur les maladies des plantes ANPP. Bordeaux,* **III** , 941-948.

Mugnier, J. ; Gouot, J.M. ; Hutt, J. ; Greiner, A. ; Chazalet, M. ; Gaulliard, J.M.Ingram, G. (1993) RPA 400727, un nouveau traitement de semences efficace contre les maladies foliaires des céréales. *Mededelingen van de Faculteit Landbouwwetenschappen Rijksuniversiteit Gent* (in press).

TRIAZOXIDE - A SEED TREATMENT FUNGICIDE FOR THE CONTROL OF SEED BORNE *PYRENOPHORA* SPECIES

S. DUTZMANN

Bayer AG, PF-F/F, Pflanzenschutzzentrum Monheim, D-51368 Leverkusen

ABSTRACT

Triazoxide is a specific non-systemic fungicide belonging to the chemical group of benzotriazines effective against seed borne *Pyrenophora graminea* and *Pyrenophora teres* on barley. With an application rate of 2 - 4 g AI / 100 kg of seed it is well qualified for completing the spectrum of activity of other fungicidal seed dressing compounds. The experience with triazoxide which is already in use in several European countries, is described and discussed.

INTRODUCTION

Cereal fungicides, both seed dressings and foliar sprays, are not only expected to provide a high level of activity but also to cover a broad spectrum of different diseases at the same time. In most cases it is not a single pathogen but a complex of them which may reduce the quality and quantity of yield. Although much effort has been made in that respect, numerous fungicidal compounds do not fully meet these requirements.

A classical gap in control of seed-borne diseases is the lack of activity of many seed dressing compounds against *Pyrenophora graminea*, the cause of leaf stripe on barley. After prohibition of mercury-containing seed dressings, leaf stripe could have become a serious problem had imazalil not been available. In this paper, triazoxide is presented as a novel benzotriazine based fungicide with a specific activity against seed-borne *Pyrenophora* diseases.

CHEMICAL AND PHYSICAL PROPERTIES

Chemical class: Benzotriazine

Chemical name: 7-chloro-3-(1H-imidazol-1-yl)-1,2,4-
                           benzotriazine-1-oxide

Common name: Triazoxide

Structural formula:

Molecular formula:	$C_{10}H_6ClN_5O$
Molecular mass:	247.7 g / mol
Appearance:	light yellow crystals
Melting point:	182° C
Vapour pressure:	20° C: 1.5×10^{-12} Pa 25° C: 5.2×10^{-12} Pa 120° C: 2.4×10^{-4} Pa
Solubility (g / l at 20° C)	water: 3×10^{-2} n-hexane: < 1 dichloromethane: 50 - 100 2-propanol: 2 - 5 toluene: 20 - 50
Stability (water): (light):	stable at ph 4 - 7 may be decomposed by light

TOXICOLOGY

Acute oral LD_{50} rats:	100 - 200 mg / kg
Acute dermal LD_{50} rats:	> 5000 mg / kg
Skin and eye irritation:	negative
Inhalation toxicity (4 hour LC_{50} rats):	0.8 - 3.2 mg / l
Teratogenicity:	no teratogenic effect to rats and rabbits
Mutagenicity:	in "in-vivo" and "in-vitro" tests no mutagenic effects

FORMULATIONS

Triazoxide can be formulated as a DS-, WS-, LS- and FS-seed treatment. Combinations with triadimenol, tebuconazole, fuberidazol, anthraquinone and bitertanol have been found to be relatively stable.

BIOLOGICAL PROPERTIES

The first tests in the greenhouse soon revealed that triazoxide is a non-systemic fungicide which is very suitable as a seed treatment. The spectrum of activity is mainly restricted to seed-borne *Pyrenophora* diseases (Table 1). Windborne *Pyrenophora*, e.g. *P. teres* is also influenced by applying a foliar spray, but due to the poor systemicity of the active ingredient, curative effects are too weak.Used as a seed treatment with the recommended rate of 2 - 4 g AI / 100 kg of seed, triazoxide perfectly controls *P. graminea* on spring and winter barley as well as seed-borne net blotch (*P. teres*). By contrast, triazoxide (like imazalil) has little effect on *Pyrenophora avenae*, the cause of stripe disease on oats.

TABLE 1. Spectrum of activity of triazoxide as a seed treatment.

Crop	Rate (g AI/100 kg seed)	Activity		
		v. good	good	poor
Barley	2 - 4	*Pyrenophora graminea*	*Pyrenophora teres* (seed-borne)	
Oats	2 - 4			*Pyrenophora avenae*
Wheat	2 - 4			*Tilletia caries*

If used with the recommended maximum rate of 4 g AI / 100 kg of seed, triazoxide displays good seed tolerance either alone or in combination with other seed dressing compounds.

TABLE 2. Efficacy of triazoxide against *Pyrenophora graminea* on winter barley (30 % disease) and on spring barley (48 % disease); Monheim field trial, 1993.

Crop	Treatment	Rate (g AI / 100 kg seed)	Efficacy (%)
Winter barley (cv. Franka)	Triazoxide (2.5 DS)	5	100
		4	100
		3	100
		2	100
	Imazalil (2.5 DS)	5	96
		4	99
		3	98
		2	96
Spring barley (cv. Asse)	Triazoxide (1.0 DS)	5	99
		4	99
	Imazalil (1.0 DS)	5	97
		4	99

As can be seen from Table 2 the performance of triazoxide is at least comparable to that of the standard compound imazalil. Even with the lowest rate of 2 g AI / 100 kg of seed, complete control of *P. graminea* on barley can be expected.

In the course of many field trials in Germany, it has been shown that leaf stripe on spring barley is slightly easier to control than on winter barley (Table 3): unfavourable growing conditions, e.g. low temperatures and unregular water supply during the phases from germination to end of tillering are known to extend the succeptible period for infection by *P. graminea*. Nevertheless, germination conditions in spring and autumn normally do not differ much from each other, thus the date of sowing alone cannot explain the difference in control. Maybe the long period of overwintering creates better conditions for systemic invasion by the fungus, while the growth of the host is slowed down or stopped. Nevertheless, triazoxide still exhibits consistently good control which can also be seen from the little variation in efficacy (figures in round brackets, Table 3).

TABLE 3. Difference in control of *Pyrenophora graminea* on spring and on winter barley.

Crop	Treatment	Rate (g AI/100 kg seed)	Control (%)	Disease (%)	Number of trials
Spring barley	triazoxide + fuberidazol	4 + 4	99.5 (98-100)*	32 (11-68)*	7
Winter barley	triazoxide + fuberidazol	4 + 4	98.8 (91-100)*	9 (1-52)*	19

* = Range

Unfavourable growing conditions may require an application rate of 4 g AI triazoxide / 100 kg of seed, e.g. for late-sown winter barley in Northern and Central Europe. However, this amount of AI may be reduced to 2 g, if the climatic conditions are more favourable (for instance in France). The same can be done, if triazoxide is combined with a seed dressing compound which has already - like tebuconazole - some activity against *P. graminea*.

In contrast to a monocyclic disease like leaf stripe on barley which can be fully eliminated by a single seed treatment, the polycyclic net blotch of barley (*P. teres*) cannot because it is the wind-borne inoculum released from stubble, which determines the severity of an epidemic. However, control of seed-borne *P. teres* is desirable since it may delay the onset of disease.

As can be seen from Table 4, both triazoxide and imazalil are able to control seed-borne *P.teres*. In combination with triadimenol an effect on early wind-borne infection may also be expected due to the high systemicity of triadimenol.

TABLE 4. Control of seed borne *Pyrenophora teres* (glasshouse test, average of 11 cultivars).

	Rate (g AI / 100 kg seed)	Control (%)
Triazoxide	5.0	95.8
	7.5	98.3
Imazalil	5.0	94.3
	7.5	95.6
Triadimenol + Triazoxide	22.5 + 7.5	99.1
Triadimenol + Imazalil	22.5 + 7.5	95.2
(Untreated mean disease level, %)		(12.1)

CONCLUSION

Triazoxide is a non-systemic fungicide for use as a cereal seed treatment. The above data demonstrate the high level of activity against *P. graminea* and seed-borne *P. teres*. As the dose required is extremely low, triazoxide is well suited for those seed dressings which exhibit a deficiency in control of barley leaf stripe. Combinations with other fungicidal seed dressings are already introduced into the German, French, Belgian, Norwegian and Spanish markets and they are under development for further European countries. Triazoxide brings new chemistry to the small group of compounds like imazalil which control *P. graminea*.

SEED TREATMENTS ON WHEAT IN THE USA USING DIFENOCONAZOLE

A. B. BASSI, Jr., D. LAIRD, L. ZANG

Ciba-Geigy Corporation, Plant Protection Division, P.O. Box 18300, Greensboro, NC, USA

N.J. LEADBITTER

Ciba Plant Protection, Basle, Switzerland

ABSTRACT

Difenoconazole is a triazole fungicide that offers broad spectrum control of seed and foliar pathogens of cereals in the USA when used as a seed treatment. In the USA it is being developed at rates of 6 to 24g AI/100kg seed.

In field trials on wheat difenoconazole has provided excellent activity against loose smut (*Ustilago tritici*), common bunt (*Tilletia caries*), dwarf bunt (*Tilletia contraversa*) and *Septoria* seedling blight (*Septoria nodorum*). In addition difenoconazole will give early season control of leaf rust (*Puccinia recondita*).

Difenoconazole also shows excellent crop safety and together with its good efficacy against a wide range of seed-borne diseases on wheat will provide a new opportunity for disease control using seed treatment in the USA.

INTRODUCTION

Seed-borne and soil-borne pathogens of wheat have become increasingly important in the USA with the advent of no-till farming and wheat mono-culture practices. The Certified Seed Program in the USA provides good quality seed to the grower but is not a 'stand alone' solution. Seed treatments are required to provide control of seed-borne diseases such as loose smut (*Ustilago tritici*), common bunt (*Tilletia caries*) and *Septoria* seedling blight (*Septoria nodorum*). Currently used products are mainly based on carboxin, triadimenol, thiram and carbendazim.

Dwarf bunt (*Tilletia contraversa*) is also a serious problem in parts of the USA i.e. Pacific North West. Due to the soilborne nature of this disease it cannot be controlled by certified seed programs. Until their ban, mercury seed treatments were used for the control of this disease. Currently there are no products registered in the USA for control of dwarf bunt.

Difenoconazole is a systemic fungicide that is highly effective against Ascomycetes, Basidiomycetes and Deuteromycetes (Ruess *et al*, 1989, Dahmen *et al*, 1992). It is currently used as a foliar fungicide in other parts of the world and will be available as a seed treatment in a number of countries. Difenoconazole is a triazole and acts on the C14 demethylation step in ergosterol synthesis.

In the USA difenoconazole has shown activity against a wide range of seed-borne diseases including dwarf bunt. Under the trade name Dividend™ it has been submitted to the U.S. Environmental Protection Agency for full registration in the USA as a seed treatment on wheat and barley. This paper gives a summary of its activity on wheat.

CROP SAFETY

Difenoconazole has been tested in field research trials for seven years in the USA. Emergence and crop vigour ratings have been evaluated to check crop safety across a wide range of environmental conditions. Difenoconazole (up to 48g AI/100kg seed) has proven to be non-phytotoxic and does not show any growth regulatory effects (data not shown).

Following treatment using the the commercial formulation of difenoconazole (3FS, 3lbs AI/gallon, 360g AI/litre) at an application rate of 48 g AI/100kg seed, seed was stored for 0, 4, 8 and 12 months at 7 and 32ºC. Crop safety was then evaluated by assessing germination after incubation in soil at 18-20ºC (Figure 1). The test showed that even after storage under adverse conditions difenoconazole at double the expected maximum application rate is safe to the seed and is generally at least equal to triadimenol (30g AI/100kg) at its standard use rate.

FIGURE 1. The influence on % germination of storage at 7 and 32 °C when wheat seed was treated with difenoconazole.

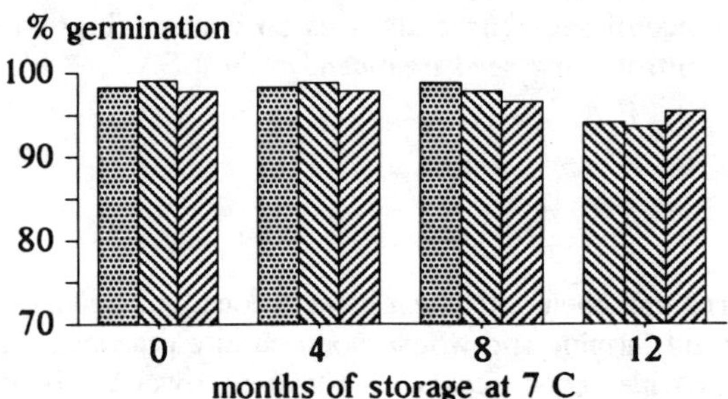

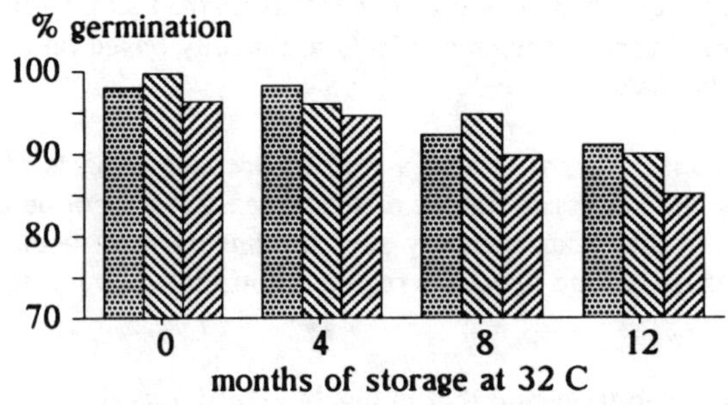

CONTROL OF DISEASES

Difenoconazole is effective against a range of diseases at rates of 6 - 24g Al/100kg depending on the disease (Table 1).

TABLE 1. Diseases of wheat controlled by difenoconazole when used as a seed treatment in the USA.

Disease	Pathogen	Rate (g Al/100kg seed)
Common bunt	*Tilletia caries*	6
Dwarf bunt	*Tilletia contraversa*	12
Loose smut	*Ustilago tritici*	12-24
Flag smut*	*Urocystis agropyri*	12-24
Septoria* seedling blight	*Septoria nodorum (seed)*	12-24
Leaf rust	*Puccinia recondita* (early season disease)	24

* data not shown

Control of *Tilletia contraversa*

Dwarf bunt is a significant problem in the Pacific North West region of the USA. No effective and commercially acceptable control of this disease is available to growers, despite the efforts of well funded, exhaustive university and industrial research programs. This lack of effective control measures has lead to a prohibition on export of soft white wheat from this area of the USA to the Peoples Republic of China which has had a zero tolerance for dwarf bunt since 1973.

In 14 Ciba field trials over a five year period (1989 - 1993) difenoconazole gave complete control of dwarf bunt at 12g Al/100kg seed. Table 2 gives a summary of data from 1989 and 1993 which show that even where high levels of disease were present, difenoconazole at 12g Al/100kg seed gave complete control. These data have been confirmed by university and USDA representatives (Goates, 1992; Sitton, 1992; Smiley, 1992).

TABLE 2. % Control of dwarf bunt in the USA using difenoconazole.

Treatment	Rate g Al/100kg seed	Year	
		1989	1993
untreated	-	(44.9)*	(81.3)**
carboxin+thiram	45 + 45	22.8	19.2
triadimenol	30	22.0	7.5
difenoconazole	12	100	100
number of trials		2	2

* number of infected plants/3.05m row   ** % incidence

Control of *Ustilago tritici*

Ustilago tritici has no zero-tolerance requirement in the USA and where control is needed a good certified seed program coupled with high levels of disease control using seed treatments is used.

In 28 trials conducted over six years difenoconazole applied at 12-24 g AI/100kg seed gave an average of 96% control of wheat loose smut. Data from 1990, 1992 and 1993 (Table 3) indicates that there is a slight dose response from 12 - 24g AI/100kg seed. At 12g AI/100kg seed difenoconazole gave more consistent and generally higher levels of control than the standard carboxin+thiram (45+45g AI/100kg seed). Where yield was taken (6 trials) the corresponding yield increase was 10% (data not shown).

TABLE 3. % Control of loose smut on winter wheat using difenoconazole. Mean data from 1990, 1992 and 1993.

Treatment	Rate	Year		
	g AI/100kg	1990	1992	1993
untreated	-	(54.2)*	(45.3)*	(28.3)*
carboxin +	45 +	69.5	37.5	94.4
thiram	45			
difenoconazole	12	91.7	94.4	95.7
difenoconazole	18	95.0	97.4	98.4
difenoconazole	24	95.0	100	99.6
Number of trials		9	1	3

* number of infected ears/plot. Plot size = 12m^2

Control of *Tilletia caries*

Both seed-borne and soilborne infections of *T.caries* can be effectively controlled by difenoconazole at 6g AI/100kg (Table 4). Control at this rate has been shown to be consistently high (98.4% control over 15 trials in four years) and equivalent to the best standards.

TABLE 4. % Control of seed and and soil-borne *T. caries* using difenoconazole.

Treatment	Rate	% infection	
	g AI/100kg	seed inoculation	soil inoculation
untreated	-	13a	20a
carboxin+thiram	45+45	0b	2.7b
triadimenol	30	0b	0c
difenoconazole	6	0b	0c

Means followed by the same letter are not significantly different according to Duncan's Multiple Range Test, P = 0.05.

<u>Control of early season foliar diseases</u>

Leaf rust (*P. recondita*) is a perennial problem in some parts of the USA. In the USA a seed treatment need only provide protection up to GS30-32 after which foliar fungicides such as propiconazole are used for its control.

In spring (GS30-32) following autumn drilling, difenoconazole (24g AI/100kg seed) will provide good control of leaf rust (Table 5) and this may be extended if disease levels remain moderate.

TABLE 5: % Control of leaf rust on winter wheat using difenoconazole as a seed treatment.

Treatment	Rate g AI/100kg	Trial			
		306 90	303 90	2 90	36 90
untreated		(9.5)*	(8.3)*	(81.3)*	(80.0)*
triademinol		46.3	67.5	10.8	15.6
difenoconazole		98.6	73.5	38.4	21.9
GS at assessment		32	31-32	31-32	31-32

* % infection in the untreated

DISCUSSION

Data collected from a wide range of sources show that difenoconazole is able to control all the most important seed-borne diseases of wheat in the USA and gives supression of some foliar pathogens. Especially important is the control of dwarf bunt for which difenoconazole is the first replacement for mercurial seed treatments.

The data also show that difenoconazole has activity against different pathogens at different rates of application (Table 1). In the USA the biological requirements for a seed treatment may differ from region to region. To reflect this difenoconazole will be recommended at a range of rates (6, 12 and 24g AI/100kg seed) depending on the needs of the region and the customer (Figure 2). This allows difenoconazole to be used at the most appropriate dose for any particular situation.

Due to its excellent broad spectrum activity and good crop safety, difenoconazole will be a useful tool for control of wheat diseases in the USA. The consistent improvements in yield and lack of negative seedling growth effects are regarded by officials as important indications that this seed treatment product is likely to be an important factor in disease management on winter wheat (Smiley, 1992).

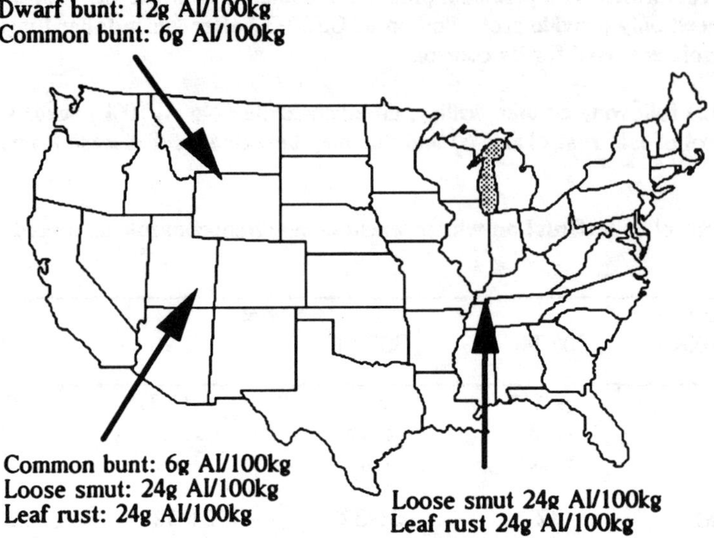

FIGURE 2. Use of difenoconazole in the USA with respect to rate and region.

REFERENCES

Dahmen, H.; Staub, T. (1992) Biologogical characterisation and uptake, translocation and dissipation of difenoconazole (CGA169374) in wheat, peanut and tomato plants. *Plant Disease*, 76, 523-526.

Goates, B.J. (1992) Control of dwarf bunt by seed treatment with Dividend 3FS. *Fungicide and Nematicide Tests*, 47, 264.

Ruess, W.; Riebli, P.; Herzog, J.; Speich, J.; James, J.R. (1989) CGA-169374, a new systemic fungicide with novel broad spectrum activity against disease complexes in a wide range of crops. *Proc. Br. Crop Prot. Conf. Pests and Diseases*, 3, 895-900.

Sitton, J.W.; Line, R.F.; Waldher, T.T.; Goates, B.J. (1992) Control of dwarf bunt of wheat with seed treatments. *Fungicide and Nematicide Tests*, 47, 273.

Smiley, R.W. (1992) 1992 Columbia Basin Agric. Res. Report. *Oregon State University Special Report* 894.

Smiley, R.W.; Uddin, W.; Kolding, M. (1992) Control of dwarf bunt with seed and foliar treatments. *Fungicide and Nematicide Tests*, 47, 274.

NEW SEED TREATMENTS BASED ON BITERTANOL AND TEBUCONAZOLE FOR SEED-BORNE DISEASE CONTROL IN WHEAT

D.B. MORRIS, A. WAINWRIGHT, R.H. MEREDITH

Bayer plc, Crop Protection Business Group, Elm Farm Development Station, Great Green, Thurston, Bury St Edmunds, Suffolk, IP31 3SJ

ABSTRACT

In trials conducted during 1988-93 on wheat the seed treatments bitertanol/fuberidazole and bitertanol/triadimenol/fuberidazole gave excellent control of bunt and of seedling blight particularly when caused by *Monographella nivalis*. Tebuconazole was effective against loose smut and both seed and soil-borne bunt, but combinations of tebuconazole with guazatine or thiram were necessary to extend the disease control spectrum to include seedling blights. These treatments offered significant improvements over modern commercial standards.

INTRODUCTION

Since the withdrawal of organomercurial seed treatments in the UK, a renewed interest has been shown in the control of seed-borne cereal pathogens and the performance of new active ingredients. This paper reports on wheat trials conducted over 6 seasons from 1988-93 with seed treatments based on bitertanol or tebuconazole. The results of trials against seed-borne diseases of barley are also reported in these proceedings (Wainwright *et al.*, 1994). Bitertanol is a broad-spectrum triazole fungicide with limited systemic activity which was first developed for application to pome and bush fruit (Birch *et al.*, 1981). Tebuconazole, also a triazole fungicide, has been developed in the UK principally for use in cereal and oilseed rape crops, and initial results obtained with spray and seed treatment formulations were reported by Wainwright & Linke (1987). Early trials with tebuconazole on wheat identified the need for an additional AI against Fusarium and applications with guazatine and thiram were included in subsequent work.

Most trials were handsown into small plots at Bayer's Elm Farm Development Station, Suffolk which enabled a large number of treatments to be tested under controlled conditions. Further experience against some of the more important pathogens was gained in 1992/3 by machine drilling trials at various sites throughout the UK.

METHODS AND MATERIALS

The treatments, which were applied as liquid or flowable formulations, are listed in Table 1. Application was by hand using glass jars for handsown trials and by mini-Rotostat for machine drilled trials. Trials were fully replicated using randomised block designs, and plot sizes were 1 m x 1 m (hand sown) and 4 m x 1.5 m (Oyjord-drilled). Naturally infected seed was used with the exception of bunt which was artificially inoculated by dusting healthy seed with fungal spores (1 or 2 g spores per kg) or by

inoculating soil using 0.5 or 1 g spores in 100 g sand per m² plot. Sowing
rates were 25 or 40 g/m² plot depending on the level of seed infection and
160 kg/ha for Oyjord drilled trials.

TABLE 1. Formulations included in trials and rates of application.

Product	Active ingredients	Application rate per 100 kg seed	
		ml	g AI
Exp.1	bitertanol/fuberidazole	150	56.25/3.45
Exp.2	bitertanol/triadimenol/fuberidazole	200	37.6/15/4.6
Exp.3	tebuconazole	120	3
Exp.4	tebuconazole	200	3
Exp.6	tebuconazole/guazatine	200	3/60
Exp.7	tebuconazole/guazatine	200	3/60
Exp.8	tebuconazole/thiram	200	3/100
Baytan® flowable	triadimenol/fuberidazole	200	37.5/4.5
Ceresol®	phenyl mercury acetate (PMA)	110	3.7
Cerevax®	carboxin/thiabendazole (TBZ)	250	90/5
Rappor®	guazatine	200	60
Vincit® L	flutriafol/thiabendazole (TBZ)	200	5/5

Sixteen hand-sown trials were located at Elm Farm Development Station:

Winter wheat - seedling blight (*Monographella nivalis*)   6 trials

            seedling blight (*Fusarium culmorum*)   1 trial

            seedling blight (*Leptosphaeria (Septoria)
                           nodorum)*   1 trial

            loose smut (*Ustilago nuda*)   2 trials

            bunt (*Tilletia caries*)   5 trials
                 (including 1 soil, 1 seed + soil-borne)

Spring wheat - bunt (*T. caries*)   1 trial

The handsown trials were generally sown late, ie November, under
conditions favouring seedling blights and bunt infection.

Two winter wheat cultivars infected with *M. nivalis* were drilled at
each of 4 sites (Norfolk, Glos., Lincs., and E. Lothian) using an Oyjord
plot drill in 1992/3 and a further 2 sites (Suffolk and Warks.) were drilled
with bunt infected seed.

Trials were assessed by counting diseased and healthy plants or ears in
each plot to obtain percent control/reduction figures. Stem base browning
symptoms on seedlings were graded and an index based on disease severity was
calculated (Table 3). Bunt and loose smut were assessed at GS 65-85 whilst
seedling diseases were assessed at GS 12-13. Plant stand was recorded at
all sites to monitor crop safety and the effect of seedling blights.

RESULTS

The results of trials are summarised in Tables 2-5. Crop stand counts reflect disease control in seedling blight trials and are given in Table 2. The results of stem base browning assessments in these trials are given in Table 3.

TABLE 2. Relative emergence (%) in seedling disease m² trials.

| Treatments | Monographella nivalis | | | | | | Fusarium culmorum | Septoria nodorum |
	FA/06 1988	FA/03 1989	FA/06 1990	FA/05 1991	FA/05 1992	FA/06 1993	FA/06 1992	FA/07 1990
bit./fub.		208	109	978	659	200	191	109
bit./triad./fub.		216	129	983	637	185	188	109
tebuconazole	306	167	107	196	185	129	189	96
tebucon./guaz.	896	198	119	974	652	177		101
tebucon./thiram				980	489	224	209	
triad./fub.	782	209	126	733	652	201	205	100
PMA	827	212	121	910				100
carboxin/TBZ				580	544	193	194	
guazatine				1058	587	216	199	
flutria./TBZ					178	104	197	
untreated (plants/m²)	(34)	(143)	(312)	(41)	(56)	(162)	(242)	(391)

TABLE 3. Reduction in stem infection (%) in m² trials.

| Treatments | Monographella nivalis | | | | | | Fusarium culmorum | Septoria nodorum |
	FA/06 1988	FA/03 1989	FA/06 1990	FA/05 1991	FA/05 1992	FA/06 1993	FA/06 1992	FA/07 1990
bit./fub.		100	96	74	89	92	89	96
bit./triad./fub.		100	100	93	98	91	95	100
tebuconazole	38	35	75	0	0	50	52	96
tebucon./guaz.	93	95	100	96	83	86		100
tebucon./thiram				77	28	88	89	
triad./fub.	64	92	91	19	30	47	92	100
PMA	68	60	37	24				99
carboxin/TBZ				0	2	18	87	
guazatine				93	74	84	23	
flutria./TBZ					0	23	56	
untreated (% infection index)	(26)	(44)	(26)	(39)	(60)	(22)	(10)	(31)

In trials that were free of seedling diseases, small reductions in crop stand and delayed emergence were usually seen with triazole treatments, though this was not observed with bitertanol/fuberidazole.

A particularly high level of bunt, 62 % infection, was recorded in FA/06 1991 (Table 4) which demonstrated the potential of soil-borne inoculum. Only 2 trials were carried out against loose smut due to limited availability of seed and complete control was achieved with all treatments containing a triazole active ingredient or carboxin, though the same seed batch had to be used in both trials.

TABLE 4. Reduction in bunt and smut infection (%) in m² trials.

Treatments	*Tilletia caries*						*Ustilago nuda*	
			(so)	(s/s)		(sp.)		
	FA/07	FA/08	FA/06	FA/08	FA/08	FA/23	FA/07	FA/07
	1988	1990	1991	1992	1993	1991	1992	1993
bit./fub.		100	100	100	100	100	100	100
bit./triad./fub.		100	100*	100*	100	100*	100*	100*
tebuconazole	99	99	100	100	93	100	100	100
tebucon./guaz.	100	100	100					
tebucon./thiram				99	98		100	100
triad./fub.	100	100	100	100	100	100	100	100
PMA		100	32					
carboxin/TBZ			54	46	87		100	100
guazatine			34	0	82	92	4	13
flutria./TBZ				100	96		100	
untreated (% infected ears)	(24)	(37)	(62)	(24)	(22)	(20)	(3)	(3)

* - rate 0.75 normal (so) - soil inoculated (s/s) - seed/soil inoculated (sp.) - spring wheat

The meaned results from Oyjord drilled trials against seedling blight and bunt are summarised in Table 5. Although disease expression varied slightly from site to site, treatment performance was similar to that observed in handsown plots.

All data were statistically analysed but lack of space prevents the inclusion of this information.

TABLE 5. Disease control in Oyjord drilled trials 1992-1993.

| Treatments | *M. nivalis* - 4 sites | | | | *T. caries* - 2 sites |
| | % Rel emergence | | % Redn infection | | % Redn infection |
	Haven	Riband	Haven	Riband	Beaver
bit./fub.	226	217	87	84	100
bit./triad./fub.	221	237	92	84	100
tebuconazole	171	161	25	20	99
tebucon./thiram	233	226	64	58	100
triad./fub.	224	232	64	33	100
guazatine	229	248	83	70	90
untreated	(22)	(21)	(52)	(50)	(53)
	(plants/m row)		(% index)		(% infected ears)

DISCUSSION

Until their withdrawal from use in 1992, the organomercury seed
treatments provided cheap and effective protection against seedling blights
caused by *Fusarium* spp. and bunt which are generally considered the most
important seed-borne diseases of wheat.

Good crop establishment may be jeopardised by seedling blights
particularly following a wet summer when seed infection is likely to be high
or when seed is drilled late. During six years of trials in the UK,
bitertanol based treatments consistently resulted in excellent crop stands
which were comparable to the best standards especially where *M. nivalis* was
present on the seed. In addition, these treatments gave the greatest
reductions in post-emergence stem browning symptoms. Tebuconazole alone was
not effective in controlling high levels of seedling blight and a
combination with guazatine or thiram was required.

It is essential to prevent bunt infection because of its ability to
build up rapidly in untreated seed stocks and to spread via soil inoculum,
the latter being cited as the cause of some recent outbreaks (Yarham, 1993).
Soil-borne bunt was only controlled well by triazole treatments; bitertanol
was surprisingly effective considering its limited sytemicity. Seed-borne
bunt was largely well controlled except in 1993 when treatments were less
reliable and carboxin and guazatine gave less than 90 % control.

Loose smut, a relatively minor disease in wheat, was completely
controlled by all triazole treatments and carboxin in these trials. Some
earlier results had indicated that bitertanol may not give consistent
control of loose smut and this led to the alternative co-formulation of
bitertanol which included triadimenol.

Against the range of seed-borne diseases of wheat, the seed treatments bitertanol/fuberidazole, bitertanol/triadimenol/fuberidazole, tebuconazole/guazatine and tebuconazole/thiram, offer a significant improvement over organomercury and modern commercial standards.

ACKNOWLEDGEMENTS

·   The authors would like to thank colleagues who have contributed to this work.

REFERENCES

Birch, P.A.; Rose, P.W.; Wainwright, A. (1981) A broad spectrum fungicide of the triazole group for use in pome and bush fruit. *British Crop Protection Conference - Pests and Diseases 1981*, **2**, 545-554.
Wainwright, A.; Linke, W. (1987) Field trials with Folicur® and Raxil® against foliar and seedborne cereal diseases, and observations on the control of rape diseases with Folicur in Great Britain (1984-1986). *Pflanzenschutz-Nachrichten Bayer,* **40**, 181-212 (English edition).
Wainwright, A.; Morris, D.B.; Meredith, R.H. (1994) New seed treatments based on bitertanol, tebuconazole and triazoxide for seed-borne disease control in barley. *These proceedings*.
Yarham, D.J. (1993) Soil-borne spores as a source of inoculum for wheat bunt (*Tilletia caries*). *Plant Pathology*, **42**, 654-656.

NEW SEED TREATMENTS BASED ON BITERTANOL, TEBUCONAZOLE AND TRIAZOXIDE FOR SEED BORNE DISEASE CONTROL IN BARLEY

A. WAINWRIGHT, D.B. MORRIS, R.H. MEREDITH

Bayer plc, Crop Protection Business Group, Elm Farm Development Station, Great Green, Thurston, Bury St Edmunds, Suffolk, IP31 3SJ

ABSTRACT

In trials on barley between 1988-1993, a seed treatment formulation containing bitertanol/triadimenol/fuberidazole was effective against loose smut, covered smut, foot rot (*Cochliobolus*), leaf stripe in spring barley and organomercury resistant leaf stripe in winter barley. Compared with triadimenol/fuberidazole, there was some improvement against leaf stripe in winter barley (organomercury sensitive), net blotch, and stem browning caused by *Monographella nivalis*.

Tebuconazole gave good control of loose smut and covered smut and was also active against leaf stripe, net blotch and foot rot (*Cochliobolus*). Co-formulation of tebuconazole with imazalil or triazoxide improved performance. Triazoxide showed outstanding activity against *Pyrenophora* spp. and in combination with tebuconazole provides a wide-spectrum treatment for barley.

INTRODUCTION

This paper reports on barley trials carried out over six seasons, 1988-1993, with seed treatments based on bitertanol or tebuconazole. Trials against seed-borne diseases of wheat are reported elsewhere in these proceedings by Morris *et al*. (1994). In discussing early results with tebuconazole on barley, Wainwright & Linke (1987) suggested the addition of a specific AI for reliable control of leaf stripe. Subsequent trials included treatments containing either imazalil or triazoxide, a new benzotriazine fungicide which is also highly specific against *Pyrenophora* spp. (Dutzmann, 1994).

Most trials were carried out as handsown plots at Elm Farm Development Station in Suffolk whilst in 1992/3 machine drilled trials were located at various sites throughout the UK.

METHODS AND MATERIALS

The treatments, which were applied as liquid or flowable formulations, are listed in Table 1. Application was by hand using glass jars for handsown trials and by mini-Rotostat for machine drilled trials. Trials were fully replicated using randomised block designs and plot sizes were 1 m x 1 m (hand sown) and 4 m x 1.5 m (Oyjord-drilled). Naturally infected seed was used with the exception of covered smut for FA/05, 1993 which was artificially inoculated using a spore suspension in water. Sowing rates were 25 or 40 g/m² plot depending on the level of seed infection and 140 kg/ha for Oyjord drilled trials.

TABLE 1. Formulations included in trials and rates of application.

Product	Active ingredients	Application rate per 100 kg seed	
		ml	g AI
Exp.2	bitertanol/triadimenol/fuberidazole	200	37.6/15/4.6
Exp.3	tebuconazole	120	3
Exp.4	tebuconazole	200	3
Exp.5	tebuconazole/triazoxide	150	3/3
Exp.6	tebuconazole/imazalil	200	3/4
Baytan® flowable	triadimenol/fuberidazole	200	37.5/4.5
Ceresol®	phenyl mercury acetate (PMA)	110	3.7
Cerevax® Extra	carboxin/thiabendazole (TBZ)/imazalil	200	60/5/4
Murganic® RPB	carboxin/phenyl mercury acetate	220	79.2/3.7
Rappor® Plus	guazatine/imazalil	220	66/5.5

Forty four trials were handsown at Elm Farm Development Station:

		winter/spring trials
Barley:	seedling blight (*Monographella nivalis*)	2[+]/ 6
	net blotch (*Pyrenophora teres*)	1 / 1
	foot rot (*Cochliobolus sativus*)	- / 5
	leaf stripe (*Pyrenophora graminea*)	9*/ 7
	loose smut (*Ustilago nuda*)	5 / 6
	covered smut (*Ustilago hordei*)	2 / -

* organomercury sensitive/resistant = 5/4
+ One trial was a spring cultivar sown in the autumn.

Four winter barley cultivars, two infected with leaf stripe and two with loose smut were drilled at each of four sites (Norfolk, Lincs., Yorks., and Kent) using an Oyjord plot drill in 1992/3.

Trials were assessed by counting diseased and healthy plants or ears in each plot to obtain percent control/reduction figures. Stem base browning symptoms on seedlings were graded and an index based on disease severity was calculated. Plant stand was recorded at all sites to monitor crop safety and the effect of seedling blights.

RESULTS

The results of individual hand-sown trials are given in Tables 2-7. Space does not allow the inclusion of crop stand data which was recorded in all trials; any reduced or delayed emergence was as normally expected in seed treatments with a triazole component. Seedling blight in barley trials was not severe even when seed was heavily infected with *M. nivalis*, and little benefit in terms of increased crop stand was obtained from seed treatments.

TABLE 2. Reduction in stem infection (%) in m² trials.

| Treatments | *Monographella nivalis* | | | | | | | |
| | W. barley | | S. barley | | | | | |
	FA/01 1988	FA/02# 1993	FA/22 1988	FA/17 1989	FA/20 1990	FA/19 1991	FA/17 1992	FA/21 1993
bit./triad./fub.		92		100	98	94	92	95
tebuconazole	0	0	0	92	96	49	7	20
tebucon./triaz.	0		0		94		31	20
tebucon./imaz.							9	24
triad./fub.	80	54	16	99	98	82	59	53
PMA	64		44	36	92	79		
carbox./TBZ/imaz.		78				81*	30	43
guazatine/imaz.		87				91	73	80
untreated (% infection index)	(12)	(9)	(62)	(17)	(10)	(7)	(19)	(32)

* - rate 1.25 normal   # - spring cultivar sown in autumn

TABLE 3. Reduction in seedling infection (%) in m² trials.

| Treatments | *Pyrenophora teres* | | *Cochliobolus sativus* | | | | |
| | W. barley | S. barley | S. barley | | | | |
	FA/03 1990	FA/23 1993	FA/20 1989	FA/21 1990	FA/20 1991	FA/18 1992	FA/22 1993
bit./triad./fub.	97	98	94	99	91	74	87
tebuconazole	72	73	48	97	64	68	72
tebucon./triaz.	94	100	73	90		75	63
tebucon./imaz.		96				83	79
triad./fub.	97	62	87	94	88	88	76
PMA	100		36	66	36		
carbox./TBZ/imaz.		98			90*	74	78
guazatine/imaz.		99			83	84	75
untreated	(2)	(12)	(33)	(42)	(38)	(33)	(28)
	(% infected plants)		(% infection index)				

* - rate 1.25 normal

Only 0.3 % covered smut was recorded in untreated plots in FA/04, 1992. In the following year an increased level of 1.3 % smutted ears was obtained with artificially inoculated seed - see Table 6.

The meaned results from Oyjord drilled trials against leaf stripe and loose smut are summarised in Table 8.

Statistical information is not included due to lack of space.

TABLE 4. Reduction in leaf stripe infection (%) in W. barley m² trials.

Treatments	Pyrenophora graminea (PMA resistant strain)				Pyrenophora graminea (PMA sensitive strain)				
	FA/02 1988	FA/04 1989	FA/04 1990	FA/03 1991	FA/03 1988	FA/05 1989	FA/02 1991	FA/02 1992	FA/03 1993
bit./triad./fub.		100	100	100		96	97	83	65
tebuconazole	100		100	100	95		85	86	30
tebucon./triaz.	100		100		100				100
tebucon./imaz.				100			97	98	72
triad./fub.	100	100	100	100	30	32	4	12	0
PMA	56	42	58	87	100	100	99		
carbox./TBZ/imaz.				100			97	99	49
guazatine/imaz.				100			99	100	80
untreated (% infected tillers)	(0.5)	(3.3)	(3.2)	(1.1)	(19)	(21)	(28)	(36)	(33)

TABLE 5. Reduction in leaf stripe infection (%) in S. barley m² trials.

Treatments	Pyrenophora graminea						
	FA/23 1988	FA/18 1989	FA/22 1990	FA/21 1991	FA/22 1991	FA/19 1992	FA/24 1993
bit./triad./fub.		100	95	98	100	100	100
tebuconazole	84	87	83	85	95	100	99
tebucon./triaz.	100	100	100			100	100
tebucon./imaz.				96	100	100	100
triad./fub.	100	100	100	98	100	100	100
PMA	47	42	0	0	0		
carbox./TBZ/imaz.				94*	100*	100	100
guazatine/imaz.				99	100	99	100
untreated (% infected tillers)	(3.0)	(9.0)	(1.3)	(6.7)	(1.5)	(5.8)	(17.7)

* - rate 1.25 normal

DISCUSSION

Leaf stripe and loose smut are generally considered to be the most important of the seed-borne diseases of barley. Seed treatments which contained organomercury alone were able to give cheap, effective control of leaf stripe and other diseases whilst carboxin could be added to extend the range to include loose smut. However, during the decade before the withdrawal of mercury, resistance in *P. graminea* was identified in the field (Locke, 1986) and clearly there was a requirement for new active compounds.

TABLE 6. Reduction in smut infection (%) in W. barley m² trials.

Treatments	Ustilago nuda					Ustilago hordei	
	FA/04 1988	FA/05 1990	FA/04 1991	FA/03 1992	FA/04 1993	FA/04 1992	FA/05 1993
bit./triad./fub.		100	100*	100*	100*	100*	100*
tebuconazole	100	99	100	100	100	100	100
tebucon./triaz.	100	100			100		
tebucon./imaz.					99		
triad./fub.	100	100	100	100	100	100	100
carboxin/PMA	100	98					
carbox./TBZ/imaz.			97	98	100	100	100
guazatine/imaz.			67	48	69	100	100
untreated (% infected tillers)	(2.6)	(9.2)	(4.9)	(10.8)	(10.2)	(0.3)	(1.3)

* - rate 0.75 normal

TABLE 7. Reduction in smut infection (%) in S. barley m² trials.

Treatments	Ustilago nuda					
	FA/18 1989	FA/21 1989	FA/23 1990	FA/22 1991	FA/20 1992	FA/25 1993
bit./triad./fub.	100	99	100	100	98*	87*
tebuconazole	100	92	86	100	90	91
tebucon./triaz.	100	100	100		93	99
tebucon./imaz.				100	91^	99
triad./fub.	100	100	100	100	99	100
carboxin/PMA		95	98			
carbox./TBZ/imaz.				98#	92	75
guazatine/imaz.				7	0	0
untreated (% infected tillers)	(0.4)	(2.2)	(1.6)	(14.3)	(5.4)	(3.8)

* - rate 0.75 normal   # - rate 1.25 normal   ^ - rate 0.67 normal

Whilst all treatments gave complete control of organomercury resistant leaf stripe on winter barley, the partial substitution of triadimenol with bitertanol led to an improvement over triadimenol/fuberidazole against the sensitive strain (Tables 4 and 5). The addition of imazalil or triazoxide to tebuconazole raised performance against leaf stripe although only the triazoxide combination was completely effective in 1993.

The seedling diseases including Fusarium (*M. nivalis*) and footrot (*C. sativus*) are usually of minor importance during the establishment of a barley crop. Against these diseases and net blotch, bitertanol conferred

TABLE 8. Disease control in Oyjord drilled trials.

Treatments	*Pyrenophora graminea* - mean of 4 sites		*Ustilago nuda* - mean of 4 sites	
	Gaulois	Flamenco	Pastoral	Panda
bit./triad./fub.	81	66	100	100
tebuconazole	85	68	100	100
tebucon./triaz.	100	100	100	100
tebucon./imaz.	99	98	100	100
triad./fub.	12	7	100	100
untreated (% infected tillers)	(14)	(22)	(8)	(7)

additional benefits over triadimenol/fuberidazole. The performance of tebuconazole against net blotch on the heavily infected cultivar Peel was clearly enhanced by the addition of imazalil and particularly triazoxide.

Tebuconazole and triadimenol formulations were very effective against loose smut, though the reliability of tebuconazole in spring barley was improved in formulations with imazalil or triazoxide. In handsown trials in 1992-93 (Table 7), reduced control may be associated with late sowing during warm, rapid-growing conditions when it is known that triazole performance can be impaired (Martin & Edgington, 1980).

For control of the range of seed-borne diseases in barley, bitertanol/ triadimenol/fuberidazole, tebuconazole/imazalil and tebuconazole/triazoxide offer improvements over existing standards. In recent years a heavy reliance has been placed on imazalil for leaf stripe control by its inclusion in a number of commercial formulations; the different chemistry and excellent efficacy of triazoxide should therefore make it a valuable addition to the limited treatment range currently available.

REFERENCES

Dutzmann, S. (1994) Triazoxide - a new seed treatment fungicide for the control of seed borne *Pyrenophora* spp. *These proceedings*.
Locke, T. (1986) Current incidence in the United Kingdom of fungicide resistance in pathogens of cereals. *British Crop Protection Conference - Pests and Diseases 1986*, **2**, 781-786.
Martin, R.A.; Edgington, L.V. (1980) Effect of temperature on efficacy of triadimenol and fenapanil to control loose smut of barley. *Canadian Journal of Plant Pathology*, **2**, 201-204.
Morris, D.B.; Wainwright, A.; Meredith, R.H. (1994) New seed treatments based on bitertanol and tebuconazole for seed-borne disease control in wheat. *These proceedings*.
Wainwright, A.; Linke, W. (1987) Field trials with Folicur® and Raxil® against foliar and seed-borne cereal diseases, and observations on the control of rape diseases with Folicur in Great Britain (1984-1986). *Pflanzenschutz-Nachrichten Bayer*, **40**, 181-212 (English edition).

COST BENEFIT ANALYSIS OF RUSSIAN WHEAT APHID (*DIURAPHIS NOXIA*) CONTROL IN SOUTH AFRICA USING A SEED TREATMENT.

M.C. VAN DER WESTHUIZEN, J. DE JAGER

Department of Zoology and Entomology, University of the Orange Free State, Bloemfontein, 9300, South Africa.

M.H. MASON, M.W. DEALL

Bayer (Pty) Ltd., Isando, 1600, South Africa.

ABSTRACT

The cost and control effectiveness of imidacloprid as the active ingredient of a seed dressing was evaluated under dry land condition in the suppression of *Diuraphis noxia* on wheat. The optimum dosage of 117 g AI or 167 g product per (deca tonne) seed provided an investment return of 4.23. The registered rate of 200 g product of Gaucho 70 WS per dt seed provides effective control of the aphid and is economically justified.

INTRODUCTION

The Russian Wheat Aphid (RWA) (*Diuraphis noxia*) is currently the most important pest of wheat in South Africa. The RWA was first recorded in South Africa in 1978 and within a few years had spread to all the wheat-producing areas in the country (Aalbersberg *et al.*, 1987). However, its major economic impact is on the wheat of the Orange Free State. One million hectares of wheat are grown in the Orange Free State, which accounts for approximately 60% of South Africa's total crop. The average yield for this region is 1.5 ton/ha. Measurements of single plant yields have shown that a reduction in yield of up to 90% can occur as a result of RWA infestations (Du Toit & Walters, 1984). The loss from RWA is estimated at between US $16 and 18 million annually.

The Orange Free State receives rainfall during the summer months, which is preserved by appropriate cultivation practices for subsequent planting of wheat in autumn. The low humidity and virtually no rainfall during the growing season favours the development of large populations of this species.

Characteristic symptoms of RWA infestation are white, reddish-purple and/or yellow streaking of infested leaves as well as inward curling of the leaf edges. The latter leads to a reduction in the photosynthetic area of the leaf. The aphids and the typical symptoms are mainly found on the new growth and in the later growth stages tend to occur on the flag leaf. Aphid colonisation may result in damage to the ears which become bent and turn white, with consequent incomplete filling of the ear (Aalbersberg, 1987).

Control strategies utilising host genetic resistance, chemical and biological methods have been used in South Africa and other areas of the world to control RWA.

Traditionally, a wide range of insecticides have been used to control RWA and in 1991 RWA resistant cultivars were introduced in South Africa for the first time. Recently an insecticide seed dressing, Gaucho 70 WS (imidacloprid), was introduced and the objective of this study was to establish guidelines for the wheat industry on the economic effectiveness of the compound in controlling RWA.

The trials were done to determine a) the influence of various dosage rates of imidacloprid on the aphid infestation rate of the crops under field conditions, b) the period of control and type of aphid suppression (tolerance or antibiosis) and c) the economic response to control measures by means of the various response curves:
- Pest infestation levels
- Cultivar growth stage response
- Cost benefit analysis
- Crop response

MATERIAL AND METHODS

The trials were conducted at the Glen Agricultural Development Institute in the Western Orange Free State. A randomised block design with five replicates was used. The plots were 2.7 m wide, consisting of six wheat rows and 10 m long. The outer row on either side was used as an infestation source and only the inner four rows were used for evaluations. Seeding was done with a Gaspardo precision planter, planting six rows, using the cultivar Scheepers '69' at a seed density of 18 kg/ha.

Treatments included an aphid free control, obtained by regular foliar insecticide application throughout the growing period and seed dressing with imidacloprid at 0, 35, 70, 105, 140 and 175 g AI/dt seed.

The following parameters for the various treatments were recorded weekly: date of infestation, number of tillers infested and growth stage (Zadoks *et al.*, 1974). Infested tillers were labelled weekly with various coloured tags indicating date (growth stage) of infestation. All infested tillers were individually harvested throughout the entire plot, while the tillers not infested were collectively yielded. Plot yields were expressed as the summed mass of these treatments. The yield and infestation parameters of the various treatments were analysed by analysis of variance and the means compared using the Student-Newman-Keuls test (Steel & Torrie, 1960). Zar's method of multiple regression analysis was used to compare slopes and elevations (Y-intercepts) of linear regressions (Zar, 1974).

The following economics of chemical control and wheat production were used in the cost benefit analysis of seed dressing as well as the calculation of economic damage and economic injury levels for foliar insecticide applications (tractor and aerial) and seed dressing:

Cost of foliar aphicide per ha	US $ 8.38
Tractor application per ha	US $ 2.55
Aerial application per ha	US $ 6.61
Gaucho 70 WS per kg	US $ 212.78
Wheat price per ton	US $ 181.38

The cost benefit analysis was based on the procedures of Flint & Van den Berg (1981) and Van der Westhuizen (1992) and the procedure of Chiang (1973) and Van der Westhuizen (1992) were used in the calculations of economic damage and economic injury levels for cereals.

RESULTS AND DISCUSSION

<u>RWA infestation in the absence of chemical control</u>

Fig. 1 shows the normal pattern of horizontal infestation (among tillers) of Scheepers '69' by the RWA in the absence of any chemical control measures. The percentage tillers infested at the various dates of infestation after planting was based on average infestation of the five plots. An exponential growth rate of 0.4563 percent tillers infested per day occurred in the absence of chemical control.

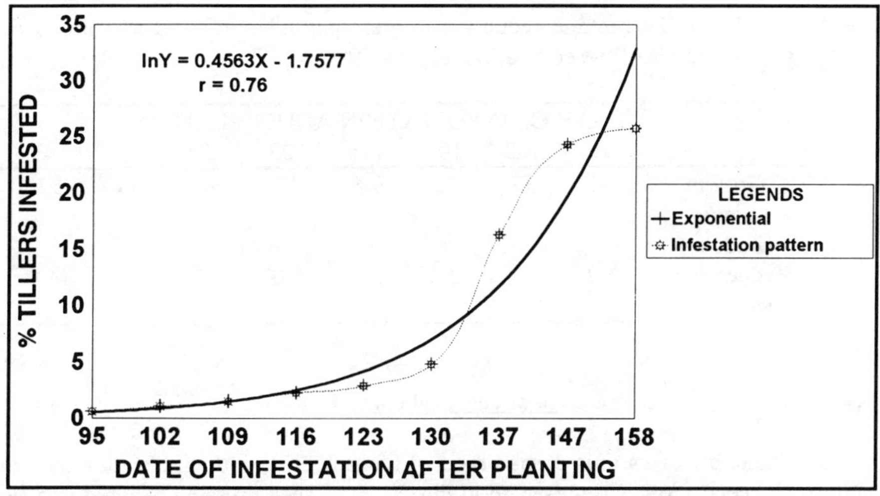

Fig. 1. Horizontal infestation of cultivar Scheepers '69' by RWA.

<u>Impact of RWA on yield in the absence of chemical control</u>

The yield of tillers infested and not infested from plants that were infested and not infested are shown in Table 1. No significant difference in yield occurred between tillers not infested from plants that were either infested or not infested ($F = 0.58$; $p > 0.05$). There was a significant difference in yield however, between infested tillers and tillers not infested ($F = 310.01$; $p < 0.01$). These results indicated that the effect on yield by the RWA feeding stays within the tiller of infestation and no reduction in yield of adjacent non infested tillers within the same plant occurred.

TABLE 1.   Inter- and intra tiller effect of RWA feeding on yield of tillers (g).

| TILLERS | INFESTATION STATUS | |
| | PLANTS | |
RWA	Infested	Not Infested
Infested	0.553	-
Not Infested	0.917	0.935

The effect of RWA infestation at various growth stages of the cultivar Scheepers '69' with regard to reduction in yield per tiller is shown in Table 2.   The growth stage response identifies the most sensitive growth stage of specific cultivars to RWA infestation. According to Table 2 the greatest impact of RWA occurs when infestation takes place at growth stage 57. Thus, the economic injury level should be based on the impact of varying infestation levels of RWA at approximately growth stage 60 of the cultivar Scheepers '69'.

TABLE 2. Percentage reduction in yield per tiller infested at various dates and growth stages of the cultivar Scheepers '69'.

| | DATE OF INFESTATION AFTER PLANTING | | | | | | | | |
	95	102	109	116	123	130	137	147	158
Growth stage (Zadoks)	30	32	33	34	37	38	43	57	69
Reduction in yield (%)	2.2	1.8	2.5	2.4	8.0	9.2	34.9	44.8	3.7

RWA infestation in the presence of chemical control

The effect of various dosage rates of imidacloprid on the infestation pattern of the cultivar Scheepers '69' is represented in Fig. 2.   The seed dressing resulted in a significant reduction in numbers of tillers infested ($F = 16.38$; $p < 0.01$).

The exponential growth of the infestation patterns achieved with the various dosages of seed treatments were determined and are shown in Table 3.   From this table it is clear that imidacloprid treatments did not affect the infestation rate ($F = 2.13$; $p > 0.05$) of RWA, but resulted in a significant decrease in the Y-intercepts of these growth curves ($F = 4.45$; $p < 0.05$).   Thus, according to these results, imidacloprid provides no suppression in infestation rate (antibiosis), but does, however, provide an increased period of total control of RWA infestation.

<u>Stimulatory effect of imidacloprid</u>

The relationship between various dosages of imidacloprid and yields of tillers not infested showed a tendency of increasing yield ($Y = 0.9292 + 9.1429 \times 10^{-5}X$, $r = 0.4180$). These increases are however not significant ($F = 1.62$; $p > 0.05$).

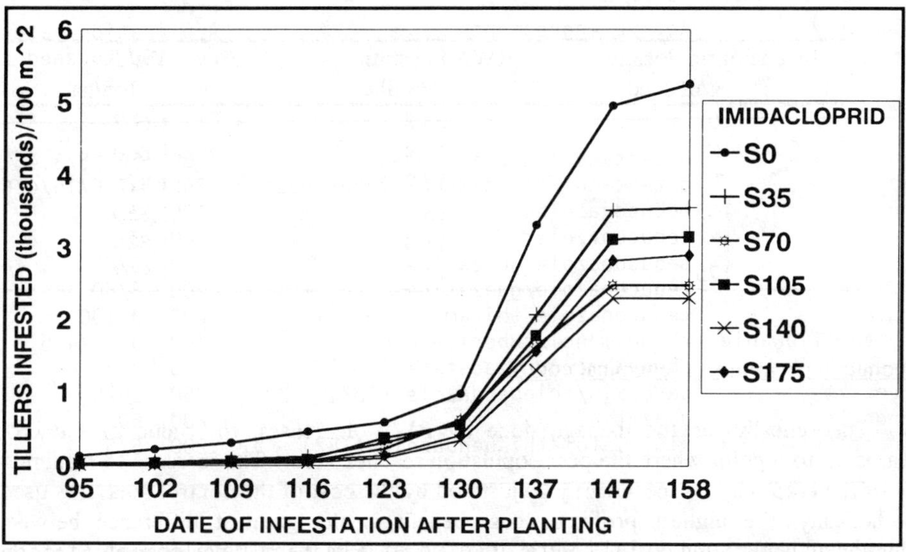

Fig. 2. Effect of various dosages of imidacloprid seed dressing (g AI/dt seed) on tillers infested with RWA.

TABLE 3. Impact of various dosage rates of imidacloprid on the growth of RWA with regard to percentage tillers infested per day.

Imidacloprid dosage g/dt seed	Exponential growth rate/day % tillers infested	Y-intercept
0	0.4563	-1.7577
35	0.4180	-2.7074
70	0.4914	-3.0506
105	0.4434	-3.0150
140	0.4482	-3.5323
175	0.4862	-3.3832

<u>Impact of RWA on yield in the presence of chemical control</u>

The effect of imidacloprid seed dressing on the mean infestation level and yields of the plots treated with various dosages is shown in Table 4. Seed dressing as a component of variance resulted in a significant decrease in final infestation ($F = 16.38$; $p < 0.01$) and increase in yield (F = 48.24; p < 0.01). The 140 g AI per dt seed treatment rate gave the best control of RWA. The highest efficacy in control and yield response lies between dosage rates of 70 and 140 g AI per dt seed.

TABLE 4. Effect of various dosages of imidacloprid on mean infestation levels and yield of cultivar Scheepers '69'.

Imidacloprid dosage g/dt seed	RWA infestation % tillers	Yield ton/ha
0	25.8	1.667
35	17.4	1.800
70	12.2	1.825
105	15.5	1.850
140	11.3	1.830
175	14.2	1.857

<u>Economic implications of chemical control</u>

Theoretically, as the damage done by RWA decreases, the value of the crop increases up to a point where the pest population reaches zero. The increase to maximum value of the crop may not necessarily be justified by the cost of the control measures used. Hypothetically, the highest profit or cost benefit is the greatest difference between production advantage and cost of control. Fig. 3 represents the cost benefit analysis for the various treatments evaluated in the trials. The cost of control was based on the amount of product used per ha. The investment return indices (IRF index) were calculated as follows:

$$\frac{(\text{Control benefit - Control cost})^3}{100}$$

Peak indices coincide with maximum return in control (Van der Westhuizen, 1992). The cost benefit analysis indicates an optimum dosage of 167 g product per dt seed with an investment return of 4.23.

The pest crop response for Scheepers '69' at the most sensitive growth stage, GS 57 (see Table 2), is shown in Fig. 4. The economic injury levels (EIL) of tractor and aerial applications of a foliar spray is indicated on the figure. Furthermore, the actual infestation of the tillers in the untreated control (S0) and the plots planted with 140 g AI imidacloprid per dt seed (S140) are indicated.

From the crop pest response curve (Fig. 4) as well as the actual RWA infestations in the untreated control, foliar application (tractor and aerial) would have been justified. However, at the seed treatment rate of 140 g AI/dt, RWA was sufficiently suppressed, so that no further corrective treatment would have been justified.

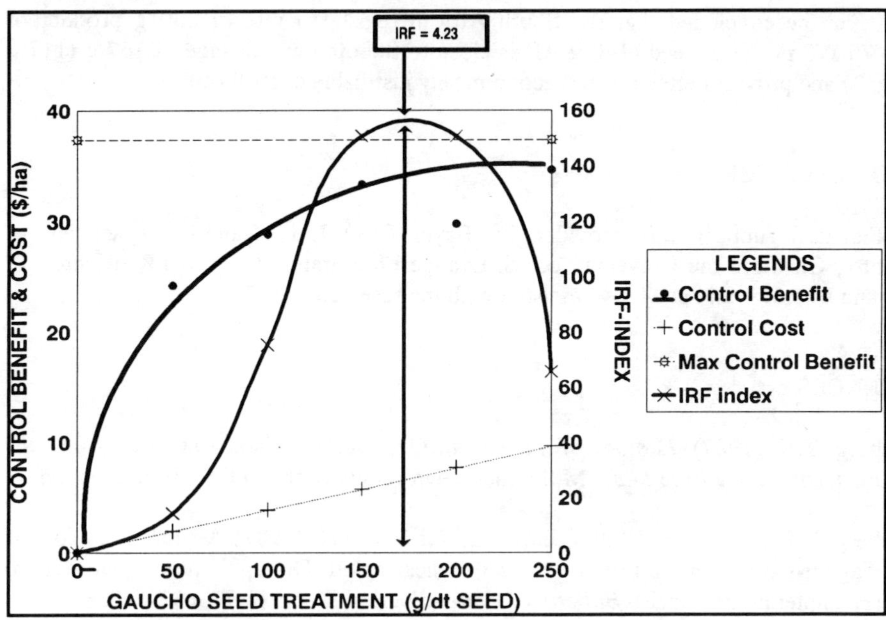

Fig. 3. Cost benefit analysis of imidacloprid seed dressing in the chemical control of RWA as pest of the wheat cultivar Scheepers '69'.

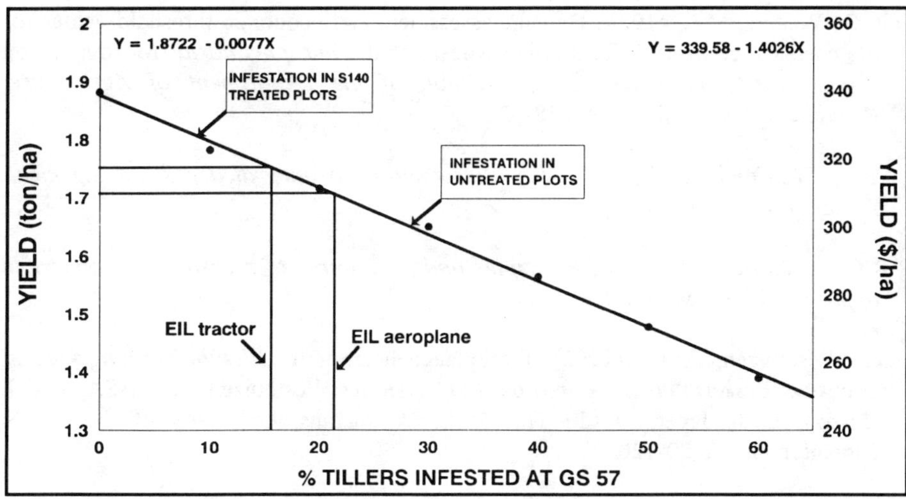

Fig. 4. Yield response of the cultivar Scheepers '69' with increasing levels of RWA infestation as well as the guidelines for corrective tractor and aerial applications of foliar aphicides.

It can be concluded that the South African registered rate of 200 g product of Gaucho 70 WS per dt of seed (140 g AI) is close to the optimum dosage of 167 g (117 g AI) (Fig. 3) and provides effective and economically justifiable control of RWA.

ACKNOWLEDGEMENTS

Financial support was provided by Bayer (Pty) Ltd., Isando, Department of Agriculture, Glen and the University of the Orange Free State. Messrs. J.R. Richter, P. Aldrich and Mrs. D. van der Merwe assisted with the research.

REFERENCES

Aalbersberg, Y.K. (1987) *Ecology of the wheat aphid* Diuraphis noxia *(Mordvilko) in the Eastern Orange Free State.* M.Sc. thesis, University of the O.F.S., Bloemfontein.

Aalbersberg, Y.K.; Van der Westhuizen, M.C.; Hewitt, P.H. (1987) A simple key for the diagnosis of the instars of the Russian wheat aphid, Diuraphis noxia (Mordvilko) (Hemiptera: Aphididae). *Bulletin of Entomological Research* 77, 637-640.

Chiang, H.C. (1973) Ecological considerations in developing recommendations for chemical control of pests: European corn borer as a model. *FAO Plant protection Bulletin* **21**, 30-39.

Du Toit, F.; Walters, M.C. (1984) Damage assessment and economic threshold values for the chemical control of the Russian wheat aphid, *Diuraphis noxia* (Mordvilko) on winter wheat. *Technical Communication of the Department of Agriculture, Republic of South Africa* **191**, 58-62.

Flint, M.L.; Van den Bosch, R. (1981) *Introduction to integrated pest management.* Plenum Press, New York.

Steel, R.G.D.; Torrie, J.H. (1960) *Principles and procedures of statistics, a biometrical approach.* McGraw-Hill, New York.

Van der Westhuizen, M.C. (1992) Insekplaagbeheer. In *Plantbeskermingskursus: Glenkovs Beskerm en Bewaar.* Eds. M.C. van der Westhuizen, AVCASA, N.C.J. Basson, J. de Jager, D. du Toit & E. Wolmarans, University of the O.F.S. Bloemfontein, pp 70-126.

Zadoks, J.C.; Chang, T.T.; Konzak, C.F. (1974) A decimal code for the growth stages of cereals. *Weed Research* **14**, 415-421.

Zar, J.H. (1974) Biostatistical analysis. In *Biological Sciences Series.* Eds W.D. McElroy & C.P. Swanson. Prentice-Hall, Englewood Cliffs, N.J, pp 228-235.

IMPACT OF SEED-COATING POLYMERS ON MAIZE SEED DECAY BY
SOILBORNE PYTHIUM SPECIES

D. C. McGEE, B. ARIAS-RIVAS, J. S. BURRIS

Seed Science Center and Departments of Plant Pathology and
Agronomy, Iowa State University, Ames, IA 50011, U.S.A.

ABSTRACT

Polymeric seed coatings were investigated as
alternatives to fungicide seed treatments of maize.
A field study was carried out in which three
polymers either alone or in combination with the
standard captan seed treatment of maize were
examined for plant emergence and infection of
nongerminated seeds by soilborne *Pythium* species.
Polymers in combination with captan were as
effective as captan alone in ensuring an adequate
stand of maize seedlings and in reducing incidence
of *Pythium* spp. seed infection. There was little
evidence that seed treatment with polymers alone
could provide protection against infection and
improve plant emergence.

INTRODUCTION

Pythium species are soil-inhabiting fungi that cause seed
rot and damping-off in a wide range of plants. Maize seeds are
particularly vulnerable to attack when subjected to cold and
wet soil conditions after planting (Johann, H. *et al*, 1928 and
Rao, B. *et al*, 1978). Effective control can be achieved by
seed treatment with the fungicide captan. However, increasing
concerns about human health risks and environmental
contamination have stimulated research for alternatives to
fungicide and reduction in fungicide dosages.

There has been considerable interest in the use of
degradable polymeric seed coatings as carriers for chemical
and biological control agents in recent years. The potential
of polymers to improve retention of the active ingredient on
seeds could lead to lower dosages of fungicides and improved
efficacy. Polymers also can regulate the rate of water
imbibition by seeds which could allow seeds to escape
infection by soilborne pathogens. It has been shown that
polymer coatings can reduce invasion of maize and soybean
Aspergillus and *Penicillium* spp. in seeds stored at high
relative humidity by reducing the rate of moisture uptake
(McGee, D. C. *et al*, 1988).

The present study investigated the potential of seed
treatment of maize with polymers, either alone or in
combination with captan, to ensure adequate plant stands and

reduce seed decay by soilborne *Pythium* spp. under field
conditions.

MATERIALS AND METHODS

Seed treatment

Polymers examined included Sacrust (30% liquid
suspension) manufactured by Sarea, Luiz, Austria; Chitosan
(2.5% liquid suspension manufactured by Nova Chem Limited,
Halifax, Nova Scotia, Canada; and Certop (30% liquid
suspension) manufactured by Cel-Pril, Manteca, California,
U.S.A. Two hybrid maize seed lots were treated with these
polymers at the rate of 125 ml/kg seed either alone or in
combination with captan (Captan 400-D)at 500 ppm a.i.
Treatments were applied in fluidized bed coater as described
previously(Burris, J. S. *et al*, 1994). Separate lots of each
hybrid also were treated with captan (Captan 30DD) at 500 ppm
a.i. and metalaxyl (Apron-FL) at 318 ppm a.i. in a Batch
Laboratory Treater (Gustafson Inc., Dallas, Texas, U.S.A.).

Field experiment

Seeds were planted in the field on April 30, 1993 near
Ames, Iowa. Individual plots comprised four rows, 5 m long and
75 cm wide, with 100 seeds planted per row at a depth of 6 cm.
A split-split-plot experimental design was used with three
replications. Main plots were the high and low vigor seed
lots, sub-plots were the polymer treatments alone or in
combination with captan, captan or metalaxyl alone, and
untreated seeds. Sub-sub-plots were evaluation dates for plant
emergence and seed colonization by *Pythium* spp. Seedling
emergence in the middle two rows was counted at 12, 16, 20 and
25 days after planting. All nongerminated seeds from a 1 m row
section in one of the outside rows of each plot were removed
at 4, 8 and 15 days after planting. Ten seeds from each plot
were randomly selected and washed with tap water for 10 min.
Seed coats and embryos were dissected aseptically from each
seed and plated separately on a medium selective for *Pythium*
spp. (Schmitthenner, 1980). Plates were incubated in the dark
at 25 C for 4 days and number of colonies of *Pythium* spp.
growing from seed coats and embryos were recorded.

RESULTS

Field emergence

Relative differences in seedling emergence between
treatments were similar at all times of evaluation, therefore
only data for the final count (25 days after planting) are
presented (Table 1). Main effects in the statistical analysis
showed that emergence of high vigor seeds at 25 days after
planting was significantly greater than that for low vigor

seeds. The emergence of seeds treated with Chitosan alone did
not differ significantly from that for seeds treated with
Chitosan combined with captan for the high vigor seed lot, but
it was significantly lower for the low vigor lot (Table 1).
For all others polymers, seedling emergence was significantly
greater when they were applied with captan than when applied
alone for both seed lots. Seedling emergence for all
captan/polymer combinations was not significantly different
from that for seed treated with captan alone or metalaxyl in
both seed lots. Polymers applied alone did not significantly
improve seedling emergence over that for untreated high vigor
seed lot, but Sacrust and Chitosan showed a small but
significant increase over untreated low vigor seeds.

TABLE 1. Effect of polymer and fungicide seed coatings on
seedling emergence and *Pythium* spp. infection of maize seeds.

Seed treatment	Seedling emergence (%)	*Pythium* spp. infection (%) of nonemerged seeds	
		Seed Coat	Embryo
HIGH VIGOR SEED			
Sacrust	65.0	53.3	26.7
Sacrust + captan	93.3	3.3	16.7
Chitosan	86.2	52.3	21.0
Chitosan + captan	86.7	10.0	6.7
Certop	77.3	3.3	10.0
Certop + captan	92.5	7.0	3.3
Captan	92.7	11.0	17.0
Metalaxyl	88.2	27.0	10.0
Untreated	80.8	20.0	20.0
LSD (P=0.05)	7.2	17.5	19.0
LOW VIGOR SEED			
Sacrust	58.5	31.0	47.7
Sacrust + captan	69.3	3.3	3.3
Chitosan	49.8	48.7	62.3
Chitosan + captan	68.3	6.7	6.7
Certop	34.7	36.7	43.3
Certop + captan	81.3	3.3	6.7
Captan	77.5	3.3	31.0
Metalaxyl	64.0	0.0	21.0
Untreated	41.3	21.0	47.0
LSD (P=0.05)	7.2	25.6	20.0

<u>Seed infection by *Pythium* spp.</u>

Relative differences in *Pythium* infection of nongerminated seeds was similar at each sampling time, therefore only data for 8 days after planting are presented (Table 1). No damped-off seedlings were detected in the experiment. Seed coat and embryo infections were consistently higher for seeds treated with polymer alone than with polymer combined with captan or captan alone for low vigor seeds. The same trend occurred for high vigor seed, but differences were significant only for Sacrust. No differences in infection of seed coats occurred for seed coated with polymer combined with captan and captan alone. Embryo infection of low vigor seeds however, was significantly less for seeds treated with polymer combined with captan compared to captan alone. No significant differences were evident between *Pythium* infection for seeds treated with the two fungicides alone.

DISCUSSION

This study has shown that treatment of maize seeds with degradable polymers in combination with captan is as effective as captan alone in ensuring adequate seedling stands and in reducing infection of seeds by *Pythium* spp. under field conditions. These effects were particularly evident for the low vigor seed lot. The test was conducted under conditions favorable for root pathogens such as *Pythium* spp. in that the soil was much wetter than normal in the four week period after planting. It is of interest that extensive *Pythium* infection occurred on nongerminated seeds, regardless of the original vigor of the lot. There was little evidence that seed treatment with polymers alone could provide protection against infection by *Pythium* spp. and improve plant emergence. Emergence was similar for seeds treated with Chitosan alone and Chitosan combined with captan, but this was seen only for the high vigor seed lot and was not reflected in *Pythium* control.

This is one of the few studies of maize seed treatments, in which pathogen control was related to seedling emergence under field conditions. It provides a starting point for further research on the mechanism of control of soilborne pathogens of maize with polymer seed treatments. While the data did not show promise for the use of polymers alone in control of seedling disease, findings were only for one growing season and this approach should not be abandoned. The fact that the addition of polymers to captan seed treatment did not compromise the efficacy of captan is an encouraging finding for further research on the use of polymers as a means of reducing the captan dosage.

ACKNOWLEDGEMENTS

Journal Paper J-15572 of the Iowa Agriculture and Home Economics Experiment Station. Project 3099.

REFERENCES

Burris, J. S.; Prijic, L. M.; Chen, Y. (1994) A small scale laboratory fluidized bed seed-coating apparatus. These Proceedings.
Johann, H.; Holbert, J. R.; Dickson, J. G. (1928) A Pythium seedling blight and root rot of dent corn. *J. Agric. Res.* **37**,443-464.
McGee, D. C.; Henning, A.; Burris, J. S. (1988) Seed encapsulation methods for control of storage fungi. In: *Application to Seeds and Soil*, T. J. Martin (Ed.) BCPC Monograph No. 39, Thornton Heath: BCPC Publications, pp.257-264.
Rao, B.; Schmitthenner, A. F.; Caldwell, R.; Ellett, C. W. (1978) Prevalence and virulence of *Pythium* species associated with root rot of corn in poorly drained soil. *Phytopathology* **68**:1557-1563.
Schmitthenner, A. F. (1980) *Pythium* species: isolation, biology and identification. In: *Advances in Turfgrass Pathology*, P. O. Larsen and B. G. Joyner (Eds.) Duluth, Minnesota, Harcourt Brace Jovanovich, pp. 33-36.

THE ROLE OF CARBOXIN + THIRAM FS IN POST-MERCURY SEED TREATMENT
FUNGICIDE STRATEGIES IN CENTRAL EUROPE

D.JACKSON, R.MARSHALL, M.J.TOMKINS
Uniroyal Chemical Ltd, Kennet House, 4 Langley Quay,
Slough, Berkshire SL3 6EH

O.NOVY
Uniroyal Chemical, Jankovcova 18, 17037 Praha 7, Czeck

Z.PAPP
Uniroyal Chemical Office, Timar Utca 20, H-1034, Budapest

J.ROMANIUK
Uniroyal Chemical, PO Box 301, Warsaw, Poland

ABSTRACT

Organomercurial seed treatments were widely used in
Central Europe particularly on small-grain cereals;
all seed treatment uses have now been terminated,
except in Romania.

Specific and broad-spectrum replacement fungicides
have been developed and the range of products now
available is reviewed.

Carboxin + thiram FS has been widely tested in Central
Europe and demonstrates broad-spectrum disease control
coupled with multi-crop use and seed safety. It is
easy to apply and has a direct stimulatory effect on
emergence and growth of seedlings. Carboxin + thiram
FS is registered internationally which will help
movement of treated seed in and out of Central Europe.

INTRODUCTION

Organomercury seed treatments became widely accepted in
Central Europe as much as in West Europe because of their
economic and effective control of important seed-borne diseases.
In spite of the effectiveness of mercury these diseases were not
eradicated altogether. Certain weaknesses in performance and
gaps in the spectrum of mercury were identified in Central Europe
as elsewhere, and development started with alternative chemicals.
Replacement products were available by the 1970's, and the
gradual prohibition of mercury seed treatments in Central Europe
started.

The importance of seed-borne diseases and the need to use
effective seed treatment chemicals continues to be recognised in
Central Europe for the same reasons as in the UK (Rennie, 1993).
Seed treatment is still the only means of controlling certain
diseases of cereals and other crops; it is a cost-effective
measure and is a key element of low-input systems and integrated

developments typical of Central Europe. Improvement of seed
health through use of seed treatments is also an important part
of the development of seed production business there.

This paper will review the modern seed treatments available
in Central Europe, and will present developments with carboxin
+ thiram formulations mainly on cereals and from Poland, Hungary,
Czech Republic, Slovak Republic and Romania.

ORGANOMERCURY STATUS IN EUROPE

Organomercury seed treatments are being prohibited in
Central Europe because of problems of acute toxicity and effect
on wildlife. Poland (1977) and Hungary (1978) were the first to
prohibit mercury use as seed treatments, with the Czech and
Slovak Republics following much later (1991) and Bulgaria and
Yugoslavia in 1992. There are derrogations in certain
Yugoslavian republics, and use is still permitted in Romania.
Some countries, such as Hungary, are particularly concerned about
safety to birds. Environmental pollution concerns, particularly
at sites of manufacture have been a strong stimulus for replacing
mercury in recent years.

NON-MERCURIAL SEED TREATMENTS

A wide range of modern fungicides is now available for use
as seed treatments, (Noon & Jackson, 1992) and can as replace
mercury. The replacement in Central Europe has been made partly
with some of these modern, often systemic, molecules and partly
with traditional commodity contact fungicides.

The newer products imported from the West, tend to be used
for certified seed production and in areas of highest disease
risk. The largest volume seed treatments are still the locally
manufactured contact fungicides.

The most commonly used mercury alternatives are listed
below (Table 1).

Typically only a few products are used on seed in most
countries because mercury replacement is relatively recent.
Hungary has the widest range of replacement products available.

Most seed treatment fungicide use is targeted against seed-
borne pathogens, with relatively little use made of systemic seed
treatments for control of foliar diseases.

Application and formulation improvements are becoming more
important as more modern chemicals become available. Until
relatively recently only powder formulations were used, applied
either dry or as slurries. Slurry application is still important
but more countries are introducing suspension concentrate (FS,
flowable) products. This development has paralleled the use of
more sophisticated application equipment e.g. Czech and Slovak
Republics, Hungary.

TABLE 1.   Seed treatment fungicides registered in
Central Europe.

Chemical Group	Chemical
Dithiocarbamates	Thiram, maneb/mancozeb/zineb
Copper salts	Copper hydroxyquinolate
Benzimidazoles	Carbendazim, fuberidazole, thiophanate methyl
Carboxamides	Carboxin
Guanidines	Guazatine
Ergosterol Biosynthesis Inhibitors	Flutriafol, triadimenol, tebuconazole, diniconazole, imazalil, bitertanol
Phthalimides	Captan

CARBOXIN SEED TREATMENTS

Carboxin was first introduced by Uniroyal Chemical as a
systemic seed treatment for control of Basidiomycete pathogens
such as *Ustilago* spp and *Tilletia* spp (Kulka & Schmeling,1987).It
is now recognised that carboxin is also active against other
pathogens such as *Pyrenophora* spp and *Fusarium* spp, although
coformulation with partner fungicides is needed for full control.
Carboxin + thiram is one of the most common mixtures used
worldwide.

Carboxin + thiram is now widely developed and registered as
a suspension concentrate formulation containing 200g/l each of
carboxin and thiram, trade name Vitavax 200FF. This formulation
has improved characteristics of uniformity of application and
adhesion, and can be applied undiluted or diluted with water
according to local conditions.

Results of efficacy and crop stimulation trials are
presented below.

RESULTS

Efficacy

Results of small-plot trials are presented for the major
seed-borne diseases. Trials were carried out with naturally
infected seed treated in the laboratory, using commercial
formulations diluted with water to give a total application
volume of 1 l/100kg seed.

<u>Bunt (*Tilletia caries*) on wheat</u>

A summary of 6 trials in 4 countries against *T.caries* is given in Table 2. The trials were all done with artificially inoculated seed (2g spores/kg) and produced a wide range of symptom levels. Carboxin + thiram FS performed as well as the reference products, with all treatments giving good control.

TABLE 2.   Efficacy of carboxin + thiram FS against *T.caries*

Treatment	Rate g ai/100kg	Slovakia 1991	Slovakia 1992	Percent control Hungary 1991	Czech 1992	Romania 1992 1	Romania 1992 2
Untreated (% infected ears)	-	(3.5)	(9.3)	(21.4)	(4.2)	(56)	(9.8)
Carboxin + thiram FS	40 + 40	100	99.9	100	-	-	-
	60 + 60	100	100	-	100	92.8	93.9
Tebuconazole	3	-	100	-	-	-	-
Organomercury	2	-	-	-	-	64.3	69.4

<u>Fusarium seedling blight</u>

Seed and seedling blight caused by *Fusarium* spp is a major problem in Central Europe, and data from a trial in Hungary on wheat cv Mv-14 are shown in Table 3.

Carboxin + thiram FS increased germination levels and emerged plants in the field, giving a final increase in yield. *Fusarium* levels on the seedlings were decreased by all products.

<u>Leaf stripe (*Pyrenophora graminea*)</u>

Barley leaf stripe is a severe disease in some areas of Central Europe, and seed treatment must be used to ensure seed certification and protect yields.

Results of five trials are summarised in Table 4; carboxin + thiram FS gave greater than 90% control at 60 + 60 g ai/100kg in 4 trials, even under conditions of high infection pressure.

<u>Seed safety and crop stimulation</u>

Carboxin + thiram FS seed treatment is very selective for most types of seed under standard germination conditions, even at high doses and after storage of treated seed.

A stimulatory effect on seedlings can often be seen, and Table 5 shows the results of carboxin + thiram seed treatment on the field emergence of winter wheat, with a strong positive effect in 2 out of 3 cultivars tested.

TABLE 3. Control of *Fusarium* seedling blight on wheat.

Treatment	Rate g ai/100kg	Percent germination	% Infected Seedlings	Plants/ m/row	Yield t/ha
Untreated control	-	90.0	18	64	4.9
Carboxin + thiram FS	50 + 50	97.5	0	77	5.2
Guazatine	70	97.5	2.5	72	5.2
Carbendazim + Cu oxyquinolate	30 + 30	97.0	0.5	80	5.1

TABLE 4. Control of leaf stripe on barley

Treatment	Rate g ai/100kg	Percent control				
		Slovakia		Romania		
		(1)	(2)	(1)	(2)	(3)
Untreated control (percent infection)	-	(3.8)	(9.5)	(22.0)	(30.3)	(9.7)
Carboxin + thiram FS	40 + 40	93.4	74.1	-	-	-
Carboxin + thiram FS	60 + 60	100	86.7	93.2	92.1	94.8

TABLE 5. Field emergence of winter wheat, Poland, 1991.

Treatment	Rate	Relative emergence		
		Cultivar A (9 trials)	Cultivar B (9 trials)	Cultivar C (9 trials)
Untreated control	-	100	100	100
Carboxin + thiram FS	60 + 60	113	110	101
Reference Products	(various)	109	-	94

One example of the effect of carboxin + thiram on early growth of seedlings is shown in Table 6, where treated seedlings produced larger root masses compared to non-treated seeds.

TABLE 6. Root growth of 3 crops treated with carboxin + thiram (Hungary, 1991, pot trial).

Treatment	Rate	Root Growth					
		Wheat		Maize		Sunflower	
		length	wt	length	wt	length	wt
Untreated control	-	100	100	100	100	100	100
Carboxin + thiram FS	60 + 60	140	145	111	143	121	114

The subsequent benefits of these early growth effects have been followed to yield in many large-scale trials throughout Europe; carboxin + thiram FS - treated seed consistently produces more vigorous crops than untreated seed, particularly under adverse growing conditions or with late sowing. For example 135 yield trials carried out in 1992 show that carboxin + thiram gave an average of 11.6% higher yields across a wide range of cereal types and growing conditions. It was noticed that the largest differences were produced on lower-input, low yielding crops.

DISCUSSION AND CONCLUSIONS

Seed treatments continue to be an important part of crop protection after mercury withdrawals in Central Europe. A range of alternative products is registered and the choice will depend on cost-effectiveness and availability. Imported systemic products are gradually replacing local contact chemicals, and improved application techniques are being used with more sophisticated formulations.

A new formulation of carboxin, carboxin + thiram FS, gives broad-spectrum disease control of small-grain cereals, is easy to apply and is very selective. Positive growth benefits have been demonstrated with the product, even in the absence of significant disease infections. This growth stimulation is thought to originate from the PGR effects of carboxin (Schmeling & Clark 1970). Coleoptile length in cereals has also been shown to be increased with carboxin treatment (Gustafson, unpub.). Faster and stronger early seedling growth is important under adverse conditions eg. late drilling, deep sowing. The combination of seed safety, stimulated growth and disease control gives an increase in crop uniformity and yield.

Carboxin + thiram FS is now registered in all countries of Central Europe on cereals and a range of other crops including maize, peas, soyabean & vegetables at a rate of 2.5 to 3l/T.

REFERENCES

Kulka, M.; Schmeling, von B. (1987) Carboxin fungicides and related compounds. *In:Modern Selective Fungicides, J.Lyr (Ed.)*, VEB Gustav Fisher Verlag Jena, pp. 119-132.
Noon, R.A.; Jackson, D. (1992) Alternatives to mercury for control of cereal seed-borne diseases. *British Crop Protection Conference - Pests and Diseases 1992*, **3**, 1127-1136.
Rennie, W.J. (1993) The need for cereal seed treatment in the UK in the post-mercury era. *Pesticide Outlook*, **4** (1), 19-14.
Schmeling, B.von; Clark, M.L. (1970) Oxathiin induced plant growth stimulation. *VII International Congress of Plant Protection*, Section B314.

THE PHENYLPYRROLES: THE HISTORY OF THEIR DEVELOPMENT AT CIBA

N.J. LEADBITTER, R. NYFELER, H. ELMSHEUSER

Ciba-Geigy Limited, Basle, Switzerland

ABSTRACT

Fenpiclonil and fludioxonil are two fungicides from phenylpyrrole chemistry that have been developed by Ciba. They are both highly active as seed treatments in a wide range of crops.

Their development at Ciba was based on the lead structure pyrrolnitrin, a natural antimycotic. Using the TOSMIC route many 4-phenylpyrroles were synthesised and tested for efficacy. The two most active compounds, fenpiclonil and fludioxonil were chosen for further development. Biological field testing, toxicological investigations, mode of action studies, formulation development, registration and marketing were all needed to bring these compounds to the market.

INTRODUCTION

In modern pest management there is a constant need for new seed treatment products. Replacements are clearly required for products that have been taken off the market or banned e.g. mercury. The usefulness of other active ingredients e.g. carbendazim has been reduced due to the development of resistance following their widespread use over a number of years. New products are needed which can control resistant pathogens. Furthermore the needs of the user and consumer can change and it is important that new products are developed which fulfil them.

The need for new products is therefore clear and ideally a modern seed treatment should combine excellent crop tolerance with a spectrum of activity that enables it to replace the older products, has low use rates and a favourable toxicology and ecotoxicology. Also new products should preferably belong to classes of chemicals which do not show cross resistance to currently used seed treatments.

It was in this environment that Ciba began the development which led to the introduction of two new active ingredients for seed treatment, fenpiclonil and fludioxonil. This paper describes the background to that development.

DISCOVERY

In the the mid 1960s pyrrolnitrin, a secondary metabolite produced by different *Pseudomonas* spp and various species from the myxobacteriales, was first isolated from *Pseudomonas pyrocinia* (Arima *et al*, 1965). Due to its interesting antifungal activity, pyrrolnitrin was developed as an antimycotic for topical application in human medicine and also served as a lead structure for pharmaceutical research (Suminori *et al*, 1969). The first use of a pyrrolnitrin analogue in plant protection was described in 1969 in a Japanese patent application (Nakanishi *et al*, 1975) and in greenhouse tests at Ciba in the late 1970s pyrrolnitrin showed interesting activity against a range of phytopathogenic fungi such as *Botrytis cinerea* and *Pyricularia oryzae*. However, pyrrolnitrin has a major drawback in that it is highly light labile (Ueda *et al*, 1982) making it unsuitable for use as a commercial fungicide.

Synthetic chemists at Ciba therefore set about the task of investigating the structural/activity relationships in attempt to identify structures that were potential commercial compounds.

Despite its apparently simple structure pyrrolnitrin proved to be a challenge for the synthetic chemists. The most useful synthesis described at that time was considered to be that described by Gosteli (1972) a method which required six steps and gave a yield of only 20%. This approach was not considered attractive for the synthesis of a large number of analogues of pyrrolnitrin. However, a rather simple process to prepare novel pyrroles had been described by van Leusen *et al* (1972) using TOSMIC (p-toluenesulfonyl methylisocyanide) as a key reagent.

Using the TOSMIC route different types of 4-phenylpyrroles were synthesized and their fungicidal activity tested. The influence of the substituents E, X and R as well as the position of X in the structure in figure 1 was tested. Results indicated that among the compounds prepared the 3-cyanopyrroles (E = -CN) had the highest activity. Furthermore this work also showed that the substituent X   should be an electron withdrawing group (in these tests -Cl, -Br, CF_3, $-OCF_2O-$) and that they should be substituted at the -2 and - 3 position on the phenyl ring. Highly active compounds were found with the 2,3-dichloro- and 2.3($-OCF_2O-$) substituted phenyl derivatives. For the N-substituted derivatives of the 3-cyano-4-phenylpyrroles it was also observed that only those with R substituents that easily hydrolysed back to the parent compound showed high fungicidal activity.

Figure 1: General formula for the 4-phenylpyrroles

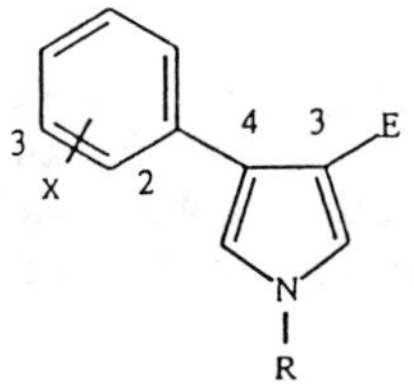

Therefore, at Ciba it was found that 3-cyano-4-phenylpyrroles with substituents in either 2- or 3- position and preferably in the 2- and 3- position of the phenyl ring proved to be most interesting. Of these compounds fenpiclonil (synthesized 1982) and fludioxonil (synthesized 1984) (Fig 2) were selected for development. Further details of this work can be found in Nyfeler & Ackermann (1992).

Figure 2: Formulae of fenpiclonil and fludioxonil

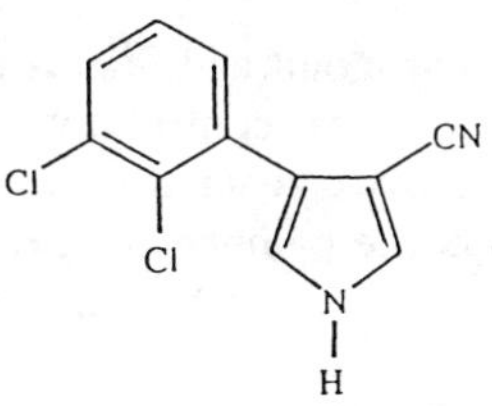

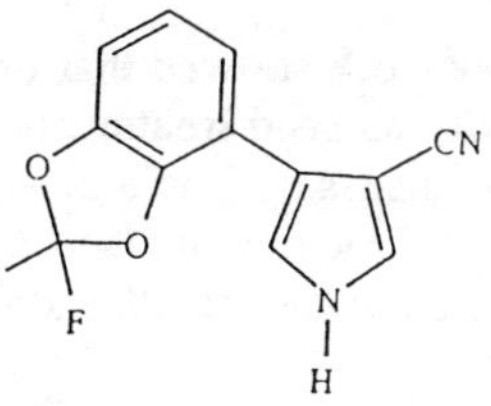

BIOLOGICAL DEVELOPMENT

Following synthesis, both compounds were rigorously screened and shown to have a similar spectrum of activity against a range of *Deuteromycete* , *Ascomycete* and *Basidiomycete* fungi when applied as either a foliar spray or as a seed treatment (Nevill *et al*, 1988; Gehmann *et al*, 1990). They were also very well tolerated by a wide range of crops.

TABLE 1. Activity of fenpiclonil (20g AI/100kg seed) and fludioxonil (5g AI/100kg seed) against seed-borne diseases of wheat and barley.

Pathogen(s)	Disease	Activity compared with mercury
Wheat		
Gerlachia nivalis	Snow mould	greater than
Fusarium spp	*Fusarium* seedling blight	greater than
Tilletia caries	Common bunt	equal to
Septoria nodorum	*Septoria* seedling blight	equal to
Urocystis agropyri	Flag smut	no data
Helminthosporium sativum	*Helminthosporium* seedling blight	greater than
Barley		
Gerlachia nivalis	Snow mould	greater than
Fusarium spp	*Fusarium* seedling blight	greater than
Helminthosporium sativum	*Helminthosporium*	greater than
Helminthosporium gramineum	Leaf stripe	equal to*
Ustilago hordei	Covered smut	no data

* fenpiclonil and fludioxonil will control mercury resistant isolates

Field trials using the phenylpyrroles as seed treatments have been carried out since the mid 1980s. So far more than 1000 trials have been initiated on an ever widening range of crops. Initial data identified that both fenpiclonil and fludioxonil were active against a broad spectrum of diseases of wheat and barley with a spectrum broader than that of mercury (Table 1, for further details see Leadbeater *et al*, 1991; Koch & Leadbeater, 1992). Investigations of dose response on cereals showed that fludioxonil is inherently more active than fenpiclonil. The level of control given by 5g ai/100kg fludioxonil being at least equivalent to that given by 20g ai/100kg fenpiclonil.

Further work showed that on other pathosystems fenpiclonil and fludioxonil, also have good activity as seed treatments. On potatoes both compounds control a wide range of seed-borne diseases (Table 2) and on peas both show activity against *Ascochyta pisi* and *A. pinodes*. Fludioxonil also has activity against seed-borne pathogens on rice, maize, cotton and a number of other crops (Leadbeater *et al*, 1991; Koch & Leadbeater, 1992).

TABLE 2. Activity of fenpiclonil (50g Al/tonne tubers) and fludioxonil (20-25g Al/tonne tubers) when used as a pre-plant application to potatoes.

Pathogen*	Disease	Activity
Rhizoctonia solani	Stem canker Black scurf	= pencycuron
Heminthosporium solani	Silver scurf	= imazalil & carbenda-zim**
Fusarium spp	Dry rot	= thiabendazole**
Polyscutalum pustulans	Skin spot	= thiabendazole

* seed-borne infections   ** will control isolates resistant to carbendazim

In situations where resistance has developed to currently used products it is important that new products have activity against isolates of pathogens insensitive to those products. In many cases the phenylpyrroles have activity against pathogens where carbendazim has commonly been used. At an early stage therefore the phenylpyrroles were tested against isolates of pathogens that had developed resistance to carbendazim. In all cases the phenyl pyrroles have been shown to be active against these resistant isolates both *in vtro* and *in vivo*. The pathogens tested so far are *Fusarium nivale* (wheat), *Ascochyta pisi* and *Ascochyta pinodes* (peas), *Gibberella fujikuroi* (rice), *Helminthosporium solani* (potatoes), *Fusarium solani* (potatoes).

MODE OF ACTION

Although mode of action studies were initiated early in the development of the phenypyrroles the mode of action is still not well understood. Studies with fenpiclonil show that there is no inhibition of respiration, chitin-, DNA-, or RNA-synthesis and it does not interfere with ergosterol or lipid biosynthesis. However, it caused a fast inhibition of mycelial growth as well as an instantaneous reduction in amino-acid uptake (Jespers & Davidse, 1990). Leroux (1991) suggested that fenpiclonil has a similar mode of action to dicarboxamides like iprodione and aromatic derivatives such as toclofos-methyl but the mode of action of these compounds is also not clearly understood and the suggested similarity has not been confirmed. Further work by Jespers *et al* (1993) has indicated that cell membrane transport processes are affected and they concluded that to their knowledge such a mechanism had not been described before for any fungicide. Further work is in progress to define more precisely the mode of action of these compounds.

TOXICOLOGY AND ECOTOXICOLOGY

Toxicological and ecotoxicological studies were started early in the development of the phenylpyrroles. Data generated from many studies has shown that the phenylpyrroles have good applicator, consumer and environmental safety.

FORMULATIONS

The development of a high quality formulation for new active ingredients is a vital part of the development process. In many cases the formulation can affect the efficacy of the chemical and also its safety to the crop. With the phenylpyrroles it was decided that water based flowable formulations (FS) should be developed to reduce the likelihood of damage to the seed which may be caused by solvent based formulations and also to provide products which are as safe as possible for the user.

Another issue regarding formulations is their ease of use. To this end FS formulations of both fenpiclonil and fludioxonil have been designed as "ready to use" formulations. For example fenpiclonil for use on wheat and barley has been formulated as a FS050 with an application rate of 400ml/100kg giving good application properties without the need for dilution. For further details of formulation development see Frank (1994).

MARKETS AND REGISTRATION

Since both compounds were shown to have broad spectrum activity against diseases on wheat and barley the first major markets were identified as the cereal growing areas of Northern Europe. Fenpiclonil was first introduced in Switzerland in 1988 and by 1990 had achieved 50% of the wheat seed treatment market. It was subsequently registered in other European countries and gained registration in the United Kingdom during 1993. The first registration for fludioxonil on cereals was achieved in France in 1993 and registrations in other European countries are expected in the near future.

As the spectrum of activity on other crops was elucidated so further markets were identified. Fenpiclonil is already registered for use on potatoes in Belgium and has been submitted for registration in a number of other countries. Although fenpiclonil will only be developed for European markets, fludioxonil will be marketed worldwide. As well as for control of diseases on cereals and potatoes it will also be used on peas, maize, cotton and rice, alone and in mixture with other active ingredients.

ACKNOWLEDGEMENTS

The authors would like to thank all those who contributed to the successful development of fenpiclonil and fludioxonil.

REFERENCES

Arima, K.; Imanaka H.; Kousaka, M.; Fukuda, A; Tamura, G. (1965) Studies on pyrrolnitrin, a new antibiotic. 1: Isolation and properties of pyrrolnitrin. *Journal of Antibiotics*, 18, 211-219.

Frank, L (1994) The life cycle of a seed treatment formulation development. *These proceedings.*

Gehmann, K.; Nyfeler, A.J.; Leadbeater, A.J.; Nevill, D.; Sozzi, D. (1990) CGA173506: A new phenylpyrrole fungicide for broad spectrum disease control. *Brighton Crop Protection Conference - Pests and Diseases 1990,* 1, 399-406.

Gosteli, J. (1972) New synthesis of the anitbiotic pyrrolnitrin. *Helv. Chim. Acta,* 55, 451-460

Jespers, A. B. K.; Davidse, L. C. (1990) *Abstract of the 4th International Mycological Congress,* Regensburg, 286.

Jespers, A. B. K.; Davidse, L. C.; de Waard, M. A. (1993) Biochemical effects of the phenylpyrrole fungicide fenpiclonil in *Fusarium sulphureum* (Schlect). *Pesticide Biochemistry and Physiology,* 45, 116-129.

Koch, E; Leadbeater, A.J. (1992) Phenylpyrroles - a new class of fungicides for seed treatment *Brighton Crop Protection Conference - Pests and Diseases 1992,* 3, 1137-1146.

Leadbeater, A.J.; Nevill. D; Steck. B; Nordmyer, D. (1990) CGA173506: A novel fungicide for seed treatment. *Brighton Crop Protection Conference - Pests and Diseases 1990,* 2, 825-830.

Leroux, P; Lanen, C; Fritz, R. (1992) Similarities in the antifungal activities of fenpiclonil, iprodione and toclofos-methyl against *Botrytis cinerea* and *Fusarium nivale. Pesticide Science,* 36, 255-261.

Nakanishi, K; Shimizu, K; Ouo, K.; Ueda, K. (1975) Japanese Patent 50,002,011 (Appl. 19.3.1969), *Chemical Abstracts,* 83, 38796.

Nevill, D.; Nyfeler, R; Sozzi, D; (1988) CGA142705: A novel fungicide for seed treatment *Brighton Crop Protection Conference 1988,* 1, 65-72.

Nyfeler, R.; Ackermann, P. (1992) Phenylpyrroles, a new class of agricultural fungicides related to the natural antibiotic pyrrolnitrin. In *Synthesis and Chemistry ofAgrochemicals,* D.R. Baker, J.G. Fenyes & J.J. Steffens(Eds), American Chemical Society pp395-404.

Suminori, U.; Kazuo, K.; Tanaka, K.; Nakamura, H. (1969) *Chem. Phar. Bulletin (Tokyo),* 17, 559-566.

Ueda, A; Nagasaki, H.; Takakura, Y.; Nishikawa, H; Nakada, A. (1982) *European Patent Application* 92, 890.

van Leusen, A.M.; Siderius, H.; Hoogenboom, B.E.; van Leusen, D. (1972) Chemistry of sulfonylmethyl isocyanides. 6. New and simple synthesis of the pyrrole ring system from Michael acceptors and (p-tolylsulfonyl) methyl isocyanide. *Tetrahed Letters,* 5337-5340.

DRESSING ZONE FORMATION, UPTAKE, TRANSLOCATION AND ACTION OF [^{14}C]IMIDA-
CLOPRID FOR WINTER WHEAT AFTER SEED TREATMENT AND UNDER THE INFLUENCE OF
VARIOUS SOIL MOISTURE LEVELS

U. STEIN-DÖNECKE, F. FÜHR, J. WIENEKE

Institute of Radioagronomy, Research Centre Jülich, 52425 Jülich, Germany

ABSTRACT

After seed treatment with [^{14}C]imidacloprid, winter wheat was cultivated at three different soil moisture levels. Higher soil moisture contents promoted the removal of ^{14}C activity from the caryopsis and distribution in the soil, and thus reduced both the potential availability of the active ingredient in the soil and also the incorporation of ^{14}C into the plant. Between 7 and 22 % of the ^{14}C activity applied per caryopsis was incorporated into the wheat shoot by the end of shooting. Up to this point in time, the concentration of active ingredient equivalents in the shoot decreased greatly as a consequence of an increasing difference between the subsequent delivery of ^{14}C and increase in new growth. A concentration gradient decreasing from the oldest to the youngest leaf for ^{14}C and imidacloprid points to an apoplastic translocation. Lower ^{14}C incorporation into the plant and lower active ingredient fractions of the ^{14}C activity in the leaf at higher soil moisture levels resulted in reduced concentrations of extractable imidacloprid in the leaf. Nevertheless, even 195 days after sowing a high toxicity to aphids was still maintained at concentrations between 12 and 110 µg of imidacloprid/kg fresh weight in the youngest leaves.

INTRODUCTION

The development of a new active ingredient, imidacloprid, means that a systemic insecticide for the seed treatment of cereals is available, with which a long-term control of aphids, and also a re-duction in infection with the Barley Yellow Dwarf Virus (BYDV) can be achieved (Schmeer *et al.*, 1990). The aphid infestation responsible for transmitting the BYDV, and thus the associated reducti-ons in yield, sometimes only takes place in spring in the case of winter wheat (Huth, 1990). This means that after seed treatment with imidacloprid sufficient active ingredient must be available in the plant or supplied from the soil even at this late point to control the aphids and prevent virus transmis-sion. In order to quantify these factors, dressing zone formation, i.e. active ingredient distribution and availability in the soil, as well as uptake, translocation and action in the plant were recorded until the spring after seed treatment of winter wheat with ^{14}C-labelled imidacloprid in an extensive study. Since it is known from studies on the uptake of triadimenol after the dressing of winter barley (Schneider, 1988) that soil moisture can have a great influence on dressing zone formation and active ingredient uptake, the present studies were carried out with three different soil moisture contents. Some of the results obtained are presented below.

MATERIALS AND METHODS

Seed Treatment

Winter wheat (Kanzler) was treated on a laboratory scale with [pyridinyl-^{14}C-methyl]imida-cloprid as a 70WS formulation, applied at 100 g a.i./100 kg seed. The analysis of 50 of a total of 850 treated caryopsis resulted in a mean ^{14}C activity of 78 kBq/caryopsis; a quantity of 50.2 µg imi-dacloprid/caryopsis was thus calculated at a specific ^{14}C activity of the active ingredient applied of 1.554 MBq/mg.

<u>Experimental Setup, Sowing, Sampling and Preparation</u>

The winter wheat was sown on 20.11.1990 in containers 1 m^2 in size filled with the topsoil of a degraded loess soil (Parabraunerde); seeds were sown every 2 cm in rows 12 cm apart at a depth of 2.5 cm. The plants were cultivated in the open air at three different soil moisture contents, amounting to 30, 40 and 50 % of the maximum water holding capacity (WHC$_{max}$) throughout the entire experiment.

The studies of dressing zone formation and ^{14}C uptake were carried out for all three soil moisture levels at three stages of development (start of tillering (21), start and end of the shooting phase (31 and 45); 111, 153 and 195 days after sowing respectively) on the basis of 10 plants and soil samples in each case. A further 10 plants served to determine the active ingredient contents and concentrations of selected leaves. The uptake and distribution of the ^{14}C activity in the plant was determined on 7 additional dates, from the 1-leafstage (11), 51 days after sowing, untill full ripeness (91), 260 days after sowing, but only for 40 % WHC$_{max}$.

Steel frames (15 cm deep, 2 x 12 cm) were used to remove the plants with their associated dressing zone covering the soil up to the adjacent dressing zones and permitting a subdivision into the region close to the grain (5 cm deep, 2 x 5 cm) and the remaining dressing zone. The plants were divided into caryopsis, root and shoot, and some of them also into individual leaves and side shoots. After determining the fresh weight, the plant samples were airdried for ^{14}C determination (<30°C) or stored at -20°C for extraction purposes. The soil samples were airdried (<30°C), weighed and homogenized for ^{14}C determination and desorption.

<u>Detection of Radioactivity and Determination of Active Ingredient</u>

^{14}C measurements were carried out for all samples by LSC (Packard, Tri-Carb^R, 460, 4530, 2500TR); solid samples were first combusted in the Packard 306 sample oxidizer. In order to determine the imidacloprid content in the plant samples, the plant material was extracted with methanol-water (80:20, v/v), the extract reduced to the aqueous fraction at 40 °C and this latter was then partitioned against ethyl acetate. Imidacloprid in the concentrated organic phase was analyzed by thin-layer radiochromatography (radio-tlc). The potential active ingredient availability in the dressing zone region close to the grain was determined by a two-stage desorption of 10 g soil with 0.01 M CaCl$_2$ solution in a ratio of 1:5. After concentrating the desorption solution, the imidacloprid fraction was detected by radio-tlc. Radio-tlc was carried out with a tlc-analyzer (Berthold, LB 2842) on silica gel plates (60F$_{254}$, Merck), developed in chloroform, acetone, glacial acetic acid and water (50:30:15:1, v/v/v/v). Imidacloprid was identified by comparing the R$_f$ values with labelled (radio-tlc, autoradiography) and unlabelled (UV detection) reference substance.

<u>Biotest with Aphids</u>

Rhopalosiphum padi and *Macrosiphum avenae* were applied in cages to the youngest, completely developed leaf of each wheat plant, whether treated with imidacloprid or not, from the 4-leaf stage (14) once a week and the mortality rates assessed after 4 days.

RESULTS

<u>Dressing Zone Formation</u>

The distribution of the ^{14}C activity in the soil was characterized by an increasing ^{14}C translocation out of the dressing zone region close to the grain into the remaining dressing zone which was intensified by higher soil moisture contents (Table 1). With values of between 45 % (30 % WHC$_{max}$) and 29 % (40 % WHC$_{max}$), 195 days after sowing, a high proportion of the ^{14}C activity of the dressing zone remained in the region close to the grain, namely a maximum of 2.5 cm adjacent to and 2.5 cm below the caryopsis. Correspondingly, the concentration of active ingredient equivalents was considerably higher than in the remaining dressing zone, irrespective of the soil moisture content, with values of between 1.1 - 1.2 mg/kg soil (111 days after sowing) and 0.3 - 0.4 mg/kg soil (195 days after sowing). As a consequence of the increasing ^{14}C translocation the active ingredient equivalents increased in the remaining dressing zone to a maximum of 0.06 - 0.07 mg/kg soil (195 days after sowing).

Table 1. Radioactivity in the dressing zone close to the grain at various soil moisture contents. Radioactivity of the total dressing zone sample = 100 %.

Days after sowing	30 % WHC$_{max}$	40 % WHC$_{max}$	50 % WHC$_{max}$
111	94	96	72
153	79	58	62
195	45	29	34

Table 2. Desorbable imidacloprid in the dressing zone close to the grain at various soil moisture contents. Radioactivity in the dressing zone region close to the grain = 100 %.

Days after sowing	30 % WHC$_{max}$	40 % WHC$_{max}$	50 % WHC$_{max}$
111	62	61	58
153	46	41	41
195	27	23	22

The potential availability of imidacloprid in the soil, measured as the desorbable fraction of active ingredient in the ^{14}C activity of the dressing zone region close to the grain (Table 2), dropped to values of between 27 % (30 % WHC$_{max}$) and 22 % (50 % WHC$_{max}$) up to 195 days after sowing, due to a reduced desorbability of the ^{14}C activity in connection with declining fractions of active ingredient in the desorbed radioactivity.

<u>Uptake, Distribution and Active Ingredient Concentrations in the Plant</u>

<u>Uptake and distribution of radioactivity at medium soil moisture</u>
At 40 % WHC$_{max}$, the ^{14}C activity in the wheat shoot rose continuously up to 19 % of the radioactivity applied on average per caryopsis or 9.5 µg of active ingredient equivalents at full ripeness (91) (Fig. 1). However, the concentration of active ingredient equivalents in the shoot during tillering and shooting dropped from 14 mg/kg fresh weight (111 days after sowing) to 0.19 mg/kg fresh weight (195 days after sowing) due to an increasing difference between the incorporation of ^{14}C and growth of fresh weight (Fig. 1). Due to the loss of fresh weight during ripening, the concentration rose to 1.0 mg of active ingredient equivalents/kg fresh weight (260 days after sowing)(Fig. 1).

The major fraction of the ^{14}C activity taken up into the plants was always translocated into the shoot and distributed there in such a way that a gradient decreasing from the oldest to the youngest leaf arose for the concentration of active ingredient equivalents. The concentrations in the oldest, senescent leaves reached values of between 20 and 80 mg/kg fresh weight, whereas in the youngest, not completely formed leaves the concentrations were in part lower than 100 µg/kg fresh weight. The youngest leaves which were completely formed and physiologically active towards the end of shooting (45) represented an exception. In this case, the concentrations were on a uniform level between 0.12 and 0.18 mg of active ingredient equivalents/kg fresh weight. At the time of full ripeness (91), the concentrations of active ingredient equivalents in the straw and chaff (1.2 - 1.9 mg/kg dry mass) were much higher than in the grain (0.04 mg/kg dry mass). On macroautoradiographies of the wheat plants an inhomogeneous ^{14}C distribution within the individual leaves with high ^{14}C concentrations at the tips, edges and in the veins can be recognized, apart from the concentration gradient between various old leaves.

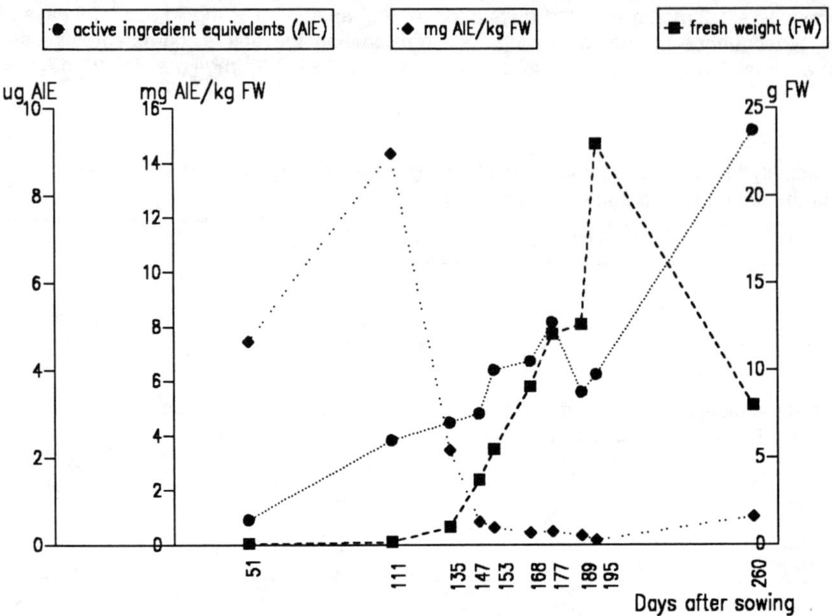

Fig. 1. Fresh weight (FW), active ingredient equivalents (AIE) and concentration of active ingredient equivalents in the wheat shoot at medium soil moisture content (40 % WHC$_{max}$).

<u>Radioactivity uptake and concentrations of active ingredient at various soil moisture contents</u>

Higher soil moisture contents caused an increased removal of ^{14}C from the caryopsis as well as reduced uptake in the plant (Table 3). Up to the end of shooting (45), between 22 % (30 % WHC$_{max}$) and 7.0 % (50 % WHC$_{max}$) of the ^{14}C activity applied per caryopsis was incorporated into the shoot. Since there was no influence of soil moisture content on the fresh weights of the wheat plants, the reduced ^{14}C uptakes at higher soil moisture contents also led to reduced concentrations of active ingredient equivalents in the shoot. As already described for 40 % WHC$_{max}$, the concentration during tillering and shooting dropped considerably at 30 and 50 % WHC$_{max}$.

The active ingredient concentrations in the wheat leaves declined considerably with increasing age of the leaves (Fig. 2) due to a decrease of the active ingredient fraction in the ^{14}C activity in the leaf. However, due to the pronounced ^{14}C concentration gradient decreasing from the oldest to the youngest leaf, the active ingredient concentrations in the older leaves were higher than those in the younger leaves at this date in spite of lower proportions of active ingredient. The influence of higher soil moisture content became apparent in the reduced concentrations of extractable imidacloprid (Fig. 2). This effect is largely based on the lower incorporation of ^{14}C into the shoot or the individual leaves, but also on reduced proportions of active ingredient in the ^{14}C activity in the leaf. 195 days after sowing, the imidacloprid concentration in the 8th leaf was between 12 µg/kg fresh mass (50 % WHC$_{max}$) and 81 µg/kg fresh mass (30 % WHC$_{max}$), whereas in the 10th leaf (flag leaf) the corresponding values amounted to 16 and 110 µg/kg fresh weight respectively.

<u>Biotest</u>

During the observation period (140 - 190 days after sowing), the insecticidal action against aphids proved to be independent of the soil moisture content, and 192 days after sowing amounted to at least 70 % against *M. avenae* and 98 % against *R. padi*, tested on the 8th - 9th leaf of the wheat.

Table 3. Radioactivity in the caryopsis, root and shoot at various soil moisture contents. Average radioactivity applied per caryopsis = 100 %.

Soil moisture content		Days after sowing		
		111	153	195
Caryopsis	30 % WHC_{max}	16	4.3	0.90
	40 % WHC_{max}	6.0	1.0	0.48
	50 % WHC_{max}	2.8	0.60	0.24
Root	30 % WHC_{max}	0.20	1.9	1.9
	40 % WHC_{max}	0.12	0.85	1.0
	50 % WHC_{max}	0.090	0.57	0.59
Shoot	30 % WHC_{max}	6.7	15	22
	40 % WHC_{max}	4.8	7.9	7.7
	50 % WHC_{max}	2.7	4.6	7.0

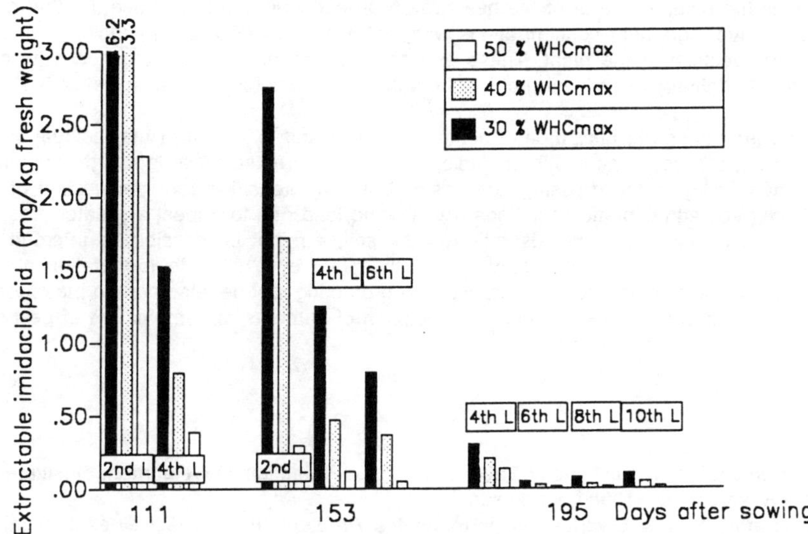

Fig. 2. Concentrations of extractable imidacloprid in the wheat leaves 111, 153 and 195 days after sowing.

DISCUSSION

A long-term protection against aphids for winter wheat by means of seed treatment requires both a sufficient subsequent supply of active ingredient from the soil or uptake into the plant as well as a distribution in the plant, by means of which effective concentrations of insecticide can be provided in all parts of the plant to be protected.

Although after seed treatment with [14C]imidacloprid the uptake of 14C-labelled compounds into the wheat plant continued even beyond the shooting stage (45) (Fig. 1) the subsequent delivery of 14C or active ingredient took place with diminishing intensity. As a result, irrespective of the soil moisture level, a decrease in active ingredient equivalent or active ingredient concentrations in the plant resulted during tillering and shooting of the wheat. Higher contents of soil moisture reduced both the time and also the level of subsequent delivery into the plants (Table 3). These relations were also observed after the seed dressing of winter cereals with [^{14}C] triadimenol (Thielert, 1984; Schneider, 1988) and were based, amongst other aspects, on a declining potential plant availability of triadimenol in the soil, which occurred due to progressive dressing zone formation and the associated adsorption, as well as the increasing active ingredient degradation. With progressing plant development, a ^{14}C translocation also took place in the dressing zone of the winter wheat treated with [^{14}C]imidacloprid (Table 1) leading to a dilution of the ^{14}C or active ingredient equivalent concentrations in the soil. In studies on desorption behaviour Fritz (1992) determined that the non-desorbable fraction is increased by a factor of 1.5 to 2 after ageing of the imidacloprid. In fact, the desorbability of imidacloprid decreased considerably during the experiment (Table 2). The lower active ingredient availability at higher soil moisture levels can be attributed to the increased dressing zone formation corresponding to the relations described above and also to degradation of the active ingredient. The observation that in the experiment the fraction of fine roots in the dressing zone was increased with lower soil moisture levels indicated that the soil moisture does not merely influence the potential active ingredient availability but rather, via root growth, the subsequent delivery of active ingredient into the plant.

The distribution of ^{14}C activity in the wheat plant was typical of xylem-mobile substances which accumulate in the plant parts with the highest transpiration so that a redistribution of the active ingredient present in the plant in favour of the new growth probably did not take place. Furthermore, as the decreasing active ingredient concentrations with leaf age show (Fig. 2), imidacloprid is subjected to a rapid metabolism in the plant. These experimental results illustrate the necessity of a continuous subsequent delivery of active ingredient from the soil for sustained insecticidal action.

Measures for improving the utilization of active ingredient and extention of the residual action after the seed dressing of winter wheat with imidacloprid must be directed towards a higher potential active ingredient availability in the dressing zone as well as an intensified root penetration of the dressing zone. Under soil and climatic conditions which could lead one to expect a greater dressing zone propagation and active ingredient adsorption in the soil, it might be possible to improve the active ingredient uptake and subsequent delivery into the plant by a suitable formulation of imidacloprid for a slow release from the seed. Possibilities of promoting root development in the dressing zone region can be found in soil working measures, addition of nutrients or combination of pesticide active ingredient and root growth regulators.

REFERENCES

Fritz, R. (1992) Personal communication. Bayer AG, Crop Protection Development, Institute for Metabolism Research, D-51368 Leverkusen, .

Huth, W. (1990) Barley Yellow Dwarf - ein permanentes Problem für den Getreideanbau in der Bundesrepublik Deutschland? *Nachrichtenblatt des Deutschen Pflanzenschutzdienstes* **42**, 33-39.

Schmeer, H.E.; Bluett, D.J.; Meredith, R.; Heatherington, P.J. (1990) Field evaluation of imidacloprid as an insecticidal seed treatment in sugar beet and cereals with particular reference to virus vector control. *BCPC, Pests and Deseases 1990, Proceedings* **1**, 29-36.

Schneider, M. (1988) Aufnahme von [^{14}C]Triadimenol über Korn und Wurzel nach Flüssigbeizung von Wintergerste: Einfluß von Bodenfeuchte und Saattermin auf Radioaktivitätsverteilung und Wirkstoffgehalt in Pflanze und Boden. Dissertation Universität Bonn. *Berichte der Kernforschungsanlage Jülich* Nr. **2253**.

Thielert, W. (1984) Aufnahme und Nachlieferung von [benzolring-U-^{14}C]Triadimenol über die Karyopse und aus den Beizhöfen nach Saatgutbehandlung von Wintergerste und Winterweizen mit einer Trockenbeizformulierung. Dissertation Universität Bonn. *Berichte der Kernforschungsanlage Jülich* Nr.**1958**.

Session 4

Seed Treatment for Non-Graminaceous Crops

Chairman	A M DEWAR
Session Organisers	M J C ASHER A M DEWAR

SEED TREATMENT USAGE ON PEAS AND BEANS IN THE UK

A.J. BIDDLE

Processors and Growers Research Organisation, Great North Road, Thornhaugh, Peterborough, Cambridgeshire, PE8 6HJ

ABSTRACT

A range of seed treatments are used in the UK to protect against several soil- and seed-borne fungal pathogens. Almost all pea seed, but only a small proportion of field bean seed is treated. Protection against damping-off, *Ascochyta* spp. and *Peronospora viciae* is achieved in peas using either single or multi-purpose products, depending on the health of the seed and the varietal susceptibility to disease. The necessity for such a range of protection in winter and spring field beans is discussed. Only limited success has been achieved in controlling root diseases in both crops and no insecticidal seed treatments are available for peas and beans to control seedling pests.

INTRODUCTION

The use of fungicidal seed treatment in peas *(Pisum sativum)* has been well established for many years. The crop is grown either for human consumption, for animal feed compounding or for seed. In 1992, vining peas, harvested green for freezing or canning occupied 44,900 ha in the UK whilst combining peas, harvested dry, occupied 78,600 ha (MAFF, 1993). Up to 75% of vining pea seed is imported, mainly from the USA, already treated with seed protectants, whilst the UK is virtually self-sufficient for combining pea seed and therefore, it is all treated in the UK.

Seed treatments are used less extensively on field beans *(Vicia faba)*. The crop is either autumn or spring planted and most of the produce is used for compounding with a small tonnage either exported or used for human consumption. The area under production in the UK in 1992 was 129,100 ha, with virtually the entire seed requirement being produced in this country.

Peas are more susceptible to seed-bed losses as a result of soil-borne fungal pathogens and the crop is also susceptible to seed-borne fungi. In the UK, seed treatments are used extensively for protection against a range of fungal pathogens.

The bean crop has expanded rapidly in recent years as a result of changes in the EEC area payment schemes. The range of varieties, particularly of spring sown beans, has been enlarged with new introductions which include types with low tannin in the seed. Hitherto, varieties were of the coloured flowered type with high seed tannin levels. Recent work has sought to evaluate any difference in susceptibility of the low tannin seeds to soil-borne pathogens.

This paper outlines the major uses of seed treatments in the pea and bean crops and discusses the possibilities for future seed and seedling protection.

SEED BED LOSSES

Pea seed, especially vining pea varieties, are particularly susceptible to pre-emergence losses caused by damping-off. The early-maturing varieties are often drilled early in the spring when soil temperatures are low and seedbeds are relatively wet (Gane & Biddle, 1978). Leakage of seed exudates under these conditions attracts *Pythium* spp. which colonise the cotyledons and become pathogenic. Dithiocarbamates are, therefore, used extensively to protect the newly geminating peas from infection.

Beans appear less susceptible to such losses. Coloured-flowered varieties often contain tannins in the seed and as such appear to be more resistant to some of the soil-borne fungal pathogens. Some new white-flowered types, however, seemed to be susceptible to fungal attack and Kantar, F., Hebblethwaite, P.D. & Pilbeam, C.J. (unpublished) found that fungicidal seed protectants were of value. With the introduction of commercially acceptable varieties of low-tannin beans, the value of seed treatments was again examined. In a series of field trials carried out by the author, between 1991 and 1993, coloured and white-flowered varieties of autumn-planted-winter and spring-planted field beans were treated with a range of seed treatments which included thiram, alone or in combination with thiabendazole, and metalaxyl. In the three years of experiments carried out at two sites, Thornhaugh and Boxworth, Cambridgeshire, there were no significant differences in plant emergence or in yield between any of the treatments or with untreated seed. The results therefore, cast doubt on the benefits of routinely treating field bean seed where protection solely from soil-borne damping-off diseases is required.

SEED-BORNE DISEASES

Ascochyta spp. remain the most common and important fungal pathogens in both peas and beans. In peas, leaf and pod spot caused by *Ascochyta pisi* is now uncommon in the UK, but it has been replaced in importance by a related disease caused by *Mycosphaerella pinodes*. Unfortunately, for the purposes of control by seed treatments, *M. pinodes*, unlike *A. pisi*, produces soil-borne chlamydospores which can infect crops later in the season, particularly during wet summers. As a seed-borne disease, *M. pinodes* causes a rot in the hypocotyl region of the pea seedling resulting in seedling death, or a more general leaf and pod spot disease.

Seed produced in the UK is often infected to a significant degree by *M. pinodes* and the proportion of seedlots containing infection is relatively high in many years (Biddle, 1986). Control of *A. pisi* and *M. pinodes* can be effected by MBC fungicides including thiabendazole (Biddle, 1981) and it is current practice to treat seed where infection is between 5% and 35%. Thiabendazole treatment for this purpose is complemented with the addition of thiram for protection against damping-off. Recent work in France has suggested that strains of *M. pinodes* may be resistant to thiabendazole. There have been no such reports in the UK, however several other fungicides are currently being evaluated as alternatives (PGRO, 1991 and 1992).

In field beans, *Ascochyta fabae* is the most serious fungal pathogen. Winter sown beans are more likely to suffer from a seed-borne infection as the disease can spread more rapidly during the wet conditions which can occur during the late winter and early spring. Following the 1992 harvest season, winter beans were more seriously infected by *A. fabae* than for many years. The results of tests carried out on 451 seed samples by PGRO in the autumn of 1992 are shown in Table 1.

TABLE 1. *Ascochyta fabae* levels in winter bean seed lots tested by PGRO in 1992

Level of seed infection (%)	All seedlots (%)	Farm-saved (%)	Certified at C_2 level (%)
< 1	69	71	67
1-5	23	25	21
5-10	5	3	7
> 10	3	1	4

Seed-borne infection of *A. fabae* is not so effectively controlled by thiabendazole seed treatment alone. Reports have shown between 60-80% reduction in disease levels (Jellis *et al*, 1988). Because field infection is also difficult to control with foliar sprays, and because infection can be severe in wet seasons, standards have been in place in the UK Field Bean Certification Scheme whereby the maximum seed infection allowed is 1% in the C_2 generation. Earlier generations have much higher standards. It has also been demonstrated that the sexual form of *A. fabae (Didymella fabae)* can develop on infected crop debris and air-borne spores may be dispersed by wind for long distances (Jellis & Punithalingam, 1991). It is, therefore, very important that the initial seed health standards be maintained. It is currently recommended by PGRO that uncertified bean seed should be treated if the level of *A. fabae* is between 1% and 3%. Seed with levels higher than this should be discarded (PGRO, 1993).

SOIL-BORNE DISEASES

Both peas and field beans can be affected by downy mildew caused by *Peronospora viciae*. At present, however, there appear to be different strains of *P. viciae* infecting the two crops (J.E. Thomas, 1993, personal communication). The fungus survives in the soil for many years and seedlings can then become infected shortly after germination. Infected pea seedlings emerge as pale, stunted plants on which the fungus produces air-borne spores. In beans, particularly spring varieties, the disease has become more common in recent years. Early seedling symptoms are not seen as commonly as with peas, but during the flowering stage, leaves can develop irregularly shaped, pale lesions which develop into larger areas

resulting in defoliation. The growing points may also become systemically
infected and further growth and pod set is reduced. There is a range of
susceptibility exhibited by both peas and field beans; the current status
of combining peas and spring beans is published in the NIAB Recommended
list of field peas and field beans (NIAB, 1993). Vining pea varieties are
tested by PGRO and Table 2 summarises the varietal susceptibility ratings
obtained from the 1992 and 1993 tests for some of the newer varieties.

TABLE 2. Susceptibility of vining pea varieties to downy mildew,
PGRO 1992-93.

Rating:	1	3	5	6	7	9
	Avola (Standard)	Caty	Deltafon	CMG282	Ambassador	Minado
	CMG 264 F	FR774	Sublima	Co400	Bastion	Solo
		Polo		Cobalt	Lambado	
		Rexado		Lynx		
		Sancho				
		Winner				

The highest ratings relate to the highest level of varietal resistance.

Seed treatment has been shown to be very effective in protecting newly
germinated pea seedlings from infection by the soil-borne source of
inoculum (Miller & de Whalley, 1981; Vulsteke & Meeus, 1985). Metalaxyl,
in combination with thiabendazole and thiram has been in general use in the
UK for several years, providing control of downy mildew and seed-borne
Ascochyta spp. (Salter & Smith, 1986). A recent introduction to the UK has
been oxadixyl, again in combination with thiabendazole and thiram.
However, whilst either treatment prevents primary infection of peas,
secondary infection from air-borne spores introduced from a neighbouring
source is not necessarily reduced. Although there is some difference in
varietal susceptibility between vining pea varieties, all seed used in the
UK is treated with either metalaxyl- or oxadixyl-based products. In
combining peas, there are more varieties which show a high level of field
resistance and therefore only those which are rated by NIAB as 6 or below
are treated routinely.

In spring beans, foliar spraying of fungicides is effective in
controlling downy mildew. However, recent work has indicated the potential
for the use of systemic fungicides as seed treatments. This form of
control may, in the long-term, be a more cost-effective option than the
current practice of crop spraying.

FOOT ROT DISEASES OF PEAS AND BEANS

Several soil-borne fungi are capable of infecting the roots or stem
bases of peas and beans either as individuals or in concert. Peas are
susceptible to *Fusarium solani* f. sp. *pisi* and *Phoma medicaginis* var.

pinodella which together cause a foot rot, especially where the crop has been grown frequently in the past (Biddle, 1983). Field beans are also susceptible to such infection, but in the last two years, 1992 and 1993, there has been a notable increase in stem base infection of spring beans attributed to *Fusarium culmorum* and *F. solani* (PGRO, 1992).

Several studies have shown the effects of seed treatment in reducing the pea disease complex in laboratory or glasshouse experiments (Gravanis, 1986; Bradshaw-Smith, 1991), but only limited reduction has been achieved in the field (Salter & Smith, 1986). Biological control agents coated onto seeds have included *Pythium oligandrum* and this also showed some effects under controlled conditions, but field tests did not confirm the earlier findings (Bradshaw-Smith *et al*, 1991).

In field beans, there appear to be varietal differences in susceptibility to the *F. culmorum/F. solani* complex (J.E. Thomas, 1992, personal communication), but in one field trial carried out at Thornhaugh in 1992, seed treatment mixtures applied to a susceptible variety failed to reduce infection significantly (Table 3).

TABLE 3. Effect of seed treatments on stem disease in spring beans, cv. Caspar, 1992

Treatment	seedling emergence (m^{-2})	% infection by stem rot	yield (t/ha)
untreated	31.4	13.0	4.39
thiram	21.1	5.3	5.75
thiram + thiabendazole	29.6	12.3	4.94
metalaxyl, thiabendazole and thiram	31.8	12.7	5.15
SED	2.8 (nsd)	4.1 (nsd)	0.4 (nsd)
CV%	11.7	53.8	12.8

SEEDLING PESTS

Both peas and beans can be attacked by the pea and bean weevil *(Sitona lineatus)* and field thrip *(Thrips angusticeps)*. Experimental work has shown that incorporated insecticide granules are more effective than sprays in reducing damage and increasing yields (King, 1981; Biddle, 1985) but the addition of an insecticide to the seed treatment has produced even better results (Baughan *et al*, 1985; Salter & Smith, 1986). However, the insecticides used experimentally had high avian toxicity and thus far, in the UK, there are no products Approved for use on peas or beans.

DISCUSSION

The use of seed treatments, either as single-purpose or multi-purpose treatments for seed and soil-borne pathogens is well established for peas in the UK.   The most common seed-borne disease, *M. pinodes*, occurs frequently during wet seasons, but is effectively controlled at present by the fungicides available.   If resistance develops in the UK, there will be an urgent need for alternative materials.   Downy mildew is also common as a soil-borne pathogen and,whilst the introduction of new varieties continues, only a small number show stable varietal tolerance and systemic acyl-analines are required for all vining pea and many combining pea varieties.

Little progress has been made in reducing root pathogens in both peas and beans and the seed treatments used currently are of little benefit in the field as they do not protect the plants beyond the seedling stage.

In beans, the requirement for seed treatment appears not to be so important.   Seedling establishment, even after winter bean seed has been ploughed-in, does not appear to be affected by seed treatment, nor do the newly introduced tannin-free varieties appear to be more susceptible to seed bed losses.   However, the improvement in control of *A. fabae* in the seed is a desirable goal and the further development of fungicides for the control of downy mildew in spring beans is also worthy of pursuit, as is the broader subject area of the use of environmentally more benign insecticidal seed treatments.

REFERENCES

Baughan, P.;   Biddle, A.J.;   Blackett, J.A.;   Toms, A.M. (1985)   Using the seed as a chemical carrier. *Symposium on appication and biology, BCPC Monograph No. 39*, Croydon: BCPC Publications.

Biddle, A.J. (1981)   Pea seed treatments to control *Ascochyta*. *Tests of Agrochemicals and Cultivars, Annals of Applied Biology* **97**, *Supplement*, No. 2, pp 34-35.

Biddle, A.J. (1983)   The foot rot complex and its effect on vining pea yield. *Proceedings 10th International Congress of Plant Protection*, 1, 117.

Biddle, A.J. (1985)   Pea pests - yield, quality and control practices in the UK.   In: *The Pea Crop - a Basis for Improvement*, P.D. Hebblethwaite, M.C. Heath and T.C.K. Dawkins (Eds), London, Butterworth.

Biddle, A.J. (1986)   Seed treatments for peas - a review. *Aspects of Applied Biology* 12, *Crop Protection in vegetables*, pp. 129-134.

Bradshaw-Smith, R. (1991)   Chemical and Biological Control of Fungal Foot Rot Pathogens of *Pisum sativum* L.   PhD Thesis,   Manchester Polytechnic.   pp 305.

Bradshaw-Smith, R;   Craig, G.D.;   Biddle, A.J. (1991)   Glasshouse and field studies using *Pythium oligandrum* to control fungal foot rot pathogens of peas. *Aspects of Applied Biology*, **27**, *Production and protection of legumes*, 347-350.

Gane, A.J.;   Biddle, A.J. (1978)   Pea establishment problems - a review. *Acta Horticulturae* **72**, 121-124.

Gravanis, F.T. (1986)   A study of the *Fusarium* foot and root rot of peas and an evaluation of certain chemicals for its control.   PhD Thesis, University of Manchester.   pp 228.

Jellis, G.J.;   Bolton, N.J.E.;   Clarke, M.H.E. (1988)   Control of *Ascochyta fabae* on *faba* beans. *Brighton Crop Protection Conference - Pests and diseases*, 895-900.

Jellis, G.J.; Punithalingam, E. (1991) Discovery of *Didymella fabae* sp nov, the teleomorph of *Ascochyta fabae* on faba bean straw. *Plant Pathology* 40, 150-157.

King, J.M. (1981) Experiments for the control of peas and bean weevil *(Sitona lineatus)* in peas. *1981 British Crop Protection Conference - Pests and Diseases*, 1, 327-331.

Ministry of Agriculture, Fisheries and Food (1993) *Agricultural and Horticultural Census: 1 June 1992, United Kingdom, Final Results*.

Miller, M.W.; de Whalley, C.V. (1981) The use of metalaxyl seed treatments to control downy mildew. *1981 British Crop Proteciton Conference - Pests and Diseases*, 1, 342-348.

National Institute of Agricultural Botany (1993) Recommended varieties of field peas and field beans 1993. *Farmers Leaflet No. 10*. NIAB, Cambridge, UK, 18 pp.

Processors and Growers Research Organisation (1991, 1992) *Annual Reports of the Processors and Growers Research Organisation*, 1991 and 1992. Peterborough, UK.

Processors and Growers Research Organisation (1993) Notes on growing field beans. *Advisory Leaflet*, Processors and Growers Research Organisation, Peterborough, UK. 11 pp.

Salter, W.J.; Smith J.M. (1986) Peas - control of establishment of pests and diseases using metalaxyl based seed coatings. *Aspects of Applied Biology*, 12, *Crop Protection in Vegetables*, 135-148.

Thomas, J.E. (1993) Personal Communication. National Institute of Agricultural Botany.

Vulsteke, G.; Meeus, P. (1985) Control of *Peronospora viciae* in peas. *Medelingen van de Faculteit Landbouwwetenschappen Rijksuniversitiet Gent* 50, 1205-1215.

CONTROL OF PESTS AND DISEASES IN SUGAR BEET BY SEED TREATMENTS

M.J.C. ASHER, A.M. DEWAR

IACR Broom's Barn Experimental Station, Higham, Bury St Edmunds, Suffolk IP28 6NP.

ABSTRACT

The current and future use of fungicides and insecticides as seed treatments to control seedling diseases and pests of sugar beet in the UK is reviewed. Thiram and hymexazol have become the standard treatments for seed- and soil-borne pathogens throughout Europe, and these chemicals continue to give good control in the UK in comparison to candidate alternatives. Two new insecticides, tefluthrin and imidacloprid are now available to UK growers as more effective alternatives to the standard methiocarb. Establishment of sugar beet in crops attacked by arthropod soils pests was as good with both insecticides as with carbamate granular insecticides. Imidacloprid also gives excellent control of foliar pests, particularly virus-carrying aphids. The benefits to the environment through introduction of these seed treatments is discussed.

INTRODUCTION

Control of pests and diseases attacking sugar beet is essential under present-day husbandry which leaves little scope for plant losses, particularly at the vulnerable seedling stage. In the UK, this has been achieved by the extensive use of pesticides; until 1992 all sugar- beet seed was treated with two fungicides and one insecticide, and additional insecticides were applied to the soil at or around drilling time to approximately 60% of crops (Asher & Payne, 1989; Winder *et al.*, 1993).

This paper examines the progress made with sugar-beet seed treatments since 1988 (Dewar *et al.*, 1988), describing the continuing search for alternative fungicides to what are now the standard thiram and hymexazol treatments, and the introduction of two new insecticides.

FUNGICIDES

Control of seed-borne *Phoma betae*

The use of the thiram steep process prior to pelleting has largely eliminated disease caused by *Phoma betae* from commercial seed sown in the UK. The additional benefit of the prolonged steep which physiologically "advances" seed, giving more rapid emergence and improved crop establishment, has been documented elsewhere (Durrant *et al.*, 1988). Despite this progress, the search has continued for possible alternative fungicides to thiram for use in the steep process, particularly for environmental reasons.

Table 1 shows the results of three trials on different sites in 1992, examining the effect of several different fungicides on the emergence of seedlings from *Phoma*-infected seed. Processed and graded seed with 67% *P. betae* infection was steeped for 12 h at 25°C in either a 0.2% aqueous suspension of thiram or a 0.1% aqueous solution of guazatine (Rappor, Dow-Elanco), iprodione (Rovral, Rhône-Poulenc), tolclophos-methyl (Rizolex, Hoechst-

TABLE 1. Percentage emergence of seedlings from *Phoma*-infected sugar-beet seed at three trial sites in 1992 following treatment with fungicides in a 12 h steep.

| | Site | | |
Treatment	Higham, Suffolk[a]	Potterhanworth[b]	Gayton, Norfolk[c]
Water	75.7	75.1	11.2
Thiram	89.6	88.3	42.1
Guazatine	84.2	82.3	48.3
Iprodione	65.1	53.4	5.0
Tolclophos-methyl	61.3	53.9	3.7
Fenpiclonil	62.5	46.7	6.0
5% L.S.D.	6.1	13.1	4.5

[a]Sown 10.4.92, counted 13.5.92; [b]Sown 9.4.92, counted 22.5.92; [c]Sown 29.9.92, counted 22.10.92

Schering) or fenpiclonil (Beret, Ciba-Geigy). A water steep was included as the control. After air-drying at room temperature, the seed was pelleted by Germains (UK) Ltd. with the standard EB3 coating but omitting all pesticides. Seed was machine drilled at three sites. At each site the six treatments were replicated six times in a randomised block or latin square design, using 12 m long single-row plots of 250 seed at *ca* 4.5 cm spacing.

Seedling emergence was generally high at the two sites sown in the spring but much lower in the autumn-sown trial (c). The performance of all treatments was compared with the water-steep treatment (without fungicides) which is known from previous experience (P.A. Payne, unpublished) to give some control of *P. betae*. Emergence was consistently better with both thiram and guazatine than with the water control whereas the other chemicals were significantly less effective, especially at Gayton. Previous work in controlled environments (P.A. Payne, unpublished) had suggested that the 0.1% concentration adopted for these fungicides (which are, in general, more soluble than thiram) was optimal for control and not phytotoxic. A range of concentrations of the active chemicals should now be studied under field conditions.

A recent survey (Dewar & Asher, in press) of seed treatments in use on sugar-beet in Europe has highlighted the popularity of thiram applied as a dust or aqueous suspension to the seed and/or incorporated in the pelleting material. Over 1.8 million ha were sown with seed treated with this fungicide in 1992.

Control of soil-borne diseases

The most important soil-borne diseases at the seedling stage in the UK are those caused by *Pythium* spp. and *Aphanomyces cochlioides* (Payne *et al.*, 1994). Since 1988, hymexazol (Tachigaren, Sankyo, Japan) has been applied in the pellet to all seed used commercially in the UK, at a rate of 10.5 g per kg seed, to control these pathogens (Payne & Williams, 1990). However, *Pythium* spp. are also very effectively controlled by a combination of the thiram steep treatment and the pelleting process. Table 2 shows the results of applying different combinations of these treatments to rubbed and graded seed that was subsequently hand-sown in soil naturally infested with *Pythium* spp. Seed which had been pelleted but which had not received any fungicide treatment, either prior to or during the pelleting process, were attacked less than the untreated control. Steeping seed in thiram suspension, but without subsequent

TABLE 2. Effect of the thiram steep and the EB3 pellet on the incidence of infection by *Pythium* spp. in naturally infested field soil at 20°C.

Treatment		
Thiram steep	EB3 Pellet[a]	% infected seedlings
—[b]	—	50.1
—[b]	+	41.2
+	—	19.8
+	+	3.9

[a] without fungicides; [b] water steep only; 5% L.S.D. = 8.7

pelleting, markedly reduced disease incidence. The combination of both treatments virtually eliminated the disease. *Pythium* spp. are known to invade seed during the early stages of germination, leading primarily to pre-emergence seedling losses. Clearly, protecting the seed with a physical barrier such as the pellet, and impregnating it with a fungicidal chemical, both contributed to the high level of disease control that was achieved.

The additional benefit of incorporating hymexazol in the pellet is the control of *A. cochlioides*, against which thiram and metalaxyl-based fungicides are ineffective (Bruin & Edgington, 1983). A recent survey (Dewar & Asher, in press) shows that hymexazol was used on sugar-beet seed on over 2.4 million ha in Europe in 1992 at rates varying from 5.6 - 28.0 g per kg seed, depending on the disease pressure in the country or region. The prevalence of *A. cochlioides* and the lack of viable alternative chemicals for its control, suggest that hymexazol will remain in widespread use on sugar beet seed for some time.

INSECTICIDES

Control of seedling pests in the soil

Since 1987, when methiocarb was the only insecticide applied to sugar beet seed pellets in the UK (Dewar *et al.*, 1988), two new active ingredients have been used in insecticides now available to growers. The first to be introduced in 1992, was tefluthrin (Force ST; Zeneca Crop Protection). Tefluthrin is a soil-stable synthetic pyrethroid with a high level of activity against soil insect pests (Jutsum *et al.*, 1986; McDonald *et al.*, 1986). It is applied to the surface of sugar beet pellets at 10 g AI per unit (1 unit = 100,000 seeds) in a micro-encapsulated formulation which allows slow release of the active ingredient, thus prolonging persistence (Marrs & Gordon, 1988). It has low solubility, but its vapour activity forms a 'seedling protection zone' around the young roots and hypocotyl, which is effective even in dry soil conditions.

In trials from 1986-1989, on sites where soil pests such as springtails (predominantly *Onychiurus* spp.), millepedes (predominantly *Blaniulus guttulatus* and *Brachydesmus superus*) and symphylids (*Scutigerella immaculata*) had been reported to cause problems in previous sugar beet crops, tefluthrin gave the best establishment in comparison to carbosulfan and furathiocarb (Dewar *et al.*, 1988; Winder, 1990). Large-scale grower trials carried out by British Sugar plc Agricultural Development staff from 1989-91 confirmed that tefluthrin performed well over a wide range of soil types giving comparable establishment to that achieved by carbamate granules such as aldicarb and carbofuran (Cook *et al.*, 1991). Similar benefits were achieved in two series of collaborative trials across Europe organised by the members of the Pests and

Diseases Group of the Institut International de Recherches Betteravieres (IIRB) in 1987 and 1988 (Dewar, 1989), and by ICI plc and SOPRA in the UK and France (Moran *et al.*, 1988).

Tefluthrin was first introduced commercially in France in 1988 and is now also available in the Netherlands, Belgium and Germany. In 1992 in Europe, it was used on a total of 160,000 ha (Dewar & Asher, in press) mostly applied at 12 g AI/unit, but as low as 4 g AI/unit in France. More recent trials conducted in France, and by members of the IIRB Pests and Diseases group, have shown that lower rates performed as well as the higher (Moran *et al.*, 1988; Muchembled, 1991; Dewar, 1992a).

This seed treatment is cheaper than the alternative granular insecticides (approximately half the price per ha) and much less active ingredient is applied to the environment. In Britain at average seed sowing rates (1.15 units/ha) only 11.5 g of active ingredient is applied per hectare compared to between 510 and 760 g AI for the various carbamates available - a substantial (at least 97%) reduction. This reduction, coupled with tefluthrin's non-soluble relative stability (Marrs & Gordon, 1988), results in very few adverse effects on non-target organisms, especially compared to sprays of the broad spectrum α-HCH (lindane) (Dewar *et al.*, 1990; Coulson *et al.*, 1990).

However, because tefluthrin is non-systemic, it has little or no effect on arthropod pests which attack young seedlings outside the soil environment. Nor does it control either beet cyst nematode (*Heterodera schachtii*) or the free-living nematodes (*Longidorus* and *Trichodorus* spp.) which cause Docking disorder (Cook *et al.*, 1991). Therefore growers with these pest problems must either continue to rely on granules or sprays, or consider the other new seed treatment, imidacloprid, at least for foliar pests.

<u>Control of soil and foliar pests of seedlings</u>

Imidacloprid (Gaucho; Bayer plc) is a member of a novel group of insecticides, the nitroguanidines, and has a different mode of action to that of organophosphorus-, carbamate- and pyrethroid-insecticides, based on an interaction with the nicotinergic acetyl-choline receptors on the post-synaptic membrane of the nerve cell junction (Diehr *et al.*, 1991). It is systemic and thus can provide a broader spectrum of control than tefluthrin. As with tefluthrin, it is applied to the outside of pelleted sugar beet seed but at a much higher rate (90 g AI/unit). It gives good control of a wide range of soil and foliar pests in sugar beet and other crops (Elbert *et al.*, 1991). Against soil pests in the UK in 1990 and 1991 imidacloprid gave poorer control of 'the soil pest complex' than tefluthrin (Table 3), but, where pygmy beetles (*Atomaria linearis*) have been a major problem, imidacloprid has performed better, particularly on the continent (Altmann, 1991; Heatherington & Bolton, 1992; Dewar, 1992a). This was probably due to the additional systemic protection afforded by imidacloprid to the cotyledons and young true-leaves of seedlings which were attacked by immigrating adult beetles (Dewar, 1991; A.Wauters in preparation).

This systemic property has resulted in good control of other foliar pests, such as the leaf-mining larvae of mangold flies (*Pegomya hyoscyami*), flea beetles (*Chaetocnema tibialis*) and the aphids, *Myzus persicae* and *Aphis fabae* (Altmann, 1991; Heatherington & Meredith, 1992; Heatherington & Bolton, 1992; Dewar *et al*, 1993; Schmeer *et al.*, 1990). Its effects on aphids are particularly noteworthy; imidacloprid has prevented or reduced aphid colonisation of beet for up to 10 weeks after sowing, giving as good or better protection than the best of the aphicidal granules, aldicarb (Dewar & Read, 1990; Dewar, 1992; Dewar *et al.*, 1993) (Fig. 1). The consequences of this latter activity have been reduced infection with virus yellows (Table 4). In laboratory experiments, transmission of the semi-persistent beet yellows virus

TABLE 3. Effects of insecticidal seed treatments (ST) or granules (Gr) on sugar-beet seedling establishment, 1990-91.

| Treatment | Rate (AI/unit or AI/ha) | Seedling establishment (%) at | | | | Mean |
		Ramsey 1990	Baston Fen* 1990	Ramsey 1991	Thorpe Tilney 1991	(4 trials)
Untreated	-	30	19	80	88	54
Tefluthrin ST	10	69	35	96	88	72
Imidacloprid ST	90	51	27	86	88	63
Aldicarb Gr	760	53	29	87	90	65
Carbofuran Gr	600	54	41	81	88	66
LSD (5%)		19.3	12.4	10.3	4.9	

*poor soil conditions at drilling

(BYV) by caged infective *M.persicae* was not prevented, but transmission of the persistent luteovirus, beet mild yellowing virus (BMYV) was substantially reduced suggesting that normal feeding behaviour was interrupted by the insecticide (Dewar *et al.*, 1992).

Imidacloprid gives the same environmental benefits as tefluthrin - lower rates of AI applied to soil, specific placement of AI around seed, and few adverse effects on non-target organisms (Pflüger & Schmuck, 1991). It is however more expensive, as would be expected for a chemical with wider activity, which may affect its sales in the market place.

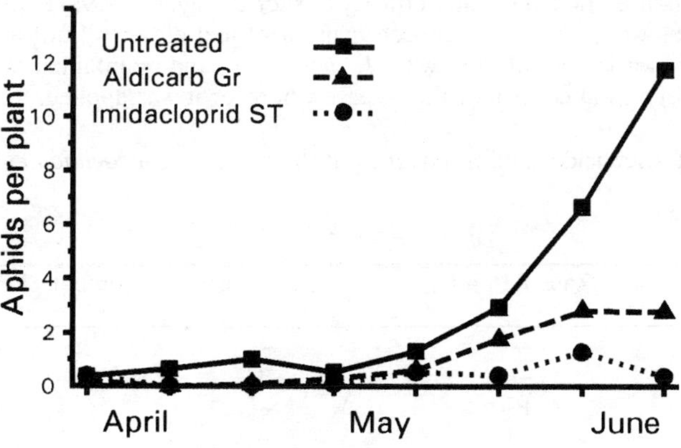

FIGURE 1. The effect of imidacloprid seed treatment and aldicarb granules on aphids in sugar beet : Sandy, Bedfordshire 1993.

TABLE 4. The effect of imidacloprid seed treatment (ST) and aldicarb granules (Gr) on the incidence of virus yellows in sugar beet in four trials in 1990 and 1993.

| | % plants infected (arcsin) at | | | | |
Treatment	Sandy 12.7.90	Clacton 26.7.90	Sandy 24.8.93	Terrington 17.9.93	Mean (4 sites)
Untreated	56.0(69)*	24.9 (18)	24.8(18)	18.8(10)	31.1(27)
Aldicarb Gr	42.7(46)	16.7(8)	22.8(15)	12.3 (5)	23.6(16)
Imidacloprid ST	29.2(24)	14.3(6)	14.7(6)	7.3(2)	16.4(8)
LSD(5%)	3.9	3.9	4.1	10.9	

*Figures in brackets are backtransformed to % plants infected.

Imidacloprid received approval for use in the UK in late 1993, and will be used extensively for the first time in the 1994 crop. It is already available in France, Belgium, the Netherlands and Finland where 50%, 62%, 12% and 2% respectively of the national crop was treated in 1993. Indeed, the uptake of the product by growers in France and Belgium has been so rapid that sales of other insecticides, such as granules and aphicide sprays, have fallen dramatically. In the UK, the use of tefluthrin in 1991 on 21% of the beet area resulted in an 18% reduction in area treated with carbamate granules and lindane sprays (Dewar & Asher, 1993). It remains to be seen what impact imidacloprid will have in 1994, but it must be a major benefit for the environment that the introduction of these new seed treatments is achieving the objective of substantial reduction in pesticide use.

INTERACTIONS BETWEEN CHEMICALS

Previous work on the incorporation of carbamate insecticides in the pellet along with the fungicide hymexazol demonstrated adverse interactions between the two chemicals, resulting in a reduction in their extractability and efficacy (Asher & Payne, 1989; Heijbroek, 1989). No adverse interactions were observed between hymexazol and either tefluthrin or imidacloprid when seeds were sown in soil infested with *A. cochlioides*, and maintained at 22°C (Table 5). Hymexazol gave very good control of the disease wherever it was applied.

Table 5. Effect of insecticide seed treatments on the control of *A.cochlioides* by hymexazol in the seed pellet.

Insecticide	Rate AI/unit	Hymexazol	Percent plants infected
—	—	—	67.4
—	—	+	2.2
Tefluthrin	10	—	62.6
Tefluthrin	10	+	6.2
Imidacloprid	90	—	50.2
Imidacloprid	90	+	2.5
LSD (5%)			9.9

CONCLUSIONS

The use of fungicidal seed treatments is now the only way of controlling seedling diseases in most countries in Europe, and use of insecticidal seed treatments is becoming more popular as better, more active chemicals become available. This trend is accompanied by a large reduction is use of granule and spray formulations, which are often applied at much higher rates. The development of seed treatments in sugar beet has thus contributed to a major reduction in pesticide use with consequent benefits to the environment.

ACKNOWLEDGEMENTS

Thanks are due to Germains (UK) Ltd and Maribo UK Ltd for carrying out trials, to Geoffrey Winder for use of unpublished data and to Glyn Williams and Lisa Read for technical assistance.

REFERENCES

Altmann, R. (1991) Gaucho - a new insecticide for controlling beet pests. *Pflanzenschutz - Nachrichten Bayer* **44(2)**, 159-174.

Asher, M.J.C.; Payne, P.A. (1989) The control of seed and soil-borne fungi by fungicides in pelleted seed. *Proceedings of 52nd Winter Congress of the Institut International de Recherches Betteravieres*, 179-193.

Bruin, G.C.A.; Edgington, L.V. (1983) The chemical control of diseases caused by zoosporic fungi. In *Zoosporic Plant Pathogens* (Ed. S.T. Buczacki), Academic Press, London, 352 pp.

Cooke, D.A.; Winder, G.H.; Prince, J.W.F.; Ecclestone, P. (1991) Tefluthrin: a new seed treatment to protect sugar beet crops. *British Sugar Beet Review* **59(2)**, 30-32.

Coulson, J.M.; Brown, R.A.; Edwards, P.J.; Lewis, F.J. (1990) The effects of tefluthrin on terrestrial non-target organisms. *Brighton Crop Protection Conference - Pests and Diseases 1990*, **3**, 975-980.

Dewar, A.M. (1989) Results of the co-operative trials on pesticides in pelleted seed, 1987-88. *Proceedings of the 52nd Winter Congress of the Institut International de Recherches Betteravieres*, 163-178.

Dewar, A.M. (1992a) The effect of pellet type, and insecticides applied to pellets on plant establishment and pest incidence in sugar beet in Europe. *Proceedings of the 54th Winter Congress of the Institut International de Recherches Betteravieres*, 89-112.

Dewar, A.M. (1992b) The effect of imidacloprid on aphids and virus yellows in sugar beet. *Pflanzenschutz-Nachrichten Bayer* **45(2)**, 423-442.

Dewar, A.M.; Asher, M.J.C. (1993) Pest and disease review of 1992. *British Sugar Beet Review* **61(1)**, 12-15.

Dewar, A.M.; Read, L.A. (1990) Evaluation of an insecticidal seed treatment, imidacloprid, for controlling aphids on sugar beet. *Brighton Crop Protection Conference - Pests and Diseases* 1990, **2**, 721-726.

Dewar, A.M.; Asher, M.J.C.; Winder, G.H.; Payne, P.A.; Prince, J.W. (1988) Recent developments in sugar-beet seed treatments. In: *Applications to Seeds and Soil*, T.J. Martin (Ed), BCPC Monograph No. 39, Thornton Heath: BCPC Publications, pp 265-270.

Dewar, A.M.; Thornhill, W.A.; Read, L.A. (1990) The effects of tefluthrin on beneficial insects in sugar beet. *Brighton Crop Protection Conference - Pests and Diseases 1990*, **3**, 987-992.

Dewar, A.M.; Read, L.A.; Hallsworth,P.B.; Smith, H.G. (1992) Effect of imidacloprid on transmission of viruses by aphids in sugar beet. *Brighton Crop Protection Conference - Pests and Diseases 1992*, **2**, 563-568.

Dewar, A.M.; Read, L.A.; Prince, J.W.F.; Ecclestone, P. (1993) Profile on imidacloprid - another new seed treatment for sugar beet pest control. *British Sugar Beet Review*, **61(3)**,5-8.

Diehr, H.J.; Gallenkamp, B.; Jelich, K.; Lantzsch, R.; Shiokawa, K. (1991) Synthesis and chemical-physical properties of the insecticide imidacloprid (NTN 33893). *Pflanzenschutz-Nachrichten Bayer*, **44(2)**, 107-112.

Durrant, M.J.; Payne, P.A.; Prince, J.W.F.; Fletcher, R. (1988) Thiram steep seed treatment to control *Phoma betae* and improve the establishment of the sugar-beet plant stand. *Crop Protection*, **7**, 319-326.

Elbert, A.; Becker, B.; Hartwig, J.; Erdelen, C. (1991) Imidacloprid - a new systemic insecticide. *Pflanzenschutz-Nachrichten Bayer*, **44(2)**, 113-136.

Heatherington, P.J.; Bolton, B.J.G. (1992) Pest control and crop establishment in sugar beet using an imidacloprid-based seed treatment. *Aspects of Applied Biology* **32**, *Production and Protection of Sugar Beet*, 65-72.

Heatherington, P.J.; Meredith, R.H. (1992) United Kingdom field trials with Gaucho for pest and virus control in sugar beet, 1989-91. *Pflanzenschutz-Nachrichten Bayer*, **45(3)**, 491-526.

Heijbroek, W. (1989) Interactions between pelleting material, insecticides and fungicides. *Proceedings of 52nd Winter Congress of the Institut International de Recherches Betteravieres*,213-220.

Jutsum, A.R.; Gordon, R.F.S.; Ruscoe, C.N.E. (1986) Tefluthrin - a novel pyrethroid soil insecticide. *Brighton Crop Protection Conference - Pests and Diseases 1986*, **1**, 97-106.

Marrs, G.J.; Gordon, R.F.S. (1988) Seed treatment with tefluthrin - a novel pyrethroid soil insecticide. In: *Applications to Seeds and Soil*, T.J. Martin (Ed.), BCPC Monograph No. 39, Thornton Heath: BCPC Publications, pp 17-23.

McDonald, E.; Punja, N.; Jutsum, A.R. (1986) The rational design of tefluthrin - a pyrethroid for use in the soil. *Brighton Crop Protection Conference - Pests and Diseases 1986*, **1**, 199-206.

Moran, A.; Painporay, G.; Cohadon, P. (1988)) Improved crop establishment in sugar beet resulting from the use of tefluthrin. *Brighton Crop Protection Conference - Pests and Diseases 1988*, **3**, 997-1002.

Muchembled, C. (1991) Development of insecticidal treatments in beet. *Pflanzenschutz-Nachrichten Bayer*, **44(2)**, 175-182.

Payne, P.A.; Williams, G.E. (1990) Hymexazol treatment of sugar-beet seed to control seedling disease caused by *Pythium* spp. and *Aphanomyces cochlioides*. *Crop Protection*, **9**, 371-377.

Payne, P.A.; Asher, M.J.C.; Kershaw, C.D. (1994) The incidence of *Pythium* spp and *Aphanomyces cochlioides* associated with the sugar-beet growing soils of Britain. *Plant Pathology* **43**, in press.

Schmeer, H.E.; Bluett, D.J.; Meredith, R.H.; Heatherington, P.J. (1990) Field evaluation of imidacloprid as an insecticidal seed treatment in sugar beet and cereals with particular reference to virus vector control. *Brighton Crop Protection Conference - Pests and Diseases* 1990, **1**, 29-36.

Winder, G.H. (1990) The development of tefluthrin as a new seed treatment for the control of soil-inhabiting pests of sugar-beet seedlings. *Brighton Crop Protection Conference - Pests and Diseases 1990*, **2**, 727-732.

Winder, G.H.; Dewar, A.M.; Dunning, R.A. (1993) Comparisons of granular pesticides for the control of soil-inhabiting arthropod pests of sugar beet. *Crop Protection*, **12**, 148-154.

REVIEW OF CURRENT AND FUTURE SEED TREATMENT USAGE IN OILSEED RAPE

D.H. BARTLETT

Uniroyal Chemical Limited, Brooklands Farm, Cheltenham Road, Evesham, Worcestershire, WR11 6LW

ABSTRACT

The seed treatments currently 'Approved' (Anon. 1991) for use on oilseed rape *(Brassica napus)* in the United Kingdom (UK) contain 7 separate active ingredients that were first patented between 17 and 52 years ago. However, in spite of their age, selection of the appropriate product will give adequate control of a) the main seed-borne diseases *Alternaria brassicae* and *Leptosphaeria maculans,* b) the main soil-borne disease, 'damping-off' *(Pythium spp.),* and c) the main seedling pests, flea beetles *(Phyllotreta spp.),* enabling the successful establishment of both winter and spring crops. It is considered desirable that, in the future, new active ingredients need to be developed to supplement or complement both thiram, for the control of 'damping-off', and gamma-HCH for the control of flea beetles.

INTRODUCTION

The 1993 oilseed rape crop is currently estimated at around 410,000 hectares (LMCS, 1993) making it the third largest arable crop in the UK. It is surprising, therefore, that combined fungicide and insecticide seed treatment products for oilseed rape have not previously been reviewed, although fungicide and insecticide seed treatments on crucifers have been discussed (Maude, 1986). The time would appear right, therefore, to carry out a review of available seed treatments.

DISEASE AND INSECT PEST CONTROL WITH SEED TREATMENTS

Target Seed-Borne Pathogens

Seed-borne pathogens of oilseed rape can be a major problem in the UK. In common with crops such as wheat, peas and linseed a cool, wet period between flowering and harvest can result in a high level of pathogens on the seed. If this infected seed is not treated and a cool, wet period follows sowing, particularly if this is late, then severe plant losses can result.

The commonest pathogen in a wet season is dark leaf spot, *Alternaria brassicae* and *A. brassicicola*, with warmer conditions favouring the development of the latter. Warm, moist conditions in May-July give rapid development, with the spores being dispersed by wind and rain splash up the plant onto the pods and which then infect the seed.

Sowing untreated, infected seed can result in reduced emergence due to plant death, similar to 'damping-off', and occasionally to wirestems (thin black stems). Infections on cotyledons and young leaves give a source of infection for subsequent leaves or adjacent plants. Serious infections of *Alternaria* have not occurred since the early 1980's, although some infection was found in 1992 on swathed crops (J.E. Thomas, personal communication). There do not appear to be any resistant varieties in commercial use in the UK.

Leaf spot and stem canker, caused by *Leptosphaeria maculans* is a very common disease of oilseed rape. The source of inoculum may be infected seed bearing pycnidia, or stubble bearing pseudothecia which release ascospores in the autumn to early spring. Two strains of *L. maculans* are recognised. Both may cause leaf spots, commonly referred to as Phoma leaf spot, but, typically, only the aggressive strain grows systemically into the leaf petiole and eventually produces a cankered stem. Several oilseed rape varieties in commercial use have moderately high levels of resistance to stem canker (Anon. 1993). Infected seed can occasionally kill seedling plants and spread the leaf spot phase to oilseed rape growing in primarily uninfected areas.

<u>Target Soil-Borne Pathogens</u>

Clubroot *(Plasmodiophora brassicae)* is potentially a serious soil-borne disease for all brassicae. It occurs in many oilseed rape growing areas, but predominantly in the west and north. There are no resistant varieties of oilseed rape, but some spring forage rapes have partial resistance.

'Damping-off' *(Pythium spp.* and *Rhizoctonia solani)* can occur in the UK but may also be due to *Alternaria, Phoma* and other soil-borne diseases. *Pythium* may occasionally cause wilting and death of young plants. *Rhizoctonia* can also cause problems at the seedling stage, but it is associated with premature ripening of more mature plants in Canada, and has been reported as causing similar symptoms in France and Germany. Its importance in the UK is not known. In Canada the *R. solani,* anastomosis groups AG2-1 and AG4, have been identified as causing pre- and post-emergence seedling damping off, seedling root rot and basal stem or foot rot (brown girdling root rot) of adult plants (Kataria and Verma, 1992). Generally, the isolates of AG2-1 are more virulent than isolates of AG4, and seedling infection by AG2-1 is favoured by cool weather, whereas warm weather is conducive to severe damping-off by AG4.

Downy Mildew *(Peronospora parasitica)* is a frequently-occurring disease of oilseed rape. The persistent resting stage in the soil infects young plants systemically and severe attacks can occasionally reduce plant population. Conidia produced from primary infections are wind dispersed and lead to further cycles of infection in cool, wet autumns. Many varieties have a high level of resistance to the disease.

Sclerotinia stem rot *(Sclerotinia sclerotiorum)* has become a widespread disease in recent years, with a wide host range, including peas, beans, linseed and potato in addition to oilseed rape. The black sclerotia develop inside the stem and can either drop into the soil, during and after harvest, or contaminate the seed. Developing seedlings can become infected by mycelium when the bottom leaf touches the ground and starts to senesce, this phase of the disease is only important when a wet autumn is associated with high seed or soil infection.

Sclerotinia is the most serious disease in France and Germany and may be a limiting factor to the increase in the growing area of oilseed rape. Some oilseed rape varieties appear to be very prone to infection. However, this appears to be almost entirely due to agronomic factors (Sweet *et al*, 1992).

<u>Target insect pests</u>

Flea beetles *(Phyllotreta spp.*, are the main species) often cause serious damage to newly-emerged seedlings of oilseed rape and other brassicas (Gratwick, 1992). The damage is usually most serious on spring-sown crops, but can occasionally affect crops sown in August. The adult beetles eat holes in the cotyledons and stems of the seedlings, beginning while plants are still below ground, and may check or kill plants, particularly in the spring, when dry weather after emergence causes wilting. Attacks can continue to the first true-leaf stage and beyond, but become progressively less damaging. It has been reported in Canada that early damage caused by *P. cruciferae* and *P. striolata* on oilseed rape delayed plant development, caused unevenness in height and maturity, reduced seed yield and raised the chlorophyll content of the seed (Lamb, 1984).

Cabbage stem flea beetle *(Psylliodes chrysocephala)* attacks only winter oilseed rape. The adult beetles move into crops in August and September and feed on leaves, causing holes and occasionally killing young plants, particularly in dry weather. Eggs are laid in the soil over several weeks and the larvae enter the plants from October to April, tunnelling in stems and leaf petioles. This flea beetle has become widely distributed.

Rape winter stem weevil *(Ceutorrhynchus picitarsus)* has a life history very similar to that of cabbage stem flea beetle. Plant attacks in September/October cause only insignificant damage, but plants severely attacked by larvae during the winter can be killed and less damaged plants stunted with a rosette like appearance.

Cabbage root fly *(Delia radicum)* is a serious pest of brassica crops in the UK. However, its effect on oilseed rape is largely ameliorated by the imperfect fit of life cycle and cropping systems. Eggs are laid in three generations - the first during mid- to late-April, the second and third generations usually overlap, so that egg laying can occur during July, August and September. The first generation can coincide with late-drilled spring, and the third generation with early-drilled winter oilseed rape. The eggs hatch into maggots, large numbers of which, feeding on roots, can virtually kill plants.

<u>Seed-borne disease control</u>

All oilseed rape seed sold by seed companies in the UK must by law be certified that it is free from disease, i.e. *Phoma* (*Alternaria* is not, at present, part of the Certification Scheme) (Anon., 1993a). Advisory tests, carried out by NIAB on brassicae seeds in 1992/93, showed that, of 39 samples tested for *Alternaria*, 73.3% were infected, and, of 36 samples tested for *Phoma*, 24.3% were infected (Reeves and Simpkins, 1993). Virtually all oilseed rape seed is treated with a fungicide/insecticide seed treatment - 94.6% was treated in 1990 (Davis *et al*, 1990).

Table 1 gives the seed treatments registered in the UK and Table 2 the seed treatments registered and hectarages grown in some of the main producing countries.

TABLE 1.   Oilseed Rape - fungicide and insecticide seed treatments registered in the U.K.

Active Ingredients (Tradename)	AI g/l or g/kg	Formu- lation	Rate/100kg Product	seed g AI	Target Diseases and Pests
Metalaxyl (Apron 350 FS)	350	FS	285ml	100	Damping-off Downy Mildew
Gamma-HCH Thiram (Hydraguard)	533 200	FS	1500ml	800 300	Flea beetle Damping-off
Gamma-HCH Thiram Thiabendazole (Hysede FL)	400 140 120	FS	2000ml	800 280 240	Flea beetle, Damping-off *Phoma*
Gamma-HCH Thiram Fenpropimorph (Lindex Plus FS)	545 73 43	FS	2200ml	1200 161 95	Flea beetle Damping-off *Phoma* *Alternaria*
Iprodione (Rovral WP)	500	WP	500g	250	*Alternaria*
Gamma-HCH Thiram Carboxin (Vitavax RS)	675 90 45	FS	2200ml	1485 99	Flea beetle Damping-off *Phoma*

TABLE 2. Seed treatment active ingredients registered on
oilseed rape in some of the main producing countries.

Country	ha (x1000) grown in 1992	Active ingredients registered
Canada	4,178	gamma-HCH/thiram/carboxin gamma-HCH/iprodione gamma-HCH/thiabendazole/thiram
Denmark	191	gamma-HCH/thiram/carboxin
France	686	isofenfos/thiram, metalaxyl, methiocarb, thiram
Germany	975	carbosulfan, isofenphos/thiram
Poland	350	carbendazim/thiram, carboxin/thiram isofenphos/thiram, thiram ·

The control of *Phoma* with carboxin has been reported
(Kharbanda, 1989); that of *Phoma* and *Alternaria* with
thiabendazole and thiram (Maude *et al*, 1984) and with
fenpropimorph, iprodione and thiram (Maude and Suett, 1986).
Further evidence that gamma-HCH/thiram/carboxin will give a
commercially acceptable level of control of A. *brassicae* comes
from a Uniroyal Chemical Ltd. contract trial carried out by NIAB
(Table 3). Here seed was used that had been artificially
inoculated during flowering in 1992. Individual seeds were sown,
in compost, in 154 modules x 6 replicates and grown in the
glasshouse. Only 3 replicates were assessed for A. *brassicae*.

TABLE 3. Oilseed rape emergence and % control of *Alternaria*
brassicae on seedlings at one true leaf.

Treatment	Rate/100 kg seed g AI	ml FP	% Emergence 19.10.92	% Control of *Alternaria* 04.11.92
Untreated (% infection)			94.9 a	(2.6) a
Gamma-HCH/thiram/ carboxin	743/99/50	1100	90.9 a	100.0 b
Gamma-HCH/thiram/ carboxin	1485/198/99	2200	93.9 a	78.3 b
Gamma-HCH/thiram/ fenpropimorph	1200/161/95	2200	93.7 a	88.9 b
LSD (P=0.05)			3.765	51.7

<u>Soil-borne disease control</u>

The control of the 'damping-off' complex *(Pythium spp)* is achieved with products containing thiram or metalaxyl. In soil inoculated with *Pythium*, seed treatment with either thiram or metalaxyl + captan gave control of pre-emergence 'damping-off' of Brussels sprout and cabbage seedlings. No post-emergence 'damping-off' occurred in these crops or in oilseed rape following treatment with metalaxyl + captan whilst post-emergence losses from untreated seed ranged from 10.2-19.4% and from thiram-treated seed from 5.7-7.4% (White *et al.*, 1984). The effect of adding metalaxyl to gamma-HCH/fenpropimorph (reduced rate of fenpropimorph) compared with standard rate gamma-HCH/thiram/fenpropimorph, resulted in further increases in the spring stand count and yield, and gave additional control of downy mildew (Smith and Margot, 1987). Metalaxyl-treated crops appeared to be more resistant to winter kill than those treated with the standard. This was believed to be related to reduced autumn downy mildew infection and/or *Pythium* control allowing improved rooting of plants going into the winter.

In Canada the control of pre-emergence damping-off and post-emergence seedling root rot caused by *Rhizoctonia solani* AG-2-1 and AG-4 with carboxin, iprodione, thiabendazole and other products has been demonstrated in growth chamber tests (Kataria and Verma, 1993).

There are no label recommendations for the control of *Sclerotinia*.

<u>Insect pest control</u>

Only one insecticide is registered in the UK for the control of flea beetles, namely gamma-HCH. Rates of use range from 800, through 1,200 to 1,485 g AI/100 kg. Contract field trials for Uniroyal Chemical Ltd., carried out by ADAS in Herefordshire and Lincolnshire on spring oilseed rape, have shown that gamma-HCH/thiram/carboxin and gamma-HCH/thiram/fenpropimorph gave significant reductions in plant damage from *Phyllotreta undulata* (Green *et al.*, 1993). Confirmation is provided by data from a further contract trial, by ADAS at Rosemaunde, Herefordshire, in spring 1993 (Table 4). Spring oilseed rape, cv. Tanto, treated at standard rate by Uniroyal Chemical Ltd., was drilled in plots 12 x 2.2m x 3 replicates on 15.4.93. Assessments were made on 18.5.93 at growth stage 1-02.

In Canada a yield increase of 61.9% was recorded on oilseed rape treated with gamma-HCH at 1,560 g AI/100kg, compared with a yield increase of 6.7% following a spray treatment of azinphos - methyl for the control of flea beetles (Westdal *et al*, 1975).

No insecticide seed treatments are currently registered for use against cabbage stem flea beetle. However, furathiocarb 25 and 50 g AI/kg seed, applied as a film coating seed treatment to oilseed rape, has given significant reduction in the number of larvae per stem at two sites (Salter and Smith, 1987).

TABLE 4. Oilseed rape emergence and damage to cotyledons from *Phyllotreta undulata*.

	Mean No. plants/m^2	Mean % plants damaged*	Mean % damage to cotyledons*
Untreated	135.6 a	74.3 b	21.0 c
Gamma-HCH/thiram/carboxin	152.3 a	45.8 a	7.8 ab
Gamma-HCH/thiram/fenpropimorph	153.1 a	51.9 a	4.9 ab
SED (12DF)	(8.63)	(8.30)	(2.85)

*data transformed using angles

Non-chemical control

Very little biological control work has been carried out. However, it has been observed that indigenous populations of *Pseudomonas fluorescens*, *Trichoderma harzianum* and the non-pathogenic binucleate *Rhizoctonia*-like forms have demonstrated a certain level of control against the virulent isolates of *R. solani* (Kataria and Verma, 1992).

Cultivar selection can be used to reduce crop susceptibility to the foliar disease *Leptosphaeria*, which must, in turn, reduce seed-borne infections. There is no published information on cultivar resistance to 'damping-off' or to flea beetle attack.

Status of current products and possible future introductions

Fungicides
Thiram, first reported in 1942 (Worthing and Hance, 1991), was not expected to survive into the 1990's. It has had its toxicology defended by a taskforce, including Uniroyal Chemical Inc., and, following a review by the FAO/WHO Joint Meeting on Pesticide Residues in 1992, the committee has re-established an Acceptable Daily Intake for thiram assuring its seed treatment uses for the present. The only alternative to thiram for the control of damping-off is, at present, metalaxyl which, while it has the advantage of giving additional control of downy mildew, does have the disadvantage of being more expensive than thiram. It is not known of any compounds being developed for this use, but it is hoped that some will be. Fenpropimorph does not appear to have any registration issues.

Several new seed treatment fungicides have been registered on cereals in recent years, but the only one which is likely to be registered on oilseed rape, for the control of *Alternaria* and *Leptosphaeria*, is fenpiclonil (Nevill *et al*, 1988) or another phenylpyrrole.

A new sequential co-application of gamma-HCH/thiram and iprodione is likely to be marketed when 'Approval' is granted.

The development of new products that would control club root and *Sclerotinia* as well as damping-off and downy mildew would be an added bonus.

Insecticides

Gamma-HCH, which was first reported in 1945 (Worthing and Hance, 1991), is no longer used in some countries such as France, Germany and Sweden. At present its future in the UK is not at risk, but it would be desirable, both from the registration point of view and to obtain a higher level of control of flea beetle, to find an alternative active ingredient. Iodofenphos is registered for this use in France and Germany, but not in the UK. The very active imidacloprid is unlikely to be developed on oilseed rape because of insufficient activity against flea beetle. However, synthetic pyrethroids are showing promise as seed treatments (Uniroyal Chemical Ltd., unpublished data).

The ideal seed treatment insecticide would be systemic to give more persistent control of *Phyllotreta spp.* and also to give control of *Psylliodes* and *Ceutorrhynchus*.

SEED COATINGS

Of all the major crops grown in the UK, oilseed rape is the one most commonly coated with coloured polymer coatings during the seed treatment process. The advantages of seed coating has been well summarised (Halmer, 1988). Several factors have accounted for their high uptake on oilseed rape including - the low seed rate/ha which reduces cost and the fact that the main product used (gamma-HCH/thiram/fenpropimorph) and also gamma-HCH/thiram/thiabendazole contain no dyes or pigments, so that the addition of a coloured coating gives a more attractive seed sample.

The future for oilseed rape seed treatments

The forecast areas of oilseed rape for 1993, 1995 and 1997 (LMCS, 1993) are given in Table 5.

TABLE 5. Hectarages of oilseed rape grown in 1991, with 1992 and forecasts for 1993-1997 expressed as a percentage of 1991 figures.

	ha (x1000) actual 1991	1992	ha expressed as % of 1991 1993	1995	1997
Denmark	276	69.2	63.0	61.6	58.0
France	735	93.3	73.7	65.2	51.6
Germany	944	103.3	64.3	46.1	38.9
UK	441	95.7	93.2	73.9	71.7
EEC total	2,444	94.6	85.1	67.8	59.4

By 1997 the EEC total hectarage of oilseed rape is forecast to fall by 40% compared with a forecast fall of 28% for the UK which is cushioned by their ability to switch to growing the cheaper spring crop. Whilst the quantity of seed treatments sold will obviously decline, their usage on virtually all the crop will no doubt continue due to their high cost-effectiveness.

ACKNOWLEDGEMENTS

I would like to thank Dr Jane Thomas of NIAB and Mr David Green of ADAS for editing, respectively, the sections on disease pathogens and insect pests.

REFERENCES

Anonymous (1991) 1991 Product Guide to Seed Treatments available in the UK, Ed. D. Soper, BCPC, 11.

Anonymous (1993) Varieties of Oilseed Crops 1993. Farmers Leaflet No. 9. NIAB, 6.

Anonymous (1993a) Oil and Fibre Plant Seed Regulations 1993. HMSO, SI 2007, 18.

Davis, M.P.; Garthwaite, D.G.; Thomas, M.R. (1990) Pesticide usage survey report 85. MAFF, 10.

Gratwick, M. (1992) Crop pests in the UK. Chapter 34 Flea beetles. Chapman and Hall, London, 173-175.

Green, D.B.; Holliday, M., Bartlett, D.H. (1993) *Tests of Agrochemicals and Cultivars*, No. **14** *(Annals Applied Biology* **122** *supplement)*, 18-19.

Halmer, P. (1988) Technical and commercial aspects of seed pelleting and film-coating. In: *Applications to Seeds and Soil*, T.J. Martin (Ed.) BCPC Monograph No. **39**, Thornton Heath: BCPC Publications, 191-204.

Kataria, H.R.; Verma, P.R. (1992) *Rhizoctonia solani* damping-off and root rot in oilseed rape and canola. *Crop Protection*, **11** (1), 8-13.

Kataria, H.R.; Verma, P.R. (1993) Efficacy of fungicidal seed treatments against pre-emergence damping-off and post-emergence seedling rot of growth chamber grown canola caused by *Rhizoctonia solani* AG-2-1 and AG-4. *Canadian Journal of Plant Pathology*, **12**, 409-416.

Kharbanda, P.D. (1989) Effectiveness of fungicides to control blackleg of canola rapeseed. *Canadian Journal Plant Pathology*, **11** (2), 193.

Lamb, R.J. (1984) Effects of flea beetles, *Phyllotreta spp.*, on the survival, growth, seed yield and quality of canola, rape and yellow mustard. *Canadian Entomologist*, **116** (2), 269-280.

LMCS Landel Mills Commodity Studies (1993). *European Agrochemical Monitor*, Proprietory Study.

Maude, R.B.; Humpherson-Jones, F.M.; Shoring C.G. (1984) Treatments to control *Phoma* and *Alternaria* infections of brassica seeds. *Plant Pathology*, **33**, 525-535.

Maude, R.B. (1986) Treatment of vegetable seeds. In: *Seed Treatment*, K.A. Jeffs (Ed) Thornton Heath: BCPC, 239-261.

Maude, R.B.; Suett, D.L. (1986) A review of the progress in the development of seed treatments for the control of *Alternaria* and *Phoma* infections of brassica seeds. *Aspects of Applied Biology*, **12**, *Crop Protection in Vegetables*, 13-19.

Nevill, D.; Nyfeler, R.; Sozzi, D. (1988) CGA 142705: A novel fungicide for seed treatment. *Brighton Crop Protection Conference - Pests and Diseases 1988*, **2**, 65-72.

Reeves, J.C., Simpkins, S.A. (1993) The incidence of some seed-borne diseases in the UK. In: *Plant health and the European single market*, D. Ebbels (Ed.), *BCPC Monograph No. 54*, Thornton Heath: BCPC Publications, 333-338.

Salter, W.J.; Smith, J.M. (1987) Furathiocarb seed coatings: potential replacements for topical pesticide applications. In: *Applications to Seeds and Soil*, T.J. Martin (Ed.), *BCPC Monograph No. 39*, Thornton Heath: BCPC Publications, 277-283.

Smith, J.M.; Margot, P. (1987) The use of metalaxyl as a seed or soil treatment. In: *Applications to Seeds and Soil*, T.J. Martin (Ed.), *BCPC Monograph No. 39*, Thornton Heath: BCPC Publications, 41-53.

Sweet, J.B.; Pope, S.J.; Thomas, J.E. (1992) Resistance to *Sclerotinia sclerotiorum* in linseed, oilseed rape and sunflower cultivars, and its role in integrated control. *Brighton Crop Protection Conference - Pests and Diseases 1992*, **1**, 117-126.

Westdal, P.H.; Romanow, W.; Askew, W.L. (1975) Annual Conference Manitoba Agronomists December 16 and 17, 1975, 55-57.

White, J.G.; Crute, I.R.; Wynn, E.C. (1984) A seed treatment for the control of *Pythium* damping-off diseases and *Peronospora parasitica* on brassicas. *Annals of Applied Biology*, **104**, 241-247.

Worthing, C.R.; Hance, R.J. (1991) *The Pesticide Manual*, **9**, 452 and 822.

DIAGNOSTIC METHODS FOR THE DETECTION OF PLANT PATHOGENS IN VEGETABLE SEEDS

C.J. LANGERAK, A.A.J.M. FRANKEN

DLO - Centre for Plant Breeding and Reproduction Research (CPRO-DLO), P.O. Box 16, 6700 AA Wageningen, The Netherlands.

ABSTRACT

Diagnostic methods to detect plant pathogens associated with vegetable seeds are applied for various reasons at different stages in the seed production chain. The reason for testing, however, may differ from that for which a particular method had been originally developed. The consequence of applying an unsuitable method is that incomplete or even false information is obtained on the health status of the seed.
In this paper an overview is presented on the various objectives of seed health testing, the principles of the available methods and the choice of the most appropriate method. Factors which might interfere with the diagnostic value of the test results are considered. Finally, some factors relevant to the correct evaluation and interpretation of the analytical data are reviewed.

INTRODUCTION

A world wide increase in the production area of vegetable seeds and a dramatic expansion of the international seed trade has enhanced the interest in diagnostic methods for detection of pathogenic fungi, bacteria and viruses associated with seeds. The occurrence of such plant pathogens in vegetable seeds has been described over the years by many authors. The most recent updated compilation of this information was presented by Richardson (1990) as section 1.1 of the Handbook on Seed Health Testing of the International Seed Testing Association (ISTA).

Reasons for analysing seeds for the presence of plant pathogens can be diverse and are related to the stage in the production process at which the seed needs to be tested. The purpose of testing, the nature of the pathogen and the level of contamination that can be tolerated will determine the methodology to be followed.

Detection methods have been described in the literature for a wide range of seed-borne pathogens. Detailed descriptions for detection of seed-borne fungi and some bacteria and viruses can be found in the Working Sheets of the ISTA Handbook on Seed Health Testing (1981). Excellent technical information for detection of bacteria in seeds is given by Saettler *et al*. (1989) and, for their identification by Schaad (1988). Serological methods which can be used for detection and identification of viral and bacterial plant pathogens were described by Hampton *et al*. (1990).

The users of seed health testing methods are often not aware of the fact that the analytical data obtained do not always provide the information which is needed in a given situation. This paper aims to evaluate the various situations which may lead to the application of diagnostic methods on seed. The benefits of testing at different stages in the seed production process, the kind of methods that are available and the limitations that

can be expected in the analysis and interpretation of the results are also
considered.

PURPOSE OF TESTING

Testing of seeds for 'health' or contamination with plant pathogens
can be performed at several stages, as illustrated in Figure 1. The reasons
for testing can vary at each stage.

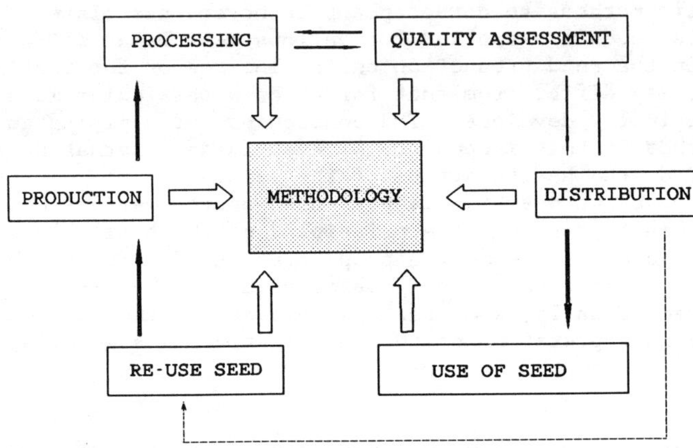

Figure 1. Scheme representing the various stages at which seed
health testing may be useful.

Production

Unripe or almost ripe seed can be collected from the seed crop during
the period the seed is formed and matures, and analysed for different
reasons. Analytical data may confirm transmission of a pathogen to the
seeds from diseased plants found in the crop itself or in adjacent fields.
Such information might be useful for curative or preventative control
measures to be undertaken prior to, or following harvest.

Pre-harvest treatment of the crop
Sprays with chemicals or biological agents may prevent further
contamination or infection of seeds with pathogenic fungi. Excessive
contamination of seed with certain pathogenic fungi, bacteria and viruses
may be prevented by early harvesting. Of course, a balance has to be struck
between the required maturity of the seed and the level of contamination
with the pathogen that can be tolerated.

Adaption of the post-harvest conditions
Quick drying of threshed or unthreshed seed may prevent further spread
of fungi or bacteria through the bulk of material. Development of embryonic
or endosperm infections which are more difficult to control may be stopped
in time. Storage at low temperatures after drying may have a similar
effect.

<u>Planning of the processing</u>
Advanced knowledge of the occurence of seed-borne pathogens in a seed lot can be used in different ways. One might avoid cross contamination between diseased and healthy seed lots during the various processes of threshing, cleaning, grading or any treatment given afterwards, especially, when one of these processes involves seeds passing through a wet or damp phase of handling. Fungi such as *Phoma* and *Septoria spp.* may release enormous amounts of conidia in water from pycnidia associated with the seed coat. Foreknowledge also allows the planning of the most appropriate curative treatment, which can be chemical, physical or biological.

<u>Planning of the sales</u>
Finally, early knowledge of the presence of a plant pathogen in a seed lot will contribute to better sales management, especially when this pathogen is subject to the phytosanitary regulations of the country of final destination of the seed.

<u>Processing</u>

It is worthwile drawing representative samples from each seed lot prior to processing, and to analyse them before further steps are undertaken. A quick visual inspection of the dry seed may sometimes indicate the presence of a pathogen. Examples of this are the occurrence of sclerotia of *Sclerotina sclerotiorum*, presence of dark coloured spots on seeds of *Phaseolus vulgaris* and *Pisum sativum* caused by *Colletotrichum lindemuthianum* and *Ascochyta spp.* respectively. In other cases, extraction or incubation procedures will be necessary for an effective diagnosis. On the basis of such diagnosis, which should preferably be a quick one, the most optimal processing can be planned. Attention can then be paid to the following substages.

<u>Cleaning</u>
Mixing of 'healthy' and 'diseased' lots prior to any cleaning process should be avoided. Additional processes can be needed for removal of sclerotia or seeds with spots or discolorations caused by pathogens. Liquid cleaning may cause smearing of certain pathogens and, under these circumstances, should not be used.

<u>Grading</u>
The prevalence of certain pathogens may be linked to sublots consisting of seeds of a certain size. It is known for example that the lightest seeds of *Cucumis melo* are more often infected with squash mosaic virus than the heavier seeds (Middleton, 1944).

<u>Storage</u>
The viability of several pathogens, for example *Septoria apiicola* in celery seed (Mujica, 1943), decreases with time. Knowledge about such contamination may postpone further steps until the risk of seed transmission after sowing has become acceptably low.

<u>Physical treatment</u>
Treatments of seed lots by heat, mostly given as a hot water treatment, may kill a target pathogen in the seed, but negative effects on the viability of the seed may be one of the other consequences. Reliable prescreening for the presence of the pathogen will avoid unnecessary loss of quality in those lots which are 'free' of the pathogen.

Physiological treatment

Treatments such as osmo- or hydropriming are often applied on vegetable seeds in order to improve the sowing value. In addition, pelleting the
seed may give further improvement. However, such expensive modification of
seeds becomes worthwhile only when there is a guarantee that no cross
contamination with pathogens takes place during one of the technological
processes. Consequently, rather sensitive health tests are needed, which
give information about such risks.

Chemical treatment

Most vegetable seeds are treated nowadays with one or more chemicals.
The choice of these chemicals should of course be targeted at the pathogens
which are associated with a particular seed lot. In practice however, all
lots of a particular seed species often get the same standard treatment
though the kind of contamination may differ in different lots. In particular the use of selective chemicals such as the benzimidazoles or, for
example, iprodione requires caution. Treatment with iprodione of carrot
seed contaminated with both *Alternaria* and *Fusarium spp.* gives lower
quality seed, as *Fusarium* will not be controlled and may even be stimulated. Treatment of such a seedlot with carbendazim causes an opposite
effect. A diagnostic method indicating the level of contamination and
specifying the fungal contaminant, at least at genus level, is a prerequisite.

Biological treatment

Interest in the treatment of seeds with biological control agents
(BCA's) has increased since the application of certain chemicals may be
prohibited in the future, in order to protect the environment. Results
however are still poor in practice. As has been mentioned for chemical
treatment, knowledge of the composition of the natural mycoflora on the
seed may be of relevance as interactions between the added BCA and the
mycoflora cannot be excluded and may vary between seed lots.

Quality assessment

Knowledge of the health status of seed lots which have reached the
stage of distribution is the most obvious reason for testing. The final
destination of the seed lot, the way it is distributed to the user and the
phytosanitary or quarantine regulations being in force determine the choice
of the method. Based on the test result it is possible that the seed may
have to be treated chemically, physically or biologically in order to meet
national quality standards or those of the potential user of the seed. In
such cases a second test may be required to verify or to measure the
efficacy of the treatment. The method used will depend on the nature of the
treatment given. When biologically active residues, e.g. fungicides,
bactericides or BCA's have been applied to the seed, traditional microbiological techniques will generally not be suitable; neither will serological
or other modern molecular techniques based on DNA or RNA analysis. A
growing-on test may solve this problem.

Distribution

Seed lots will generally meet the quality standards of the marketing
company, and those of the Nation or State in which they are finally
processed and made ready for sale. Additional, and often different,
requirements for certain pathogens may arise at the moment export takes

place. The method of processing and the treatment that was applied will
then determine whether testing for the presence of such pathogens is still
possible and reliable. Analytical data from tests carried out at an earlier
stage during seed production and processing may be useful to convince the
client that the seed lot meets all requirements set by phytosanitary or
quarantine regulations.

<u>Use</u>

Health tests are normally not necessary once the seed has reached its
destination, viz. the retailer in an importing country or the grower of a
vegetable crop, unless a check has to be carried out by law in order to
meet the requirements for an Import Permit. Again, difficulties will arise
when the pathogen is detectable in such a situation and it cannot be proved
that the seed has been treated in such a way that the risk of disease
transmission is negligible. It is often not realised, however, that the
sensitive methods based on serological and molecular techniques that are
now available are not suitable as they do not necessarily indicate survival
of the target pathogens. The only reliable test would then be a culture
method or a growing-on method, showing specific symptoms of the disease
after imbibition of the seed.

<u>Re-use</u>

Decisions have to be made at this stage about the use of seed for
further multiplication or for long term storage in genebanks. Consequently,
the requirements set for disease 'freedom' should be as stringent as possible. Health tests performed in earlier stages may already have given enough
information; otherwise, very sensitive and if possible, non-destructive
tests have to be carried out to confirm the absence of the target pathogen.

PRINCIPLES OF DETECTION METHODS

Methods for the detection of seed-borne fungi, bacteria and viruses
have been reviewed by Neergaard (1977) and Agarwal & Sinclair (1987).
Official testing procedures for detection of fungi can be found in the
Rules of the ISTA (International Seed Testing Association, 1993). These
methods are recommended for official certification purposes provided that
the seeds submitted for testing are not treated with chemicals and have
been obtained according to a standardised sampling procedure. Other methods
for fungi and for a number of seed-borne bacteria and viruses are presented
in a series of separate Working sheets (ISTA, 1981- onwards). They have
been evaluated by the Plant Disease Committee of the ISTA for reproducibility. In a number of sheets an indication is given of the diagnostic value
of the method(s) described.

<u>Direct inspection of the seed or seed washings</u>

The most simple procedure by which seed infection through fungi,
bacteria or viruses can be determined is by inspection of the dry seed with
the naked eye or with the help of a microscope. One may inspect for
presence of sclerotia, specific discoloration or malformations. Fungi with
characteristic conidia or spores can also be detected on dry seed or in
seed washings. One should realise, however, that symptomless seed may still
carry slight and latent infections.

<u>Growing-on tests</u>

Another generally applicable procedure for a number of fungi, bacteria and viruses is the so-called growing-on test in which seeds are incubated in an artificial substrate, sand or soil in order to induce specific symptom development in the seedlings. This kind of test may also be needed to confirm, for example, the efficacy of an eradicative seed treatment against a bacterium or a virus. The possibility that some seeds may be unable to germinate because of lethal infections should not be overlooked.

<u>The most common tests for fungi, bacteria and viruses</u>

Most commonly applied methods for fungi are based on incubation of the seed on paper substrates (blotter tests) or on nutrient agar. These procedures are not suitable for detection of bacteria or viruses. These pathogens are now detected most commonly by serological and/or a biochemical techniques as part of the test procedures which are described below.

The test procedure for detection of seed-borne bacteria can include the following steps: a) extraction of subsamples of whole or disrupted seeds in water or buffered solution, b) isolation of the target bacterium from the seed extract by plating the extract or dilutions of it on (semi) selective media, c) identification of suspect colonies via biochemical tests or a serological test with a specific antiserum, and d) testing the pathogenicity of positively identified isolates.

Detection of seed-borne viruses involves a serological test on either extracts of ground seeds or on sap obtained from seedlings raised from seeds. In the first type of extracts both active and inactivated virus particles will be detected; in the latter case information is also obtained on the transmission of active virus, provided remnants of the seed coats are separated from the seedlings prior to extraction. For confirmation of the identity of the virus, additional methods have been reviewed by Agarwal & Sinclair (1987). Newer techniques, such as application of the polymerase chain reaction (PCR), are still awaiting widespread application.

CHOICE OF THE DETECTION METHOD

The choice of which diagnostic method is best suited to analysing a seed lot or sample depends primarily on defining the identity of the target pathogen(s), selecting the optimum sampling procedure for detecting of the pathogen(s) and defining tolerance levels. The sub-sampling procedure, with respect to the number of seeds to be analysed individually or pooled in sub-samples, has to be adjusted to the tolerance level set for the target pathogen (Geng *et al.*, 1983). Choice of a method of detection and sampling also depends on whether qualitative and /or quantitative information is required, e.g. presence or 'absolute' (i.e. 'presumed)' absence of a certain pathogen in a seed lot. Information has to be available on how analytical data can be evaluated and interpreted, what kind of conclusions can be drawn from them and what actions have to be taken as a result. The choice of a method may also depend on the skill of the analysts running the tests. Apart from the purpose of the test, the available methods and their principles, it is important to know what logistic and technical facilities are needed to carry out the planned tests (Langerak *et al*, 1988).

FACTORS INTERFERING WITH THE DIAGNOSTIC VALUE OF THE TEST RESULT

The use of any method must be based on relevant research or practical experience, giving information on how effectively the pathogen in question can be detected. Moreover, one should know how well results can be reproduced and interpreted, recognising the factors which might have been influencing the final test result.

History of the seed lot

At any time, independently of the stage at which a seed crop or lot is sampled for seed health testing, information should be available about: the origin (i.e. the location of production) and the climatological conditions during seed formation, maturation, and at harvest. Similar information is needed on the environmental conditions during transport, storage, processing and treatment of the seeds. Furthermore, it is relevant to have information on any chemical, physical, physiological or biological treatment given to the seed from the moment it is formed. All these factors may influence the location of the pathogen in the seed, the level of contamination and the prevalence of seed infection in the lot, and the chance of detecting the target pathogen.

Sampling and storage of the samples

Samples drawn from a seed lot must be representative of that lot. Directions for correct sampling procedures are given in the Rules for seed testing of the ISTA (1993). One of the prerequisites is that the lot is homogeneous. The size of the sample must be sufficient to ensure that the test results give relevant information. The size of the working sample and the number of seeds taken from it for testing must be closely related to the tolerated infection percentage. Statistical handbooks may be consulted in order to find the optimal sample size. It is very important to dry samples with a high moisture content when the period between sampling and testing exceeds several days, otherwise moulds associated with such seeds may become active. As a consequence, detection of the target pathogen may become difficult or even impossible.

Technical factors

Several factors of technical origin can cause variation in test results. The seeds and associated micro-organisms may respond differently under variable incubation conditions. Therefore, equipment needed for seed health testing must be clean and sometimes even sterile in order to avoid contaminating the material to be incubated. Supervision on correctness of settings for temperature, moisture, light, pressure and cleanness must take place according to a well defined schedule. General guidelines for installations and instruments used in microbiological studies are given in ISO 7218 of the International Organization of Standardization (ISO, 1985).

Materials such as water, filterpaper, sand, soil, agar, nutrient media, chemicals, plates, glassware etc. must be of a standard quality and need regular checks for purity and eventual toxicity for both the seeds and the seed-borne organisms. As the quality of tap water may vary over the year and will, in general, differ from laboratory to laboratory, it is recommended that deionised or distilled water is used. Test material of a biological nature such as antisera, test plants or reference seed samples need special attention as the risk of quality loss is rather high. Regular checks on quality have to be carried out according to a fixed schedule.

Personnel factors

One of the most important factors in performing seed health tests is the human one. The skills of the personnel determine the success and standardisation of tests. This specially holds for methods in which assessment of characteristics of a biological, and thus variable, nature such as symptoms, colonies of bacteria or fungi and fructifications of fungi are required. It is recommended that such tests be carried out at least in duplicate and are assessed by two analysts. Criteria should be established with respect to accuracy and performance of the test.

EVALUATION AND INTERPRETATION OF TEST RESULTS

It is important that in any laboratory where seed health tests are carried out, comparable results are obtained in repeated testing. A detailed protocol that can be followed has been described by Sheppard in "Guidelines for collaborative study procedures to validate characteristics of a method of analysis" (1993, unpublished), which was adapted from the "Guidelines for collaborative study procedures in chemical analysis" (Anonymous, 1989).

Repeatability and reproducibility of test results

The reproducibility of test methods may vary and different methods may become more or less sensitive to test conditions. It is essential therefore to repeat testing on well-stored reference samples in order to evaluate the precision of test results. The cause of variation should be found and eliminated if reproducibility is lower than originally described and established for the method.

Statistical treatment of analytical data

Regardless of the kind of data that are obtained, a statistical treatment is necessary before a final report on the test result is presented. Results can be expressed as the percentage infected or contaminated seeds, the number of fungal spores or colony-forming units per gram or number of seeds tested, the occurrence or non-occurrence of the pathogen in a working sample or number of subsamples, etc. In the latter case MPN tables can be used to estimate the number of infested seeds by the most probable number method (Taylor & Phelps, 1984). For official seed health testing, it is usually necessary to state the maximum infection level for a negative result at a given probability (e.g. P = 0.05). It is also important to note in a test report the size of the working sample i.e. the number of seeds tested, and the method of statistical analysis.

Incorrect, improper, illusory analytical data

Certain methods, especially those used for detection of low infection levels of bacteria or viruses may sometimes give false positive and false negative values. Examples of how to interpret these have been worked out by Sheppard *et al.* (1986).

DISCUSSION

High quality standards are required for the successful marketing of
vegetable seeds worldwide. An important element of quality is the health
status, the foundation of which is established during the seed production
but may be improved subsequently by technological processes. Diagnostic
methods by which the presence of plant pathogens in seed can be detected
are mostly used at the stage when seed lots are ready for sale. There are
several other stages at which application of such methods would be justified, thereby contributing to more efficient quality control management
between the time the seed is produced and used. Figure 1 and the comments
made on the possible reasons for testing, make it clear that obtaining
health data before harvest can be used for planning and organising the
technological processes which follow after harvest. Data obtained during
the various processing stages will help to avoid unnessary spoilage of seed
through cross contamination between and within seed lots. Moreover, such
data could also be used to choose the best seed treatment if needed.

Another aspect requiring comment concerns the kind of method to be
chosen at a particular stage of processing. It has been emphasized that the
choice of method depends on the purpose of testing, the nature of the
pathogen and the principle of the method. A great diversity of methods is
available currently and new ones are still being developed, mostly based on
molecular techniques. Nevertheless, it has been pointed out that special
attention should be given to the interpretation of test results obtained by
the modern methods as they often do not give information about the viability of the inoculum detected. The latter aspect is importance if the target
organism is not considered as a quarantine pathogen and a certain level of
contamination can be tolerated. Very sensitive, specific and fast methods
may be required to detect the survival and presence of living inoculum of
the pathogen after a seed treatment, provided the aim of the treatment was
to kill the inculum or to reduce the transmission rate to acceptable
levels.

Tests may give false negative or false positive results (Sheppard *et
al*, 1986). For example, heavy contamination of seeds with saprophytic fungi
or bacteria may mask the presence of a pathogen when plating the seed or
seed extracts on filterpaper or nutrient agar (de Tempe & Limonard, 1973;
Franken *et al*., 1991). Also, seeds of different cultivars may respond in
different ways to the various treatments. Furthermore, residues of chemicals on the seed, applied either on the seed crop or after harvest during
any stage of the processing, may also interfere with the result of the seed
health test (Franken *et al*., 1993).

ACKNOWLEDGEMENTS

Thanks are due to Dr. Raoul Bino for critically reading the manuscript.

REFERENCES

Agarwal, V.K.; Sinclair, J.B. (1987) *Detection of seedborne pathogens,
Principles of seed pathology, volume II*, CRC Press, Inc, Boca Raton,
Florida, USA, 29-76.

Anonymous (1989) Guidelines for collaborative studies, *Journal of the Association of Analytical Chemists*, **72**, 694-704.

Anonymous (1993) Seed Health Testing in: Annexes to chapter 7 of International Rules for Seed Testing, 1993, *Seed Science and Technology*, **vol. 21**, supplement, 211-216.

Franken, A.A.J.M.; van Zeijl, C.; van Bilsen, J.G.P.M.; Neuvel, A.; de Vogel, R; van Wingerden, Y; Birnbaum, Y.E.; van Hateren, J.; van der Zouwen, P.S. (1991) Evaluation of a plating assay for *Xanthomonas campestris* pv. *campestris*, *Seed Science and Technology*, **19**, 215-226.

Franken, A.A.J.M.; Kamminga, G.C.; Snijders, W.; van der Zouwen, P.S.; Birnbaum, Y.E. (1993) Detection of *Clavibacter michiganensis* ssp. *michiganensis* in tomato seeds by immunofluorescence microscopy and dilutionplating, *Netherlands Journal of Plant Pathology* **99**, 125-137.

Geng, S.; Campbell, R.N.; Carter, M.; Mill, F.J. (1983) Quality control programs for seedborne pathogens, *Plant Disease* **67**, 236-242.

Hampton, R.; Ball, E.; De Boer, S. (1990) *Serological methods for detection and identification of viral and bacterial plant pathogens. A laboratory manual*, APS Press, the American Phytopathological society, St. Paul, Minnoseta, USA, 389 pp.

International Organization for Standardization (1985) *ISO 7218, Microbiology General guidance for microbiological examinations*, International organization for standardization, Geneva, Switzerland, 14 pp.

International Seed Testing Association (1981 -) *ISTA Handbook on seed health testing, section 2, Working sheets, each dealing with one pathogen on one host*, Zürich, Switzerland.

International Seed Testing Association (1993) International Rules of Seed Testing 1993, *Seed Science and Technology* **21**, Supplement, Zürich, Switzerland, 288 pp.

Langerak, C.J.; Merca S.D.; Mew, T.W. (1988) Facilities for seed health testing and research, Proceeedings of the International Workshop on Rice Seed Health, 16-20 March, 1987, pp 235-246.

Middleton, J.T. (1944) Seed transmission of squash mosaic virus, *Phytopathology*, **34**, 405.

Mujica, F. (1943) La septoriosis del Apio. *Simiente*, **12**, 81.

Neergaard, P. (1977) Seed Pathology, Volume I and II, The MacMillan Press LTD, London, England, 1187 pp.

Richardson, M.J. (1990) *An annotated list of seed-borne diseases, Section 1.1.*, *ISTA Handbook on Seed Health Testing*, International Seed Testing Association, Zürich, Switzerland.

Schaad, N.W. (1988) *Laboratory guide for identification of plant pathogenic bacteria*, American Phytopathological Society, St. Paul, Minnosota, USA.

Saettler, A.W.; Schaad, N.W.; Roth, D.A. (1989) Detection of bacteria in seed and other planting material, 122 pp., APS Press, The American Phytopathological Society, St. Paul, Minnesota, USA.

Sheppard, J.W.; Wright, P.F.; DeSavigny, D.H. (1986) Methods for the evaluation of EIA tests for use in the detection of seed-borne diseases, *Seed Science and Technology*, **14**, 49-59.

Taylor, J.D.; Phelps, K. (1984), Estimation of percentage seed infection, *Report on the first International Workshop on Seed Bacteriology*, Angers, France, 12-14.

Tempe de, J.; Limonard, L. (1973) Seed-fungal-bacterial interactions, *Seed Science and Technology*, **1**, 203-204.

SEED-BORNE PATHOGENS OF LINSEED IN THE UK

P.C. MERCER

Plant Pathology Research Division, Department of Agriculture for
Northern Ireland, Newforge Lane, Belfast BT9 5PX

ABSTRACT

The commonest seed-borne pathogens of linseed in
the UK are *Alternaria linicola*, *Botrytis cinerea*, and
Fusarium spp. Their incidence is dependent on a
number of factors, the most important of which is the
weather. Most of the pathogens are found only on the
outside of the seed and do not infect the embryo; they
are thus relatively easily controlled by seed-
treatments at the seedling stage. However, some
pathogens, particularly *B. cinerea*, may also be
transmitted in other ways and can attack the growing
crop independent of their incidence on the seed and
may require further control measures. Choice of
cultivar may also influence the incidence of seed-
borne disease.

INTRODUCTION

Linseed in the UK suffers from a number of diseases which
can be seed-borne (Mercer *et al.*, 1991). The damage caused to
the emerging seedling is dependent on a combination of the
incidence of the pathogens on the seed and the weather
conditions at the time of sowing (Fitt *et al.*, 1991). Although
little can be done about the latter factor, the incidence of
seed-borne pathogens can be reduced by the judicious choice of
seed or by seed-treatment. This paper describes the background
to current practices and suggests ways that they could be
improved in the future.

PATHOGENS

Those pathogens most commonly isolated from linseed seed in
the UK (Fig. 1) are *Alternaria linicola*, *Botrytis cinerea* and
Fusarium spp. (mostly *F. avenaceum* and *F. culmorum*). Others
less frequently isolated are *Phoma exigua* var. *linicola*,
Colletotrichum linicola, *Mycosphaerella lini* and *Fusarium
oxysporum* f.sp. *lini*. The incidence of pathogens is largely
determined by the weather during the period of capsule
maturation, being considerably higher if this is wet as in
1987/88 (Fig. 1). The effect of weather is also reflected in
regional differences in pathogen incidence, *A. linicola*, for
example, being readily isolated from seed every year in Northern
Ireland, while sometimes being at a relatively low incidence in
the drier south east of England. Conversely, *F. oxysporum* f.sp.
lini, which requires relatively high soil temperatures, is a
major problem on the continent, is occasionally found in the

south east of England, but has not been recorded in Northern
Ireland in recent years.

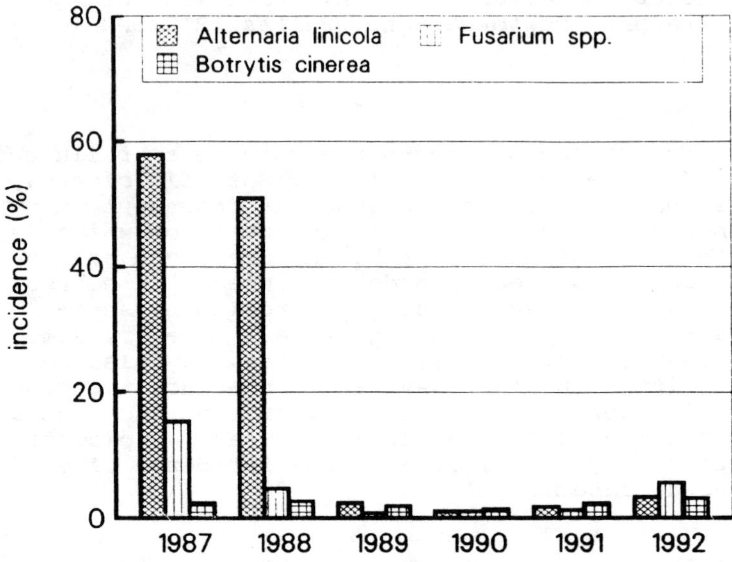

FIGURE 1. Mean pecentage incidence (when present) of seed-
borne pathogens on UK linseed seed from 1987 - 1992.

EPIDEMIOLOGY

<u>Pathogen transmission</u>

     Although the pathogens above have been classified as seed-
borne, many of them may also be transmitted in other ways.
Botrytis cinerea is a ubiquitous soil- and air-borne pathogen
and can infect the growing linseed crop each year independent of
the health-status of the seed, if weather conditions are
suitable (Mercer *et al.*, 1991). Similarly, *Fusarium* spp. have
been readily isolated in N. Ireland from lesions on roots of
linseed plants grown in the field from seed free of pathogens
(P.C. Mercer, unpublished). *Phoma exigua* var *linicola* has been
observed to cause severe damage in experimental plots in
N. Ireland (Mercer & Hardwick, 1993a) even though there was no
indication of its presence on the seed (P.C. Mercer,
unpublished). There was also evidence from a trial in
N. Ireland in 1993 of a low, but significant level of
transmission of *A. linicola* via the soil (P.C. Mercer,
unpublished), even though it is clear that the main method of
transmission is via the seed.

<u>Position of pathogens on the seed</u>

Most of the important pathogens of linseed capable of being seed-borne in the UK, are found in the seed coat (Mercer & Hardwick, 1991) where they appear to be present as resting hyphae. There is little evidence for the presence of spores or other propagules. Nor is there much evidence for fungal colonisation of the embryo. Ultrastructural studies have shown that the resting hyphae are located mainly in cells underlying the outermost gelatinous layer (P.C. Mercer, unpublished). From here they can rapidly resume growth, as the seed imbibes water, and grow out to infect the erstwhile sterile seedling.

<u>Host/pathogen interaction at germination</u>

The degree to which the pathogen is successful at colonising the seedling is dependent on weather conditions at the time of sowing. Low temperatures favour the pathogens at the expense of the host, while at higher temperatures, the host is able to grow sufficiently vigorously to "escape" what are generally relatively weak pathogens. An experiment carried out in a heated and an unheated glasshouse in N. Ireland with samples of seed with different levels of *Alternaria linicola* showed a higher incidence of the pathogen on the roots of seedlings germinated in the unheated glasshouse compared with the heated (Fig. 2). Percentage germination was also significantly poorer in the unheated house, although in this instance it was not significantly correlated with the percentage

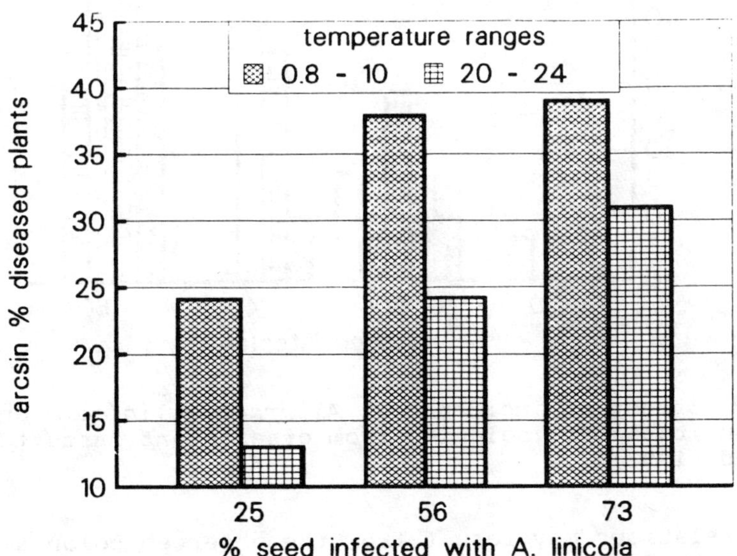

FIGURE 2. Effect of temperature at germination and percentage of seed infected with *Alternaria linicola* on the incidence of *A. linicola* on roots of 15 cm high seedlings.

incidence of *A. linicola* on the seed. However, this can occur
under field conditions - in a trial in N. Ireland in 1988 there
was a reduction in emergence of at least 50% resulting from
using seed with an incidence of 57% *A. linicola* (Mercer &
Hardwick, 1991). This led to a consequent drop in yield of 15%.

<u>Effect of seed-health on the growing crop</u>

        Although much of the effect of seed-borne pathogens is
observed at crop emergence, the growing crop is also subject to
attack from a range of pathogens, some of which are also capable
of being seed-borne, *e.g. Botrytis cinerea, Fusarium* spp. and
Phoma exigua var. *linicola*. There generally appears to be
little correlation between incidence of the pathogen in the seed
and incidence in the growing crop. However, disease-assessment
of a linseed trial at growth stage 50 (Freer, 1991) in
N. Ireland in 1993 (P.C. Mercer, unpublished) showed that the
incidence of *Alternaria linicola* on untreated seed could be
correlated positively with its later incidence on roots, stem-
bases and leaves, although not capsules (Fig. 3). An assessment
one week later indicated that there might even be a slight
correlation with capsule-colonisation.

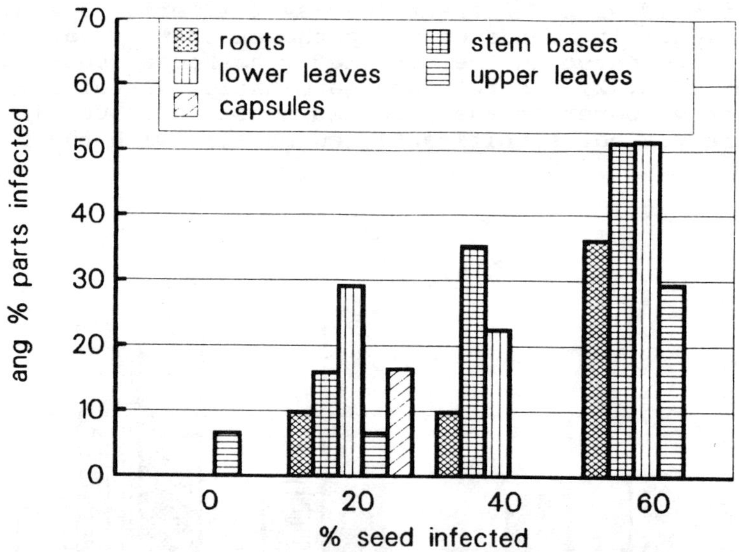

FIGURE 3. Effect of incidence of *Alternaria linicola* on seed of
linseed on its later isolation from other plant parts (GS 50,
N. Ireland, 1993).

These correlations may result from the observed colonisation
pattern of this pathogen, moving from lower to upper plant parts
with time (Fig. 4) but without large numbers of spores being
produced until the end of capsule production (Mercer *et al.*,
1992).

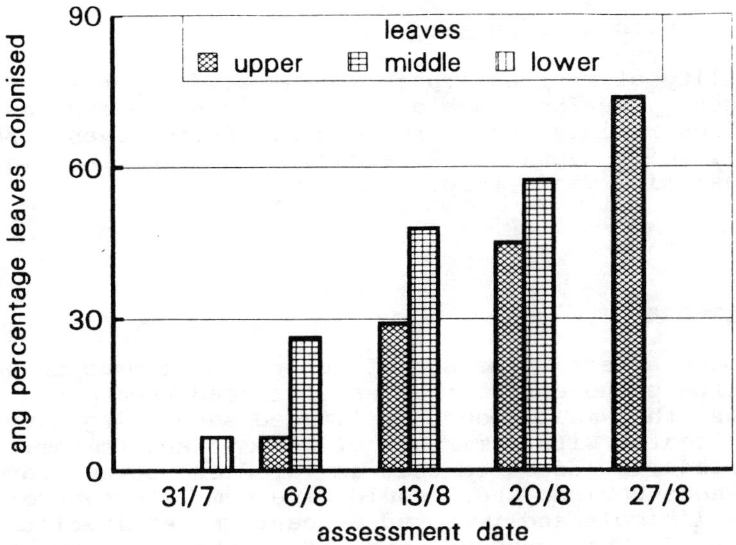

FIGURE 4.   Progressive colonisation of leaves by *Alternaria
linicola* on linseed plants in N. Ireland in 1991.

Infection of seed by pathogens

Seed-borne pathogens enter capsules and seeds as these
organs mature. Passage appears to be either through the
capsule's walls or central stalk. The means whereby the
actively growing pathogen becomes converted into resting hyphae
embedded in the seed's gelatinous layer is largely unknown.

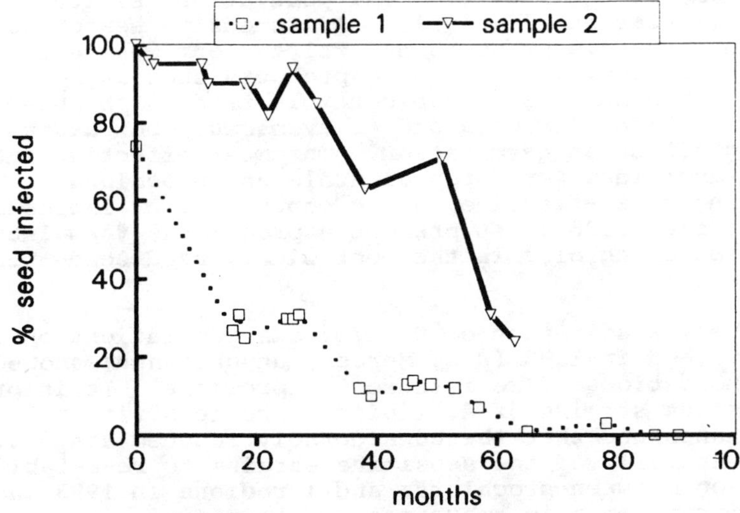

FIGURE 5.   Effect of time on *Alternaria linicola* viability in
seed (initial infections of samples: 1 - 74%; 2 - 100%).

<u>Viability of pathogens on seed</u>

Viability of resting hyphae in the seed depends strongly on the pathogen. The incidence of *Botrytis cinerea* and *Fusarium* spp. declines rapidly over a few months after harvest (Mercer *et al.*, 1991), while hyphae of *Alternaria linicola* can remain alive for at least five years (Fig. 5).

CONTROL

<u>Seed-treatments</u>

The most effective method of control of damage to seedlings by seed-borne pathogens is the use of a seed-treatment. In the early 1980s, the small amount of linseed seed being sown was generally treated with a mixture of benomyl and thiram. However, research indicated that an iprodione powder formulation (Rovral, Rhone-Poulenc Ltd.) could give complete control of *Alternaria linicola* and improved percentage germination (Mercer *et al.*, 1985). This product then became the industry standard.

However, at that point *A. linicola* was perceived as the most important seed-borne pathogen and research was concentrated on its control. In 1985, some seed had, as well as *A. linicola*, a relatively high incidence of *Fusarium avenaceum*, which because of its suppression by *A. linicola*, frequently only became apparent following application of iprodione (which controlled *A. linicola* only). Control of the *F. avenaceum* required the further addition of benomyl (Mercer & McGimpsey, 1987).

Resistance by *A. linicola* to iprodione appeared first in 1986 and spread rapidly so that by 1988, 85% of seed samples had at least some of their *A. linicola* population resistant to iprodione (Mercer *et al.*, 1991). An intensive search was then made for alternative products and thirty-four were examined (Mercer & Hardwick, 1993b). Some products such as fenpropimorph/benomyl and propiconazole/tridemorph showed good control of both *A. linicola* and *F. avenaceum*, but had an inhibiting effect on germination. The most effective and least damaging fungicides were propiconazole and prochloraz, the latter being more effective in the control of *A. linicola* (Mercer *et al.*, 1988). At present a prochloraz formulation (Prelude, Schering plc) is the most widely used seed-treatment on UK linseed.

However, examination of *A. linicola* populations on random samples of seed in 1993 (P.C. Mercer, unpublished) showed no further indications of resistance to iprodione. It is probable that resistant strains of *A. linicola* are less fit than sensitive ones and that the considerably reduced usage of iprodione has allowed the sensitive strains to re-establish. A comparison between prochloraz and iprodione in 1993 showed similar improvements in emergence and reductions in *A. linicola* over an untreated control (Fig. 6).

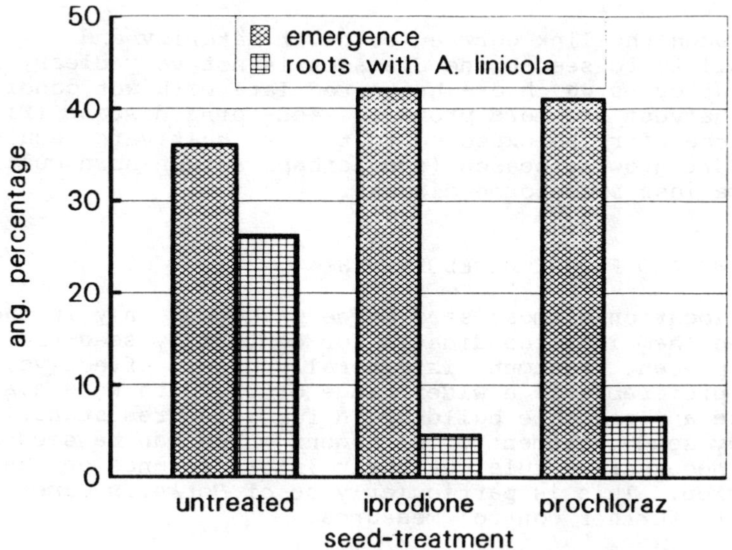

FIGURE 6. Effect of seed-treatment on emergence and root infection (GS 15) by *Alternaria linicola* following the use of seed which had an incidence of 60% *A. linicola* (N. Ireland, 1993).

Crop spraying to reduce seed-borne disease

The application of fungicide sprays to the growing crop has been shown to reduce the incidence of some seed-borne pathogens, *e.g. A. linicola* (Fitt & Ferguson, 1990; Mercer & Ruddock, 1993) and *B. cinerea* (Mercer & Hardwick, 1991). However, the reduction in incidence of *A. linicola* is variable and frequently not below the 5% level required for seed-certification. Further, although fungicide sprays do reduce *B. cinerea* on the growing crop and can increase yields, the tendency for badly infected capsules to fall to the ground may sometimes result in a negative relationship between the incidence on the crop and on the seed (Thomas *et al.*, 1993). The economics of linseed, at present, will only allow for a single fungicide spray and it is not yet clear when this is most effectively applied (Mercer & Hardwick, 1991).

Use of cultivars to control seed-borne disease

Some cultivars are more resistant to some of the seed-borne diseases than others. For example, the incidence of *A. linicola* on the capsules and seeds of cv. Andro was much lower than that on other cultivars in a trial in N. Ireland in 1992 (Mercer & Ruddock 1993). There are also indications of cultivar effects in susceptibility to *B. cinerea* (Thomas *et al.*, 1993), although, as observed above, incidence on the growing crop does not always correlate with that on the seed.

Although the link between cultivar maturity and
susceptibility to seed-borne diseases is not very clear, it does
appear that crops which are harvested late with wet conditions
prior to harvest are more prone to seed-borne disease (Fig. 1).
It would therefore be expected that those cultivars requiring
only a short growing season (and perhaps autumn-sown cultivars)
would have less seed-borne disease.

CONCLUSIONS AND FUTURE DEVELOPMENTS

The location of most seed-borne pathogens only in the seed
coat makes them ideal candidates for control by seed-treatment.
Although present treatment is generally highly effective, it
would be preferable if a wider range of products were available
to obviate any possible build-up in fungicide-resistance.
Control by seed-treatment of pathogens which can be seed-borne
does not necessarily rule out their later presence on the
growing crop. This is particularly so of *Botrytis cinerea* which
may require further control measures.

In the future, varietal control may be more widely used for
disease control in linseed, and biocontrol seed-treatments may
also become available. However, a major requirement for the
evolution of the most effective disease-control measures is a
much fuller understanding of the epidemiology of the seed-borne
pathogens.

REFERENCES

Fitt, B.D.L. & Ferguson, A.W. (1990) Responses to pathogen and
    pest control in linseed. *Brighton Crop Protection
    Conference - Pests and Diseases 1990*, **2**, 733-738.
Fitt, B.D.L., Ferguson, A.W., Dhua, U. & Burhenne, S. (1991)
    Epidemiology of *Alternaria* spp. on linseed. *Aspects of
    Applied Biology, Proceedings of Conference on Production
    and Protection of Linseed*, **28**, 95-100.
Freer, J.B.S. (1991) A development stage key for linseed (*Linum
    usitatissimum*). *Aspects of Applied Biology, Proceedings of
    Conference on Production and Protection of Linseed*, **28**, 33-
    40.
Mercer, P.C. & Hardwick, N.V. (1991) Control of seed-borne
    diseases of linseed. *Aspects of Applied Biology,
    Proceedings of Conference on Production and Protection of
    Linseed*, **28**, 71-78.
Mercer, P.C. & Hardwick , N.V. (1993a) Methods of control of
    diseases of linseed. *Proceedings of Crop Protection in
    Northern Britain 1993*, 165-170.
Mercer, P.C. & Hardwick, N.V. (1993b) Chemical control of seed-
    borne diseases of linseed in the UK. *Proceedings of the
    International Organisation for Biological Control (West
    Palearctic Region) Rennes, France 1992* (in press).
Mercer, P.C. & McGimpsey, H.C. (1987) Joint control of
    Alternaria linicola and *Fusarium avenaceum* on the seed of
    linseed. *Tests of Agrochemicals and Cultivars (Annals of
    Applied Biology 110*, Supplement) **8**, 54-55.

Mercer, P.C. & Ruddock, A. (1993) Effects of cultivars and
     iprodione on the incidence of *Alternaria linicola* in
     capsules and seeds of linseed in Northern Ireland. *Tests
     of Agrochemicals and Cultivars (Annals of Applied Biology
     122, Supplement)* **14**, 146-147.
Mercer, P.C., McGimpsey, H.C. & Ruddock, A. (1988) The control
     of seed-borne pathogens of linseed by seed-treatments.
     *Tests of Agrochemicals and Cultivars (Annals of Applied
     Biology 112, Supplement)* **9**, 30-31.
Mercer, P.C., Hardwick, N.V., Fitt, B.D.L & Sweet, J.B. (1991)
     Status of diseases of linseed in the UK. *Home-Grown
     Cereals Authority Research Review OS3*, 76 pp.
Mercer, P.C., McGimpsey, H.C., Black, R. & Norrie, S. (1985)
     The chemical control of *Alternaria linicola* on the seed of
     linseed. *Tests of Agrochemicals and Cultivars (Annals of
     Applied Biology 106, Supplement)* **6**, 56-57.
Mercer, P.C., Ruddock, A., Fitt, B.D.L & Harold, J.F.S. (1992)
     Linseed diseases in the UK and their control. *Brighton
     Crop Protection Conference - Pests and Diseases 1992,* **3**,
     921-930.
Thomas, J.E., Mercer, P.C. & Ruddock, A. (1993) Incidence of
     Botrytis cinerea on linseed cultivars at sites in England
     and Northern Ireland. *Tests of Agrochemicals and Cultivars
     (Annals of Applied Biology 122, Supplement)* **14**, 150-151.

MERCURY-BASED SEED TREATMENTS FOR CONTROL OF BACTERIAL BLIGHT IN COTTON

M.A.T. POSWAL[‡]

Department of Crop Protection, Ahmadu Bello University, Samaru- Zaria, Nigeria.

ABSTRACT

The evaluation of mercury based compounds in Nigeria for bacterial blight control commenced in 1953. Disease incidence, in some of the trials, were reduced by as much as 93%. Formulations of a mixture of phenyl mercuric acetate (PMA) and ethyl mercuric chloride (EMC) (3% or 5% Hg), were then recommended for treating all cottonseed used by farmers. However, these mercurials exhibited high mammalian toxicity and phytotoxicity. Cuprous oxide (45% Cu), a less effective bactericide, was also recommended. Based on trials conducted between 1967 and 1970, bronopol (12%), a less poisonous and effective bactericide, was recommended as an alternative to the mercurial and copper formulations. Recent experiments have found liquid formulations containing 30% or 60% of 2-(thiocyanomethylthio)-benzothiazole (TCMTB) and acid-treatment to be as effective as bronopol in disease control. The problems that were associated with the use of mercury based seed treatments by Nigerian cotton farmers are discussed.

INTRODUCTION

Bacterial blight (<u>Xanthomonas</u> <u>campestris</u> pv. <u>malvacearum)</u> is the major disease of cotton in Nigeria. Annual yield loss is estimated at 10 -20% (Dransfield, 1965). Integrated management strategy for the disease involves the combined use of field sanitation, partially resistant cultivars and chemical seed treatment. The evaluation of seed treatment chemicals commenced in 1953 at the now Institute for Agricultural Research, Samaru, with particular attention being paid to their efficacy against bacterial blight, cost effectiveness and levels of both mammalian toxicity and phytotoxicity (Dransfield, 1968). This paper reports the contributions of mercury based seed treatment chemicals and their alternatives in bacterial blight control and cotton production in Nigeria and highlights some of the socio-economic and technical problems associated with their use.

[‡]Present address: Department of Agronomy, University of Fort Hare, Private Bag X1314, Alice 5700, South Africa

MATERIALS AND METHODS

Over 60 different seed treatment chemicals were evaluated between 1953 and 1967 (Dransfield, 1968). Organo-mercurial dust formulations tested included: PMA - phenyl mercuric acetate + EMC - ethyl mercuric chloride (Agrosan 3W, 3% Hg), PMA + EMC (Agrosan 5W, 5% Hg), MC - mercuric chloride + MI - mercuric iodide (Abavit B, 8.4% Hg), PMU - phenyl mercury urea (Abavit B red, 1.0% Hg), M - mercury (1.25%) + D - dieldrin (75%) (Dieldrex A (1.25% Hg), PMA (Leytosan P, 3.0 - 7.6% Hg), MMS - methoxyethyl mercuric silicate (Leytosan M, 3.0 - 7.6% Hg) and EMC + PMA (Leytosan E, 3.0 - 7.6% Hg). The three organo-mercurial liquid formulations assessed were MMN - methyl mercury nitrile (Agrosol, 1.0% Hg), CMG - cyanomethyl - mercury - guanidine (Panogen, 1.5% Hg) and PMAA - phenyl mercuric ammonium acetate (Mist-O-Matic, 4% Hg). Copper formulations tested were cuprous oxide (Shell Cu_2O, 50% Cu) and cuprous oxide (Cuprocot, 45% Cu). The dust and liquid mercury formulations were applied at a dosage rate (weight of chemical: weight of seed) varying from 1: 100 - 1: 400. The dosage rate of the copper formulations ranged from 1: 100 - 1: 300. Machine delinted (fuzzy) Samaru 26C seed was used in 1953, and Samaru 26 J seed subsequently. To ensure proper seed coating, seeds and chemicals were vigorously shaken in closed containers for 5 minutes. Replicated field trials were usually sown (June or July) in randomized block designs (Dransfield, 1968). Plots consisted of one, two or three ridges (10m x 0.9m). Six seeds of each treatment were sown per hill/stand, using a within-row spacing of between 15 - 45 cm. Parameters assessed for each treatment were stand counts, seedling emergence, and bacterial blight incidence. Proportional reduction in bacterial blight - a measure of the efficiency of each treatment - was expressed as the mean ratio of the percent diseased plants in treated plots to that of the untreated plots.

Trials between 1967 and 1972 were conducted with non-mercurial seed treatment chemicals. Two new dust formulations, Bronocot (12% bronopol) and Unicot (40% cufraneb), were compared with standard recommended seed treatment chemicals, PMA + EMC (3% Hg) and cuprous oxide (45% Cu), based on the 1953 - 1967 experiments (Dransfield & Beeden, 1974). The new chemicals were evaluated at a dosage rate of between 1: 100 and 1: 250 w/w. Seeds of Samaru 26J were used. Experimental layouts were generally similar to those of the 1953 - 67 trials. However, yield trials were conducted on larger plots (up to 200m²). Parameters evaluated were germination, disease incidence and severity, and yield of seedcotton.

Over ten non-mercurial liquid and dust/powder chemicals were compared with bronopol in laboratory, glasshouse and field tests, between 1980 and 1989 (Poswal & Erinle, 1987; Poswal et al., 1992). The effect of acid-delinted seeds, with or without chemical seed treatment, was also evaluated. Machine - and acid-delinted seeds of Samaru 71, Samaru 72 and Samaru 77 were used. All new chemicals were tested at the manufacturers recommended rates. Liquid formulations were applied, as slurries, by firstly dispersing the recommended rates in quantities of water equivalent to 2.0% of the weight of seed treated. Cotton seeds and the chemicals were placed in polythene bags and agitated until the seed was uniformly coated. The procedure of Cross (1962) for acid-treatment of machine-delinted seed was used. Seed-seedling parameters evaluated in the laboratory and glasshouse tests were germination and seedling emergence, radicle damage, root length and seedling height. Seedling emergence, disease incidence and severity, and yield of seedcotton were parameters assessed in the field.

RESULTS AND DISCUSSION

The relative efficacy of some of the over 60 seed treatment chemicals tested between 1953 and 1967 is presented in Table 1. The dust mercurials, represented by the PMA + EMC, PMA and MMS formulations were the most effective, compared to the non-mercurials (cuprous oxide). A reduction in the application rates from 1: 100 to 1: 400 lead to reduced bacterial blight control. PMA + EMC (5% Hg) at the rate of 1: 150, was adopted as the standard commercial treatment until 1966 (Dransfield, 1968).

TABLE 1. Summary of the relative efficiency of some of the seed treatment chemicals evaluated between 1953 and 1967.

Seed treatment	Dosage rate	Rank order	Percentage* efficiency
PMA + EMC (5% Hg)	1 : 150	1	96.4
	1 : 200	3	96.0
	1 : 250	32	86.8
	1 : 300	30	87.6
PMA + EMC (3% Hg)	1 : 150	12	93.7
	1 : 250	43	73.3
PMA	1 : 150	5	95.6
MMS	1 : 150	6	95.2
EMC + PMA	1 : 150	13	93.1
M + D (1.25% Hg)	1 : 150	26	89.2
	1 : 100	4	95.8
CMG (1.5% Hg)	1 : 200	15	91.9
	1 : 300/400	33	86.3
	1 : 100	11	94.2
PMAA (4% Hg)	1 : 200	21	90.4
	1 : 300/400	40	79.6
	1 : 150	38	80.5
Cuprous oxide (45% Cu)	1 : 200	46	66.4
	1 : 150	45	69.2
Cuprous oxide (50% Cu)	1 : 250	56	48.7

* Percentage efficiency = 100 - (Mean ratio x 100); Mean ratio = mean of ratio of percent diseased plants in treated to untreated plots).
Source: Dransfield (1968)

In 1966 the rate of application of PMA + EMC (5% Hg) was reduced to 1: 200, and disease incidence varied from 2.7% to 4.8% and from 96.7% to 100% for the untreated control. By 1967, PMA + EMC (3% Hg) (1: 150) replaced PMA + EMC (5% Hg) (1: 200) due to the comparable level of disease control recorded between 1965 and 1967. The PMA + EMC (3% Hg) dosage rate was further reduced to 1: 170. None of the liquid mercurial treatments were recommended for commercial use. This was due to the high degree of phytotoxicity expressed by MMN (1.0% Hg) and CMG (1.5% Hg). No phytotoxic effects were observed with PMAA or with PMA + EMC formulations. In addition, machines for treating liquid treatments became clogged and choked by damp lint. Consequently the objective of testing liquid treatments as alternatives to the dust chemicals, given their health hazards during the seed ginning and treatment process, was abandoned. Cuprous oxide (45% Cu), a less efficient but safer treatment, was also recommended.

By 1966, copper treatments were increasing in cost and the disadvantage of organo-mercurial treatments became widely known and a matter of concern to the Nigerian government. Intensive research between 1967 and 1970 with substitute formulations, containing other active ingredients, lead to the recommendation of bronopol. Differences between bronopol and PMA + EMC (3% Hg) were not significant; however, PMA + EMC (3% Hg) consistently gave higher seed germination and lower disease incidence and severity (Table 2). Similarly, increase in seedcotton yields, over the control, were 10.5%, 11.7% and 10.2% for PMA + EMC (3% Hg), cuprous oxide (45% Cu) and bronopol, respectively (Table 3). In five year trials, (Dransfield & Beeden, 1974) showed that cufraneb (40%) at a dosage rate of 1: 150 was as effective as the mercurial and bronopol formulations, and more effective than the copper treatments (Table 4). Furthermore, no evidence of phytotoxicity was indicated. As a result of its low mammalian toxicity (acute oral LD_{50} = 2700 mg/kg) compared to those of PMA + EMC (3% Hg) (30 - 50 mg/kg) and bronopol (400 mg/kg), cufraneb was added to the list of recommended seed treatments.

TABLE 2. Bacterial blight in large-scale trials at Samaru in 1968-70

Seed treatment	Dosage rate (wt/wt)	Germination (%)	Disease incidence (%)	Mean disease score (0-5 Scale)
PMA + EMC (3% Hg)	1 : 150	67	0.7	0.01
Bronopol	1 : 150	59	1.3	0.03
	1 : 160	59	2.2	0.05
"	1 : 170	60	2.5	0.05
"	1 : 180	59	3.0	0.06
"	1 : 190	58	3.9	0.09
"	1 : 200	57	4.5	0.10
"	1 : 225	60	6.9	0.17
"	1 : 250	60	8.2	0.19
Untreated control	–	61	69.2	2.05

Source: Dransfield (1971)

Poswal & Erinle (1987) highlighted certain problems associated with the practice of producing and distributing treated cotton seed in Nigeria. The use of dust chemicals consistently creates potential health hazards, while poor storage and handling of treated machine-delinted seed may also lead to excessive loss of the chemicals before planting. As a result of the bulk packaging (in 25 - 50kg jute or polypropylene bags), there is frequent wastage and misuse of treated seed. Most disturbing, however, is the practice of washing and feeding Bronocot treated seeds to livestock by many farmers. The defunct Nigerian Cotton Board then began investigations into the feasibility of replacing the old plantector and drum ginning machines with mechanical and slurry seed treaters. Consequently, the evaluation of non-mercurial liquid bactericides, Busan 30 and Busan 72, containing 30% and 60% of TCMTB, respectively, was initiated in 1980. When compared to bronopol, no significant differences were recorded (Table 5). This preliminary trial suggested that TCMTB (30%) is a potentially good alternative to bronopol (Poswal & Erinle, 1987).

TABLE 3. Yields of seedcotton in 1968 and 1969 trials (sown mid- June and sprayed against insects), combined over five locations.

Seed treatment	Yields (kg/ha)	
	Mean	% increase over control
12% Bronopol	985	10.2
45% Copper	999	11.7
3% Mercury	988	10.5
Control	894	–

Source: Dransfield (1971)

TABLE 4. Relative efficiency of cufraneb for bacterial blight control at Samaru in 1972.

Seed treatment	Germination (%)		Disease incidence (%)	Mean disease score (0-5 scale)
	7 days	13 days		
PMA + EMC (3% Hg)	71	98	3.4	0.06
Cufraneb	80	97	1.8	0.04
Bronopol	72	98	3.5	0.08
Untreated	77	98	65.3	2.32
S.E ±	6.4	1.2	2.44	0.051

Source: Dransfield & Beeden (1974)

Laboratory, glasshouse and field trials with non-treated and chemically treated acid-delinted seed lots gave significantly higher seed germination and seedling emergence, and more vigorous seedlings than machine-delinted seeds. The incidence and severity of bacterial blight was also reduced and seed cotton yields increased (Table 6).

TABLE 5. The effect of TCMTB (3%) and TCMTB (60%) on the development of bacterial blight in 1980.

Bactericidal treatment	Dosage rate (A.I.kg/ seed)	Germination (%)	Disease incidence (%)	Disease severity (0-5 scale)
TCMTB (60%)	9.3 ml	58.2	23.2	0.42
TCMTB (30%)	18.0 ml	58.5	19.4	0.30
Bronopol	20.0 g	59.6	17.8	0.29
Untreated	–	53.1	74.1	1.80
L.S.D. (0.05)		N.S.	6.84	0.16

Poswal & Erinle (1987)

It is hoped that cotton production in Nigeria could be enhanced if farmers are issued with acid-delinted and chemically treated seeds in small polythene or polypropylene bags, commensurate with their yearly requirements. This should eliminate some of the problems and misuse associated with machine-delinted seed (Poswal <u>et al</u>., 1992).

TABLE 6. Effect of delinting method on bacterial blight incidence/ severity, seedling emergence and seedcotton yield.

Delinting method	Bacterial blight		Seedling emergence (%)	Seedcotton yield (kg/ha)
	Incidence (%)	Severity (0-5 scale)		
Machine	86.3	1.86	42.3	1347
Acid	51.8	0.84	64.0	1656

Source: Poswal (1987)

REFERENCES

Cross, J.E. (1962). An improved method of acid-treating small quantities of cotton seed. *Empire Cotton Growing* Review, **39**, 205.

Dransfield, M. (1965). Cotton seed dressing in northern Nigeria. *Samaru Research Bulletin*, **48**, 261-265.

Dransfield, M. (1968). Seed dressing trials on cotton in northern Nigeria, 1953-1967. *Samaru Miscellaneous Paper*, 26, 1-22.

Dransfield, M. (1971). Seed dressing trials to control bacterial blight of cotton (*Xanthomonas malvacearum*) in the northern States of Nigeria. *Proceedings of the Sixth British Insecticide and Fungicide Conference*, 225-230.

Dransfield, M.; Beeden, P. (1974). A new cotton dressing for northern Nigeria. *Samaru Miscellaneous Paper*, **46**, 1-9.

Poswal, M.A.T. (1987). Survival of *Xanthomonas campestris* pv. *malveacarum* and performance of seedlings from commercially ginned cottonseed as influenced by length of storage. In: *Plant Pathogenic Bacteria*, E.L. Civerolo, A. Collmer, R.E. Davis and A.G. Gillapsie (Eds), Netherlands: Nijhoff Publishers, pp. 741-745.

Poswal, M.A.T.; Erinle, I.D. (1987). Comparative study of the efficacy of Busan 30, Busan 72 and Bronocot for the control of bacterial blight of cotton in northern Nigeria. In: *Plant Pathogenic Bacteria*, E.L. Civerolo, A. Collmer, R.E. Davis and A.G. Gillapsie (Eds), Netherlands: Nijhoff Publishers, pp. 976-981.

Poswal, M.A.T.; Atangs, P.A.; Akpa, A.D. (1992). Laboratory and glasshouse evaluation of seed treatment chemicals in relation to some seed-seedling parameters in cotton. *Seed Science and Technology*, **20**, 69-76.

FILM-COATING OF LEEK SEEDS WITH INSECTICIDES: EFFECTS ON GERMINATION AND ON
THE CONTROL OF ONION FLY (*DELIA ANTIQUA* (MEIGEN))

A. ESTER

Research Station for Arable Farming and Field Production of Vegetables (PAGV),
8200 AK Lelystad, The Netherlands

R. DE VOGEL

Nunhems Zaden BV, 6080 AA Haelen, The Netherlands

ABSTRACT

Field experiments were carried out in 1991 and 1992 to assess the control of the
onion fly (*Delia antiqua*) in a winter leek crop (*Allium porrum* L.) by film-coating the
seeds with insecticides. Plants were raised in a seed-bed.
The efficacy of benfuracarb, carbofuran, imidacloprid and isofenphos at two
rates as seed film-coatings was compared to a conventional application of chloor-
fenvinphos to the seed-bed plus two spray applications with carbufuran. Germination of only benfuracarb film-coated seed was comparable to the untreated controls in all tests. Control of onion fly by benfuracarb applied as a filmcoating was as
effective as the conventional application.
This seed treatment will reduce the necessary amount of insecticide for control of
the onion fly in a winter leek crop by 98% to 3 g per 100 m^2 seed-bed.

INTRODUCTION

Increasing costs and increasing concern about the environmental impact of insecticides
have produced a need to apply insecticides more economically and efficiently (Halmer, 1988). In
controlling the cabbage root fly (*Delia radicum*) a significant reduction in the necessary quantity of
insecticide was achieved by applying the insecticides as a film-coating to cabbage seed (Ester &
De Moel, 1992). This paper reports on controlling the onion fly (*Delia antiqua* (Meigen)) in leek with
a reduced amount of insecticides. Onion seeds filmcoated with the insecticide benfuracarb have
been on the market in The Netherlands for some years. Narkiewicz-Jodko (1991) reported successful control of onion fly in onion by applying carbosulfan or isofenphos as a seedcoating. In a
leek crop the onion fly is mainly a problem in seed beds where plants are raised at high densities.
Generally transplanting to the production field takes place approximately 12 weeks after sowing.
The onion fly attacks the seedling by hollowing out the basal part of the plant resulting in its
collapse. Due to the high plant density in the seed bed, neighbouring plants are easily attacked
too, resulting in patches of collapsed plants. In the production field plant spacing is much wider
and the onion fly is not a real problem. Therefore the effects of a seed film-coating were investigated after sowing in seed beds, according to common practise. A preliminary report has been
published (Ester, *et al* 1992).

MATERIALS AND METHODS

All experiments were carried out with the winter leek variety Porino. Film-coating of the
seed was done by SUET (Saat- und Erntetechnik, Eschwege, Germany) using a fluidised bed film-
coating technique. The film-coating contained polymers to give a dust free product. In order to
obtain the same amount of insecticide per seed, rates are expressed per unit of seed, with one
unit equalling 250,000 seeds. All treatments contained the same amount of fungicide, namely

thiram at 1 g AI per unit of seed, except the isofenphos/thiram treatment. For this treatment a combined formulation of isofenphos and thiram has been used (powder for dry seed treatment DS 40/10%, containing 40% isofenphos and 10% thiram). The thiram rates were 0.9, 1.4 and 3.6 g AI/unit of seed for the 7, 11 and 14 g AI rates of isofenphos respectively. Untreated seeds were treated as a film-coating with thiram only. Four insecticides were applied at different rates. Benfuracarb (wettable powder WP 40%) was applied at 20, 27 or 40 g AI per unit seed, carbofuran (suspension concentrate 500 SC) at 27 g AI per unit of seed, imidacloprid (water dispersible powder for slurry treatment, 70% WS) at 14 or 28 g AI per unit of seed and isofenphos at 7, 11 or 14 g AI per unit of seed.

All germination tests in the laboratory were carried out in silver sand (sand test) or in a peatbased potting compost (soil test), using 3 replications of 100 seeds. Trays were placed in germination cabinets at 15°C night and 20°C day. The percentage of normal plants emerging was assessed after 16 days (sand test) or 15 days (soil test).

Insecticide efficacy and/or emergence experiments were carried out at four field sites in The Netherlands with a history of onion fly attack: Rijsbergen (1991; 1992), Berkel-Enschot (1992), Tollebeek (1992) and Haelen (1992). The control treatment in the seed bed consisted of a soil treatment of chlorfenvinphos at 6 l AI per ha and two spray applications with carbofuran at 4.4 l AI per ha. The soil treatment was sprayed and incorporated just before sowing, followed by crop sprayings six and twelve weeks later. Each replicate included one plot without insecticide treatment i.e. film-coated seeds with thiram only. All the experiments were sown with a "Nibex" hand-sowing machine and carried out as a randomized block design.

At Rijsbergen site, where the soil was sandy, the treatments were randomized within three replicates (1991; 1992). Each plot consisted of 12 rows (11 cm between rows) of 4.5 m length in 1991 and 13 rows of 2.50 m length in 1992. The seeds were sown in mid-April in both years.

At Berkel-Enschot site, also a sandy soil, the treatments were randomized within four replicates. Each plot consisted of 14 rows (15 cm between rows) of 3.25 m length. The seeds were sown in mid-April 1992.

At Tollebeek site, a marine loam soil, the treatments were randomized within three replicates. Each plot consisted of 12 rows (14 cm between rows) of 3.75 length. The seeds were sown at the end of April 1992.

At Haelen site, another sandy soil, trials were especially designed to assess field emergence of the insecticide-coated seeds compared to seeds without any insecticide. For each treatment 200 seeds were hand-sown in one row of 3 m length. A fully randomized block design was used with 4 replicates. The seeds were sown at the end of April 1992. The number of emerged plants in each row was assessed four weeks after sowing.

Field emergence on the seedling-bed was assessed by counting six rows of one meter per replicate at Rijsbergen in 1991. In 1992 field emergence was not assessed except at Haelen, due to a defect with the sowing machine.

The damage to all the crops by the onion fly was assessed regularly between six and eleven weeks after sowing by observing the percentage of collapsed plants.

The statistical analysis was performed with the statistical package Genstat.

RESULTS

<u>Germination</u>
Film-coated seeds without insecticide showed no reduced germination or field emergence compared to non-filmcoated seeds (data not shown). Seeds film-coated with benfuracarb and carbofuran in both 1991 and 1992 did not differ in percentage normal plants in laboratory germination tests carried out in 1992 as compared to the control film-coated seeds. Imidacloprid at 28 g AI and isofenphos at 11 g AI and 14 g AI/unit of seed significantly lowered the percentage of normal plants in the sand test in comparison with untreated film-coated seeds. In the soil test only isofenphos-treated seeds showed a significantly lower percentage of normal plants (Table 1.).

TABLE 1. Laboratory germination of filmcoated winter leek seeds in 1992. Percentage normal plants after 16 days in the sand test and after 15 days in the soil test.

Insecticides	Rate g AI/unit	Sand test 1992[**]	Sand test 1992	Soil test 1992
untreated	-	92	93	91
benfuracarb	20	93	92	92
	30	-	92	90
carbofuran	27	-	91	-
imidacloprid	28	78	87	90
isofenphos	7	89	89	85
	11	-	87	-
	14	85	-	-
LSD (p=0.05)		4.8	4.8	4.6

[**] seed treated in 1991, tested one year later.

In 1991 only benfuracarb at 40 g AI/unit of seed showed significantly lower field emergence compared to the control film-coated seeds. In 1992 only the field emergence of benfuracarb at 20 g AI/unit of seed and imidacloprid-treated seed was not significantly different from the untreated seed (Table 2). Because of a defect in the sowing machine a reliable assessment of the field emergence could be made only at the Haelen site.

TABLE 2. Field emergence of winter leek seeds. Number of seedlings per meter row length four weeks after sowing in Rijsbergen (1991) and Haelen (1992).

Insecticides	Rate g AI/unit	Number of plants 1991	Number of plants 1992
film-coating			
untreated	-	32	82
benfuracarb	20	30	78
	30	-	73
	40	27	-
imidacloprid	14	33	-
	28	32	80
isofenphos	7	34	68
	14	29	-
soil + crop treatment			
chlorfenvinphos + carbofuran	6+4.4[*]	33	-
LSD (p=0.05)		3.2	4.7

[*] g AI per hectare.

<u>Efficacy</u>

Both the soil plus crop treatment and the seed film-coatings reduced the onion fly damage significantly compared to the untreated control, except at the Rijsbergen site in 1992 (Table 3). In this trial carbofuran showed no difference from the untreated control after 10 weeks. Although the percentages of damaged plants were lower for all other seed coatings, these results were not significant. Control of the onion fly by all seed coatings was not significantly different from the standard soil plus crop treatment, except for the carbofuran coating at the Rijsbergen site in 1992.

TABLE 3. Efficacy of insecticides applied as a seed coating for onion fly control in leek. Percentages of damaged plants 14 weeks (1991) and 10 weeks (1992) after sowing.

Insecticides	Rate g Al/unit	Rijsbergen 1991	Rijsbergen 1992	Berkel Enschot 1992	Tollebeek 1992[***]	Tollebeek 1992
film-coating						
untreated		1.6	7.5	13.5	13.6	13.6
benfuracarb	20	0.03	-	-	0.1	-
	30	-	0.9	0.0	-	0.0
	40	0.03	-	-	-	-
carbofuran	27	-	10.6	1.3	-	0.8
imidacloprid	14	0.10	-	-	-	-
	28	0.03	0.6	0.9	-	0.1
isofenphos	7	0.03	0.3	0.1	0.0	0.0
	11	-	0.4	0.1	-	0.2
	14	0.03	-	-	-	-
soil + crop-treatment						
chlorfenvinphos +						
carbofuran[**]	6+4.4	0.03	0.4	0.0	0.0	0.5
LSD (α = 0.05)		0.70	7.39	3.76	5.26	5.26

[**] Control treatment g Al per ha.
[***] Seed treated 1991 and sown 1992.

DISCUSSION AND CONCLUSION

The aim of insect control is to avoid economic crop losses. This research demonstrates that, for the control of onion fly in leek, this can be achieved by seed coatings with insecticides. Of the compounds tested, carbofuran failed to give sufficient protection in one trial, possibly due to a longer onion fly flight. Benfuracarb, isofenphos and imidacloprid reduced the amount of onion fly attack to levels comparable to those achieved by the current standard treatment in The Netherlands, a soil treatment with chlorfenvinphos and two crop treatments with carbofuran.

In the laboratory tests phytotoxicity has been observed for the isofenphos seedcoatings at high rates and also at the lowest rate of 7 g Al/unit of seed in the field trials in 1992 .
As imidacloprid also showed a reduced laboratory germination, only benfuracarb at the lowest rate of 20 g Al/unit seed always gave adequate control without phytotoxicity.
Film-coating leek seeds with benfuracarb at the rate of 20 g Al/unit of seed illustrates clearly the possibility to reduce the amount of insecticides required to control the onion fly in a leek crop.
Compared to the conventional application, of soil treatment plus two spray applications, a seed-application will reduce the necessary amount of insecticide by 98% to 3 g per 100 m^2.

In The Netherlands, currently benfuracarb at the rate of 20 g Al/unit of seed has received a clearance to be used as a seed coating of leek. This method of control is recommended as part of the integrated pest management scheme for leek.

REFERENCES

Ester, A.; Embrechts, A.; Vogel, R. de; Schouten, K. (1992) Tegen zaadcoating kan uievlieg niet op. *Groenten en Fruit/Vollegrondsgroenten*, 44, 10-11.
Ester, A.; Moel, C.P. de (1992) Controlling cabbage root fly in Brussels sprouts by filmcoating seeds with insecticides. *Proceedings Experimental & Applied Entomology, NEV Amsterdam*, 3, 181-190.
Halmer, P. (1988) Technical and commercial aspects of seed pelleting and film-coating. In: *Applications to Seeds and Soil*, T.J. Martin (Ed.), *BCPC Monograph No. 39*. Thornton Heath: BCPC Publications, pp. 191-204.
Narkiewicz-Jodko, J. (1991) Effect of Marshal 25 ST Carbosulfan in control of onion fly *Hylemya antiqua* Meig. and Carrot fly *Psila rosae* F. *Mededelingen van de Faculteit Landbouwweten-schappen van Rijksuniversiteit Gent, 56/3b* 1143-1150.

ERADICATION OF *FUSARIUM* FROM OIL PALM BY SEED TREATMENTS.

J. FLOOD, R. MEPSTED, S. TURNER & R.M. COOPER

School of Biological Sciences, University of Bath, Claverton Down, Bath BA2 7AY

ABSTRACT

Fusarium oxysporum f.sp. *elaeidis* can be present naturally on or within oil palm seeds and a low proportion of infested seed gives rise to infected plants. Vacuum infiltration and soaking for 7 days with captafol or a formulation of prochloraz plus carbendazim effectively eradicated the pathogen. Seed treatment should be used as a precaution to reduce the possibility of the disease being introduced into previously disease free regions.

INTRODUCTION

Vascular wilt, caused by *Fusarium oxysporum* f.sp. *elaeidis* (*F.o.e.*) is the most serious disease of oil palm in West Africa (Turner, 1981). The disease appears to be absent from major producing areas in S.E. Asia such as Malaysia and Indonesia but has occurred in Brazil (Van de Lande, 1984) and Ecuador in 1986 (Renard & de Franqueville, 1989). The origin of these outbreaks is not known but seed-borne transmission is likely; the pathogen can be present along with *Fusarium solani* on the seed coat (Locke & Colhoun, 1974) or on the kernel surface inside seeds (Flood *et al.*, 1990). Also, isolates of the pathogen from Ivory Coast, Brazil and Ecuador are vegetatively compatible, according to results with nitrate non-utilising (*nit*) mutants which may indicate a common origin; seeds from Ivory Coast were used to plant some of the affected plantations in South America(Flood *et al.*,1992). The pathogen can survive normal routine seed processing (soaking for 7days at 25°C, heating for 75 days at 39°C followed by further soaking at 25°C for 7days)
and contaminated seed can give rise to infected plants albeit at a low frequency. Consequently, an effective fungicidal seed treatment was required which could be incorporated with normal seed processing.

EXPERIMENTAL

Natural contamination of oil palm seed by *F.o.e.* can vary considerably between seed consignments and between individual seeds (Flood *et al.*, 1990) and thus, seeds from susceptible crosses were artifically inoculated (Flood *et al.*,1994) to allow quantification and statistical analysis.

Quantification of *Fusarium oxysporum* on oil palm seeds

Following inoculation, levels of *F. oxysporum* on seed coats and kernel surfaces were quantified as mean colony forming units (cfus) per shell or per kernel using dilution and plating techniques (Flood *et al.*,1990).

Fungicide treatments

A suspension of captafol (Sanspor 50% a.i. ICI Agrochemicals) was prepared (1 g a.i./litre water plus 0.1ml Tween 20) and applied as either a 7 day soak (change of fungicide every day) or as a vacuum infiltration

treatment. In the latter treatment, captafol plus seeds were placed in a
vacuum chamber (Edwards-Pearce freeze drier) in which the air pressure was
reduced until the liquid began to boil (1000 - 1200 Pa). This process was
repeated three times with alternating repressuration. Following vacuum
infiltration, the seeds were soaked for 7 days with a daily change of
captafol.

Seeds were also vacuum treated with a suspension (1 g a.i./litre water
plus 0.1ml Tween 20) of benomyl (Benlate 50% a.i., Dupont) using the method
described above and then soaked for 7 days with a daily change of benomyl.

After fungicide treatment, the mean cfus of *Fusarium oxysporum* per
seed were determined and compared with those obtained after seed had
received a routine 7 day soak with water. Vacuum infiltration plus soaking
with captafol eradicated *F. oxysporum* from shells and kernels (Table 1) but
soaking alone with this fungicide failed to eradicate the pathogen from
kernel surfaces. Benomyl significantly reduced *F.oxysporum* from seed coats
and kernels but it failed to eradicate it.

TABLE 1 Levels of *Fusarium oxysporum* on seed coats
and kernels following fungicide treatments.

Treatment	Mean cfus/shell	Mean cfus/kernel
Captafol soak (daily change of fungicide for 7 days)	0^b	155^{ab}
Vacuum infiltration with captafol plus 7 day soak	0^b	0^b
Vacuum infiltration with benomyl plus 7 day soak	22^b	54^b
Water soak (daily change of water for 7 days)	2340^a	274^a

cfus = colony forng units (Flood *et al.*, 1989).
Values represent a mean of 14 replicates.
Within each column, values with the same letter are not
significantly different at 1% level using STD test for
non-parametric data (Sokal & Rohlf, 1981).

Captafol and a formulation of prochloraz and carbendazim (Sportak Alpha,
Schering Agrochemicals) were chosen for further experimentation as a post-
heat treatment in order that fungicide treatments could be incorporated
into routine seed processing protocols. Hence, the fungicides (1g
a.i./litre water plus 0.1ml Tween 20) were vacuum infiltrated into

inoculated seeds following heat treatment. The seeds were soaked for 7 days
(daily change of fungicide) and air dried. They were cracked aseptical
and shells and kernels plated onto *Fusarium* selective medium and incubated
for 14 days at 25°C. Following incubation, the presence of *F. oxysporum* on
seeds and kernels was determined and compared with seeds which had received
heat treatment and water soaking only. Vacuum infiltration with both
fungicides eradicated *F.oxysporum* from shells and kernels (Table 2).

TABLE 2 Presence of *F.oxysporum* on heat- treated inoculated seeds following
fungicide vacuum infiltration.

Treatment	Shell	Kernel
Control (7 day soak in water)	45^{+} a	45 a
Vacuum infiltration plus 7 day soak with captafol.	0 b	0 b
Vacuum infiltration and 7 day soak with prochloraz plus carbendazim.	0 b	0 b

+Values represent results from 50 replicate seeds.
Within each column, values followed by the same letter
are not significantly different using x^{2} analysis (< 0.01).

Effect of fungicide treatment on seed germination and on plant development
Artifically inoculated seeds which had been heat treated and then
treated with either fungicide were incubated in plastic bags at 25°C for
approximately 3 weeks. Germination rates of 60 seeds treated or not treated
with fungicide were very similar: 70% for fungicide treated seeds as
compared to 68% for untreated seeds (P>0.05 using x^{2} analysis). No
abnormalities in root or shoot development were evident following fungicide
treatment either when the young seedlings were transplanted into trays or
at the 1-2 leaf stage when the seedlings were transferred to individual
pots. All plants appeared to grow normally throughout the 9 month period of
the experiment.

DISCUSSION

Vascular wilt pathogens such as *Verticillium* and *Fusarium* are
generally considered as soil-borne fungi; their dissemination is mostly by
water or wind erosion of soil, irrigation or movement of man and equipment.
Consequently, their disease spread is ususally confined to a restricted
area. However, long distance movement can also occur through the movement

of infected vegetatively propagated material or seed and many of the *formae speciales* of *Fusarium oxysporum* have also been demonstrated to be seed-borne (Gambogi, 1983).

F. oxysporum f. sp *elaeidis* can be naturally present on the seed coats and kernel surfaces of oil palm seeds (Locke & Colhoun, 1974; Flood *et al.*, 1990) but embryo infection has not yet been demonstrated. Oil palm seeds routinely undergo heat treatment (at 39°C) to induce germination and high temperatures have proved to be effective seed treatments for the removal of several other *formae speciales* e.g. *lycopersici* (Besri, 1978). However, although heat treatment drastically reduced the populations of *F. oxysporum* and *F. solani* on oil palm shells and kernels, neither of these fungi were eradicated (Flood *et al.*,1994).

Also, *F. oxysporum* was reisolated from stem base tissue of 2 out of 60 plants grown from seeds inoculated with *F o e* and these plants developed characteristic vascular necrosis (Flood *et al* .,1994). Thus, with a low initial inoculum (mean 7 cfus per kernel and 40 cfus per shell) infected plants were produced, albeit at a low frequency. Although direct comparisons are difficult to make, contamination levels of up to 5 x 10^3 cfus per seed and up to 100 cfus per kernel have been reported from naturally contaminated seeds (Flood *et al.*, 1990).Thus, the pathogen may be introduced into new areas in this way as is likely to have already occurred in South America.

Consequently, as a precautionary measure, fungicide treatment of seeds is required and eradication of the pathogen on kernel surfaces is essential.

Vacuum infiltration of benomyl effectively reduced *Fusarium oxysporum* on both oil palm seed coats and kernels but it failed to eradicate these fungi. Haware *et al.* (1978) similarly reported that benomyl alone reduced *F. oxysporum* f. sp. *ciceri* from chickpea seed but it failed to eradicate the pathogen. In contrast, vacuum infiltration with captafol completely eradicated *F. oxysporum* from oil palm seed coats and kernels and had no phytotoxic effects on seed germination or seedling development, but captafol has been withdrawn from recommended use. Vacuum infiltration plus a 7 day soak with a formulation of prochloraz plus carbendazim was shown to be as effective as captafol and could be used for batches of seed exported from Africa to countries which are currently wilt-free such as Malaysia, Indonesia, Papua New Guinea and India.

ACKNOWLEDGEMENTS

The authors wish to thank Unilever Plantations Group for funding JF and RM. We would also like to thank staff at Joint Research Scheme (Plantation Lever au Zaire and Societe au Zaire) Binga, Zaire for their co-operation.
The novel seed treatment described in this paper is the subject of pending patent protection filed by Unilever.
The experiments were conducted under Licence PHF 275A/17 (3) from the Ministry of Agriculture, Fisheries and Food.

REFERENCES

Besri,M. (1978) Phases de la transmission de *Fusarium oxysporum* f.sp.
    lycopersici et de *Verticillium dahliae* par les semences de
    quelques varieties de tomate. *Phytopathologische
    Zeitschrift* **93**,148-163.

Flood, J.; Cooper, R.M.; Lees, P.E. (1989) An
    investigation of pathogenicity of four isolates of
    Fusarium oxysporum from South America, Africa and
    Malaysia to clonal oil palm. *Journal of Phytopathology*,
    124, 80-88.

Flood, J.; Mepsted, R.; Cooper, R.M. (1990) Contamination of
    oil palm pollen and seeds by *Fusarium* spp. *Mycological
    Research*, **94**, 708-709.

Flood, J.; Whitehead, D.S.; Cooper, R.M.(1992) Vegetative
    compatibility and DNA polymorphisms in *Fusarium oxysporum*
    f.sp.*elaeidis* and their relationship to isolate virulence and
    origin. *Physiological and Molecular Plant Pathology*,**41**,201-215.

Flood,J.;Mepsted,R.; Cooper.R.M.(1994) Population dynamics of *Fusarium*
    species on oil palm seeds following chemical and heat treatments.
    Plant Pathology (in press).

Gambogi,P.(1983) Seed transmission of *Fusarium oxysporum*: epidemiology
    and control. *Seed Science & Technology*, **11**, 815-827.

Haware, M.P.; Nene,Y.L.; Rajashwari,R.(1978) Eradication of *Fusarium*
    oxysporum f. sp. *ciceri* transmitted in chickpea seed. *Phytopathology*,
    68,1364-1367.

Locke, T.;Colhoun, J.(1974) *Fusarium oxysporum* f. sp. *elaeidis* as a seed
    -borne pathogen. *Transactions of the British Mycological Society* **60**,
    594-595.

Renard, J.L.; De Franqueville, H. (1989) Oil palm vascular wilt.
    Oleagineaux, **44**, 341-347.

Sokal,R.R.; Rohlf,F.J.(1981) *Biometry:The Principles and Practice of
    Statistics in Biological Research*. (2nd Edition) New York: Freeman and
    Company.

Turner, P.D.(1981) *Oil Palm Diseases and Disorders*.Oxford:
    Oxford University Press.

Van de Lande,H.(1984) Vascular wilt disease of oil palm (*Elaeis
    guineensis Jacq.*) in Para Brazil. *Oil Palm News*, **28**, 6-10.

FIELD EMERGENCE OF PEAS AS AFFECTED BY SEED QUALITY AND FUNGICIDE SEED TREATMENTS.

P.S.R. KOSTERS

S&G Seeds B.V., P.O. Box 26, 1600 AA Enkhuizen, the Netherlands

ABSTRACT

Laboratory germination and field emergence of pea (*Pisum sativum L.*) was studied in three seed lots using fungicide film-coating. The best field emergence was obtained with the lot which showed the best results in the laboratory germination tests, and which also had the lowest electroconductivity. The level of field emergence equalled the laboratory germination of the untreated seed if the seeds were film-coated with the fungicide combinations of oxadixyl + cymoxanil + carbendazim or metalaxyl + thiabendazol + thiram. The use of oxin-copper had no significant effect on field emergence if added to the mixtures. Thiram or thiram + oxin-copper did not lead to good field stands when used on the weaker lots.

INTRODUCTION

Field emergence of pea is important in relation to field establishment and final yield (Trawally *et al.*,1984). Emergence cannot be predicted from the standard germination test (Duczmal & Minicka, 1989). Field emergence is determined by seed quality , seed-borne pathogens and soil conditions (biotic and abiotic). The complex interactions between these factors make it impossible to study them separately. Fungicide seed treatments may contribute to a good field stand by protecting the germinating seed from seed borne and soil borne pathogens.

Several different seed treatments are commercially available for peas (F. Haquin, 1987). In this study we tested the effect of fungicides applied in a film-coating on laboratory germination and on field emergence. Root rot development was not assessed as several reports indicate that chemical control does not effectively control soil borne root rot (Kraft, 1982; van Loon & Oyarzun, 1988).

MATERIALS AND METHODS

Seed treatment

The three pea lots used in this study were produced in central France in 1992, and were selected for their differences in the levels of germination and the levels of seed borne organisms. There were three cultivars, Florado (lot A), Spartan (lot B) and Ninado (lot C); all these varieties had wrinkled seeds. Thousand seed weight was 174 g for lot A, 158 g for lot B and 105 g for lot C. Seeds were tested for levels of seed borne infection by "Ministere de l'Agriculture et de la fôret service de la protection des vegetaux" Angers (France) in september 1992. Seeds were film-coated with 6 fungicide combinations (table 1) in batches of 400 g using a laboratory fluidised bed coater. Polymers were used to give a uniform and dust free product.

The composition of the fungicide formulations used was; Pulsan TS Pepite (40% oxadixyl, 16% cymoxanil), Bavistine (50% carbendazim), Apron combi 453 (233 g/l metalaxyl, 120 g/l thiabendazol, 100 g/l thiram), Aatiram 75 (75% thiram), Quinolate Pro FL (120 g/l carbendazim, 120 g/l oxin-copper), Quinolate 400 (400 g/l oxin-copper). Oxin-copper was used as a split factor in the trial.

TABLE 1. Fungicide seed treatments of pea seeds in gram AI per kg of seed.

Treatment AI	1	2	3	4	5	6
oxadixyl	0.60	-	-	0.60	-	-
cymoxanil	0.24	-	-	0.24	-	-
carbendazim	1.00	-	-	0.30	-	0.30
metalaxyl	-	0.80	-	-	0.80	-
thiabendazol	-	0.41	-	-	0.41	-
thiram	-	0.34	0.30	-	0.34	0.30
oxin-copper	-	-	-	0.30	0.30	0.30

Germination tests

Seeds were germinated in 4 replications of 50 seeds according to the ISTA method (sand/perlite at 20°C) and in a cold test (sand/perlite for 10 days at 8°C followed by 20°C). Seeds were sown in the field by hand in a completely randomised design in 7 replications of 100 seeds on April 1, 1993. Observations on emergence were done on 20 days after sowing (total plants), 26 days after sowing (total plants) and 50 days after sowing (normal plants). The soil was a sandy clay in Enkhuizen. Mean day temperatures were below 10 °C during the first three weeks after sowing.

Conductivity was determined in a bulk conductivity test with 2 replicates of 50 seeds at 20 °C for the untreated seeds.

RESULTS

Levels of seed infection are given in table 2. Laboratory germination under standard conditions (table 3) was not affected by the fungicide treatments for lot B and lot C. For lot A, treatments 1 and 4 showed lower germination than the control; treatments 5 and 6 germinated better. In the cold test all fungicide treatments had a negative effect on germination for lot B; germination of lot C was not affected. The effects for lot A were variable; compared to the control germination was lower in treatment 4, and better in treatments 2,3 and 5.

TABLE 2. Percentage fungus infected seeds in the seed lots used in the trials.

fungus	lot A	lot B	lot C
Botrytis cinerea	9	6	10
Penicillium spp.	66	7	0
Alternaria spp.	2	27	23
Stemphyllium spp.	1	3	5
Mycosphaerella pinodes	1	15	0.5

Other pathogens detected in the test (*Fusarium spp.*, *Trichoderma spp.*, *Phoma medicaginis*, *Cladosporium spp.*) were at levels of 1% or below for the 3 lots.

TABLE 3. Percentage germination in sand/perlite at 20°C and in a cold test (10 days at 8°C followed by 20°C), and field emergence of pea seed lots A, B and C with the different seed treatments.

| | Treatment (see TABLE 1) | Germination at 20 °C | | Cold test | Field emergence | |
		7 days normal	10 days normal	15 days normal	20 days total	50 days normal
Lot A	1	54	70	49	72	85
	2	54	75	65	63	80
	3	56	78	65	60	64
	4	50	71	43	63	81
	5	81	87	71	63	81
	6	73	90	80	40	44
	control	63	80	51	28	28
Lot B	1	82	88	69	66	82
	2	75	83	70	63	81
	3	68	82	73	63	72
	4	74	82	71	62	82
	5	71	85	73	63	78
	6	76	84	71	59	64
	control	71	79	79	43	40
Lot C	1	99	99	93	84	96
	2	98	98	91	82	93
	3	99	99	94	85	95
	4	95	96	87	83	97
	5	96	97	90	84	93
	6	97	97	92	80	91
	control	98	98	90	74	72
LSD at p=0.05		6.5	6.3	5.1	8.1	6.1

Field emergence was improved by all fungicide treatments. Compared to the controls, field emergence was increased from 28% to 85% for lot A, from 40% to 82% for lot B and from 72% to 97% for lot C. For lot A and lot B the best emergence was obtained with treatments 1, 2, 4 and 5. For lot C all treatments gave the same level of field emergence. The correlation between laboratory germination and field emergence was only 0.477.

Conductivity, as determined for the untreated seeds, was 34.81 μs/cm (lot A), 32.58 μs/cm (lot B) and 17.82 μs/cm (lotC).

DISCUSSION

No correlation existed between standard laboratory germination or the cold test and field emergence. For lot A the highest level of germination in the laboratory test (treatment 6) resulted in the lowest field emergence. A very high correlation was obtained between conductivity and field emergence for the untreated seed (r^2 = 0.978). For all lots, treatments 1,2,4 and 5 gave good field stands. For the best lot (lot C), all fungicide treatments gave the same level of field emergence. It is striking that the germination level of the untreated seed in the laboratory (10 days, 20 °C) was always matched by the field emergence if the seed was correctly treated with fungicides.

The high incidence of *Penicillium* on lot A could be due to unfavorable post harvest weather conditions which lead to low vigour.

Even with weak lots good field stands were obtained if seeds were treated with the oxadixyl/cymoxanil/carbendazim or metalaxyl/thiabendazol/thiram mixtures. Addition of oxin-copper to these mixtures did not improve stands any further. To obtain seeds correctly treated with these mixtures, film-coating techniques are needed for their application.

REFERENCES

Duczmal, K.W.; Minicka, L. (1989) Further studies on pea seed quality and seedling emergence in the field. *Acta horticulturae,* **253,** 239-246.

Haquin, F. (1987) Traitements de semences: ça bouge aussi pour les oléoprotéagineux. *Semences et Progrès,* **53,** 3-12.

Kraft, J.M. (1982) Field and Greenhouse Studies on Pea Seed Treatment. *Plant disease,* **66,** 798-800.

van Loon, J.J.A.; Oyarzun, P. (1988) Zaadinfectie bij droge erwten, een potentiele bron van verspreiding van voetziekten in de erwtenteelt. *Gewasbescherming,* **19 (2),** 51-60.

Trawally, B.B.; Noonan, M.J.; Close, R.C. (1984) The effect of seed treatment on plant establishment, downy mildew and yield of peas. *Proceedings of the 37th New Zealand weed and pest control conference,* 163-166.

EFFECT OF THE PERIOD BETWEEN SOWING AND TRANSPLANTING ON CABBAGE ROOT FLY (*DELIA RADICUM*) CONTROL IN BRASSICAS WITH CHLORPYRIFOS FILM-COATED SEEDS

P.S.R. KOSTERS, S.B. HOFSTEDE

S&G Seeds B.V., P.O. Box 26, 1600 AA Enkhuizen, the Netherlands

ABSTRACT

Cabbage root fly (*Delia radicum*) control with chlorpyrifos film-coated seed has been succesfully practised in the Netherlands for the last 2 years. The influence of the period between sowing and transplanting (27 to 133 days) on the resulting control of cabbage root fly was assessed for cauliflower and Brussels sprouts at two sites in the Netherlands. Infection rate was extremely high at both sites. The percentage of dead plants was the lowest for the chlorpyrifos film-coated seed in all cases. The root damage index (RDI%) was lower for chlorpyrifos film-coated seed than for the control in all cases. Control by chlorpyrifos film-coating is effective for a period of up to 133 days between sowing and transplanting. Longer periods were not tested.

INTRODUCTION

Cabbage root fly (*Delia radicum*)is an important pest in horticultural brassicas. Film-coating seeds with chlorpyrifos gives reliable control of cabbage root fly and reduces the amount of insecticide needed (Kosters *et al* 1993; Ester *et al* 1993). Growing practice for brassicas, and more specifically for cauliflower and Brussels sprouts, varies considerably in the Netherlands. The time between sowing and transplanting can vary from 4 to 18 weeks (de Moel, 1993). Cabbage root fly control by chlorpyrifos also depends on soil and weather conditions if applied onto the soil at the moment of transplanting (Rouchaud *et al*, 1989). With the much reduced quantity of chlorpyrifos used in a film-coating compared to field treatment this dependency may be greater.

In this study we investigated the effect of a period of 27 to 133 days between sowing and transplanting of plants in the field on control of cabbage root fly by chlorpyrifos film-coated seed. Two fields with different soil types and expected high levels of cabbage root fly attack were used.

MATERIALS AND METHODS

<u>Treatments</u>

Seeds of the cauliflower variety Lindurian and Brussels sprouts variety Oliver were film-coated with fungicides only or with fungicides and insecticide. The standard commercial treatments as registered in the Netherlands were used (Table 1). Three treatments were compared on all sowing dates; fungicide control, chlorpyrifos film-coating and granulate treatment with 1 g of 5% chlorpyrifos granules per plant (0.05g a.i./plant) applied at transplanting to plants raised from fungicide-treated control seed.

TABLE 1. Film-coating treatments used in the experiments.

Treatment	AI	treatment rate (AI)
fungicide control	thiram	2 g/kg seed
	carbendazim	1 g/kg seed
	iprodione	5 g/kg seed
chlorpyrifos film-coating	thiram	2 g/kg seed
	carbendazim	1 g/kg seed
	iprodione	5 g/kg seed
	chlorpyrifos	0.096 g/1000 seeds

Field experiments

Plant raising

Plants were sown on six different occasions, viz December 3, January 5, January 29, February 19, March 11 and March 26 which gave differences in the time between sowing and transplanting in the field from 27 to 133 days (Table 2). Plants at all sowing dates were sown and raised in modules with a peat-based potting compost by a commercial plant raiser in the Netherlands, according to standard Dutch horticultural practice.

TABLE 2. Sowing dates and transplanting dates for the cauliflower and Brussels sprouts.

Field site	Sowing dates	Transplanting date	Period between sowing and transplanting (days)
Westmaas	cauliflower: 3/12; 5/1; 29/1; 19/2.	April 15	133, 100, 76, 55
	sprouts: -- ; 5/1; 29/1; 19/2.		--, 100, 76, 55
Prinsenbeek	cauliflower and sprouts:	April 22	
	5/1; 29/1; 19/2; 11/3; 26/3.		107, 83, 62, 42, 27

Field trials

Field sites were situated at Westmaas, which had a light clay soil, and Prinsenbeek, which had a sandy soil. Westmaas was selected for its known history of early cabbage root fly infection, Prinsenbeek for its light and drought-sensitive soil. Cauliflower and Brussels sprouts were transplanted separately by hand in randomised complete blocks combined over sowing dates. The trial had 5 replications of 65 plants at Westmaas and 4 replications of 72 plants at Prinsenbeek. Number of dead plants was assessed by eye at Westmaas 36, 41 and 48 days after transplanting (DAP), and at Prinsenbeek 21, 29, 35 and 41 DAP.

Mean root damage index (RDI%) was assessed 43 DAP in Westmaas and 49 DAP in Prinsenbeek. Ten complete root systems per replication were harvested and, after washing the surface area of root cortex, damage was assessed by eye. Each root system was then ascribed to one of six categories (0% damage, 1-10% damage, 20-30% damage, 40-60% damage, 70-80% damage, 90-100% damage). The mean root damage index was calculated by multiplying the number of plants in each category by the mean value of that category, adding all of these results together and then dividing by the total number of plants assessed per plot

(Lole, 1992). As an example, a 30% damage level in cauliflower results in considerable crop loss (Long, 1992).

RESULTS

Cabbage root fly attack at 43 DAP for Westmaas and 49 DAP for Prinsenbeek, and the RDI% are presented in figures 1 to 4. The LSD is presented for p=0.05.

<u>Westmaas</u>

Cauliflower
Film-coating with chlorpyrifos, and the granulate treatment reduced the percentage of dead plants 48 DAP and the RDI% compared to the controls. The RDI% for the last sowing date was lower for the granulate treatment than for the film-coating treatment (Fig. 1).

Brussels sprouts
Percentage dead plants in the controls was much lower for Brussels sprouts than for cauliflower. The RDI% was reduced to the same levels by the chlorpyrifos film-coating and by the granulate treatment, except on the last sowing date when plants from the granulate treatment were more extensively damaged (Fig.2).

<u>Prinsenbeek</u>

Cauliflower
The percentage dead plants 49 DAP was very high in the controls on all sowing dates. This percentage was reduced to almost zero by the chlorpyrifos film-coating treatment on all five sowing dates, which was lower than the percentage with granulate treatments. No difference in the RDI% was found between the film-coating and the granulate treatments for the first three sowing dates, for the last two sowing dates the RDI% for the granulate treatment was lower (Fig.3).

Brussels sprouts
Both percentage dead plants and RDI% were lower for the chlorpyrifos film-coating and the granulate treatment than in the control on all sowing dates. For some sowing dates control by the chlorpyrifos film-coating was better than with the granulate treatment (Fig.4).

DISCUSSION

Cabbage root fly attack was extremely high in the Netherlands in 1993 (Anon, 1993). This allowed an excellent but severe test of the cabbage root fly control methods. The number of plants killed by cabbage root fly in the plots with chlorpyrifos film-coated seed was very low and was unaffected by sowing date. The RDI% of cauliflower was higher in Westmaas when the sowing date was close to the transplanting date. This may be due to plants from these last dates being smaller and thus more vulnerable to attack. This also is illustrated by the higher RDI% for the controls of both cauliflower and Brussels sprouts on the last sowing date. In Prinsenbeek the RDI% for cauliflower exceeds 30 for both chlorpyrifos film-coating and granulate treatments on all sowing dates.

Surprisingly the RDI% for granular treatments tended to be higher than for film-coating. Dead plants which were selected for RDI assessment were rated as 100% damaged which led to a higher RDI%. The higher RDI% for the granular treatments may be because egg laying by cabbage root flies had already started before the transplanting date. Plants from film-coating were protected from the moment of sowing while the effect of granular treatment only starts after the insecticide has leached from the granules. Granulate application often leads to slight phytotoxic effects on the plants which delays establishment. This short period may be critical for good control. To understand the different control mechanisms from film-coating or granulate application combined field work and analysis of chemical residues in the plant is needed. Analytical work in 1993 at Zaadunie indicated that chlorpyrifos was present in the plant roots from film-coated seed at the moment of transplanting for all sowing dates (personal communication).

In this work it was proven that cabbage root fly control with a film-coating at 0.96 g chlorpyrifos per 1000 seeds gave comparable or better control than the standard granular treatment with 50 g chlorpyrifos per 1000 plants. The effect was irrespective of the period between sowing and transplanting. It was also shown that this year the level of control obtained in cauliflower at one site was not sufficient to reduce the RDI% to a low enough level to avoid economic losses. In years with heavy infestation a second treatment in the first period after transplanting may be necessary if granulate or film-coating is used. However, it was reported that in practice in the Netherlands in 1993 better control was obtained with the film-coating treatments than with the granulate treatments (Anon, 1993, Vader, 1993).

REFERENCES

Anon. (1993) Zaadcoating by spruitkool geeft afdoende bescherming. *Groenten+Fruit*, **no. 25**, p. 13.

Ester, A; Hofstede, S.B.; Kosters, P.S.R.; de Moel, C.P. (1993) Film-coating of cauliflower seed (*Brassica oleracea L. var botrytis L.*) with various insecticides to control the cabbage root fly (*Delia radicum*). *Crop Protection* (in press).

Kosters, P.S.R.; Hofstede, S.B.; Ester, A. (1993) Effects of film-coating Brussels sprouts seeds with various insecticides on germination and on the control of cabbage root fly (*Delia radicum*). In: *Fourth international workshop on seeds: basic and applied aspects of seed biology, Vol.3.*, Daniel Come & Francoise Corbineau (Eds.), Université Pierre et Marie Curie, Paris, 1081-1086.

Lole, M.J. (1992) Control of cabbage root fly on brassicas with 'Gigant' seed dressing. ADAS Contract report C001/158.

Long, E. (1992) Dressing beats cabbage root fly. *Grower*, **188**, (8), 24- 25.

Moel de, C.P. (1993) Teelt van bloemkool. PAGV teelthandleiding no. 51. Proefstation voor de Akkerbouw en de Groenteteelt in de Vollegrond, Lelystad, the Netherlands.

Rouchaud, J; Gustin, F; van de Steene, F.; Pelerents, C; De Proft, M.; Seutin, E.; Vanparys, L.; Benoit, F.; Ceustermans, N. (1989) Soil metabolism of insecticides and their protection efficiency. *Revue de l'Agriculture*, **42**, 1309-1326.

Vader, R. (1993) Resultaat legt sceptici het zwijgen op. *Groenten+Fruit*, **no. 31**, p. 9.

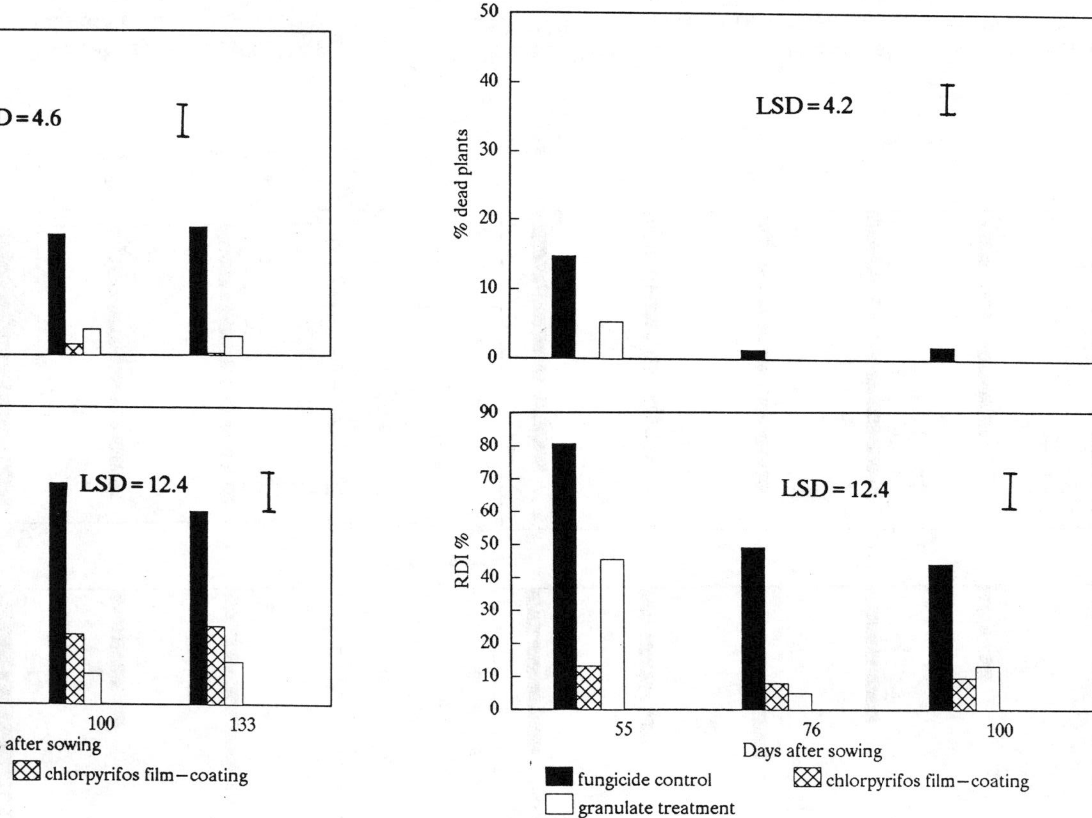

Figure 1. Percentage dead plants from cabbage root fly 48 DAP and RDI% 43 DAP for cauliflower at Westmaas. Transplanting date April 15, 4 sowing dates.

Figure 2. Percentage dead plants from cabbage root fly 48 DAP and RDI% 43 DAP for Brussels sprouts at Westmaas. Transplanting date April 15, 3 sowing dates.

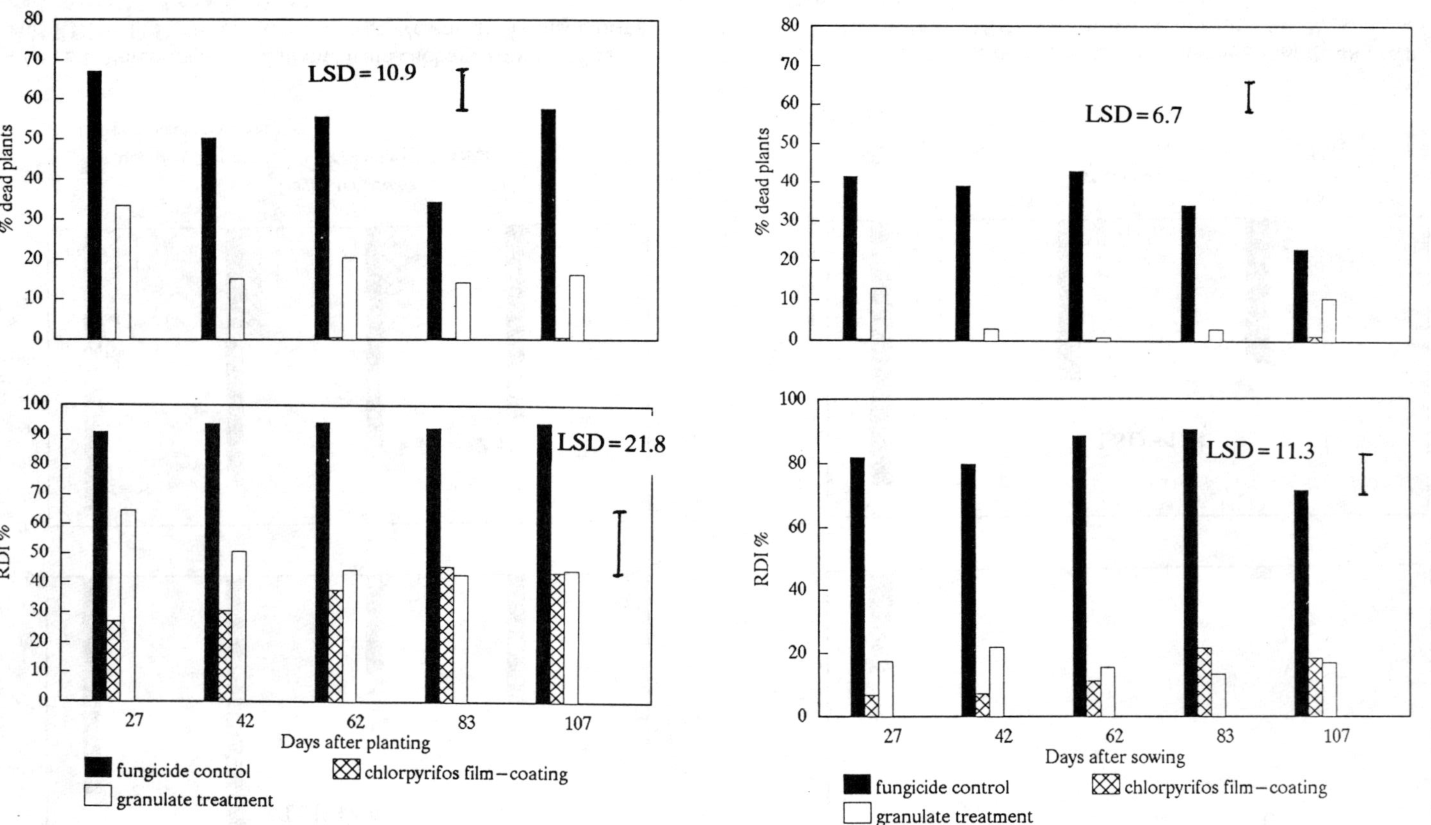

Figure 3. Percentage dead plants from cabbage root fly 35 DAP and RDI% 49 DAP for cauliflower at Prinsenbeek. Transplanting date April 22, 5 sowing dates.

Figure 2. Percentage dead plants from cabbage root fly 35 DAP and RDI% 49 DAP for Brussels sprouts at Prinsenbeek. Transplanting date April 22, 5 sowing dates.

Session 5
Potatoes and Bulbs

Chairman and
Session Organiser   G A HIDE

USE OF SEED TREATMENTS IN THE INTEGRATED CONTROL OF POTATO TUBER DISEASES

S. J. WALE

Scottish Agricultural College, 581 King Street, Aberdeen, AB9 1UD

ABSTRACT

Fungicide treatment of seed potatoes is an important integral part of a strategy for control of tuber diseases. It must be considered through all stages of seed multiplication. Choice of fungicide and correct timing of application depend on an assessment of disease risk. This risk can be assessed from a thorough knowledge of storage diseases but requires to be interpreted in a practical context.

INTRODUCTION

Whilst the ware grower converts the seed he buys to a saleable crop in one season, seed growers have to multiply a stock up over several seasons. In doing so there is the potential each year for ingress of disease both in the field and during storage. For a grower to produce healthy seed suitable for top quality ware production, he has to adopt an appropriate strategy to minimise or prevent tuber disease from the initial multiplication (VTSC) to the point of sale.

Seed tuber fungicide treatment is only one element in any strategy, for it must be integrated with husbandry techniques. Whilst the use of a fungicide treatment requires planning from the start of seed multiplication, flexibility is also needed to cope with changing disease risks and practical problems that arise each season. The choice of fungicide, time of application and method of application are all extremely important for successful control of seed tuber diseases.

The range of diseases which can affect tuber health in Great Britain is wide. In this paper consideration will be given primarily to storage fungal diseases for which fungicide seed tuber treatments are available (gangrene - *Phoma foveata*; dry rot - *Fusarium solani var. coeruleum*; skin spot - *Polyscytalum pustulans*; silver scurf - *Helminthosporium solani*; black dot - *Colletotrichum coccodes* and black scurf - *Rhizoctonia solani*).

For an effective disease control strategy to be developed, a knowledge of the life cycle and epidemiology of each pathogen is crucial. Tuber storage diseases have been reviewed by a number of authors (Boyd, 1972; Hide & Lapwood, 1992). In this paper essential features of each of the fungal storage diseases listed above are reviewed and the husbandry factors that reduce them are summarised in Table 1. Thereafter, the use of seed treatments in the integrated control of disease will be discussed.

ESSENTIAL FEATURES OF THE BIOLOGY OF STORAGE DISEASES

<u>Silver Scurf</u>

During seed multiplication the main source of inoculum is the seed tuber, soil-borne contamination being considered a minor source of the fungus. *H. solani* is easily detected in dirt and dust in potato stores (Hide, 1992) and initial contamination of stocks derived from mini-tubers or VT clones is likely to occur in store.

Infection of progeny tubers can occur pre-harvest and/or post-harvest. Pre-harvest infection is probably dependant on suitable conditions for spread and infection. In general, early harvest results in limited pre-harvest infection (Hide, 1992). Infection can occur in the crucial period immediately post-harvest when air humidity and temperatures can be high and moisture may be present on tuber surfaces. Infection can continue during storage, given relatively high temperatures and humidity. The concept of 'dry-curing' has been introduced as a method of minimising early post-harvest infection by maintaining unfavourable conditions for infection (Hide & Boorer, 1991).

Pre-harvest infection can result in symptoms already being expressed at harvest, but usually symptom expression follows a period of storage. Expression can be delayed, or minimised, by holding tubers at constant low temperatures (eg <4°C) (see Table 3) but after a period of cool storage, once tubers are placed in a warm environment, symptoms can rapidly develop.

<u>Skin spot</u>

Skin spot is a 'latent' disease, symptoms developing several months after infection has occurred. At harvest it is not possible to visually assess the extent of infection. Like silver scurf, the main source of inoculum is probably the seed tuber although inoculum has been found capable of surviving in soil. During multiplication from pathogen-free material, VT clones or mini-tubers, limited soil-borne inoculum may be responsible for initial infection of stocks. However, viable spores of *P. pustulans* are readily detectable in store dust (Carnegie *et al.*, 1978) and contamination of stocks during storage is probably the major method responsible for ingress of the pathogen (Carnegie & Cameron, 1987).

Inoculum of *P. pustulans* can multiply after planting as the fungus invades stem bases, roots and stolons and sporulates readily from these lesions. This explains why correlation between contamination of daughter tubers and that on seed tubers can be poor. Before harvest tuber infection, usually of eyes, is possible but the majority of infection occurs at or after harvest. Infection is very dependent on environmental conditions. Prolonged low temperatures (4°C) and high humidity, prolonging tuber dampness, lead to extensive infection. Dry curing minimises infection. There is potential for infection throughout storage if suitable conditions for infection occur, hence tubers in refrigerated stores with temperatures as low as 1-2°C may be at greater risk.

Early harvest of seed crops reduces the incidence of infection. This is due to less opportunity for pre-harvest infection in the soil, a lower inoculum build up on the harvested tubers and conditions more consistently suitable for rapid drying of tubers after lifting.

<u>Dry rot</u>

In Scotland, dry rot has been reported and observed more frequently in recent years. This might be a consequence of widespread use of 2 amino-butane which does not control this disease, a series of dry seasons or a general move to earlier lifting. Of the pathogens associated with dry rot, *F. solani var. coeruleum* is most prevalent in Scotland. *F. sulphureum* and *F. avenaceum* are rarely isolated.

Inoculum is both soil-borne (Carnegie & Cameron, 1990) and carried on the seed tuber. Initial contamination of early generations of seed may come from the soil, or from store dust (Carnegie, 1992). Infection is dependent on damage to tubers, either directly through physical wounding during handling or indirectly through lesions of other pest and diseases (eg. slugs, common and powdery scab, gangrene). The greater the damage and/or the higher the inoculum, the greater the risk of infection.

In store, symptoms develop after a latent period. Frequently, symptoms develop after grading in January to March, implying that the most important period of infection follows damage on the grading line. However, trials by Carnegie *et al.* (1986) indicate that fungicide treatment into store gives the greatest control of dry rot. This implies that lesions inflicted at harvest are most important. Lesions may remain latent until the stimulus provided by movement during grading encourage symptom expression. However, infection through wounds inflicted at grading can occur.

<u>Black dot</u>

Black dot is not a common disease on Scottish seed potatoes, although the similarity in appearance to silver scurf may have resulted in it being overlooked. Major factors contributing to a lower incidence in seed are probably the enforced 5 year minimum interval between seed potato crops and generally earlier harvest. Low levels of inoculum on seed and relatively limited soil inoculum appear to result in a limited build up of the disease during multiplication. Dry curing reduces the incidence and severity of black dot especially when used in association with early harvest. Irrigation tends to increase the disease but this is unusual on seed crops in Scotland.

<u>Gangrene</u>

Soil-borne *P. foveata* is unlikely to be a source of inoculum for seed potato crops (Carnegie & Cameron, 1990). Diseased tubers or tubers contaminated from dust and dirt in the store (Carnegie & Cameron, 1987) are the principal sources of inoculum. The fungus develops on stems after senescence or chemical desiccation. Spores from these infected stems and from rotted mother tubers contaminate progeny tubers. This contamination increases the later tubers are havested. Infection is dependent on wounding, and may thus occur at any time tubers are handled. The disease exhibits a long latent period and symptoms following infection at harvest may not develop for several months. Sometimes grading is the stimulus for the latent infection to develop. Curing, after damage has been inflicted, can reduce infection. Conditions that are unfavourable for wound healing thus favour the disease.

Black scurf

Although some post-harvest development can occur most infection occurs
pre-harvest. As resting structures, the black sclerotia of *R. solani* are
found only on the tuber surface and do not usually affect the seed tuber in
store. Once sprouting has begun and if conditions favour growth of mycelum,
infection of sprout initials can occur. Eyes are rarely killed by the
fungus and secondary sprouts can develop. Invasion of eyes can continue
after planting. Infection of established stems can occur resulting in stem
canker symptoms. Fungicide treatment prior to planting can limit infection
of sprouts and subsequent development of black scurf on progeny tubers.
However, soil-borne inoculúm is not controlled by seed tuber treatments and
black scurf can develop on progeny tubers from this source.

**Table 1. Summary of husbandry factors that reduce (+) tuber fungal storage
disease**

Husbandry factor	Gangrene	Skin spot	Dry rot	Silver scurf	Black dot	Black scurf
Cultivar resistance	+	+*	+*	+*	+*	+*
High seed tuber health	+	+	+	+	+	+
Early haulm destruction and lifting	+	+	+/-	+	+	+
Good skin set	+	-	+	-	-	-
Minimising damage	+	-	+	-	-	-
Dry curing	+	+	+	+	+	-
Minimising condensation during storage	+/-	+	+/-	+	+/-	-
Use of fungicide seed treatments at:						
harvest	+	+	+	+	-	-
grading	+/-	+	+/-	+	-	+
pre-planting	-	+	-	+	-	+
Store, box and machinery hygiene	+	+	+	+	+	-
Grading warm tubers	+	-	+	-	-	-

+* disease resistance ratings not recorded but cultivar differences exist

DISCUSSION

The need for and choice of fungicide treatment(s) during storage each
year requires a risk analysis. Routine treatment will be needed if:
a) there is a high disease risk for a particular cultivar
b) there is a prevalent disease problem on the farm
c) there were significant disease levels on the seed planted

d) there is a disease which would influence marketability, even at low
   levels
Other factors need to be considered. These include: seasonal or husbandry
factors that could increase the risk of disease, such as late harvesting;
the practical potential for fungicide application and the likely
effectiveness and risk of resistance following application. If the disease
risk is perceived as low and good husbandry is followed, routine fungicide
use is not necessary.

Fungicide seed treatments constitute an important element of seed production
but their use should be considered in relation to husbandry and the time of
application before selecting the fungicide to use.

<u>Husbandry practices and their relation to fungicide use</u>

   Fungicides are not substitutes for good husbandry, indeed they work
most effectively only in conjunction with good husbandry practices. Hide
(1992) demonstrated that for the surface blemish diseases, silver scurf and
black dot, not only were there significant individual effects on infection
by seed treatment (applied prior to planting) and husbandry factors such as
harvest date and dry curing but also significant interactions between
factors. Thus, for example, the effect of curing was greater in earlier
harvested crops.

   Evidence from studies on almost all storage diseases demonstrate that
early haulm destruction and early harvest minimises disease risk. This is
because lower levels of inoculum are present on harvested tubers. The later
the harvest, the greater the inoculum and the higher the disease pressure.
Consequently, the task facing a seed treatment in a later harvest is much
greater.

   There are associated practical reasons why a late harvest increases the
risk of disease. Soils are often wetter as the autumn proceeds and more
earth adheres to the tubers during harvesting. This decreases the efficacy
of fungicide treatment either by spray applications or as a fumigant. Also,
with later harvests, soil and tuber temperatures have fallen with dual risks
of increased tuber damage and slower wound healing.

   Dry curing is another factor that can have a major influence on most
tuber storage diseases. Keeping tubers dry during the early phase of storage
is essential if infection is to be minimised at this critical time. Early
lifting usually ensures tubers are above $10^{\circ}C$ entering store and wound
healing, so essential to prevent gangrene and dry rot, proceeds rapidly. At
temperatures above $10^{\circ}C$ there is usually no need to permit tuber
temperatures to rise for curing. Rather, curing will proceed sufficiently
fast even where ventilation to dry the tubers and maintain dry conditions is
applied continuously. When tubers are lifted later in cooler conditions,
not only is wound healing slower but drying is less efficient because the
moisture holding capacity of the air is lower.

   Rapid drying of tubers entering store, particularly when they are wet,
has been complicated by the move by seed growers from bulk to box storage.
In the latter it is less easy to pass air positively past all tubers. Of the
many ventilation systems available, few have the potential to pass air
through boxes and achieve very rapid drying of tubers.

Infection by *H. solani*, *P. pustulans* and *C. coccodes* can occur
throughout storage provided conditions are suitable for infection. When the
tuber temperature has fallen to that desired for long term storage (3-4°C),
the effect of moisture is less important. The aim, however, must be to
continue to minimise condensation through judicious ventilation. The impact
of the microclimate around tubers on disease development is poorly
understood and deserves further investigation.

Store, box and machinery hygiene is a little studied area of potato
production. Confirmation that store dust harbours spores of many fungal
pathogens indicates a potential for rapid contamination of healthy stocks.
A move to positive ventilation and maintenance of dry conditions throughout
storage has resulted in dust becoming a major problem, particularly during
grading. The benefits of thorough cleaning of stores and boxes are
uncertain but removal of as much dust as possible should reduce tuber
contamination. Regular cleaning of the grading line is also important but
often difficult to achieve. There have been no studies on the spread of
fungal spores on grading lines. The value of disinfectants to clean grading
lines, stores or boxes has not been evaluated.

Fungicide timing

At harvest. For all diseases that develop in store, infection can occur
at or shortly after harvest. Therefore, this is probably the most crucial
time for fungicide application to control them. Carnegie *et al.* (1986)
demonstrated that thiabendazole was less effective against gangrene, skin
spot and dry rot the later it was applied after harvest. Even if infection
occurs later in storage, the presence of fungicide on the tuber surface from
application immediately post-harvest would limit it. However, despite the
importance of harvest treatments, there are practical difficulties for
growers applying fungicides at this time.

There are several opportunities to apply spray fungicides. The first
one is on the harvester at the point where they tumble into a hopper.
However, treatment on the harvester has rarely proved effective. Another
opportunity comes if tubers are split dressed or pre-cleaned before storage
and a final opportunity is when they pass up an elevator in store before
placing in boxes. The current, widely adopted, practice of lifting straight
into boxes minimises handling but restricts the opportunity for these last
two options. The restriction imposed on many fungicides, that they can only
be applied to seed tubers, has resulted in an increase in the practice of
split grading into store. Into store separators are expensive but do
present a clear opportunity for seed-size tubers to be treated with
fungicide and additionally offer practical advantages to rid tubers of
excess soil and reduce the proportion of lifted tubers that need to be
stored as seed. However, there is the disadvantage that extra handling
increases damage thereby increasing the risk of invasion by wound
pathogens. When lifting in wet conditions the use of into store separators
can be totally impractical.

At grading. Effective fungicide treatment at grading is far easier to
achieve because fungicide can be targeted at seed to be sold rather than all
the crop, soil contamination on the tuber surface has been removed and
roller tables provide an even supply of tubers for efficient fungicide spray
applications. Fungicide treatment at grading will only have limited effect
on disease initiated at harvest. But because grading can result in wounding
of tubers, fungicide applications at this time will reduce any subsequent

invasion by wound pathogens. However, treatment at grading should be
primarily intended for reduction of inoculum on seed to minimise spread to
progeny tubers after planting.

Conditions at grading can play an important role in the risk of
infection by fungal wound pathogens. When tubers are graded at temperatures
close to 3-4°C, damage will be greater than if they are warmed prior to
grading. The curing of wounds is also slower. However, whilst warming prior
to grading is a desirable objective it can be difficult to achieve.

Pre Planting.   Fungicide treatment just prior to planting is also
intended to reduce inoculum and spread to progeny tubers. Most of the
fungicides used at this time are dusts applied to seed in the hopper of a
planter. Recently, application of fungicides in a spray applied to tubers as
they are collected by the planter cups has been investigated.

Choice of fungicide

Of the fungicide options available (Table 2), experience would indicate
that 2-aminobutane is the most consistently effective fungicide against
gangrene and skin spot. Trials support this contention (Graham *et al.*;
1981). This is not due to greater efficacy but to the fact that, by
fumigation, the fungicide is usually better distributed over and absorbed
into the surface of tubers (Malone, 1986). 2-aminobutane does have a
restricted spectrum of activity. In particular, it does not control dry rot
and where this disease is present or a variety is susceptible an additional
treatment or alternative fungicide is advisable.

Those fungicides that control *R. solani* alone are generally not
applied into store, although provided only the seed fraction is treated and
tubers are free from soil effective control of tuber-borne inoculum is still
possible. Of those combinations of fungicides where one component controls
R. solani, there is also a presumption against applications at or soon after
harvest for the same reasons.

Of those fungicides that control diseases that develop in store,
thiabendazole is the cheapest option. For application into store, because
it has approval for seed and ware, there is no requirement to split grade at
harvest. However, the use of thiabendazole alone results in resistant
strains of *H. solani* (Hide *et al.*, 1988; Burgess *et al.*, 1993), and to a
lesser extent of *P. pustulans* (Carnegie & Cameron, 1992) developing. Thus
the use of this fungicide is <u>not</u> advised at any time during seed
multiplication. Where this strategy is followed ware growers should be able
to use thiabendazole, their only fungicide option into store, without the
risk of exacerbating infection by these two pathogens.

For varieties where silver scurf does not affect the marketability of
the final crop and where skin spot resistance is high, the option to use
thiabendazole would seem valid. However, because store dust can readily
contaminate other varieties in a store the spread of resistant isolates to
varieties where this disease is of greater importance could be rapid and is
clearly undesirable.

There is, therefore, a presumption in favour of into store treatment
using 2-aminobutane, imazalil alone or imazalil in combination with
thiabendazole. The effect of the fungicide mixture on development of
isolates of *H. solani* or *P. pustulans* resistant to thiabendazole is not

Table 2.   Seed potato fungicide tuber treatments

Active ingredient(s)	Product	Formulation[1]	How applied	For use on		Application timing			Relative disease control - good (++) or moderate (+)						
				Seed	Ware	Har	Gra	Pre	GAN	SSP	DRO	SSC	BDO	SCA	BSC
2-aminobutane	CSC 2-aminobutane	SL	Fumigant	+	–	+	–	–	++	++	–	+	–	–	–
thiabendazole	Storite Clear Liquid	SL	Spray	+	+	+	+	–	++	++2	++3	++2	–	–	–
	Storite Flowable	SL	Spray	+	+	+	+	–	++	++2	++3	++2	–	–	–
	Tecto Dust	DP	Dust	+	+	+	–	+	++	++2	++3	++2	–	++	+
thiabendazole + imazalil	Extratect Flowable	SC	Spray	+	–	+	+	–	++	++	++	++	–	+	–
	Seedtect	DP	Dust	+	–	–	–	+	–	++	–	++	–	++	+
imazalil	Fungazil 100SL	SL	Spray	+	–	+	+	–	++	++	++	++	–	–	–
iprodione	Rovral WP	WP	Spray	+	–	–	+	+	–	–	–	–	–	++	++
tolclofos-methyl	Rizolex Flowable	FS	Spray	+	–	+	+	–	–	–	–	–	–	+	++
	Rizolex	DP	Dust	+	–	–	–	+	–	–	–	–	–	+	++
pencycuron	Monceren Flowable	FS	Spray	+	–	+	+	–	–	–	–	–	–	+	++
	Monceren DS	DP	Dust	+	–	–	–	+	–	–	–	–	–	+	++
pencycuron + imazalil	Monceren IM	DP	Dust	+	–	–	–	+	–	–	–	++	–	+	++

1) SL-Soluble concentrate, DP-Dustable powder, SC-Suspension concentrate, WP- Wettable powder, FS-Flowable concentrate

2) Insenstive strains have been detected 3) Insensitive strains of *F. sulphureum* have been detected

4) GAN-Grangrene, SSP-Skin spot, DRO-Dry rot, SSC-Silver scurf, BDO-Black Dot, SCA-Stem canker, BSC-Black scurf

clear. Limited evidence (Burgess *et al.*, 1993) has suggested that the
mixture does not encourage resistance although this is still under
investigation.

A delay in emergence has been recorded with a number of fungicide
treatments, especially dusts applied pre-planting. Such effects are usually
transient. However, where two fungicide applications are made, then the
liklihood of delayed emergence is increased (Table 3). This effect occurs
relatively infrequently and in a strategy trial in progress by SAC where
sequences of fungicide treatments are being compared over several years, no
major delays in emergence using treatments into store and at grading have
been recorded. The benefits or otherwise of sequential fungicide treatments
over the passage of several years of seed multiplication and the effect on
production of a ware crop are not yet fully understood.

Table 3. SAC Potato Fungicide Strategy trial 1989-91: Effect of fungicide
treatments at harvest, grading and pre-planting alone or in combination on
emergence and silver scurf development on the subsequent crop stored at 4°C
or 8°C for six months. Aberdeenshire.

Treatment			Emergence (%)		S. scurf - April 91 (% surface area)	
Harvest	Grading	Pre-planting	5/6/90	25/6/90	4°C	8°C
Nil	Nil	Nil	48.0	95.5	6.6	25.2
TBZ+Im	Nil	Nil	36.5	96.0	2.6	12.6
TBZ+Im	Nil	T-Methyl	18.5	98.0	1.3	4.5
Nil	TBZ+Im	Nil	46.5	98.5	4.1	11.5
Nil	TBZ+Im	T-methyl	18.5	94.5	5.9	19.8
S.E.D.			6.41	1.96	1.49	6.11

TBZ+Im - Thiabendazole + Imazalil; T-methyl - Tolclofos-methyl

CONCLUDING REMARKS

In general, the more years of multiplication there are the more likely
inoculum has built up and the greater the risk of fungal storage disease.
Surveys of seed stocks have shown that the percentage of tubers contaminated
by *P. foveata* and *P. pustulans* was greater in lower grade stocks than for
VTSC (Carnegie, 1992). This was not true for *H. solani* where contamination
was similar at all grades. Every effort to minimise inoculum build up is
needed, therefore, throughout seed multiplication. Thus during
multiplication of VTSC fungicide treatment each year is recommended. The
potential for rapid contamination by *H. solani* has encouraged VTSC growers
to keep each years production in separate stores and to mainatain high
hygiene standards.

To consistently produce high quality, high health status seed potatoes, attention to detail is important. In particular this will mean early haulm destruction and lifting but the financial returns must be guaranteed for a crop which is intended primarily for seed. Otherwise the grower will be tempted to leave the crop to bulk up and produce a ware fraction. Such a 'dual purpose' crop presents a greater disease risk and difficulties for selection and application of fungicide seed treatments. In the future, more rational use of seed treatments will come when rapid means of measuring inoculum levels on harvested tubers are possible and risk assessments rely less on judgement.

REFERENCES

Boyd, A.E.W. (1972) Potato Storage Diseases. *Review of Plant Pathology* 51, 297-321.

Burgess, P.J.; Wale, S.J.; Oxley, S.J.P.; Lang, R.W. (1993) Fungicide treatments of seed potatoes: .strategies for the control of silver scurf (*Helminthosporium solani*). *Proceedings Crop Protection Northern Britain 1993*, 319-324.

Carnegie, S.F. (1992) Improving seed tuber health. *Aspects of Applied Biology, 33, Production and protection of potatoes,* 93-100.

Carnegie, S.F.; Adam, J.W.; Symonds, C. (1978) Persistence of *Phoma exigua var. foveata* and *Polyscytalum pustulans* in dry soils from potato stores in relation to re-infection of stocks derived from stem cuttings. *Annals of Applied Biology,* **90**, 179-186.

Carnegie, S.F.; Cameron, A.M. (1987) Contamination of healthy seed tubers by *Polyscytalum pustulans* and *Phoma exigua var. foveata* in potato stores. *Potato Research,* **30**, 79-88.

Carnegie, S.F.; Cameron, A.M. (1990) Occurrence of *Polyscytalum pustulans, Phoma foveata* and *Fusarium solani var. coeruleum* in field soils in Scotland. *Plant Pathology,* **39**, 517-523.

Carnegie, S.F.; Cameron, A.M. (1992) Resistance to thiabendazole in isolates of *Polyscytalum pustulans* (skin spot) and *Fusarium solani var. coeruleum* (dry rot) in Scotland. *Plant Pathology,* **41**, 606-610.

Carnegie, S.F.; Ruthven A.D.; Lindsay, D.A. (1986) Control of fungal storage diseases on seed tubers by fungicide application at grading. *Aspects of Applied Biology,* **13**, *Crop protection of sugar beet and crop protection and quality of potatoes,* part II, 273-278.

Graham, D.C.; Hamilton, G.A.; Quinn, C.E.; Ruthven, A.D. (1981) Summary of results of experiments on the control of the potato tuber fungal diseases, gangrene, dry rot and skin spot with various chemical substances, 1966-80. *Potato Research,* **24**, 147-158.

Hide, G.A. (1992) Towards integrated control of potato storage diseases. *Aspects of Applied Biology,* **33**, *Production and protection of potatoes,* 197-204.

Hide, G.A.; Boorer K.J. (1991) Effects of drying potatoes after harvest on the incidence of disease after storage. *Potato Research,* **34**, 133-137.

Hide, G.A.; Hall, S.M.; Boorer K.J. (1988) Resistance to thiabendazole in isolates of *Helminthosporium solani*, the cause of silver scurf disease of potatoes. *Plant Pathology,* **37**, 377-380.

Hide, G.A.; Lapwood, D.H. (1992) Disease aspects of potato production. In: The Potato Crop, P. Harris (Ed.), Chapman & Hall, London, pp. 403-437.

Malone, J. (1986) Interactions of the fungicide 2-aminobutane with potato tubers. PhD. Thesis, University of Edinburgh.

POTATO TREATMENT: GETTING IT OUT OF YOUR SYSTEM

THE WORK OF THE BCPC POTATO TREATER GROUP

G.H. INGRAM (Convener) and members of the Group

c/o Rhône-Poulenc Agriculture, Fyfield Road, Ongar, Essex CM5 0HW.

SUMMARY

The BCPC Potato Treater Group was conceived about 3 years ago, and since its birth has gathered together representatives of a wide range of disciplines, who have an interest in potatoes. This interest is primarily concerned with optimizing the application of agrochemicals to seed or ware potatoes with various agrochemical products and mechanical techniques.

To this end, a series of information sheets and guidelines have been produced covering different aspects of treatment such as:-

Calibration of equipment (No. 1)
Roller table design improvement (No. 2)
Tuber disease control factors (No. 3)
Roller table efficacy (No. 4)
Tuber treatability scores (No. 5)
Disease assessment guidelines (No. 6)
Residues and deposit analyses guidelines (No. 7)

Liaison between the Group and training organisations, ATB and NPTC, has resulted in a new module PA12 for the treatment of crops on conveyor systems.

To encourage standardization and improvement in the interpretation of results of trials, etc. guidelines have been issued for the recording of both disease incidence/severity and levels of product deposits.

Future projects include:-
- the significance of surface soiling of tubers
- application volumes, the facts and myths
- improved treater systems and tuber feed systems
- information on products available and use

INTRODUCTION

The BCPC Seed Treatment Working Party has an interest in the treatment of potatoes, particularly seed tubers, and has published "Guidelines for the effective and safe treatment of potatoes" (Scott, 1990). It has long been recognised that potato treatment was overall rather "hit and miss", with each component working well, but normally unco-ordinated, resulting in widely varying quality of treatment and tuber deposit levels.

Approximately three years ago various interested parties, representing the main sectors of the potato industry, set up the BCPC Potato Treater Group in consultation with their own sectors of the industry. The BCPC Seed Treatment Working Party acted as the umbrella organization. The core Group is composed of:-

A.C. Cunnington	Sutton Bridge Expt. Stn.	representing	Potato trials/testing
P.A. Dover	ADAS (Secretary)	"	ADAS
G.A. Hide	Rothamsted Expt. Station	"	Research Organizations
R.T. Hubbard	Team Sprayers	"	Sprayer Manufacturers
G.H. Ingram	Rhône-Poulenc (Convener)	"	British Agrochemical Association
D.C. McRae	S.C.A.E.	"	Scottish Agricultural College
J.A. Nightingale	Dalgety Produce	"	National Assn. of Seed Potato Merchants
R.M.J. Storey	Potato Marketing Board	"	Potato Marketing Board
W.R. Threapleton	Golden Wonder	"	Potato Processors Assn.
A. Toon	Peal Engineering	"	Agricultural Engineers

From time to time, specialists are co-opted as required for specific skills, e.g. members of the disciplines represented.

The objectives of the Group can basically be summarized as "getting the best performance out of tuber treatment systems", such as :-

- Enhance the standards of potato treatment (initially wet liquid systems).
- Draw on the experience of all sectors of the "Potato" industry, to establish their needs and target objectives for future work.
- Promote educational material for plant operators, growers, seedsmen and merchants, etc.
- Optimize the use of the systems currently available.

- Stimulate investigations into the detailed problems of potato treatment and novel systems.
- Enhance the safety to operators and the environment, when using potato treaters.

It is not the objective of the Group to specifically recommend one product or one piece of equipment, but to give guidance for the optimum utilisation of products and equipment to the benefit of the industry as a whole.

THE WORK OF THE GROUP

The Group is involved in many aspects of potato treatment and, having identified factors which have a significant impact on the efficacy/application of products, produce information sheets. These have been given wide circulation at potato demonstrations, in newsletters and via various company information distribution channels.

Calibration

This was one of the first areas in which the Group had an impact. It was recognised that growers and seed houses did not necessarily match equipment calibrations and several misconceptions abounded. Typical problems were:-

- incomplete coverage of the roller table by tubers.
- arbitrary adjustment of spray pressure with different table loadings.
- inaccurate calibration of tables and sprayers omitting go-down time, box changing and using average hourly or daily work rates.

Coupling these problems easily led to over 50% reductions from target doses. Information Sheet No. 1 was issued on calibration of systems.

Tuber Size Implications

Work commissioned with Scottish Centre for Agricultural Engineering, by joint funding from industry sources, quantified the large variation in tuber surface areas due to tuber size, and the variations of tuber numbers, tuber weight and tuber volume for a range of sizes across the major UK varieties. Once the mechanical application problems are resolved, then dose rate versus tuber surface area relationships can be examined, and these may impact on product recommendations, and application volumes.

Another aspect of grading was examined, using video techniques by Writtle College. These investigations clearly showed that tubers within a narrow size band greatly improved rotation on roller tables, otherwise movement was disrupted resulting in uneven spray coverage. In addition it was found that one misshapen tuber caused abnormal movement or indeed no movement in a large area around it, out of all proportion to the tuber itself.

Thus a general conclusion was that for optimum spray deposit, even graded, well shaped tubers were necessary. A key to improve roller table efficacy was drawn up by Writtle College, and can be used to give a % score which indicates the optimization of tuber presentation to sprays. This is presented as Information Sheet No. 4.

The general conclusion was that for optimum treatment conditions evenly graded crops with no misshapen tubers gave the best chance of improving product targeting, dose implication being a subject of further debate.

Roller Tables

Under the auspices of the Group, SCAE have been investigating the potential for enhancing tuber movement on roller tables by table design, and have concluded that for treatment of seed grade tubers a treatment table that is purpose designed for such crops should be considered (Information Sheet No. 2). This would feature smaller rollers than typically used on closer centres, as shown in the diagram, which improves coverage of the tables by tubers, thus reducing the amount of product lost onto rollers. An example of the concept was produced for the Group by Peal Engineering for a recent demonstration and attracted considerable attention and several tables for "seed" treatment have now been commercially produced. The potential for lost (or off target) spray can be very high, as shown in the following diagram.

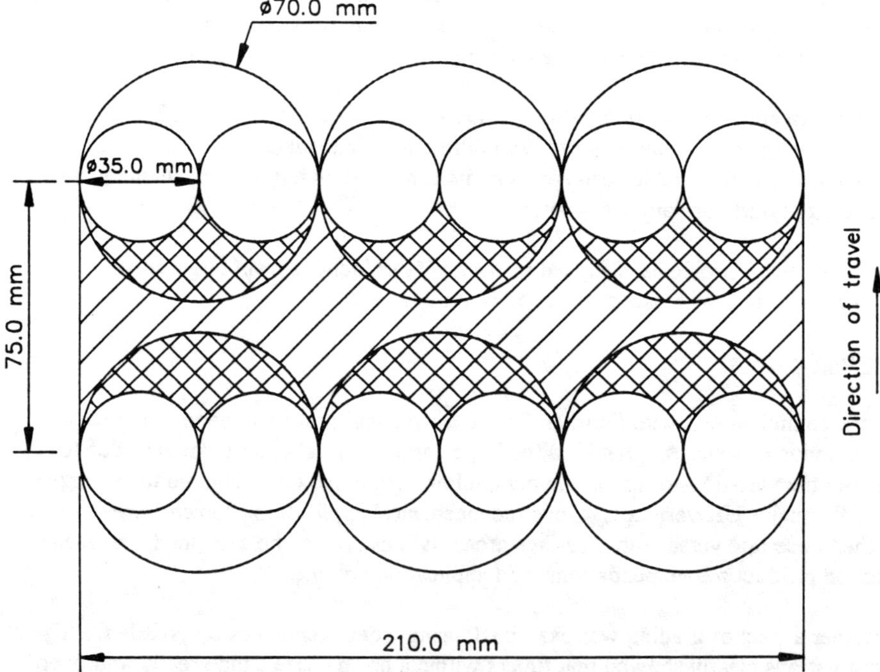

The space occupied by two sizes of potatoes (spherical shape) on rollers at 75mm pitch

7-27% of spray falling on the shaded area is wasted when spraying 70mm potatoes, depending on tuber shape. 53-63% of spray falling on the shaded area plus cross hatched area is wasted when spraying 35mm potatoes of various shapes.

In the future it is proposed to look at roller surfaces to ensure rotation even when the tubers are wet or soiled; at present smooth rollers acting like skid pans with the tubers turning unevenly, or not at all. This work would also look at the impact of tuber cleanliness on tuber rotation.

Spray Timing and Disease Development

It is important to optimize application timing for the highest level of disease control. The Group's disease specialist has drawn up Information Sheet No. 3 to aid growers, advisors, and others in selecting the best treatment timing for the target diseases on their seed tubers.

As a further aid, particularly for advisory and education purposes, the specialist has drawn up a series of disease life cycles which will shortly be available through the Group. These detail the life cycles for the major tuber diseases and will improve the understanding of potato workers for improved control measures and the importance of coverage and timeliness.

Tuber Descriptive Score

It was agreed by several areas of the industry that a means of categorising tuber condition would be advantageous. Reference to the suitability for spraying or for storage and as a means of long term reference for trials or seed lots, would assist in upgrading standards in both trials and commercial situations and in the interpretation of levels of control (or failure). Accordingly a score system has been defined by the Group's advisors and issued as Information Sheet No. 5. The key is described as follows, and can be expressed as a unique 3 digit number.

Condition of Tubers

Tubers should be assessed for each of the following factors:

A.	**Surface Moisture**	B.	**Soil Thickness***	C.	**Soil Cover %**
0	Dry	0	Clean (washed)	0	< 5
1	Damp	1	Thin	1	5-10
2	Moist	2	Medium	2	10-25
3	Wet	3	Thick	3	> 25

* A thin soil layer is one which should be penetrated by fungicide.

Treatability score	Optimal	0 or 1	in all factors
	Acceptable	2 (or lower)	in any factor
	Inadvisable	3	in any factor

A score of 110 would be ideal for treatment, etc. whereas 233 would be unacceptable for any use.

Operator and System Safety

This is an area of concern to the industry as a whole. It was found operators need to be qualified only for treatment of the seed crop, whereas considerable quantities of ware crop is treated for direct public consumption.

The ware industry representatives on the Group considered that for obvious reasons staff need the same or higher standard of qualifications .

Discussions took place with training representatives and as a result of the Group's detailed collaboration with the National Proficiency Test Council, a test module has been agreed within the Scheme, specifically for crop treatment on conveyor systems. This will be issued as module PA12.

Concern over anomalous requirements for various sectors of the industry, and in various regulations such as labelling, has prompted the Group to take these up with Pesticide Safety Directorate and these are currently ongoing.

Standardization of Terms and Methods

It soon became apparent that there was considerable variation in terminologies and methodologies within the potato industry, for both commercial and technical aspects. The Group has therefore undertaken a consolidation of the various details, including disease assessment, product deposits and product residues. This has been done in close consultation with the appropriate sectors of the industry and regulatory authorities.

As a result, two guidelines have been issued which it is hoped will result in improved standardization. These are available from the Group as:-

-   Guidelines for the assessment of potato tuber diseases.
    These detail the assessment of flesh and skin diseases by using the most widely accepted methods and include expression of data, sample sizes and sample preparation.

-   Definition of objectives and methodology for the analysis of agrochemical products on tubers.
    These highlight the difference between chemical deposits and chemical residues. They detail aspects of the sample preparation and selection that are required for these two different objectives.

It is hoped that these harmonizing exercises will improve exchange of data and information, and will minimize confusion caused by a multiplicity of terms.

OTHER GROUP INITIATED PROJECTS

Tuber Soiling

An area that has stimulated considerable discussion, with little hard fact, is the significance of soiling of tubers. We are encouraging some research projects into the real importance of loose skin, soil and dust in the retention (or not as the case may be) of products on the tubers, with reference to both seed and ware crops. The work aims to quantify the problem and look for solutions and simultaneously dispelling some myths.

It is currently thought that this could be a major source of loss of product post-treatment during storage and handling.

Volume of Application

There is a trend towards using medium volumes, e.g. 0.75-1.0 L/T for spraying. Over the years there have been several trends, e.g. high volume 3-5 L/T to get total coverage, medium volume 1.5-2.5 L/T for good coverage, 0.05-0.1 L/T as a concentrate spray for ease of use, minimum refilling and minimum tuber wetting. There have been many comments regarding the inducing of rots by high spray volumes and the merits of coverage and deposit levels. However, there is little real data to support or refute the various hypotheses.

Therefore the Group has encouraged and supported moves to elucidate the problem in research projects, one of which is being carried out by the Potato Marketing Board.

Treater Feed Systems

It is recognised that even and regular tuber flow onto the treatment table is important to optimize (a) tuber flow, (b) treatment quality and (c) economical use of products. Work is needed to develop the "feed-on" systems.

Improved Treatment Systems

The Group is keen to encourage collaboration between companies in different disciplines, and to develop systems of treatment that will bring together the optimum means of presenting a product to a potato. As such we have facilitated communication in some areas, resulting in improved treatment systems and novel approaches to the market in terms of specialist application equipment and treatment lines.

THE FUTURE

The Group is keen to stimulate work in many different disciplines that will generally improve the standards of potato treatment.

We welcome feed-back from the Industry on their current problems, areas which need research, and indeed the success of the current lines of communication.

It must be remembered that we are self-funding, and much of the information is derived from projects set up as a direct result of either a general support request, or because a college or company has the necessary facilities available. Therefore we would welcome any proposals for future projects, offers of assistance in carrying these out, and for future information sheets, their production and circulation (which currently stands at approximately 4,000 per issue, indicating good uptake).

ACKNOWLEDGEMENTS

The author wishes to thank the other members of the Group for their ideas, inspirations and support in the wide-ranging activities of the Group. The advice and suggestions from others in the Potato Industry is also acknowledged.

REFERENCES

Scott, G.C. (1990) Guidelines for effective and safe application of potato tuber treatments. Farnham: BCPC, 23pp.

GREEN-CROP-HARVESTING, A MECHANICAL HAULM DESTRUCTION METHOD WITH POTENTIAL FOR DISEASE CONTROL OF TUBER PATHOGENS IN POTATO

P.H.J.F. VAN DEN BOOGERT, P. KASTELEIN, A.J.G. LUTTIKHOLT

DLO-Research Institute for Plant Protection (IPO-DLO), P.O. Box 9060
NL-6700 GW Wageningen, The Netherlands

ABSTRACT

Green-crop-harvesting (GCH) is a recently developed mechanical potato haulm destruction method that helps to adequately prevent virus transmission to the progeny tubers. It provides excellent healing conditions for tuber skin damage, reduces many important tuber diseases, and provides an opportunity to control diseases on the progeny tubers in soil. Recent data on biocontrol of black scurf (*Rhizoctonia solani*) and soft rot and black leg (*Erwinia* spp.) in GCH are presented. A miniaturized model system of GCH was used to screen antagonists and their compatibility with fungicides for broad disease control. Control of black scurf was improved in GCH by application of the mycoparasite *Verticillium biguttatum*. *Erwinia* spp. also were effectively controlled by fungal and bacterial antagonists. *V. biguttatum* is tolerant of Oomycete-specific fungicides and several potential antagonists of *Erwinia* spp, so mixtures of microbial and chemical agents could be used for broad disease control in GCH.

INTRODUCTION

To prevent virus infection, Dutch growers of seed potato crops use to kill the foliage shortly after mass flights of virus-transmitting aphids begin. Simultaneously with mechanical harvesting of seed tubers, chemical and mechanical haulm killing became a standard procedure to prevent transmission of virus and *Phytophthora infestans* to the tubers. The current method to kill green potato haulms is to pulverize the foliage and then either pull the haulms or spray them with a herbicide. Black scurf caused by *Rhizoctonia solani* is an important problem in seed potato production and is strongly promoted by chemical haulm destruction and, to a lesser extent, haulm pulling (Dijst *et al.*, 1986). Tuber exudates and volatile compounds from decaying plant parts favour black scurf formation (Dijst, 1990), so the release of tuber-derived exudates should be prevented and decaying plant parts should not accumulate near tubers. A method that brings about a spatial separation between tubers and the remaining plant parts combines haulm destruction and black scurf prevention. Research on harvest methods and prevention of black scurf has led to green-crop-harvesting (GCH) which involves the following (Bouwman *et al.*, 1990):

1. removal of potato haulms by cutting off at soil level or by pulling;
2. digging up the progeny tubers;
3. placing the tubers back on a new soil bed;
4. covering them with soil to obtain new ridges for an *in situ* maturation period.

Photo courtesy supplied by Ir. A. Bouman (IMAG-DLO) and Dr. L. J. Turkensteen (IPO-DLO)

FIG. 1. Green-crop-harvesting equipment: haulm stripper in the front; tuber lifter in the middle; roto discs to built new GCH-ridges. The two containers on the back of the lifter serve to treat the tubers with antagonists and/or fungicides.

GCH is an effective haulm destruction method (Mulder *et al.*, 1992) and the specially designed diggers provide an acceptably low level of skin damage (Bouwman *et al.*, 1990) so that the incidence of some important tuber diseases compared favourably with that following chemical haulm destruction (Table 1).

TABLE 1. Effect of green-crop-harvesting (GCH) on potato tuber disease incidence compared with chemical haulm destruction (CHD) methods (Mulder *et al.*, 1992).

Tuber disease (pathogen)	Effect of GCH on disease incidence compared with CHD
Black scurf (*Rhizoctonia solani*)	decrease
Gangrene (*Phoma exigua* var. *foveata*)	decrease
Silver scurf (*Helminthosporium solani*)	indifferent
Soft rot and black leg (*Erwinia* spp.)	indifferent
Tuber late blight (*Phytophthora infestans*)	indifferent/slight increase[*]

[*] after haulm stripping/after haulm pulling, respectively.

GCH also provides the opportunity to apply control agents directly to progeny tubers after the first lifting between steps 2 and 3. For successful introduction of antagonists in GCH, the control agents must be proven to be i) effective against the target pathogen *in situ*, ii) compatible with other co-applied chemical and biological control agents for broad disease control, and iii) mass-producible at economic rates. To further reduce the dependency on chemical control agents recent research has been aimed at i) isolating and screening of antagonists for effectiveness against potato tuber pathogens *in situ* during GCH and ii) testing for compatibility with chemical control agents for broad disease control. In this review we concentrate on recent data obtained with antagonists of *R. solani* and *Erwinia* spp.

GREEN-CROP-HARVESTING AND CONTROL OF BLACK SCURF

A miniaturized model system for GCH

A bioassay had to be developed to screen potential antagonists for their suppressive properties under realistic GCH conditions. From *in situ* observations in the GCH ridges the components to be recognized as necessary are: immature progeny tubers and (diseased) potato plant parts. These may be mimiced in a miniaturized system for GCH ridges - the mini tuber system (MTS). The MTS consists of plastic containers (12 cm diam; 11 cm height) filled with 250 g natural soil and 20 mini potato tubers sandwiched between two layers of soil. Mini tubers (1 cm diam) were obtained from *in vitro* propagated potato plants (cv Bintje) grown in pots filled with compost-perlite mixture for 6 weeks. At two-week intervals mini tubers were harvested to provide uniformly sized and physiologically young tubers throughout the year. The amount of plant fragments and densities of test organisms (*R. solani* and antagonists) applied either to soil or on mini tubers were adjusted according to the experimental need. Each treatment was replicated 4 times. Soil moisture content was adjusted to and kept at 50 % water holding capacity.

Requirements for black scurf development in MTS

Below-ground plant parts left in the GCH-ridges and colonized saprophytically are thought to be the main inoculum source of facultative pathogens, including *R. solani*. The effects of plant residues on black scurf development were studied extensively in MTS. For this, mini tubers were incubated in soil amended with 1% (w/w) pulverized potato plant parts. Perlite-borne sclerotia of *R. solani* were mixed into the soil at a density of 500 particles kg^{-1}. Black scurf development on the tubers was periodically assessed and its amount was expressed as a sclerotium index (SI; 0-100). The SI was calculated as follows.

$$SI = 25(1N_1+2N_{2\text{-}5}+3N_{6\text{-}10}+4N_{>10})/N_{tot},$$

in which $N_{1,2\text{-}5,6\text{-}10,>10}$ = Number of tubers with 1, 2-5, 6-10 and > 10 sclerotia per tuber;

N_{tot} = total number of tubers

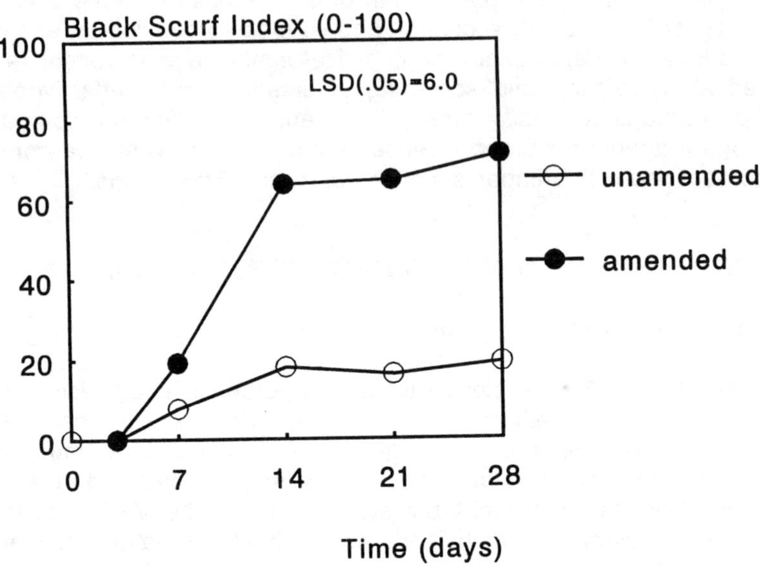

FIG. 2. Development of black scurf (*R. solani*) on mini tubers in soil unamended or amended with potato plant fragments.

Black scurf development was strongly enhanced in the presence of potato plant fragments (Fig. 2). This development of black scurf is essentially similar to that on potato tubers in the field after chemical or mechanical haulm destruction (Mulder & Roosjen, 1984). There was a linear correlation between the amount of plant residue added to soil and black scurf development in MTS. Pieces of stolon or stem added to soil caused stimulation of black scurf whilst other plant parts, such as roots or leaves caused little or no stimulation. Stem pieces from different potato cultivars gave similar black scurf responses; stem pieces from other plants caused no black scurf response (wheat), slight stimulation (peas) or marked stimulation (tomato) similar to potato. Thus, both the concentration of plant material and its source influence black scurf development on potato tubers. In the practice of GCH, part of the below-ground plant material remains present in the newly built GCH-ridges, especially after haulm stripping, so additional control measures are needed to suppress black scurf on the progeny tubers.

<u>Screening of fungal antagonists of *R. solani.*</u>

In an exploratory study on *R. solani*-suppressive soils, attention was drawn to mycoparasitic fungi as possible biocontrol agents of black scurf. *Verticillium biguttatum* proved to be the most effective one following seed tuber inoculation; other mycoparasites such as *Trichoderma* spp., *Gliocladium* spp. and *Hormiactis fimicola* were less effective (Van den Boogert & Jager, 1984). Although *V. biguttatum* has potential for practical application as a seed tuber inoculant (Jager *et al.*, 1991), its relatively high minimum temperature (> 13°C) limits its feasibility for use in Dutch conditions. However, at the time of GCH (last two weeks of July), the soil temperature is more favourable to *V. biguttatum* than the temperature at planting time. Some of the mycoparasites in the IPO-DLO collection, including *V. biguttatum* M73, were assayed for potential to control black scurf in MTS. For this, the mini tubers were dipped into conidial suspensions (10^7 conidia ml^{-1} water) following a two-week incubation period in soil amended with potato plant fragments.

TABLE 2. Effect of tuber inoculation with different antagonistic fungi on black scurf incidence as assayed in MTS after 14 days of tuber maturation.

Fungal inoculant	Black scurf index (0-100)
none	81.1 a
+ *V. biguttatum* (M73)	34.1 b
+ *Trichoderma* sp. (IVT-10)	85.8 a
+ *T. viride* (IPO-1811)	86.4 a
+ *T. harzianum* (IPO-1812)	89.5 a
+ *Gliocladium roseum* (IPO-1813)	80.6 a
+ *G. nigrovirens* (IPO-1815)	89.6 a
+ *G. roseum* (IPO-M171)	82.7 a

Data followed by the same letter are not significantly different at P=0.01

Tuber-inoculation with spores of *V. biguttatum* significantly reduced the black scurf incidence, whereas no other tested mycoparasite did so (Table 2). When individual sclerotia of *R. solani* from the *V. biguttatum* treated series were checked for viability by plating onto water agar they did not yield hyphae of *R. solani*, whilst sclerotia from the *Trichoderma* or *Gliocladium* series were still viable. The data from MTS thus provided evidence of the antagonistic potential of *V. biguttatum* under experimental conditions.

Dosage-response of *Verticillium biguttatum* on black scurf

The most feasible option for applying control agents in GCH is by spraying the progeny tubers. From a practical and economic point of view, 400 l spray liquid ha^{-1} is the upper limit for current spray equipment. Dosage-response tests were designed accordingly. Preliminary experiments on the efficiency of the spray equipment on GCH machines showed that about 10% of the spray liquid was deposited onto the tubers. Apart from any loss to the air the rest contacted the soil and might thus indirectly affect the tubers. Technical improvements in the spray methodology would seem worthwhile. To find out at what spore density of *V. biguttatum* effectively suppresses black scurf under GCH conditions, dosage-response experiments were done in MTS and in the field. For this purpose, tubers were either dipped into a spore suspension for MTS or sprayed with a spore suspension in GCH. The field experiments were laid out in a randomized block design with 4 replications. Per plot (20 x3 m) 100 tubers were rated for black scurf infestation.

TABLE 3. Dosage-response of *V. biguttatum* on black scurf in MTS and in the field after 14 and 20 days of tuber maturation, respectively.

Concentration of *V. biguttatum* spores ($ml-1$; equivalent of 400 l ha^{-1})	Black scurf index	
	MTS	Field
0	73 ab	24.2 a
3.2×10^4	70 ab	17.9 a
1.6×10^5	80 a	43.9 b
0.8×10^6	65 b	25.3 a
4.0×10^6	46 c	7.9 cd
2.0×10^7	33 d	2.2 d

Data followed by the same letter(s) in each column are not significantly different P = 0.05

A concentration of 4.0×10^6 spores ml^{-1} in 400 l ha^{-1} gave significant and worthwhile control of black scurf in MTS as well as in the field (Table 3). For current field experiments 400 l suspension liquid (tap water) with a spore density of 1.0×10^7 spores ml^{-1} are normally used. Spores of *V. biguttatum* can be cultured on common mycological agar media and to produce 4×10^{12} spores for application to 1 ha, a surface grown culture equivalent to 12 m^2 is required; there is need to develop the technology for inoculum production.

Compatibility of *V. biguttatum* with chemical control agents

When potato foliage was infected by *P. infestans*, haulm pulling led to a higher incidence of tuber blight (Mulder *et al.*, 1992). No increase of tuber attack was observed after application of chlorothalonil, mancozeb or cymoxanil at a recommended dosage for foliar application in GCH (Mulder *et al.*, 1992). However, these compounds might be incompatible with *V. biguttatum* in broad disease control. We investigated this in MTS and in the field. Potato tubers were treated with spores of *V. biguttatum* and the chemical components in mixed suspensions. Mini tubers used for MTS were dipped into the spore suspension and tubers used for field experiments were sprayed with suspension. Dosages of spores and fungicides were adjusted to equivalents of 4×10^{12} spores ha^{-1} and the recommended chemical dosage for foliage application in 400 l water (Table 4).

TABLE 4. Effect of tuber inoculation with *V. biguttatum* alone and with chemical control agents used for tuber late blight control on the incidence of black scurf (*R. solani*) in MTS and in the field after 14 and 20 days tuber maturation, respectively.

Tuber treatment (commercial product; a.i. kg ha^{-1}	Black scurf index	
	MTS	Field
Water (400 l)	50 a	29.3 a
V. biguttatum (4×10^{12} spores in 400 l water)	28 b	17.2 b
+ cymoxanil (a.i.; 0.20)	32 bc	18.8 b
+ propamocarb (Previcur; 1.44)	34 bc	17.6 b
+ thiabendazole (Lyrotect; 1.00)	35 bc	18.0 b
+ fluazinam (Shirlan; 0.20)	43 ac	nt
+ chlorothalonil (Daconil; 1.50)	44 ac	28.1 a

Data followed by the same letter(s) in each column are not significantly different at P=0.05
a.i. = active ingredient
nt = not tested

The Oomycete-specific fungicides cymoxanil and propamocarb did not affect the antagonistic action of *V. biguttatum*. In combination with the broad-spectrum fungicides chlorothalonil or fluazinam, the antagonistic action of *V. biguttatum* was completely suppressed. Nevertheless, the broad-spectrum fungicide thiabendazole, currently used to control fusarium dry rot and silver scurf in storage, appeared to be compatible with *V. biguttatum*.

GREEN-CROP-HARVESTING AND CONTROL OF BACTERIAL DISEASES

<u>Requirements of bacterial diseases</u>

GCH involves the lifting of immature tubers, which are prone to damage. As skin wounds are avenues of entry for soft-rot *Erwinia* spp., it was expected that GCH would lead to a considerable increase in bacterial infection. Therefore, field experiments were done on the effect of the current chemical method of killing green potato haulm on infestation of progeny tubers with *Erwinia carotovora* subsp. *atroseptica* (Eca) and *E. chrysanthemi* (Ech). The experiments involved seed potato crops with or without supplementary inoculum of bacteria. Inoculum consisted of tubers that had been induced to rot by strains Eca (IPO-161) or Ech (IPO-502) and then buried in the ridges after pulverization of the foliage. Non-inoculated tubers were used in plots without enhanced contamination. Infestation of tubers with Eca and Ech was assessed by semi-selective isolation from peel extracts and immunofluorescence colony staining (IFC; Van Vuurde & Roozen, 1990). With IFC, the bacterial colonies in agar isolation plates were stained with FITC-conjugated antibodies raised against strains of Eca or Ech to enumerate the *Erwinia* colonies. During the 1990 and 1991 growing seasons GCH did not lead to marked increase in infestation with Eca or Ech.

Fungal and bacterial antagonists of *Erwinia* spp.

When the mycoparasitic fungi *G. roseum* (IPO-1813) and *T. harzianum* (IPO-1812) were used to control dry rot (*Fusarium sulphureum)* a remarkable reduction of tuber soft rot by *Erwinia* was observed (Mulder *et al.*, 1992). This observation led to exploratory field trials on the control of Ech (trial 1) and Eca (trial 2) in potato crops with mixtures (400 l ha^{-1}) of *Trichoderma* spp. (10^9 spores ml^{-1} in trial 1 or 10^6 spores ml^{-1} in trial 2) or *Gliocladium* spp. (10^8 spores ml^{-1}) or fluorescent *Pseudomonas* spp. (10^9 cells ml^{-1} in trial 1 or 10^{10} cells ml^{-1} in trial 2). Tubers were harvested after two and seven weeks after haulm destruction in trial 1 and 2, respectively (Table 5).

TABLE 5. Effect of tuber inoculation with fungal and bacterial antagonists on the numbers of *Erwinia chrysanthemi* (Ech in trial 1) and *E. carotovora* subsp. *atroseptica* (Eca in trial 2) on potato peel after a two-week tuber maturation period in GCH ridges.

Inoculant	*Erwinia* count (log10 ml^{-1} peel extract)	
	Ech	Eca
None	1.13 a	0.00 a
Erwinia spp. (Ech in trial 1; Eca in trial 2)	2.71 b	1.68 a
+ *Pseudomonas* spp. (WCS 358;IPO-a,b)	1.00 a	1.78 a
+ *Trichoderma* spp. (IVT-10;IPO-1811,1812)	1.43 a	0.46 a
+ *Gliocladium* spp. (IPO-1813, 1815)	1.42 a	nd

Data followed by the same letter in each column are not significantly different at
P = 0.05
nd = not determined

In trial 1 all three groups of antagonists prevented the colonisation of Ech from inoculated tubers to progeny tubers. However, trial 2 failed to show significant differences among experimental treatments. Apparently the antagonistic interactions are very specific or are influenced by site variables. It should be noted that antagonistic organisms in these trials were not selected for specific antagonism against either Eca or Ech, so the success in trial 1 holds promise for control of pathogenic *Erwinia* spp. by microbial antagonism. Currently, antagonists of *Erwinia* spp. are being isolated and screened for effectiveness and compatibility with *V. biguttatum* in MTS. So far, most of the organisms applied in trials for the control of Eca and Ech are fully compatible with *V. biguttatum* as assayed in MTS. Hence, future MTS and field experiments may show whether mixtures of antagonists can control both bacterial diseases and black scurf.

CONCLUDING REMARKS

MTS is a helpful tool in screening for antagonists under realistic conditions of GCH. *Verticillium biguttatum* is a promising biological control agent of black scurf in seed potatoes and, since it is a specific antagonist of *R. solani* (Van den Boogert *et al.*, 1989), additional means are needed for broad spectrum disease control. Some Oomycete-specific fungicides were shown to be compatible with *V. biguttatum* and in fields with high risk of tuber blight, mixtures of these with *V. biguttatum* could be applied in GCH to prevent tuber blight and black scurf on progeny tubers. In addition, a promising outlook was established for biological control of bacterial diseases with fungal and bacterial antagonists. The obligately mycoparasitic nature of *V. biguttatum* appears to ensure that it is compatible with a variety of bacterial and fungal antagonists. One of the most challenging ways to promote biological control is the development of mixtures of antagonists that control an array of plant pathogens under different environmental circumstances.

Green-crop-harvesting is an acceptable alternative to the current chemical haulm destruction method and, if it can be integrated with the use of microbial antagonists to control tuber diseases, it will satisfy the increasing demands for reduction of agrochemical inputs in agriculture.

REFERENCES

Bouwman, A.; Mulder, A.; Turkensteen, L.J. (1990) A green crop lifting method as a new system for lifting potatoes. Abstracts of Conference Papers of the 11th Triennial Conference of the EAPR, Edinburgh, Scotland, UK. pp. 386-387.

Dijst, G. (1990) Effect of volatile and unstable exudates from underground potato plant parts on sclerotium formation by *Rhizoctonia solani* AG-3 before and after haulm destruction. Netherlands Journal of Plant Pathology **96**, 155-170.

Dijst, G.; Bouwman, A.; Mulder, A.; Roosjen, J. (1986) Effect of haulm destruction supplemented by cutting off the roots on the incidence of black scurf and skin damage, flexibility of harvest period and yield of seed potatoes in field experiments. Netherlands Journal of Plant Pathology **92**, 287-303.

Jager, G.; Velvis, H.; Lamers, J.G.; Mulder, A.; Roosjen, J. (1991) Control of *Rhizoctonia solani* in potato by biological, chemical and integrated measures. Potato Research **34**, 269-284.

Mulder, A.; Roosjen, J. (1984) Control of *Rhizoctonia solani* in seed potato crops by soil treatments with tolclofos-methyl, furmecyclox and pencycuron. In: Abstracts of Conference Papers of the 9th Triennial Conference of the EAPR, Interlaken, Switzerland. pp. 107-108.

Mulder, A.; Turkensteen, L.J.; Bouman, A. (1992) Perspectives of green-crop-harvesting to control soil-borne and storage diseases of seed potatoes. Netherlands Journal of Plant Pathology **98** (supplement 2), 103-114.

Van den Boogert, P.H.J.F.; Jager, G. (1984) Biological control of *Rhizoctonia solani* on potatoes by antagonists. 3. Inoculation of seed potatoes with different fungi. Netherlands Journal of Plant Pathology **90**, 117-126.

Van den Boogert, P.H.J.F.; Reinartz, H.; Sjollema, K.A.; Veenhuis, M. (1989) Microscopic observations of the mycoparasite *Verticillium biguttatum* with *Rhizoctonia solani* and other soil-borne fungi. Antonie van Leeuwenhoek **56**, 161-174.

Van Vuurde, J.W.L.; Roozen, N.J.M. (1990) Comparison of immunofluorescence colony staining in media, selective isolation on pectate medium, ELISA and immuno-fluorescence cell staining for detection of *Erwinia carotovora* subsp. *atroseptica* and *E. chrysanthemi* in cattle manure slurry. Netherlands Journal of Plant Pathology **96**, 75-89.

FUNGAL AND NEMATODE PATHOGENS OF *NARCISSUS*: CURRENT PROGRESS AND FUTURE
PROSPECTS FOR DISEASE CONTROL

C. A. LINFIELD

Plant Pathology and Weed Science Department, Horticulture Research
International, Wellesbourne, Warwick, CV35 9EF, UK.

ABSTRACT

The major limiting factor for the increased export of ornamental
bulbs from the United Kingdom is disease control. *Narcissus*, the
most important ornamental bulb crop in the U.K. is susceptible to
several diseases whose control would offer increased export
opportunity. Control of these may be achieved by modification of
crop husbandry practices, chemical control and the use of field
resistant cultivars. Fungicide insensitivity is an increasing
problem especially in the control of *Fusarium oxysporum* f.sp.
narcissi and *Penicillium* spp. Future opportunities for disease
control lie in maximising control by crop husbandry practices, the
use of alternative chemicals, biological and integrated control,
pest and disease forecasting systems and the use of resistant and
indexed stocks.

INTRODUCTION

The United Kingdom ornamental bulb and corm industry is dominated by
one species; the daffodil. The United Kingdom is the world's largest
producer of daffodils with current production at 4,000 ha per annum (Table
1). This area probably accounts for over half of the world *Narcissus*
production. The U.K. has a good export trade in *Narcissus* currently valued
at £4 M. The large-bulbed *Narcissus* cultivars are suited to mechanised
handling of high planting densities on large farms where cultivation is
compatible with cereal and potato growing. This has done much to aid the
success of the crop in the U.K. As the *Narcissus* crop has increased in the
U.K., so the tulip crop has declined from 360 ha in 1982 to 111 ha in 1992
(Table 1). The species making up the remainder of the 194 ha of field-
grown flower bulbs in the U.K. in 1991 mainly comprise iris, anemone, lily
and gladiolus. The concepts applied to *Narcissus* disease control are
equally applicable to minor bulb crops and they are not considered
separately here. Diseases are one of the major limiting factors to bulb
production and consideration of their control is essential for cost
effective management. The development in recent years of the export trade
in bulbs and bulb flowers has emphasised the importance of good disease
control. Bulb diseases can be controlled or their effect on quality and
yield minimised in several ways; the modification of crop husbandry
practices to limit disease development, use of disease resistant or
tolerant cultivars, elimination or reduction of the pathogen at source and
the use of fungicides/nematicides or other control agents.

Specific environmental conditions often favour development of bulb
diseases and in some cases crop husbandry practices can do much to
eliminate or aggravate the situation. For example, depth of planting and
time of planting and lifting can influence disease incidence. Careful

handling of bulbs, thus reducing wounding and bruising, prevents invasion
by opportunistic fungi. In store, humidity and temperature regimes can be
optimised to minimise disease progression.

Daffodil cultivars vary in their susceptibility to basal rot, some are
highly resistant while the most commonly grown cultivars are very
susceptible. All cultivars seem to be highly susceptible to neck rot.
Many factors influence the choice of cultivar grown and the most resistant
are not always the most suitable. In recent years, virus-free stocks have
been released to the industry and a programme of breeding for resistance to
Fusarium oxysporum f.sp. *narcissi* is in progress.

Elimination of the pathogen at source or reduction of inoculum level
provides a further approach to control. Fungi such as *Fusarium* are able to
survive in the soil for many years and soil sterilisation of hectares of
daffodil fields is not practical or cost effective. However, rotations of
six to seven years can reduce the inoculum level. Many bulb diseases are
carried on the outside of the bulb and the planting of good pathogen-free
stocks will do much to reduce disease development during the growing
season. Removing affected plants during the growing period will also limit
the build up of diseases in a stock.

Fungicides may be used on bulbs as protectants and or eradicants and,
in some instances, provide the only effective method of disease control.
They are most commonly used as foliar sprays and bulb dips. Insensitivity
to fungicides is a problem which has increased in recent years. Under
selection from fungicide use, insensitive variants will over a period of
time, begin to account for a greater proportion of the fungal population,
thus resulting in ineffective disease control. In particular, insensitive
strains of *Penicillium, Botrytis* and *Fusarium* are increasingly a cause for
concern. Recently, there have been several studies on the potential use
of biological control agents (BCAS) for control of bulb diseases.

Table 1. Total areas (ha) of field-grown bulbs and allied crops in England
and Wales[*]

Bulb crop	1982	1983	1984	1985	1986	1987	1988[**]	1989	1990	1991
Narcissus	3675	3505	3616	4013	3721	3814	4047	3826	3961	3972
Tulip	360	368	282	269	235	193	190	166	127	111
Iris	56	48	36	36	31	28	26	24	32	20
Anemone	150	39	39	45	40	36	31	30	27	18
Others	109	125	155	153	155	153	157	157	157	156
Total	4350	4085	4128	4516	4182	4224	4451	4203	4304	4277

[*] Source: MAFF Basic Horticultural Statistics for the United Kingdom, July
1993.
[**] ADAS interpolations due to cancellation of October 1987 census.

FOLIAR DISEASES

The most serious foliar disease "smoulder", caused by *Botrytis narcissicola*, is currently controlled by regular sprays with MBC fungicides, dicarboximide fungicide or chlorothalonil. In years when "leaf scorch", caused by *Stagonospora curtisii*, is also a problem dithiocarbamate fungicides can be incorporated in the control programme. In the first year of a two year crop, spray programmes continue after flowering; in the second year sprays are targeted in the early part of the growing season to avoid prolonged green leaf retention. Strains of *Botrytis* insensitive to MBC fungicides are now common and strains insensitive to dicarboximide fungicides and chlorothalonil have also been reported. No cultivars are resistant to these two diseases. By avoiding the planting of infected bulbs and removing affected plants, the level of infection within a crop can be reduced.

BASAL ROT AND EELWORM

Basal rot, caused by the fungus *Fusarium oxysporum* f.sp. *narcissi*, continues to be the most serious problem faced by the bulb industry. The correct management of bulbs in the dry bulb phase is of crucial importance for the maintenance of disease free stocks.

The basal rot pathogen occurs in most soils where *Narcissus* have been grown and, although the fungus is specific to this host, it has also been recorded in soils where cultivated *Narcissus* have never been previously grown. The fungus produces three types of spores: micro-conidia, macro-conidia and chlamydospores. It is the latter that enable the fungus to survive for many years in soil and on the outside of bulbs.

Root infection and disease development are closely related to temperature. The fungus is active over the range 5° - 35°C but with an optimum for activity at 25°C. Bulb to bulb spread of the disease in soil is favoured by the high planting densities currently used. Studies on fenland soils demonstrated the distance the fungus can spread in a field. Healthy bulbs were planted, touching or at a distance of 10, or 20 cm from infected bulbs. By the end of the first season 60% of the bulbs which were in direct contact with infected bulbs were rotting. At 10 cm distance 27% of bulbs were infected, and at 20 cm distance 6% of the bulbs were infected, which was a similar incidence to infection in control plots.

Complete control of basal rot is difficult to achieve mainly because of the ability of the fungus to survive in soil for prolonged periods. However, it is possible through a combination of cultural and chemical measures to significantly reduce the level of disease in a stock. Rotation should be as long as possible, preferably 8-10 years. Bulbs should be lifted and dried early to avoid exposure to warm soil temperatures. Careful handling of bulbs and removal of diseased bulbs in the field and during lifting will reduce potential sources of inoculum. Bulbs should be stored at temperatures not exceeding 17°C before planting to reduce in-store spread of the pathogen. Planting depths of 12-15 cm are favourable to shallower planting as this avoids the warm late summer/autumn soils.

The chemical control of basal rot is closely connected to that of the stem and bulb nematode (*Ditylenchus dipsaci*). Unlike the basal rot

pathogen this pest also attacks tulips, gladioli and snowdrop. Soil and
debris cleaned from bulbs are the primary source of infection. Debris from
a stock of bulbs with 11% infestation yielded 210,000 nematodes per 1000
bulbs (Hesling, 1967). Of the major cultivars grown none are highly
resistant. As with basal rot, control involves rigorous inspection,
correct rotation, removal of affected plants and chemical treatment.

Hot water treatment (HWT) at 44.4°C for 3 h is the only method of
controlling the stem nematode. The addition of 0.5% formalin ensures that
free swimming nematodes are also killed and additionally kills spores of
Fusarium thus preventing bulb to bulb contamination in the tank. The
addition of a benzimidazole fungicide to the HWT tank offers some
protection against *Fusarium* infection after planting. When HWT is carried
out within 10 days of lifting disease control is considerably improved,
however this practice often causes flower damage in the first year unless
bulbs are pre-warmed.

Providing a stock is not infested with eelworm a cold formalin dip
within 48 h of lifting can greatly reduce the level of disease. More
effective but also more expensive dips in benzimidazole fungicides can be
utilised at this stage.

In recent years it has been shown that control is further improved by
using a double-fungicide treatment i.e. a post-lifting dip in either
formalin or thiabendazole plus a pre-planting dip with thiabendazole in the
HWT. When applied annually, disease levels in a stock were substantially
reduced. However in a conventional biennial crop it could take up to six
years to achieve the 1% disease level required for export.

Approval has recently been granted for an alternative treatment using
prochloraz (as Sportak 45) either in the HWT tank or as a cold dip with or
without the addition of captan. Alternating this treatment with
thiabendazole may offer some protection against the build up of fungicide
insensitive populations of *Fusarium* and *Penicillium* sp.

ALTERNATIVES TO FORMALDEHYDE

Formaldehyde vapour is an irritant and, whilst the vapour levels near
HWT tanks are within safety limits, working conditions in these areas are
often unpleasant. Doubts about the mammalian toxicity of formalin have
been raised in the USA and Germany, and stricter control on its production
and use are being considered in these countries (Zell, 1984; Pearce, 1984).
Because of these considerations, and the fact that formaldehyde fails to
control *F. oxysporum* totally, alternatives have been sought. The aim of
the tests was to assess the efficacy of compounds in killing *F. oxysporum*
chlamydospores, and to determine whether they were phytotoxic to *Narcissus*
bulbs (Linfield, 1991).

Compounds were tested at 44.4°C for up to 3 h (Table 2). At rates of
0.25% active ingredient and above, a glutaraldehyde formulation (Cidex)
killed all spores within 160 min. Peratol, a hydrogen peroxide and
peracetic acid-containing formulation, gave 100% kill after 80 min at a
concentration of 0.5%. A thiabendazole formulation (Storite Clear Liquid)
killed all chlamydospores, but would not be expected to exert an effect on
the stem and bulb nematode. Decon 90 failed to control the fungus

adequately. Formaldehyde (as 0.5% formalin) gave good, but less than
complete control.

TABLE 2. Effect of chemicals on the survival (%) of chlamydospores of
Fusarium oxysporum f.sp. *narcissi* at 44.4°C.

| Time (min) | Concentration (%) of chemical | | | | | | | |
| | formalin | | Cidex | | Peratol | | Decon | |
	0.5	2.5	0.05	0.5	0.5	1.3	1.0	10
10	3.19	2.74	1.13	0.10	1.59	0.07	81	32
20	0.11	0.08	0.80	0.08	0.68	0.0	81	22
40	0.06	0.05	0.59	0.01	0.45	0.0	68	14
80	0.05	0.03	0.42	0.01	0.0	0.0	57	12
160	0.01	0.01	0.33	0.0	0.0	0.0	52	10
320	0.01	0.01	0.02	0.0	0.0	0.0	53	3

All the products were tested for their effect on the flower
development of *Narcissus* cvs. St. Keverne, Golden Harvest, Ice Follies and
Carlton. Bulbs had HWT at 44.4°C for 3 h in a solution containing 0.5 or
2.5% formalin. After treatment during the first week of August, bulbs were
dried and planted in mid-September. The treatment was repeated, with
formalin being replaced by the test products at the rates shown in Table 2.
The 2.5% formalin treatment led to flower malformation and corkiness of the
bulb base plate. The other four chemicals tested had no adverse effect on
flower quality and yield. Further work is in progress to determine the
effect of these four chemicals on the stem and bulb eelworm, and their
ability to control the fungus in an infested stock. Undoubtedly one of the
main factors will be their relative cost compared with formalin.

NECK ROT

In recent years neck rot has increasingly become a cause of narcissus
bulbs failing export inspections. Neck rot starts at the top of the bulb
in the old scape and surrounding tissues. After harvest the tissue becomes
progressively rotten and typically chocolate brown in colour if the
infection is caused by *Fusarium oxysporum* f.sp. *narcissi*. In other cases a
gingery rot of much slower progression is observed, caused by *Penicillium
hirsutum*. In some instances, a non-spreading form of neck rot is seen,
caused by physical damage or non-invasive pathogens. In a survey of crop
husbandry practices conducted over several years (Linfield, 1990) it was
found that "top bashing" tripled the percentage rejection when compared
with standard flailing. The time between flailing and lifting can also be
important, with an increase in neck rot the longer the bulbs are left in
the ground.

Experiments have shown that moisture is a pre-requisite for *Fusarium*
infection to take place (Table 3). Application of a fungicide at flailing
greatly reduces disease incidence. Practically this is difficult and a dip

or spray within two days of flailing would probably suffice. *Penicillium* neck rot has been little investigated although the disease has greatly increased in recent years. This fungus is frequently seen in large quantities on otherwise healthy bulb stocks and its prevalence seems to be linked with increased fungicide insensitivity.

TABLE 3. Effect of flailing-off leaves, inoculation with *Fusarium oxysporum* f.sp. *narcissi* and fungicide treatment on the development of *Narcissus* neck rot.

Treatment		% Neck Rot
Flower stalk	Inoculated with	
Left uncut	Untreated	0
Cut	Water	5.1
Cut	*Fusarium* spores[*]	27.4
Cut	*Fusarium* spores + thiabendazole[**] after 48 h	8.6
Cut	Thiabendazole + *Fusarium* spores after 48 h	8.0
Cut	Thiabendazole	1.2

[*] Spore suspension (1 x 10^6) applied as 2.5ml/bulb
[**] Applied as 1.0ml/bulb of 5ml/l Storite Clear

HOST RESISTANCE

The use of disease indexed or disease resistant stocks is an effective way of controlling disease development especially when combined with other control measures. For the fungal foliar pathogens there are few sources of resistance and the removal of diseased plants together with chemical control is essential.

About 20 viruses are known to infect *Narcissus* although many are not common. The most damaging viruses are aphid borne: narcissus yellow stripe (NYSV), narcissus degeneration virus (NDV) and narcissus white streak virus (NWSV). Yields can be seriously reduced by virus infection; 10-35% has been recorded in stocks affected with NYSV and NWSV (Asjes, 1990). Propagation schemes of virus-tested material offer an excellent opportunity for the improved health of bulb stocks. Virus-tested (VT) stocks of over 30 *Narcissus* cultivars have been produced and many of these have been released to the U.K. industry (Brunt, 1985). In Scotland, VT bulb production and certification schemes are also in place (Rankin, 1986). One problem is the rate at which virus-free bulbs become re-infected, but the wider use of VT stocks to achieve increased yield and flower quality and a reduction in export rejection is in the long term interest of the grower.

Among *Narcissus* cultivars and species there is considerable variation in susceptibility to basal rot; many of the most commonly grown commercial cultivars are very susceptible but are grown for other qualities. Significant reduction of the disease in the long term will be best achieved by growing cultivars resistant to *Fusarium*. In other crops, resistance to

Fusarium oxysporum has been found amongst the wild relatives, and this is
also the case with *Narcissus* (Linfield, 1992). When species were screened
for resistance to the pathogen it was found that accessions were either
wholly resistant (bulbs remaining uninfected), partially resistant (a
variable portion of the bulbs infected) or susceptible (all the bulbs
infected). *N. pseudonarcissus*, the species from which many of the
cultivars are derived was susceptible. Those species found to be resistant
are now being used as parents in a breeding programme at Horticulture
Research International. Progeny from these crosses will be tested to
determine the heritability of the resistance. In addition, cultivars with
tolerance of the disease, observed in the field i.e. St Keverne, are being
crossed with large-yellow trumpet types to produce Division 1 bulbs with
resistance to basal rot. In trials, continuous variation of percentage
bulb survival between progenies suggests a polygenic mode of inheritance;
there is no evidence of maternal inheritance. Parental general combining
ability is highly significant and accounts for much of the difference in
survival between progenies (Bowes *et al*., 1992). By its nature this type
of selection programme is long term, a period of 5-7 years is required to
progress from seed production to the production of a bulb that will flower.
Similar selection programmes are underway in Holland for resistance to
Fusarium of tulips, gladioli and lily (van Eijk & Eikelboom, 1983;
Straathof *et al*., 1993).

MACROPROPAGATION

The slow rate of multiplication of *Narcissus* can be enhanced by the
use of the chipping technique. Bulbs are divided into a number of wedge-
shaped longitudinal segments, normally 8-16 from each bulb. These
propagules (chips), on incubation in a suitable moist substrate, produce
adventitious bulbils from the abaxial scale surface close to the junction
with the base plate. *Narcissus* bulbils grow to flowering size in three to
four years, and the rate of bulb multiplication is greatly increased. Such
a method of enhanced multiplication is important for the propagation of
virus-tested, disease resistant and other stocks. Whilst on a small scale
results have been satisfactory, commercial exploitation of the technique
has shown the need for adequate disinfection of chipped tissues and there
have been reports of benomyl-insensitive strains of fungi associated with
Narcissus tissue.

The major fungal pathogens in large scale chipping have been *Rhizopus*
spp., largely due to the heat generated when large volumes of vermiculite
are wetted. This can be avoided by ensuring the vermiculite is cool before
chip incubation. Commonly found fungi on bulbs include *Penicillium* spp.
Trichoderma spp. and *F. oxysporum* f.sp. *narcissi*. Control of all fungi in
chipping is important because saprophytes otherwise become established and
invade through cut surfaces to compete with the developing bulb for
available nutrients. The most effective fungicide for use during chipping
was captafol until its withdrawal in 1990. Since then a wide range of
compounds have been tested. This included a systemic fungicide effective
against benzimidazole-insensitive strains of *Penicillium*, but which was
phytotoxic to the bulbils. Other fungicides tested include thiram,
chlorothalonil, thiabendazole and captan.

Whilst all these fungicides gave some benefit over no-treatment, only
captan proved to be as effective as captafol on the numbers and weights of

bulbils produced. Combining captan with other broad-spectrum fungicides
may further enhance yield and disease control.

After bulbil formation a pre-planting dip in 1% captan or 1%
zineb/maneb plus a systemic compound has been found to improve skin
quality, counteract basal rot and often results in higher yields.

INSENSITIVITY TO FUNGICIDES

Insensitivity to fungicides has become an increasing problem in bulb
production in recent years. Forms of a fungus insensitive to one fungicide
are usually also insensitive to related fungicides with the same mode of
action. In some instances isolates have been found that are insensitive to
fungicides belonging to more than one type . Benzimidazole insensitive
isolates of *Penicillium* from *Narcissus* have been found extensively in the
UK, and in Holland resistant forms have been obtained from lily and iris.
Benzimidazole insensitive strains of *F. oxysporum* f.sp. *gladioli* (fusarium
yellows of gladiolus) have been found in Holland and recently insensitive
strains of *F. oxysporum* f.sp. *narcissi* to the related fungicide
thiabendazole have also been discovered.

The use of fungicide mixtures may reduce the build up of insensitive
strains and a benomyl/mancozeb mixture has given good results in trials
(ADAS, 1986).

BIOLOGICAL CONTROL

Biological control of *F. oxysporum* formae speciales has proved
successful (Locke *et al.*, 1985; Sneh *et al.*, 1985). The potential for
integrated or biological control of basal rot was reported by Langerak
(1977) who found no colonisation by *F. oxysporum* f.sp. *narcissi* in the
rhizosphere of fungicide treated bulbs when the fungicide levels decreased
to non-toxic levels. Roots were colonised by *Trichoderma* spp. and
Penicillium spp. The same fungi inhibited growth of the pathogen *in vitro*
and *in vivo*, so enhancing the control provided by fungicides alone. Beale
& Pitt (1990) identified several soil- and bulb-borne micro-organisms
antagonistic to the pathogen, some of which were tolerant to thiabendazole.
In pot trials, *Minimedusa polyspora* and a *Streptomyces* sp. reduced the
incidence of disease by over 50%. In a 1-year field trial these
antagonists and *Trichoderma viridae* reduced infection and increased bulb
yields. The integrated use of *T. viridae* with thiabendazole gave bulb
yields significantly greater than either *T. viridae* or thiabendazole alone
whilst the integrated use of *M. polyspora* and thiabendazole resulted in
significantly greater flower numbers and significantly fewer infected bulbs
than either treatment alone. Further work has indicated that in some years
the antagonists are adversely affected by environmental conditions in the
field and their ability to be effective in a 2-year crop remains to be
established.

A formulation of *Streptomyces griseoviridis* (Mycostop), already
available in several countries, has been shown in trials in the UK to be
effective against *Fusarium* in cyclamen. Its ability to control basal rot
in *Narcissus* is currently being evaluated. Initial results indicate that
the organism effectively reduces the incidence of infection over a 1-year

period.

There is clearly potential for the biological and integrated control of *F. oxysporum* f.sp. *narcissi*. To establish the full potential of antagonists, efficacy tests over a 2-year period are required. Antagonists that are ecologically suited to the rhizosphere of bulbs are more likely to be successful over this time period, and more effective at excluding the pathogen from its niche. Compatibility with pesticides used in bulb production must be considered.

FUTURE PROSPECTS

Future disease control in bulbs will be dependant on good crop husbandry practices, the availability of effective control agents and disease resistant stocks. The production of *Narcissus* as an annual crop, instead of employing the normal biennial system, has been shown to significantly reduce basal rot and increase the effectiveness of fungicidal control measures. A fundamental change such as this may be worthwhile where a susceptible cultivar is grown for export or where small expensive stocks are grown. Availability of an alternative to formalin, for use in the HWT system, that gives total control of *Fusarium* and is effective against eelworm is still required. In addition to *Fusarium* and eelworm control such a chemical needs to be compatible with other fungicides used in bulb dips, of low cost, usable in early HWT without causing flower damage and easy to dispose of without the risk of environmental damage.

Virus-tested and indexed stocks are currently available in small quantities; their continued use and maintenance will do much to reduce the spread of virus infection. Breeding for resistance to *F. oxysporum* f.sp. *narcissi* is in progress and flowering size bulbs are now available. Release of material to the industry and its subsequent multiplication will provide further opportunities for improvement of stocks. With an increased concern over the continued large scale use of pesticides, biological and integrated control continues to offer an attractive alternative. Several micro-organisms have been identified as antagonists of *Fusarium* and their ability to control diseases in a biennial crop should be assessed. Much will depend on appropriate formulation and their ability to survive a range of environmental conditions over several seasons.

Research on pathogens of *Narcissus* and their control at HRI is conducted with funding from the Ministry of Agriculture Fisheries and Food (England & Wales).

REFERENCES

ADAS, (1986) Control of diseases of bulbs. Booklet 2524. Ministry of Agriculture, Fisheries and Food. (Publications), Alnwick, 63 pp.
Asjes, C.J. (1990) Production for virus freedom of some principal bulbous crops in the Netherlands. *Acta Horticulturae*, **266**, 517-529.
Beale, R.E. ; Pitt, D. (1990) Biological and integrated control of *Fusarium* basal rot of *Narcissus* using *Minimedusa polyspora* and other micro-organisms. *Plant Pathology*, **39**, 477-488.
Bowes, S.A.; Edmondson, R.N.; Linfield, C.A.; Langton, F.A. (1992). Screening immature bulbs of daffodil (*Narcissus* L.) crosses for

resistance to basal rot disease caused by *Fusarium oxysporum* f.sp. *narcissi*. *Euphytica*, **63**, 199-206.

Brunt, A.A. (1985) The production and distribution of virus-tested ornamental bulb crops in England: principles, practice and prognosis. *Acta Horticulturae*, **164**, 153-161.

Hesling, J.J. (1967) The distribution of eelworm in a naturally-infested stock of *Narcissus*. *Plant Pathology*, **16**, 6-10.

Langerak, C.J. (1977) The role of antagonists in the chemical control of *Fusarium oxysporum* f.sp. *narcissi*. *Netherlands Journal of Plant Pathology*, **83**, 365-381.

Linfield, C.A. (1990) Neck rot disease of *Narcissus* caused by *Fusarium oxysporum* f.sp. *narcissi*. *Acta Horticulturae*, **266**, 477-481.

Linfield, C.A. (1991) A comparative study of the effects of five chemicals on the survival of chlamydospores of *Fusarium oxysporum* f.sp. *narcissi*. *Journal of Phytopathology*, **131**, 297-304.

Linfield, C.A. (1992) Wild *Narcissus* species as a source of resistance to *Fusarium oxysporum* f.sp. *narcissi*. *Annals of Applied Biology*, **121**, 175-181.

Locke, J.C.; Marois, J.J.; Papavizas, G.C. (1985) Biological control of *Fusarium* wilt of greenhouse grown chrysanthemums. *Plant Pathology*, **69**, 167-169.

Pearce, F. (1984) Coming to terms with a carcinogen. *New Scientist*, **104**, No. 1426, 14-15.

Rankin, R.A. (1986) Narcissus certification in Scotland. *Acta Horticulturae*, **177**, p. 578.

Sneh, B.; Agami, O.; Baker, R. (1985) Biological control of *Fusarium*-wilt in carnation with *Serratia liquefaciens* and *Hafnia alvei* isolated from the rhizosphere of carnation. *Phytopathologische Zeitschrift* **113**, 271-276.

Straathof, Th.P.; Jansen, J.; Loffler, H.J.M. (1993) Determination of resistance to *Fusarium oxysporum* in *Lilium*. *Phytopathology*, **83**, 568-572.

van Eijk, J.P.; Eikelboom W. (1983) Breeding for resistance to *Fusarium oxysporum* f.sp. *tulipae* in tulip (*Tulipa* L.) 3. Genotypic evaluation of cultivars and effectiveness of pre-selection. *Euphytica*, **32**, 505-510.

Zell, R. (1984) Campaign against formaldehyde grows. *New Scientist*, **103**, No. 1420, p. 6.

THE USE OF IPRODIONE AND IMAZALIL FOR DISEASE CONTROL OF SEED POTATO
TUBERS.

D. A. JAMES, S. HIGGINBOTHAM.

Rhône-Poulenc Agriculture Limited, Fyfield Road, Ongar, Essex, CM5 OHW

ABSTRACT

Iprodione and imazalil have been extensively used in the UK and in Europe for the control of most
seed-borne diseases of potato tubers, when applied prior to planting. The use of new formulations
suitable for the latest application equipment is described. Benefits of iprodione for optimising tuber
grade are considered.

INTRODUCTION

Seed potatoes in store are vulnerable to a range of seed-borne diseases which can affect both the
vigour of the seed, as well as the quality of progeny potatoes. These diseases include silver scurf
(*Helminthosporium solani*), skin spot (*Polyscylatum pustulans*), dry rot (*Fusarium* spp.), gangrene
(*Phoma foveata*). In addition black scurf and stem canker (*Rhizoctonia solani*) causes problems in the
growing crop. Increasingly, seed potatoes are treated with a fungicide during storage to inhibit the
growth of these pathogens in store and to prevent the infection of progeny (Hide & Cayley, 1981;
Hide *et al.*, 1987).

In 1976 imazalil developed by Janssen Pharmaceutica, was found to be active against silver scurf,
skin spot, gangrene and dry rot (Cayley *et al.*, 1981; Cayley *et al.*, 1983), giving a viable alternative to
thiabendazole, with which some resistance problems were occurring (Hide *et al.*, 1988). Imazalil was
launched in the UK as an emulsifiable concentrate by Rhône-Poulenc during 1988 and an improved
aqueous concentrate formulation is currently marketed as FUNGAZIL 100 SL.

Iprodione was discovered in 1971 by Rhône-Poulenc, and is active in potatoes against *Rhizoctonia
solani* (Cayley & Hide, 1980; Hide & Cayley, 1982; Hide & Sandison, 1985). This disease can reduce
yields by affecting foliar development and by pruning stolons, reducing tuber numbers. Also it can
reduce the quality of potato skin by development of black clusters of sclerotia. Traditionally,
treatments have been made prior to planting to control the disease on the mother tuber and reduce the
infection of the daughter tubers. However, the disease does develop mycelia in store (personal
communication by Dr. S. J. Wale) and there is a case for eliminating the disease in store as well as
control in the field. The benefits of this treatment for optimising target grade production have only
recently been fully realised (Hardwick & Bevis, 1987). Wettable powder formulations such as Rovral
WP have drawbacks in terms of handling particularly with the new ULV/LV potato treatment
equipment, where there is less agitation; a ready to use product is preferred. Thus a more suitable
500g /l flowable formulation of iprodione, EXP80038 has been developed. The use of imazalil and
iprodione together controls all major seed-borne potato diseases, in store and post-planting.

MATERIALS AND METHODS

<u>Formulations used;</u>

Active ingredient	g A.I./t	Product	Application Rate
Imazalil	10	Fungazil 100 SL	100 ml/t
Iprodione	100	EXP80038	200 ml/t
Thiabendazole	40.5	Storite Flowable	90 ml/t *
Tolclofos-methyl	125	Rizolex Flowable	250 ml/t

* used as a dip only

<u>Dip tests</u>

<u>Treatment</u>

Tubers were treated by immersing them in fungicide suspensions for five minutes. Commercial formulations were made up in water to produce the required concentrations of active ingredient.

<u>Gangrene (*Phoma foveata*)</u>

P.foveata was cultured in 2% malt extract solution and after ten weeks the cultures were macerated and diluted with soil and water to give slurries of two inoculum levels, Pentland Crown tubers were immersed in the soil slurry, allowed to dry overnight and then given four cut and crush wounds. After fungicide treatment the tubers were stored as three replicates of fifty tubers at 5 C.

<u>Black scurf and stem canker (*Rhizoctonia solani*)</u>

In one test King Edward tubers with black scurf were treated with fungicides and slices of tissue bearing sclerotia incubated at 15 C. In a second test, ten moderately infected tubers of each of three varieties (Wilja, Maris Peer and Pentland Dell) were washed and then treated with fungicide. After drying, tubers were potted in a peat/sand compost and incubated at 15 C until emergence. The degree of stem canker infection was recorded after four weeks on a 0-4 scale.

<u>Skin spot (*Polyscytalum pustulans*)</u>

Three replicates of ten King Edward tubers affected by skin spot were fungicide treated, and dried before potting in a peat/sand compost. Pots were incubated at 15 C until emergence and then grown on for a further nine weeks. Stem base infection was recorded on a 0-3 scale.

<u>Silver scurf (*Helminthosporium solani*)</u>

Three replicates of twenty severely affected (>30% cover) Maris Piper tubers were washed and treated. After drying the tubers were incubated at 15 C for six weeks when *H. solani* conidia were collected by washing and centrifuging. Numbers of conidia were counted using a haemocytometer. In a second test, tubers with discrete silver scurf lesions were treated and three replicates of ten dry tubers were stored at 10 C in sealed plastic bags. The increase in lesion diameter was measured.

<u>Field Trials</u>

Two trials each using the same two seed stocks, Maris Piper and Pentland Squire, were carried out during 1990/1, Both stocks were infected with black scurf and silver scurf, Pentland Squire was more severely infected. In 1991/2 a third experiment was carried out on a stock of Désirée with a very low level of black scurf infection. Treatments were varied between sites and all trials were in the Edinburgh area.

Application was carried out using electrostatically charged ULV equipment (Microstat) with a total spray volume of 900 ml per tonne of seed, and treatments were applied over a roller table at a throughput of 4 tonnes per hour. Seed was stored under controlled temperature during the winter, and tubers were generally at "eyes open" to early chitting at treatment. Treated tubers were stored pre-planting under the normal conditions.. The seed was planted as randomised block field trials with four replicates and the husbandry of the trials was as recommended for local conditions. Numbers of stolons and numbers pruned were assessed and were scored as slight (1-33% stem affected), moderate (34-66%) or severe (>67%) and the results expressed as an index.

$$\text{Index} = \frac{(1 \times \% \text{ slight stems}) + (2 \times \% \text{ moderate stems}) + (3 \times \% \text{ severe stems})}{100}$$

At harvest, yields in each grade were taken and samples of fifty tubers were assessed for skin disease as occurrence (% tubers infected) and severity (% tuber surface area infected). These assessments were repeated after six months storage at 4°C and 8°C .

RESULTS

<u>Dip tests</u>

All treatments except iprodione gave good control of gangrene and reduced dry rot incidence. (Table 1)

TABLE 1. Incidence of disease on wounded tubers immersed in solutions of different chemicals (% wounds infected after 12 weeks)

Treatment	gAI/litre	P.foveta (1)		F.solani (2)		F.sulphureum (2)	
		Low	High	Low	High	Low	High
Untreated		65	58	35	72	20	28
Iprodione	18	4	27	34	67	24	34
Imazalil + Iprodione	2 + 14	0	3	0.3	3	2	2
	3 + 21	0	0	1	2	2	2
Imazalil	2	1	2	1	4	0	1
Thiabendazole	2	0	2	0	2	0	0

(1) Assessed after 12 weeks.   (2) Assessed after 8 weeks.

R.solani sclerotia germination and hyphal growth were significantly decreased with iprodione. Other active ingredients were less effective. Iprodione treatment prevented development of stem canker.   Stem base browning caused by *Polyscylatum pustulans* was reduced by imazalil and in its mixtures with iprodione (Table 2).

TABLE 2. The effect of different chemicals on seed-borne diseases

Chemical	g.AI/litre	R.solani (%)		P.pustulans (%)
		Germination	Stem canker	stems
Untreated		65	29	47
Iprodione	18	2	0	23
Imazalil Iprodione	2 + 14	0	0	2
	3 + 21	0	0	0
Imazalil	2	34	30	2
Thiabendazole	2	19	12	15

Both imazalil and iprodione greatly reduced conidial production by . Imazalil with or without iprodione had the greatest effect in reducing the increase in lesion size. Thiabendazole was ineffective (Table 3).

TABLE 3. Number of *Helminthosporium solani* conidia following fungicide treatment and increase in lesion size, 10 weeks after treatment.

Chemical	AI/litre	Number (x 10^4)	% increase
Untreated		187.7	55
Iprodione	18	17.7	27
Imazalil + Iprodione	2 + 14	13.9	24
	3 + 21	6.3	17
Imazalil	2	7.3	10
Thiabendazole	2	157.0	48

<u>Field Trials</u>

Stem canker was greatly decrease by iprodione and tolclofos methy in both cultivars. However with Pentland Squire stolon pruning was not proportionally decreased (Table 4).

TABLE 4. Stem canker development in growing crops, July 1991

Treatment	Timing		Pentland Squire		Maris Piper	
	Autumn	Spring	stem canker Index	% stolon Pruning	stem canker Index	% stolen Pruning
Untreated			34.3	36	8.3	25.3
Tolclofos methyl		x	1.3	20	0	0
Iprodione		x	1.0	27	1.3	0
Imazalil +	x		4.3	15		
Iprodione		x				

The largest yield of Pentland Squire tubers was achieved by the combination treatment of imazalil applied in the autumn, followed by iprodione applied in spring, this was achieved by a marked increase in the yield of seed grade tubers. In Maris Piper, total yield was not affected but yields of tubers 35-55mm were significantly increased (Table 5).

TABLE 5. Yield of tubers (t/ha)

| Treatment | Timing | | Pentland Squire | | | Maris Piper | | |
	Autumn	Spring	Total Yield	Yield 35-55mm	Yield >55mm	Total Yield	Yield 35-55mm	Yield >55mm
Untreated			48.8	8.2	40.0	41.0	17.6	22.8
Tolclofos methyl		x	48.7	10.5	37.7	45.7	21	21.2
Iprodione		x	48.5	11.6	36.6	43.1	24.3	17.7
Imazalil	x		50.4	9.6	40.3			
Imazalil + Iprodione	x	x	52.6	14.7	37.7			

After harvest and on removal from long term storage, iprodione treatments had significantly fewer tubers with black scurf compared with untreated. Fewest affected tubers were from seed treated with imazalil in the autumn and with iprodione at chitting (Table 6).

TABLE 6. Assessment of black scurf before and after long term storage (4°C)

| Treatment | Timing | | Pentland Square | | Maris Piper | |
	Autumn 1990	Spring 1991	In store Oct. 1991	Out store April 1992	In store Oct. 1991	Out store April 1992
Untreated			32 (1.2)	26 (1.1)	17 (0.8)	8 (0.4)
iprodione		x	13 (0.6)	5 (0.3)	2 (0.1)	0 (0)
imazalil + iprodione	x	x	6 (0.3)	6 (0.4)		

Treating seed with iprodione increased total yield and yield of 35-55mm but decreased the yield of large tubers (Table 7).

TABLE 7. Yield of Tubers (t/ha)

| Treatment | Timing | | | Desirée | |
	Atumn	Spring	Total Yield	Yield 35-55	Yield >55mm
Untreated			43.3	28.5	14.4
imazalil	x		42.0	24.9	16.0
iprodione		x	45.5	31.5	13.4

After storage silver scurf was decreased by treating seed with imazalil although iprodione also had a useful effect. The severity of skin spot was reduced by imazalil. (Table 8).

TABLE 8. Silver scurf and skin spot after long term storage (4°C), May 1993

Treatment	Timing		Desirée	
	Autumn 1991	Spring 1992	% tubers infected skin spot	(% severity) silver scurf
Untreated			72 (11.7)	36 (12.6)
imazalil	x		59 (7.8)	16 (2.5)
iprodione		x	61 (11.2)	23 (5.1)

ACKNOWLEDGEMENTS

The authors would like to thank S.M.Hall and G.A.Hide of Rothamsted Experimental Station for results of dip tests, S.J.P. Oxley, P.J. Burgess S. Bowen of SAC for the field trials and G.I. Ingram of Rhône-Poulenc Agriculture for assistance in the preparation of this paper.

REFERENCES

Cayley, G.R.; Hide, G.A. (1980) Uptake of iprodione and control of storage disease Potato Research 25, 89-97.

Cayley, G.R.; Hide, G.A.; Read, P.J.; Dunne, Y. (1983) Treatment of potato seed and ware tubers with imazalil and thiabendazole for control of silver scurf and other storage diseases. Potato Research 26, 63-173.

Cayley, G.R.; Hide, G.A.; Tillotson, Y. (1981) The determination of imazalil on potatoes and its use in controlling potato storage diseases. Pesticide Science 12, 103 (1981) Tests of Agrochemicals and Cultivars (Annals of Applied Biology 97, Supplement), No. 2, 24-25.

Hide, G.A.; Cayley, G.R. (1982) Chemical techniques for control of stem canker and black scurf (*Rhizoctonia solani*) disease of potatoes. Annals of Applied Biology 100, . 105-116.

Hide, G.A.; Sandison. J.P. (1985) *Tests of Agrochemicals and Cultivars* (Annals of Applied Biology 106, Supplement), No. 6, 58-59.

Hide, G.A.; Hall, S.M.; Boorer, K.J. (1988) Resistance to thiabendazole in isolates of *Helminthosporium solani*, the cause of silver scurf disease of potatoes. Plant Pathology 37, 377-380.

Hide, G.A.; Read, P.J.; Sandison. J.P.; Hall, S.M. (1987) *Tests of Agrochemicals and Cultivars* (Annals of Applied Biology 110, Supplement), No. 8. 72-73.

Hardwick, N.V.; Bevis, A.J. (1987) Proceedings Crop Protection in Northern Britain 1987, pp. 153-158.

A COMPARISON OF TECHNIQUES AND DEVELOPMENTS IN THE APPLICATION OF PENCYCURON
FORMULATIONS TO SEED POTATOES

A.C. ROLLETT, D.M. ROBERTS, D.B. MORRIS

Bayer plc, Crop Protection Business Group, Elm Farm Development Station,
Great Green, Thurston, Bury St. Edmunds, Suffolk, IP31 3SJ

ABSTRACT

Pencycuron was successfully applied to potato tubers using a wide
range of application techniques in tests carried out from 1988 to
1993. Pre-planting treatments made with the flowable formulation
using roller table applicators achieved efficiences in the range
40-70 % of target dose. Scope for obtaining the higher levels was
illustrated where improved target area, electrostatic charging or
higher water volumes were used. Treatments applied at planting
using dry powder formulations gave efficiences of 45-60 % of
target dose but tended to be more variable. The use of dispenser
systems for both formulation types at planting gave comparable
results and offered significant practical benefits.

INTRODUCTION

The fungicide pencycuron has shown particular activity against black
scurf (*Rhizoctonia solani*) on potatoes and trials commenced in the UK during
1985 with dry seed treatment (DS) and suspension concentrate formulations.
The techniques used before 1988 were described by Rollett *et al.* (1988) and
biological results reported by Adam & Malcom (1988).

During 1988, work commenced with the application of the flowable seed
treatment (FS) formulation of pencycuron using commercial in store
equipment.

This paper reports on the efficiency, in terms of chemical loading, of
the various application techniques from 73 tests carried out between 1988
and 1993. In addition to comparing existing equipment some modifications
have been tested including dispenser systems for use at planting.

MATERIALS AND METHODS

All tests were carried out with seed potatoes, mostly graded 35-55 mm,
and with the pencycuron formulations given in Table 1.

TABLE 1. Formulation details and application rates.

Trade name	Active ingredient	Formulation	Rate of Use
Monceren® DS	pencycuron	12.5 % DS	200 g/100 kg seed
Monceren® flowable	pencycuron	250 g/l FS	60 ml/100 kg seed

<u>Pre-planting application</u>

Pencycuron FS was applied by various applicators (Table 2) mounted over standard inspection roller tables with a roller pitch of c. 70 mm. Some work was also done using a table with a narrow roller pitch of 60 mm to improve the target area (Fig. 1). The roller tables were fully loaded and the applicator output and roller speed were adjusted to achieve the required dose.

TABLE 2. Pre planting application techniques.

Applicator	Atomiser	Volume l/t
Fischer	Single spinning disc	0.6
Mantis LK250	Twin spinning disc	0.6
Microstat	Single 'charged' spinning disc	0.6
Mistral	Twin fluid sonic nozzle	2 - 3
Hydraulic	Hollow cone 65-80°	2 - 3

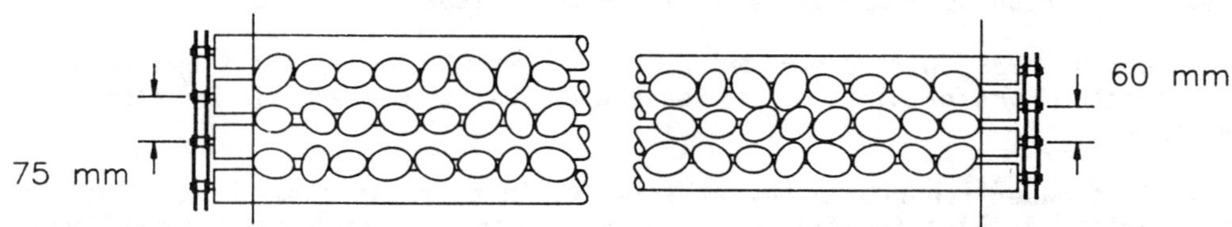

FIG 1. Effect of roller pitch on potato target area.

<u>At planting application</u>

Tests were performed using various makes of cup feed automatic planters and a Cramer Minor was used for much of the initial work.

Pencycuron DS was applied in the planter hopper by the standard layering technique using 1, 2 or 3 chemical layers during filling with potatoes (Fig. 2). A simulation of this technique as used in trials was achieved by treating tubers in a polythene bag (Rollett *et al*., 1985).

On-planter dispensers, including a powder applicator and a liquid applicator, metered chemical dose to the tuber pick-up area (Fig. 2). The liquid applicator dispensed pencycuron FS at the standard rate in volumes of 8-13 l/t through 65-80° cone nozzles.

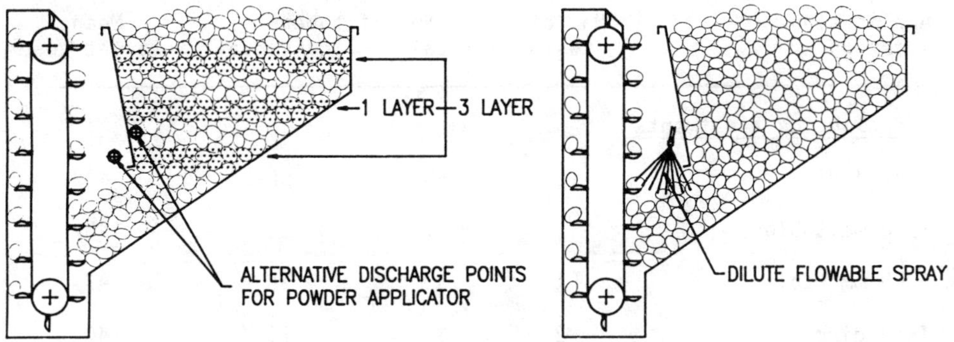

FIG 2. On-planter application positions.

Tuber sampling and analysis

Tubers were usually sampled systematically when the equipment had reached a steady state either from the treatment table or the planter cups to give 5 lots of 5 or by sampling 10 tubers from the treated bulk. They were placed individually or in fives in polythene bags and frozen prior to analysis.

Chemical loading analysis was based on the extraction of the Ceres red dyestuff, present in both formulations at a standard rate, from frozen tubers with dichloromethane. Absorption of the extract was determined at a wavelength of c. 500 nm using a Compur mini photometer. From the extract volume, weight of the tuber and photometer reading the tuber loadings were calculated.

RESULTS

A summary of the tuber loading results obtained from the tests conducted with the different application methods is given in Table 3. This data gives the total number of tubers analysed and the overall mean loading expressed as a percentage of the target dose.

The distribution of loadings on individual tubers is shown in Table 4 following an analysis of the data according to Tukey (1977). The median value of the data is similar to the overall mean value and 50 percent of the results lie within the box.

The results were obtained from different numbers of tests on different seed lots and at different times therefore they should be interpreted with care.

Biological data were obtained from some of the treated potatoes but this has not been included in this report. In general differences between treatments were small which reflects on the high activity of pencycuron against black scurf.

TABLE 3. Tuber pencycuron loading results.

Application method	No. of tests	No. of tubers Total	Individual	Mean % target dose
Pre planting treatments				
Single disc	12	165	90	44
- 70 mm table	1	30	15	43
- 60 mm table	1	30	15	47
Twin disc	2	35	20	40
Microstat	6	150	20	72
Mistral	2	51	10	57
Hydraulic	6	105	30	58
At planting treatments				
In bag	2	49	49	55
Single layer	3	75	75	46
Double layer	3	75	30	47
Triple layer	6	150	105	65
Powder dispenser	10	243	98	43
Spray nozzle	21	590	65	48

DISCUSSION

All results were obtained from carefully supervised tests and therefore the overall achievement of a tuber loading of 40-70 % of target dose was considered high. In practice tuber treatment loadings would often be expected to be lower depending on uniformity and cleanliness of seed and operation of the equipment.

Roller table applicators depend on the table being fully and evenly filled with potatoes, however with a 75 mm pitch roller only c. 40 % of the sprayed area is covered with tubers (Anon, 1992). The target area can theoretically be increased by 20 % using rollers with a reduced pitch (Fig. 1). A preliminary test comparing two tables with different roller pitches and using a spinning disc applicator showed a small improvement from 43 to 47 % of target dose. Further developments on the feeding and distribution of tubers on roller tables are being pursued.

TABLE 4. Distribution of tuber pencycuron loadings from individual tuber samples - Tukey analysis.

Application method	Percentage of target dose				
	0	50	100	150	(>200)

Pre planting treatments

Single disc	⊦----[*]-----⊣ o
- 70 mm table	⊦-----[*]----⊣
- 60 mm table	⊦--[*]----⊣
Twin disc	⊦--[*]------⊣ ⟨□⟩
Microstat	⊦--[*]------------⊣
Mistral	⊦----[*]⊦⊣
Hydraulic	⊦--[*]-------------⊣ ⊕ ⊕

At planting treatments

In bag	[*]------------⊣ o
Single layer	⊦--⊦[*]-------------⊣ o o o ⟨□⟩
Double layer	⊦--[*]-----------⊣
Triple layer	⊦-------[*]--------------------⊣ o ⟨o⟩
Powder dispenser	⊦--[*]-------------------⊣ o o □ ⟨□⟩
Spray nozzle	⊦--------[*]--------⊣ ⊕ □

* - median value, box - 50 % of value, o - outlier value, □ - extreme value

Spinning disc applicators are widely used for application on roller tables as they use undiluted product and give even distribution across the table and therefore tuber surfaces. Very consistent results were obtained in the 40-50 % of target range. Losses tended to occur due to a build-up of soil on the rollers and some excessively overloaded tubers were found when the product collected along the table edge.

The use of electrostatic charging vastly improved deposition of chemical (72 %) but increased variability. The reliability of this method in practice is a problem due to the difficulty in maintaining an even charge distribution. Higher water volumes using hydraulic or Mistral applicators also improved chemical loading (57-58 %). Dilution resulted in a more complete cover of the tuber surface and reduced losses on the rollers as the greater moisture allowed secondary redistribution.

Applications at planting have the advantage that tubers are treated as they are planted and consequently chemical losses prior to planting are eliminated. Furthermore any of the applied chemical not carried on the seed tuber can be incorporated close to the tubers in the furrow. Tuber loadings may therefore underestimate the effective dose.

Powder formulations can be effectively distributed in the planter hopper using the layering technique with more layers tending to give better tuber loadings (61 % of target). This compared well with the "in bag" experimental technique (55 % of target).

On-planter dispenser systems are likely to be restricted to planters with cup feed spacing mechanisms since a certain amount of movement of the tubers in the pick-up area is necessary to give even distribution of chemical on the tuber surfaces. Both powder and liquid applicators gave tuber loadings comparable to other techniques, 46-51 % of target, whilst saving the operator time and reducing chemical exposure.

The scope for improving tuber chemical loadings by improved roller table treatment arrangements have been demonstrated and this could result in improved product performance or lower application rates. On-planter application techniques may not be improved so easily but dispenser systems offer significant practical benefits.

ACKNOWLEDGEMENTS

The authors thank their colleagues who participated in the experimental work, Mr C.P. Pill for the analytical work and farmers and co-operators, in particular Mr R.P. Baker, for supplying seed, treatment facilities and trials sites.

REFERENCES

Adam, N.M.; Malcom, A.J. (1988) Control of *Rhizoctonia solani* in potatoes in the UK with pencycuron. *British Crop Protection Conference - Pests and Diseases 1988*, **3**, 959-964.
Anon (1992) Potato Working Group information sheet no 2. BCPC Publications.
Rollett, A.C.; Roberts, D.M.; Malcom, A.J.; Wainwright, A. (1988) Techniques for the application of pencycuron to control black scurf (*Rhizoctonia solani*) on potatoes. In: Applications to seeds and soil, T.J. Martin (Ed.), *BCPC Monograph no 39*, Thornton Heath: BCPC Publications, pp. 363-370.
Tukey, J.W. (1977) Exploratory data analysis. Reading, Mass Addison-Wesley.

CHEMICAL SEED TREATMENT FOR FUNGAL DISEASE CONTROL IN PROGENY TUBERS

A.C. CUNNINGTON

Potato Marketing Board, Experimental Station, Sutton Bridge, Spalding, Lincs. PE12 9YB.

ABSTRACT

Seed potatoes were sprayed with a range of fungicidal treatments immediately after harvest and/or before traying up for sprouting using hydraulic or electrostatic applicators.
Imazalil treatment resulted in some delays in the emergence of crops but this did not result in any effect on saleable yield (i.e. 45-85mm fraction).
At harvest, silver scurf incidence on the progeny was reduced by treating with fungicides containing either imazalil or prochloraz-manganese complex. Development of the disease during subsequent storage of the progeny was suppressed but not prevented by the treatments in comparison with the untreated control.

INTRODUCTION

There has in recent years been an increasing need to address the problems of skin blemish diseases, notably silver scurf (*Helminthosporium solani*) and skin spot (*Polyscytalum pustulans*), with an integrated control strategy (Hide, 1992). Such a strategy would include measures like the choice of clean seed, early harvesting of the crop and the use of dry curing in store. A further area in which control of these diseases can be addressed is that of chemical seed treatment.

However, the use of fungicides for the control of silver scurf, in particular, has been hampered by the development of resistance by the pathogen to one of the key products used, thiabendazole (TBZ) (Hide *et al.*, 1988) and by a lack of new products to use as alternatives to TBZ or the specialist applied 2-aminobutane.

This trial, started in 1988, was designed to evaluate the use of some new chemical treatments which are either under development or have been recently introduced to the commercial market and to look at their effects on disease levels at harvest and, subsequently, after about six months' storage of the progeny crop grown from the treated seed.

EXPERIMENTAL DETAILS

The treatments used in the trial are detailed in Table 1. Chemicals were applied using hydraulic sprayers fitted with hollow cone nozzles except in year 3 when a Microstat electrostatic machine was used. Volumes applied were 2 l/t and 720 ml/t with the hydraulic and Microstat respectively. In each year, untreated tubers were included as controls.

Treated seed was chitted prior to planting in late March/April each year at two sites. One site (i.e. grower) was used throughout the three years of the trial (site A). The other site was at a different location in each season (sites B, C and D in 1989, 1990 and 1991 respectively).

Locations and soil types for each of the sites were: Site A,Cambs., peaty clay loam; Site B, Lincs., medium silt; Site C, Cambs., silty clay loam; Site D, Cambs., clay loam over chalk.

TABLE 1. Chemical treatments employed and application dates

Year / seed stock / treatment no. /chemical	Application date	
Year 1 (crop year 1989): Maris Piper SE2		
1. Prochloraz Mn / tolclofos-methyl	November 1988	(at lifting [L])
	January 1989	(pre-chitting [PC])
2. Imazalil [a] + iprodione [b]	November 1988	[L]
	January 1989	[PC]
3. Imazalil / thiabendazole [d]	November 1988	[L]
	January 1989	[PC]
Year 2 (crop year 1990): Maris Piper SE1		
1. Prochloraz Mn / tolclofos-methyl	September 1989	
2. Imazalil [c]	September 1989 + January 1990	
3. Imazalil / thiabendazole [d]	September 1989 + January 1990	
Year 3 (crop year 1991): Maris Piper SE1		
1. Prochloraz Mn / tolclofos-methyl	November 1990	
2. Imazalil [c]	November 1990 + January 1991	
3. Imazalil / thiabendazole [d]	November 1990 + January 1991	

a: Fungaflor C, b: Rovral, c: Fungazil (Rhône Poulenc Agriculture)   d: Extratect (MSD Agvet)
January applications made at Sutton Bridge Experimental Station; all others carried out in Scotland at seed source. Untreated control = Treatment 4.

Emergence assessments (1990, 1991 only) were made on the growing crop. Crops were grown to the normal commercial practice of each grower and harvested using standard commercial equipment on the following dates:

	1989	1990	1991
Site A	18 October	25 October	8 October
Site B/C/D	8 October (B)	19 October (C)	7 October (D)

Yield measurements were made on each crop at harvest at Sites A, C & D. At site B (year 1) yield could not be accurately measured as only part of each of the 0.2 ha plots was harvested for storage at Sutton Bridge Experimental Station.

Prior to storage, disease incidence was assessed on 6 x 10kg samples taken from each treatment. Crops were then cured at 10-12°C for 7-14 days and then the temperature decreased to 5-6°C. Storage was undertaken in a commercial box store and a further 6 x 10kg samples per treatment were placed in nets within the one-tonne boxes. Two nets were placed in each of the three boxes stored. The store was unloaded after 6-7 months and assessments made on the netted samples of quality and disease incidence.

Disease was assessed on a rating scale according to the surface area affected. Grades were <2%, 2-10%, 10-25% and over 25% and the rating was calculated using the mid-points of each category (1,6, 17.5 and 62.5 respectively) to give a mean incidence ranging from 0 to 62.5. Tests in year 3 also included assessments on additional sub-samples for pre-pack quality on a 1-5 scale where 1 represented top quality and 4 poorest quality suitable for pre-packing. A score of 5 was unacceptable.

RESULTS

Chemical deposition data are summarised in Table 2. Target rates for a single application were: prochloraz Mn 50g/t; tolclofos-methyl 125g/t; imazalil 10g/t; iprodione 100g/t and thiabendazole 40g/t.

Emergence data (Table 3) showed that treatments 2 and 3 caused delay in emergence; both treatments were applied twice to seed. In 1991, when dormancy had broken early, this resulted in additional sprout damage compared with treatments 1 and 4. Tubers with 20, 53, 56 and 12% damaged sprouts were recorded on 15 February for treatments 1 to 4, respectively.

TABLE 2. Chemical deposits on seed (mg/kg detected)

Year /treatment	Prochloraz Mn	Tolclofos-methyl	Imazalil	Iprodione	TBZ
1989					
1 L	4.8	17.9	-	-	-
2 L	-	-	2.2	12.0	-
3 L	-	-	4.3	-	19.6
1 PC	na	-	-	-	na
2 PC	-	-	5.0	-	-
3 PC	-	-	na	-	na
4 +	-	-	nd	-	-
1990					
1	1.2	2.0	-	-	-
2 (x2)	-	-	3.1	-	-
3 (x2)	-	-	1.3	-	8.8
4 +	-	-	-	-	-
1991					
1	1.9	8.9	-	-	-
2 (x2)	-	-	21	-	-
3 (x2)	-	-	25	-	77
4 +	-	-	0.1	-	-

– = not tested; nd = not detected; + = untreated control; na = not available; x2 = two applications.

Effects of the chemicals on yield, however, were not always consistent with these early effects on sprouting and emergence. Yield data for site A for the three seasons (Fig. 1) showed that the effects of treatments on the total yield and the proportion of ware size potatoes varied although, in each year, treatment 2 gave a slightly lower yield than the untreated control. Also, with the exception of treatment 1 in year 1, no seed treatment increased total yield compared with the control; the proportion of ware size tubers was sometimes increased. At sites C and D, there were no trends in the yield of the saleable ware (45-85mm) fraction (Fig. 2).

TABLE 3. Plant emergence (%) [++]

Year	Treatment	Site A	Site C	Site D
1990	1	91	72	-
	2	70	51	-
	3	82	60	-
	4	95	79	-
1991	1	93	-	100
	2	100	-	96
	3	87	-	87
	4	100	-	97

[++] assessed by minimum of 3 x 25 plant counts per treatment on the following dates:
1990: Site A, 24 May; Site C, 25 May. 1991: Site A, 11 June; Site D, 12 June

The primary objective of applying seed treatments was to control silver scurf. However, in two of the three years, silver scurf levels were generally low and only in 1990 was the incidence of the disease greater than trace level. Significant effects of treatments on silver scurf were detected in both 1990 and 1991 but the effects were inconsistent between sites (Table 4). In 1990, each of the three chemicals significantly reduced silver scurf compared with the control at site C, but not at site A. In 1991, a similar significant response was measured at site A but not site D although there were no differences between chemicals. Similarly, there was no significant effect on pre-pack out-turn (Table 5) in 1991 when the incidence of other fungal diseases was negligible at intake; this had also occurred in the two previous seasons.

After storage, progeny tubers of untreated seed usually had the highest incidence of the disease (Table 4). There was a significantly greater amount of silver scurf than in the imazalil treatment in all three seasons (mean data) and all treatments had a significantly better pre-pack score after storage in 1991/92 than untreated (Table 5). Throughout the three years, levels of other fungal diseases were low so that no effects of the seed treatments on black scurf, skin spot and black dot were determined.

TABLE 4. Silver scurf mean incidence (rating) at intake & after storage

Year /treatment		Intake			After storage		
1989/90		Site A	Site B	Mean .	Site A	Site B	Mean
1 L		1.0	1.0	1.0	2.6	6.0	4.3
2 L		1.0	1.1	1.1	5.4	6.8	6.1
3 L		1.0	1.0	1.0	3.2	6.1	4.7
1 PC		1.0	1.0	1.0	5.6	6.6	6.1
2 PC		1.0	1.0	1.0	6.3	6.3	6.3
3 PC		1.0	1.1	1.0	6.2	8.7	7.5
4		1.0	1.0	1.0	5.2	10.4	7.8
sig		ns	ns	ns	***	**	**
LSD	P<0.05				1.8	2.4	1.5
	P<0.01				2.5	3.3	2.0
	P<0.001				2.6		
1990/91		Site A	Site C	Mean	Site A	Site C	Mean
1		9.2	2.7	6.0	41.5	9.6	25.6
2		9.5	3.4	6.4	28.6	11.8	20.2
3		16.1	2.4	9.2	36.4	8.6	22.5
4		14.4	8.4	11.4	47.9	15.7	31.8
sig		ns	**	*	*	ns	*
LSD	P<0.05		2.6	3.5	12.8		6.9
	P<0.01		3.8				
1991/92		Site A	Site D	Mean	Site A	Site D	Mean
1		0.1	1.2	-	5.9	4.8	5.3
2		0.2	0.1	0.2	2.7	2.4	2.5
3		0.1	0.0	0.1	0.3	1.3	0.8
4		1.5	1.2	1.3	9.1	9.9	9.5
sig		***	ns	**	**	***	***
LSD	P<0.05	0.6		0.7	4.8	2.7	2.6
	P<0.01	0.8		0.9	6.7	3.8	3.5
	P<0.001	1.2				5.3	4.7

Fig 1 Saleable yield (t/ha) 45-85mm : Site A

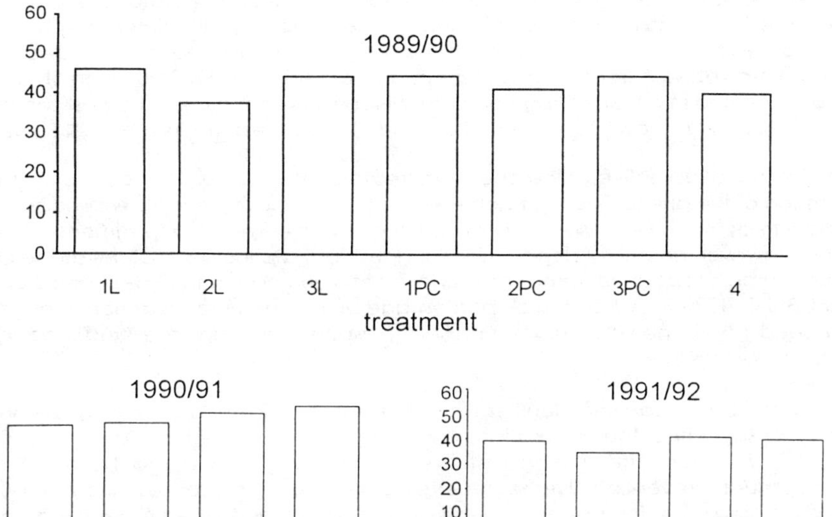

Fig 2 Saleable yield (t/ha) 45-85mm : other sites

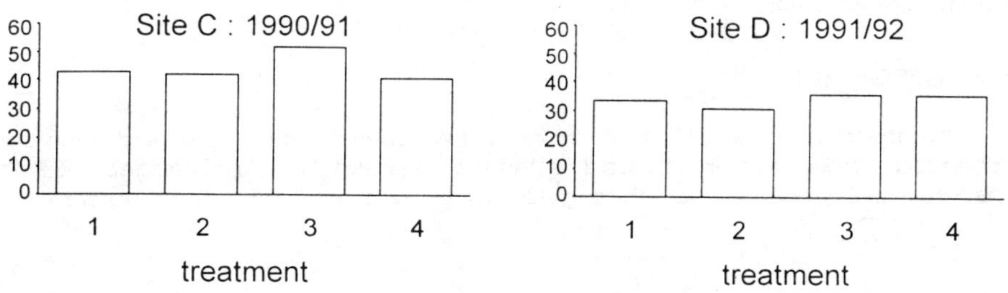

TABLE 5. Pre-pack out-turn, 1991/92 mean data (2 sites)

	Intake	After storage
Prochloraz Mn / tolclofos-methyl	2.9	2.8
Imazalil	2.9	2.8
Imazalil / TBZ	2.9	2.7
Untreated	2.8	3.1
sig	ns	**
LSD P<0.05		0.17
P<0.01		0.23

DISCUSSION

The use of seed treatment chemicals in this trial had slight detrimental effects on crop emergence and on yield. This could be attributable to either a phytotoxic effect of the active ingredient imazalil or, more probably, to the double application of the two treatments which caused more physical damage to the sprouts, as confirmed in 1991. Effects on yield were most marked with treatment 2, imazalil, and especially in 1989/90 where its use with iprodione resulted in a high proportion of small tubers (<45mm) in the harvest sample. Iprodione has been shown to increase the yield of seed sized tubers (Oxley & Bowen, 1993).

The major objective of the chemical treatments was silver scurf control. Total suppression of the disease was not achieved but a lower level was generally observed where chemical treatments were used, although differences between the treatments were very variable.However, chemical deposit data suggest that a principal factor was the chemical application process. Sometimes deposits were not high enough to expect good disease control. But, in 1990/91, when a high percentage of the chemical treatment was deposited on the seed, results showed a highly significant response although the incidence of silver scurf at intake was low.

Despite application difficulties which have been shown to be largely attributable to machine design rather than operation (Anon, 1992), it is apparent that all three products used in this work were effective against silver scurf on progeny tubers at harvest. Some were less successful in preventing silver scurf development during subsequent storage but this was not unexpected since the presence of any inoculum, either on the progeny crop or from adjacent stored crops could provide a source of infection. Because of this risk of cross-contamination, results after storage should be treated with caution but are a realistic illustration of the situation likely to be encountered in commercial storage.

The chemical seed treatments tested, exhibited good disease control characteristics, particularly with regard to silver scurf, but problems with application probably limited their success. Although the chemicals did not prevent disease development during storage of the progeny, their relative effects seen at intake were generally still evident after storage. Some benefits of seed treatment were therefore carried through to the stored ware crop, which will be sold for consumption.

ACKNOWLEDGEMENTS

The assistance of SOAFD, East Craigs and Wedderspoon Processes Ltd., who carried out chemical applications in Scotland for this trial, is gratefully acknowledged. MSD Agvet, Rhône Poulenc Agriculture and Schering Agriculture, who each provided fungicide, are also thanked.

REFERENCES

Anon. (1992) Evaluation of application methods for treatment of seed and ware with fungicides. In: *Sutton Bridge Experimental Station Annual Review 1992* A.C.Cunnington (Ed.), Oxford: Potato Marketing Board, pp. 20-22.
Hide, G.A.; Hall, S.M.; Boorer, K.J. (1988) Resistance to thiabendazole in isolates of *Helminthosporium solani*, the cause of silver scurf in potatoes. *Plant Pathology*, **37** (3), 377-380.
Hide, G.A. (1992) Towards an integrated control of potato storage diseases. *Aspects of Applied Biology*, **33**, 197-204.
Oxley, S.J.P.; Bowen, S.A. (1993) Fungicide strategies to control fungal diseases on tubers. *Abstracts from the 12th Triennial Conference of the European Association for Potato Research, Paris, France, 18-23 July 1993.* pp. 107-108.

CONTROL OF GANGRENE, DRY ROT, SKIN SPOT AND SILVER SCURF ON STORED SEED
POTATO TUBERS BY IMIDAZOLE AND PHENYLPYRROLE COMPOUNDS

S. F. CARNEGIE, A. M. CAMERON, D. A. LINDSAY

Scottish Agricultural Science Agency (SASA), East Craigs, Craigs Road,
Edinburgh, EH12 8NJ

ABSTRACT

The efficacy of imidazole and phenylpyrrole fungicides in
controlling storage diseases of potato tubers was compared with
an established fungicide treatment, 2-aminobutane applied as a
gas. In the trial incorporating 3 stocks, the fungicides were
applied using an electrostatic sprayer. Best control of
gangrene and skin spot was achieved with 2-aminobutane or with a
formulated mixture of thiabendazole and imazalil. This mixture
gave complete control of dry rot caused by <u>Fusarium solani</u> var.
<u>coeruleum</u> while fenpiclonil and a mixture of prochloraz and
tolclofos-methyl also gave good control. The largest reductions
in silver scurf were achieved following applications of
fenpiclonil or the mixture of thiabendazole and imazalil.

INTRODUCTION

Storage diseases are a major problem for seed growers who are aiming
to present their customers with blemish-free tubers. The most important
are two latent diseases: gangrene, a fungal rot, caused by <u>Phoma foveata</u>
and skin spot, caused by <u>Polyscytalum pustulans</u>, affecting both viability
of eyes and skin appearance. Dry rot <u>(Fusarium solani</u> var. <u>coeruleum</u>) can
cause severe rotting of tubers after grading late in the storage season
leading to blanking and uneven emergence in the field. Silver scurf
(<u>Helminthosporium solani</u>) has become increasingly important for the ware
grower marketing washed, pre-packed tubers and this has increasingly led
to requests for seed tubers with minimal infections of the pathogen.

Until 1991 the main fungicides used in Scotland to control storage
diseases of potato were 2-aminobutane (2-AB) applied as a gas and
thiabendazole (TBZ) applied as a spray. While gangrene and skin spot are
controlled by 2-AB, the fungicide only reduces silver scurf and does not
control dry rot (Graham <u>et al</u>., 1973). Its approval, as a fungicide under
the Control of Pesticides Regulations 1986, is now limited to seed, making
its usage less attractive to growers who have to separate seed and ware
before applying the fungicide. TBZ can be applied to both seed and ware
crops; however isolates of <u>P.pustulans</u> and <u>H.solani</u> resistant to TBZ have
been found in Scottish seed tubers (Hide <u>et al</u>., 1988; Carnegie & Cameron,
1992). Treating tubers bearing resistant isolates with TBZ is likely to
result in poor disease control (Hall & Hide 1992; Carnegie, 1992).

Alternative fungicides are being developed and some of these have
recently gained full commercial approval for use on potato tubers
(imazalil) or are likely to do so in the near future (prochloraz,
fenpiclonil). However, in field trials with these fungicides, the
predominant disease has been silver scurf followed by skin spot with only

a few reports of rot diseases (Cayley _et al._, 1983; Oxley _et al._, 1990; Burgess _et al._, 1993). This paper reports the results of a trial made in 1992-93 to determine the efficacy of these fungicides applied after harvest to tubers of three stocks selected to produce a range of diseases.

MATERIALS AND METHODS

In 1992 seed tubers of a stock (stock 1) of cv. Pentland Squire classified Super Elite 1 were obtained from a farm in Perthshire where dry rot had occurred the previous year. The tubers were machine harvested on 27 September, sized over riddles and one tonne of the 35-55 mm fraction was delivered to East Craigs on 29 September. Tubers were clean and dry and spray applications were made on the day of delivery.

Tubers of cv. Pentland Squire were also produced at SASA Gogarbank Farm by planting seed from two stocks. In stock 2 most tubers bore gangrene lesions and were sparsely affected by skin spot. From stock 3 in which skin spot was common, tubers for planting were mixed in the ratio of two tubers bearing dry rot lesions to one rot-free. Tubers were planted on 6 May and haulm killed by applying diquat dibromide (Reglone, ICI) on 24 August. Tubers were harvested on 7 October by single row digger into 0.25 tonne boxes and spray applications made on the same day. Tubers were damp with soil adhering.

The fungicides tested were imazalil (Fungazil 100SL, Embetec), thiabendazole + imazalil (Extratect, Merck Sharp & Dohme Ltd), prochloraz manganese chloride complex + tolclofos-methyl (Sporgon 50 + Rizolex Flowable, Schering), fenpiclonil (proposed name Gambit, Ciba Geigy) and 2-aminobutane (Chemical Spraying Co Ltd). The first four fungicide formulations were applied as sprays using an electrostatic sprayer (Microstat, Horstine Farmery Ltd) on a roller table with a throughput of 5 tonnes/hr. The intended application rates (mg AI/kg) were 40 TBZ, 10 imazalil, 50 prochloraz, 124 tolclofos-methyl and 50 fenpiclonil. The fumigations with 2-AB were made on 14 October (stock 1) and 30 October (stocks 2 and 3) using a 1 tonne capacity chamber and a dosage rate of 200mg/kg. For each treatment there were 4 replicate batches, each of 80 tubers. Each batch was kept in a numbered net bag in a 0.25 tonne box.

On 7 January tubers were passed over a reciprocating riddle, rots counted and removed, and remaining tubers stored until a final assessment of rots was made in late March. The identity of the pathogens causing rots was confirmed by placing a small sample of pieces of tissue from edge of rots on 2% malt agar. Fifty tubers from each replicate were washed and assessed for % surface area affected by skin spot and silver scurf (Carnegie _et al._, 1994).

Tubers for fungicide deposit analysis were stored at $4^{o}C$ and tests made on a representative sample of opposite quarters from at least 5 unwashed tubers. Deposits of 2-AB were extracted by steam distillation and determined by hplc using fluorescence detection as described by Hunter & Lindsay (1981), except that automated on-line derivatisation with orthophalaldehyde was used. TBZ, prochloraz and tolclofos-methyl deposits were extracted with ethylacetate, and imazalil and fenpiclonil with acetone. These fungicides were determined directly by reverse phase hplc using fluorescence detection for TBZ and ultra-violet diode array detection for the remainder.

RESULTS

Deposits of imazalil were similar when applied alone or in the mixture (Table 1). The proportion of fungicide deposited on the tubers relative to the intended dose was 44% for imazalil, 38% for TBZ, 34% for prochloraz, 43% for tolclofos-methyl and 58% for fenpiclonil. Much less 2-AB was present on tubers from stock 1 (3%) than from the other two stocks (11%).

TABLE 1. Fungicide deposits (mg/kg) on unwashed tubers of three stocks treated with fungicide after harvest.

	Stock number		
Fungicide	1	2	3
2-aminobutane	5.7	24.8	19.8
imazalil	4.0	5.1	2.9
thiabendazole + imazalil	15.0 + 5.3	13.6 + 4.4	17.0 + 4.5
prochloraz manganese complex + tolclofos -methyl	18.7 + 58.9	19.4 + 59.7	13.6 + 41.9
fenpiclonil	32.8	28.7	25.7

Gangrene developed on stock 2 only (Table 2). All fungicides reduced the incidence of rots; 2-AB was most effective giving a 90% reduction and fenpiclonil least effective giving a 33% reduction. Dry rot developed on the untreated tubers of all stocks and was most prevalent on stock 3 (Table 2). Virtually all rots were caused by <u>F.solani</u> var. <u>coeruleum</u>. On all stocks the incidence of dry rot was greater following treatment with 2-AB than with no treatment. The mixture of TBZ and imazalil gave 100% reduction in dry rot whereas imazalil alone was ineffective in controlling dry rot on stocks 1 and 3 but was 100% effective on stock 2. The mixture of prochloraz and tolclofos-methyl, and fenpiclonil also gave good control with reductions ranging from 78 to 100%.

TABLE 2. Percentage (degrees) tubers affected by gangrene (_Phoma foveata_) and dry rot (_Fusarium solani_ var. _coeruleum_) on three stocks in relation to fungicide treatment after harvest.

Fungicide	Gangrene	Dry rot		
	2	1	2	3
Nil	15.5	4.7	6.6	16.9
2-aminobutane	1.6	11.5	8.8	27.6
imazalil	6.4	6.9	0	13.9
thiabendazole + imazalil	4.0	0	0	0
prochloraz manganese complex + tolclofos -methyl	6.6	1.6	0	3.8
fenpiclonil	10.4	0	1.6	2.2
SED (32 DF)	2.81	1.72	2.18	3.19

Skin spot was most severe on stock 2 (Table 3). The mixture of TBZ and imazalil gave the best control of skin spot on all stocks (66-86% reduction). Control with 2-AB was similar on stocks 1 and 3 but less (48%) on stock 2. The other three fungicides gave smaller reductions in skin spot. Silver scurf on stock 1 was almost three times as severe as on the other stocks. The mixture of TBZ and imazalil gave best control although the reduction, 55-75%, was smaller than that achieved for skin spot. Fenpiclonil was slightly less effective than the mixture of TBZ and imazalil; however both imazalil and prochloraz/tolclofos-methyl mixture reduced the severity of silver scurf only on stock 1.

TABLE 3. Surface area (index) affected by skin spot and silver scurf on three stocks in relation to fungicide treatment after harvest.

Fungicide	Skin spot			Silver scurf		
	1	2	3	1	2	3
Nil	4.4	13.0	6.7	41.5	14.8	9.3
2-aminobutane	1.3	4.4	1.4	31.1	11.6	7.2
imazalil	2.2	6.8	2.3	27.2	15.0	9.4
thiabendazole + imazalil	1.5	1.8	1.5	10.2	6.2	4.2
Prochloraz manganese complex + tolclofos- methyl	3.1	8.1	3.5	24.9	14.7	8.8
fenpiclonil	3.6	7.8	3.8	15.6	7.3	8.2
SED (32 DF)	0.80	0.80	0.66	2.67	1.22	1.45

DISCUSSION

Our results show that overall the best control of the four diseases was achieved by applying the mixture of TBZ and imazalil. This formulation was as effective as 2-AB in controlling gangrene. Furthermore this mixture provided complete control of dry rot whereas the application of 2-AB appeared to enhance its development. Carnegie et al., (1990) also reported a stimulation in dry rot associated with 2-AB treatment which was partly related to the treatment causing necrosis of damaged tissue allowing entry of the pathogen. However, this type of chemical damage of the tuber skin was not seen in this experiment.

Imazalil and the mixture of prochloraz and tolclofos-methyl tended to give the same degree of control except for dry rot; this mixture gave good control on all stocks but imazalil was ineffective on 2 stocks. Both fungicides reduced silver scurf but were not as effective as fenpiclonil or the mixture of TBZ and imazalil. Leadbeater & Kirk (1992) reported that fenpiclonil was effective in controlling tuber rotting by F.solani var. coeruleum after planting. Our results show that the fungicide is effective in controlling dry rot in store but is less effective in controlling gangrene. Also fenpiclonil was slightly less effective in controlling skin spot than imazalil whereas Leadbeater & Kirk (1992) found that the fungicides were similar in their effectiveness.

In trials where tubers were dipped in suspensions of the mixture of TBZ and imazalil, the ratio of TBZ to imazalil deposited on the tuber was in the range of 1:1 to 1:2 (Carnegie et al., 1994). In these trials where the formulation was applied as a spray the ratio was 3:1 matching the ratio of the formulated product. It would thus appear that the method of application can effect the amount of fungicide deposited on the tuber. This emphasises the importance of measuring chemical residues on the tubers in any experiments evaluating the efficacy of fungicides.

ACKNOWLEDGEMENTS

We thank Miss E Sharp for the fungicide residue analyses, Mr I Nevison of the Scottish Agricultural Statistics Service for the statistical analyses and the chemical companies for supplying the fungicides.

REFERENCES

Burgess, P. J.; Wale, S. J.; Oxley, S. J. P.; Lang, R. W. (1993) Fungicide treatment of seed potatoes: strategies for the control of silver scurf (Helminthosporium solani). Proceedings of Crop Protection in Northern Britain, pp 319-324.

Carnegie, S. F. (1992) Improving seed tuber health. Aspects of Applied Biology, 33, 93-100.

Carnegie, S. F.; Cameron, A. M. (1992) Resistance to thiabendazole in isolates of Polyscytalum pustulans (skin spot) and Fusarium solani var. coeruleum (dry rot) in Scotland. Plant Pathology, 41, 606-610.

Carnegie, S. F.; Ruthven, A. D.; Lindsay, D. A.; Hall, T. D. (1990) Effects of fungicides applied to seed potato tubers at harvest or after grading on fungal storage diseases and plant development. Annals of Applied Biology, **116**, 61-72.

Carnegie, S. F.; Cameron, A. M.; Hide, G. A.; Hall, S. M. (1994). The occurrence of thiabendazole-resistant isolates of Polyscytalum pustulans and Helminthosporium solani on potato seed tubers in relation to fungicide treatment and storage. Plant Pathology, in press.

Cayley, G. R.; Hide, G. A.; Read, P. J.; Dunne, Y. (1983) Treatment of potato seed and ware tubers with imazalil and thiabendazole for the control of silver scurf and other storage diseases. Potato Research, **26**, 163-173.

Graham, D. C.; Hamilton, G. A.; Quinn, C. E.; Ruthven, A. D. (1973) Use of 2-aminobutane as a fumigant for control of gangrene, skin spot and silver scurf diseases of potato tubers. Potato Research, **16**, 109-125.

Leadbeater, A. J.; Kirk, W. W. (1992). Control of tuber-borne diseases of potatoes with fenpiclonil. Brighton Crop Protection Conference - Pests and Diseases 1992, **1**, 657-662.

Hall, S. M.; Hide, G. A. (1992) Fungicide treatment of seed tubers infected with thiabendazole-resistant Helminthosporium solani and Polyscytalum pustulans for controlling silver scurf and skin spot on stored progeny tubers. Potato Research, **35**, 143-147.

Hide, G. A.; Hall, S. M.; Boorer, K. J. (1988) Resistance to thiabendazole in isolates of Helminthosporium solani; the cause of silver scurf disease of potatoes. Plant Pathology, **37**, 377-380.

Hunter, K.; Lindsay, D. A. (1981) High-pressure liquid chromatographic determination of 2-aminobutane residues in potatoes. Pesticide Science, **12**, 319-324.

Oxley, S. J. P.; Lang, R. W.; Wale, S. (1990). Control of silver scurf and skin spot by fungicide seed treatment. Proceedings of Crop Protection in Northern Britain 1990, pp 351-357.

Session 6
Biological Seed Treatments

Chairman	R B MAUDE
Session Organisers	J W HENFLING
	R B MAUDE

MECHANISMS OF PROTECTION OF SEED AND SEEDLINGS BY BIOLOGICAL SEED TREATMENTS: IMPLICATIONS FOR PRACTICAL DISEASE CONTROL

G. E. HARMAN, E. B. NELSON

Departments of Horticultural Sciences and Plant Pathology, Cornell University, Geneva and Ithaca, NY 14456 and 14850, respectively.

ABSTRACT

Seed treatments with biological control agents have been extensively tested, and are beginning to be used commercially. However, there are substantial gaps in our knowledge of mechanisms by which various bacterial and fungal biocontrol agents control plant pathogenic fungi. Good evidence exists for the role of antibiotics or siderophores in seed protection, but other mechanisms probably also occur. Evidence for and against mycoparasitism, competition for iron or key stimulants of microbial propagule germination, and attachment of bacterial cells to hyphae are presented. Criteria for, and progress in, mechanistic studies with various organisms will be considered. The implications for these various mechanisms in the development of biological seed treatment formulations and methods for successful biocontrol will be considered.

INTRODUCTION

Seed treatments with fungal or bacterial biocontrol agents have been proposed, tested, and are likely to be used soon for commercial plant disease control (Harman, 1991). They may be used to control seedborne plant pathogens, control soilborne seed attacking fungi, or as a method of introducing agents that will colonize newly formed roots or other plant parts (Harman et al., 1989; Harman, 1990; Vannacci and Harman, 1987). However, performance of biological seed treatments has been variable. One important component of performance in any biocontrol system is an understanding of the mechanism by which control occurs.

While many, if not most, biocontrol organisms may suppress pathogens through several mechanisms (Ownley *et al.* 1992), principal mechanisms of biocontrol exist and are likely to be different between various organisms. Three mechanisms have been proposed most frequently for biocontrol, as follows (Baker, 1986): 1. Production of antibiotic substances; 2. Competition for space or nutrients; and 3. mycoparasitism. In addition, there are at least two other mechanisms that have been proposed for biocontrol agents applied to seeds, and that are likely also to occur in other application/delivery methods, as follows: 4. Adsorption/catabolism of specific plant metabolites that are necessary for stimulation of propagules of pathogenic fungi (Nelson, 1992); and 5. Attachment of bacteria to fungal hyphae by lectin interactions that inhibit further fungal growth (Nelson *et al.* 1986).

Biocontrol processes are partially determined by the intrinsic ability of biocontrol organisms to grow on and colonize seed surfaces, which is a property we will describe hereafter as spermosphere competence. In addition, the apparent spermosphere competence of any microorganism is strongly affected by the method of application and by the physiological, ecological, and edaphic interactions that occur immediately after treatment and/or planting of the seed. Similarly, the ability to colonize root surfaces (rhizosphere competence) is also determined by both the genetic and physiological characteristics of the organism in question and also by the various environmental factors such as pH, temperature, and water potential. Consequently, if a biocontrol agent has appropriate genetic and physiological attributes to provide effective seed or root disease control, the method of seed treatment may strongly influence the success or failure of the disease control strategy. Logically, seed treatment efficacy can be made more effective if mechanisms of disease control are known, so that the seed treatment composition and method of application can provide maximum levels of the critical metabolites or propagule growth at an appropriate time.

Timing is a fundamentally important factor in any biological seed treatment method. Not only do some pathogens attack seeds very quickly after planting, but competitive microbial populations proliferate rapidly and may successfully compete or otherwise prevent growth of the biocontrol agent (Hubbard *et al.* 1983). Determination of biological control mechanisms were, of necessity, first determined by in vitro interaction studies, usually on artificial media. Whereas such studies indicated the kinds of mechanisms that may occur between biocontrol agents and target pathogens, such studies cannot prove that similar mechanisms occur in vivo. We consider that the only definitive studies that prove or disprove specific mechanisms of action require a molecular version of Koch's postulates. In such studies, mutant strains are prepared that are deficient in the particular character to be considered. In addition, the gene coding for the specific character is isolated, and added back to the deficient strain. The various strains so constructed are tested for biocontrol ability under realistic conditions, and in this way, the contribution of any gene, and presumably its gene product, can be determined both qualitatively and quantitatively.

This paper will critically examine our knowledge of specific mechanisms of biocontrol by various microbial agents. Further, we will consider the temporal interactions of bioprotectants, seeds, pathogens, and other microorganisms around planted seeds, and how such interactions affect our understanding of mechanisms of seed protection. Finally, we will relate our understanding of mechanisms and temporal relationships to strategies and formulation of biological seed treatments, and indicate how further studies may enhance efficacy and usefulness of this method of seed and seedling protection.

TEMPORAL RELATIONS OF PATHOGEN-SEED-ANTAGONIST INTERACTIONS

For seed-rotting pathogens, early events in seed germination are particularly important in determining the success or failure of seed infections. Generally, the first 12-48 hr of seed germination and seedling development are critical to longer-term plant health. This phenomenon is particularly notable for seed-rotting *Pythium* species. Some critical time periods for particular pathogens and biocontrol agents are as follows (primarily as summarized by Harman and Stasz, 1986).

Time after Sowing Seed	Activity
0-4 h:	Propagules of *Pythium* spp. begin to germinate. Bacterial biocontrol agents metabolically active
4-24 h:	Seed coats are infected by *Pythium* spp. The number of infected seeds increases linearly over this time period (Taylor *et al*, 1991).
5-12 h:	Propagules of fungal biocontrol agents begin to germinate on seed surfaces.
5-60 h:	Chlamydospores of *Fusarium* spp. germinate and infect seeds.
24-40 h:	Embryos of seeds are infected by *Pythium* spp.

These representative data indicate that seeds are at risk very soon after planting, and that if bioprotectants are to protect them, they must react quickly. Clearly there is a window of vulnerability that must be closed if biocontrol of soil-borne seed and seedling pathogens are to be controlled. Further, biocontrol agents must grow if they are to be effective, and pathogens must proliferate if they are to infect their hosts. Therefore, the nature of the nutritive or stimulatory materials released from seeds, which are energy sources for both beneficial and deleterious organisms, must be considered.

Our current knowledge of pathogen stimulants was recently reviewed (Nelson, 1990) and will only be updated here to include more recent findings. Our most complete knowledge of pathogen stimulants from seeds comes from studies of *Pythium* spp. Conclusions drawn from many previous studies of propagule germination among *Pythium* spp. have led to the suggestion that carbohydrates and amino acids are the primary exudate components responsible for initiating *Pythium*-seed interactions in nature (see Nelson, 1990), although sporangia of *Pythium* species clearly respond to other stimuli such as volatiles (Nelson, 1987) and can be manipulated in ways that eliminate their responses to sugars and amino acids while maintaining their responses to seed exudates (Nelson and Craft, 1989). For example, sporangia produced on most synthetic culture media germinate readily in response to sugars and amino acids present in cotton seed exudate, but fail to respond to these same molecules when produced on metabolically active plant tissue such as diseased seeds and radicles or on a lecithin-containing mineral salts medium. Regardless of how sporangia are reared, they remain fully germinable in response to unfractionated cotton seed exudate (Nelson and Craft, 1989). The inability of plant-produced sporangia to germinate in response to sugars and amino acids under some conditions in vitro raises questions about the types of compounds responsible for stimulating sporangium germination in soil.

The active components of cotton seed exudates stimulatory to *P. ultimum* sporangia consist of unsaturated fatty acids and triglycerides of these fatty acids (Ruttledge and Nelson, submitted). These stimulatory molecules are present in the exudate as soon as seeds begin imbibing water and reach maximum levels around 4 hr after imbibition begins. Interestingly, some degree of unsaturation is required for activity. All of the unsaturated fatty acids tested to date are effective sporangium germination stimulants and active at concentrations of at least 100 ug/ml. On the other hand, none of the saturated fatty acids tested to date have shown stimulatory activity at those concentrations.

The activity of unsaturated fatty acids in stimulating fungal spore germination has been demonstrated before. Papavizas and Adams (1969) showed that endoconidia and chlamydospores of *Thielaviopsis basicola* did not germinate in response to saturated fatty acids such as stearic or palmitic acid, but did germinate in response to linolein and linolenic acid. Additionally, Harman *et al* (1978) showed that unsaturated fatty acids such as oleic, linoleic, and linolenic acid stimulated germination of *Alternaria alternata* conidia while saturated fatty acids such as stearic and palmitic acid, were ineffective. They have proposed (Harman *et al*, 1980; Harman *et al*, 1978) that the volatile peroxidation products of unsaturated fatty acids may be the active germination stimulants in fungi, since as little as 200 µg/l 2,4-hexadienal in aerial solution stimulated germination of *A. alternata* conidia. Although this compound was not stimulatory to *F. oxysporum* f.sp.*pisi* chlamydospores, relatively high concentrations (400nl/petri dish) of trans, trans-2,4-nonadienal were stimulatory (Harman *et al*, 1980). These observations, coupled with the fact that volatiles from seeds and decaying plant tissues are stimulatory to a range of soilborne pathogens (Gilbert and Linderman, 1971; Gilbert and Griebel, 1969; Gorecki *et al*, 1985; Linderman and Gilbert, 1969; Linderman and Gilbert, 1975; Nelson, 1987; Norton and Harman, 1985; Paulitz, 1991) supports the notion of fatty acids and volatile peroxidation products as being important stimulants of propagules of seed-rotting pathogens.

SEED-BIOPROTECTANT INTERACTIONS AND MECHANISMS OF BIOCONTROL

With the background presented above, it is possible to discuss the mechanisms by which biocontrol may operate and to place this discussion in an ecological framework. The biocontrol agents to be discussed will be bacterial strains in the genera *Enterobacter*, *Pseudomonas*, and *Serratia*, and fungi in the genera *Gliocladium* and *Trichoderma*. These organisms have been more extensively studied than other organisms, particularly as seed treatments. The discussion that follows is not exhaustive, but will provide specific examples of biocontrol mechanisms.

Antibiotics and toxicants

Very compelling evidence indicates that some microbes exert biocontrol ability through production of toxic substances. Howell and his coworkers (Howell, 1987; Howell and Stipanovic, 1983) have demonstrated that biocontrol ability of seed treatments with *Gliocladium virens* is primarily mediated through the production of antibiotics. If these antibiotics were eliminated through the production of antibiotic deficient mutants, the organisms largely lost their biocontrol ability. These studies have been extended and corroborated by other workers; however, these substances may not act alone. Di Pietro *et al* (1993) discovered that the endochitinase produced by this organisms can synergistically enhance activity of the *G. virens* antibiotic gliotoxin, and has the effect of making target fungi more sensitive to the antibiotic.

Bacteria also produce antibiotics effective in biocontrol. Definitive studies in which genes have been identified that code for antibiotics, together with transposon mutagenesis to elucidate their role have been conducted (Farrand *et al*, 1985; Gutterson *et al*, 1986; Jones *et al*, 1986; Keel *et al*, 1992; Laville *et al*, 1992; Slota and Farrand, 1982; Thomashow and Weller, 1988; Vincent *et al*, 1991; Voisard *et al*, 1989). In nearly all of the studies in which bacterial genes involved in fungal suppression have been cloned, they have been involved in antibiotic biosynthesis (e.g. oomycin A, phenazine-1-carboxylic acid, pyoluteorin, pyrrolnitrin, 2,4-diacetylphloroglucinol, hydrogen cyanide) in *Pseudomonas fluorescens*. In these bacteria, antibiotics clearly play an important, and in some cases, definitive role.

Competition: Siderophores

Probably the only conclusive research demonstrating a role for competition for a specific substance comes from the research on siderophore competition for iron. A few studies have also demonstrated that siderophore biosynthesis in *P. fluorescens* plays a role in pathogen suppression (Duijff *et al*, 1991; Loper, 1988), whereas a few studies have found siderophores to play little or no role in these processes, particularly with *Pythium* species (Hamdan *et al*, 1991; Keel *et al*, 1989; Paulitz and Loper, 1991).

Competition: Inactivation of germination stimulants

The first 6-12 h of germination is a key period of vulnerability to *Pythium* infection; after that, developing seedlings become less susceptible to infection (Nelson, *et al.*, 1986; Maloney and Nelson, *unpublished*). The suppressive effects of *E. cloacae* on *Pythium* behavior have been observed as reductions in *Pythium* colonization of bacterized seeds as compared with untreated seeds. These reductions occur as early as 4-6 h after sowing seed (Maloney and Nelson, unpublished). One possible hypothesis to explain the rapid reduction in *Pythium* response and infection of germinating seeds in the presence of *E. cloacae* is that *E. cloacae* can inactivate molecules in seed exudates that *P. ultimum* requires for propagule activation and/or germination. In the absence of these signal molecules, *P. ultimum* cannot establish pathogenic interactions with its host. Growth of various strains of *E. cloacae* on 4-hr cotton seed exudate results in dramatically reduced levels of sporangium germination when cell-free exudate solutions are assayed for stimulatory activity (Nelson, 1992; van Dijk and Nelson, unpublished). Likewise, when seeds are coated with strains of *E. cloacae*, sporangia do not germinate in response to exudate stimulants released into and extracted from sand or unsterilized soils (Nelson, 1990). Further, wild-type strains of *E. cloacae* eliminate the stimulatory activity of linoleic acid in as little as 4 hr (van Dijk and Nelson, unpublished). However, transposon-induced mutants of *E. cloacae* have been identified that fail to inactivate the activity of linoleic acid after 24 h and these same mutant strains no longer protect cucumber seeds from infection by *P. ultimum* (Maloney and Nelson, unpublished). It is likely that in this system, stimulant inactivation is an important mechanism of seed rot suppression.

In other systems, stimulant inactivation appears to be playing a role in biological control activity. Elad and Chet (1987) found a significant correlation between the ability of various bacterial strains to inhibit *Pythium* seed rot of cucumber and their ability to inhibit oospore germination of *P. aphanidermatum*. Effective bacterial biocontrol agents inhibited oospore germination by as much as 57% while ineffective biocontrol strains inhibited oospore germination by only 13-20%. There was no direct interaction between the bacteria and oospores, and the ability of effective strains to inhibit oospore germination was not related to the production of inhibitory metabolites. They speculated that bacterial strains catabolized exudate components responsible for stimulating oospore germination. In other studies, pea, cotton, and soybean seeds evolved significantly lower levels of ethanol and acetaldehyde during germination when treated with *Enterobacter cloacae*, *Trichoderma harzianum*, or *Pseudomonas putida* as compared with untreated seeds (Gorecki *et al*, 1985; Nelson, 1990; Paulitz, 1990). The volatiles released from treated seeds were less stimulatory to sporangium germination of *P. ultimum* than were volatiles from untreated seeds (Gorecki *et al*, 1985; Nelson, 1990). In addition, Ahmad and Baker (1988) observed reductions in sporangium germination of *P. ultimum* in the presence of *Trichoderma*-treated seeds as compared with untreated seeds. Thus, there is a substantial and growing body of evidence indicating that competition for germination stimulants may play an important role in biocontrol.

An explanation of the mechanisms of biocontrol by *E. cloacae* is likely to occur soon. A mutant library has been constructed using a mini-TN5/phoA plasmid (de Lorenzo *et al*, 1990); the phoA reporter gene makes it particularly well-suited to detection of gene production expressed on the outer surface of the plasma membrane. Five mutants possess altered biocontrol phenotypes, and one no longer protects seeds against rots (Maloney *et al*, 1994). A 16 kb clone has been discovered that completely restores biocontrol activity; this gene likely provides a global function such as nutrient transport of sensing. Elucidation of the role of this gene should indicate the mechanism of action of this biocontrol agent.

Mycoparisitism and production of cell wall degrading enzymes

Mycoparisitism is a complex process by which biocontrol fungi may attack pathogenic fungi and involves the following steps, as inferred from in vitro studies (primarily from (Chet, 1987): (1) the biocontrol fungi grow tropically toward the target fungi, (2) hyphae of the biocontrol fungi bind to lectins on the surface of the target fungi via attachment of carbohydrate receptors on the surface of the biocontrol fungus to lectins on the target organism, (3) cell wall degrading enzymes are produced that attack the target fungus and destroy its integrity; these enzymes have recently been shown to be complex mixtures of synergistic proteins that act together against pathogenic fungi (Lorito *et al*, 1993a), and (4) appressoria-like structures are produced that apparently initiate penetration of the target fungus by the bioprotectant. While mycoparasitic structures have been observed on *Trichoderma*-treated seeds (Hubbard *et al*, 1983), Lifshitz *et al* (1986) considered this mechanism to be unlikely to protect seeds against *P. ultimum*. They based this hypothesis on the fact that (a) mycoparasitic structures were rarely observed, and (b) infection by this pathogen occurs too rapidly for growth of *Trichoderma* and mycoparasitism to occur on the seed surface. These considerations may well be correct, however, the time required for *P. ultimum* to breach the seed coat may indeed provide the requisite time for mycoparasitism to occur. Definitive results concerning the role of mycoparasitism by *Trichoderma* in control of a range of pathogens should be available soon. The genes for cell wall degrading enzymes are being isolated (Hayes *et al*, 1994), and the first chitinase deficient mutants have been prepared (Harman and Hayes, 1993). Once a series of mutants deficient in specific cell wall degrading enzymes have been prepared, as well as strains to which the genes have been restored, we will be able to definitively assess the role of the enzymes in biocontrol.

Chitinolytic enzymes from *Serratia marcescens* can play a role in biocontrol. A gene (ChiA) from this bacterium was inserted into the nonbiocontrol agent *Escherichia coli*, and the transgenic bacterium achieved biocontrol ability (Shapira *et al*, 1989). Similarly, *T. harzianum* was transformed with plasmids that resulted in integration of ChiA from *S. marcescens* into the *T. harzianum* genome (Haran *et al*, 1993). The gene was under control of the CaMV 35S promoter and so the heterologous enzyme was produced constitutively. The transformed strains were more capable of overgrowing *Sclerotium rolfsii* in vitro than the original strain from which it was derived. Further, it produced wider lytic zones in the area of contact with the pathogen than the wild type. Both the wild type and the transformed strains had similar growth rates and conidiation levels. These results indicate that the biocontrol ability of this strain was improved by transformation.

Adherence of *E. cloacae* to *P. ultimum*

One of the more conspicuous traits of *E. cloacae* in its interaction with *Pythium* spp. is its ability to adhere to hyphae. Empirical relationships between adherence of *E. cloacae* to hyphae of *P. ultimum* and biological control properties in the bacterium have been established (Nelson *et al*, 1986). However, pretreating cell suspensions of *E. cloacae* with various mono-, di-, and trisaccharides, as well as certain amino sugars, or a-linked glucosides, prevents cells from attaching to intact hyphae or agglutinating hyphal fragments. Addition of these same sugars also eliminates the ability of cells to inhibit fungal growth. On the other hand, pretreatment of cells with certain other monosaccharides, methylated sugars, or α-linked glucosides does not interfere with the ability of *E. cloacae* to attach to hyphae and inhibit the growth of *P. ultimum* (Nelson *et al*, 1986). The same carbohydrates that block binding of *E. cloacae* to *P. ultimum* hyphae also block ammonia production by *E. cloacae* (Howell *et al*, 1988) and allow cells to disperse through water films adjacent to *P. ultimum* hyphae (Maloney and Nelson, unpublished). These results suggest that bacterial adherence to hyphal cell walls might involve the binding of a fimbrial adhesin to specific sugar residues, possibly glucosides, that are associated with the fungal cell wall. Analogous interactions between E. coli and animal cells have been described (Duguid and Old, 1980).

Microscopic studies of the interactions between bacteria and live mycelium revealed bacterial cells clustered primarily around hyphal tips which were devoid of organized and streaming cytoplasm, around tips which were sealed off from their hyphae by septa, or along hyphae which contained cytoplasm that was no longer streaming and was withdrawn from the mycelial cell wall (Maloney & Nelson, unpublished). Only rarely were bacteria clustered in large numbers along hyphae which had actively-streaming cytoplasm. These observations have led us to speculate about the nature of the *Pythium* cellular changes occurring as a result of the close interaction of *E. cloacae* with hyphal tips. It is unclear from our preliminary experiments whether *E. cloacae* cells bind preferentially to regions of hyphae devoid of cytoplasm, or whether binding of *E. cloacae* results in the observed loss of cytoplasm and hence viability.

Although relationships between bacterial adherence and biological control activity appear to firm, molecular genetic analysis of these interactions has not supported the contention that adherence properties in *E. cloacae* are in any way related to biological activity. After screening a mutant library of *E. cloacae* transconjugants, several have been identified from three types of adherence assays, that are deficient in adherence properties. However, all of these strains have remained suppressive to *Pythium* seed rot of cucumber (Maloney & Nelson, unpublished).

IMPLICATIONS FOR DEVELOPMENT OF BIOLOGICAL SEED TREATMENTS

In spite of the efforts described here, as well as information from other studies and reviews, it is apparent that biological seed treatments can still be dramatically improved. This improvement can logically come from enhancement of the biocontrol agents themselves, and from developments in seed treatment methodology. These will be discussed separately.

Genetic improvement

Once a full understanding of the genetic basis for mechanisms of biocontrol agents is in hand, we will have a series of "molecular bullets" for improvement of biocontrol agents. Such genes may be used for other purposes such as producing disease resistant plants. Key to making rapid progress are understanding (a) of synergistic interactions between biocontrol organisms, and various biological and chemical components of plant disease control, and (b) use of promoters that change development or regulatory control of metabolites critical for biocontrol. We recently have discovered high levels of synergy between enzymes from *T. harzianum* and various other organisms or materials. Chitinolytic and glucanolytic enzymes from *T. harzianum* are strongly synergistic; each individual protein requires 40 to 150 ug/ml to achieve ED50 levels, while for combinations of enzymes, less than 2 ug/ml of total protein is required to achieve the same effect (Harman *et al*, 1993; Lorito *et al*, 1993a; Lorito *et al*, submitted). Further, these same enzymes are synergistic with *E. cloacae*. If levels of both enzymes and bacteria too low to have an effect are mixed with the test pathogen are added to the pathogen *Botrytis cinerea*, dramatic effects can be seen. First, within a few minutes the bacterium binds to hyphae of the pathogen,; binding ordinarily does not occur in this medium. In the presence of the enzyme, but not in its absence, proliferation of the bacteria occurs on the hyphal surface and within 24 hr the target fungus is nearly completely destroyed (Lorito *et al*, 1993b). Further, these same enzymes are strongly synergistic with chemical fungicides; the presence of the enzyme may increase the sensitivity of target fungi to the fungicide by more than 100-fold (Lorito *et al*, 1993c). Other synergistic combinations will no doubt be determined. Addition of such synergistic combinations of enzymes to appropriate biocontrol organisms (*E. cloacae* is an obvious choice) should result in much more potent biocontrol agents. As noted earlier, the first few such strains have already been produced, and even more effective strains should follow rapidly (Haran *et al*, 1993, Shapira *et al.*, 1989).

Changes in regulatory control of specific gene products also are likely to be effective. For example, chitinolytic enzymes are usually only produced by bioprotectants after the two organisms come into contact. If genes coding for these enzymes are placed under the control of constitutive rather than inducible promoters, the temporal relationships of pathogen and protectant are likely to be improved in favor of the biocontrol agent. Promoters active at different developmental or in response to various metabolites are available and a number of strategies can be devised for use of such promoters for improvement of biocontrol agents.

Improved biocontrol seed treatment strategies

A number of strategies to improve temporal, edaphic, or competitive interactions of seed treatment bioprotectants have been developed. Concepts and practices of such treatments recently have been reviewed (Harman, 1991; Jin *et al*, 1992; Taylor and Harman, 1990; Taylor *et al*, submitted). The level of improvement, which can usually be attributed to an apparent enhancement of spermosphere competence, can be dramatic. For example, if *T. harzianum* is used to treat cucumber seeds and is applied as a slurry, biocontrol efficacy is marginal. However, if the organism is applied to the seed in a nutritive base, and then an inert layer placed over this (double coating), biocontrol efficacy is considerably enhanced. Further, if the seeds are then incubated for a few days at 100% relative humidity, the bioprotectant grows and colonizes the seed surface. Such seeds can then be dried and nearly perfect biocontrol can be obtained. Similar excellent results can be obtained with the use of solid matrix priming (SMP) with either fungal or bacterial biocontrol agents. In SMP, seeds are treated and then added to moistened substrate that provides sufficient moisture for microbial growth but not for seed sprouting. After a few days, biocontrol microbes proliferate by an order of magnitude and colonize the seed coat. Additives to seed treatments, such as specific food bases or materials to control pH can also be useful.

However, biological seed treatments can and must be further improved. Even though such treatments as double coating and SMP provide excellent biological control, the processes are too labor-intensive, and hence to expensive, for many crop seeds, including most agronomic crops. Ideally, biological seed treatments should meet the following criteria: (1) provide excellent and reliable control of pathogens on a wide range of crop seeds, (2) be inexpensive and simple to apply, (3) have no deleterious effects upon seed viability during storage, and (4) the biocontrol agent must retain viability and efficacy during storage for over one year at room temperature. This latter property is primarily a function of the production and formulation technique (Jin *et al*, 1992; Harman, unpublished) employed, but no components of the seed treatment formulation can adversely affect viability of the biocontrol agent.

Are such treatments possible? The authors feel that the answer is definitely yes. Seed treatment techniques described above indicate that *T. harzianum* and, to a lesser extent, *E. cloacae*, can be effective on a wide range of crop seeds, even when existing strains are employed. Seed treatments that enhance rapid growth and development, i.e. that increase apparent spermosphere competence, which leads directly to improved rhizosphere competence in strains possessing this property (Harman, 1991) can provide the necessary level of enhancement of biological seed treatments. Microbial nutrition added to the seed treatment is likely the key to development of more economical and successful treatments. Such development is likely to arise from a knowledge of the nutritive requirements of the agent of interest, and from the nature of seed exudates from various species. To illustrate this concept, we can consider the relative efficacy of *T. harzianum* and *E. cloacae* on different seeds. *T. harzianum* is quite effective, even as a slurry treatment on snap bean seeds, but requires a relatively complex double coating or SMP treatment to achieve similar efficacy on cucumber seeds. Conversely, *E. cloacae* is effective on cucumber, but has little or no effect on snap beans (Nelson *et al*, 1986). For *T. harzianum*, the difference lies on its ability to colonize these two seeds. The fungus readily colonizes bean seeds, but rather poorly and slowly colonizes cucumber seeds. Seed exudates from beans are richer and support growth of *T. harzianum* much better than do exudates from cucumbers. *E. cloacae*, on the other hand, is blocked from hyphal attachment by the level of sugars from snap beans, and this apparently prevents its biocontrol activity on this seed type. Clearly, however, a greater knowledge of the qualitative and quantitative differences in seed exudates, can assist in developing inexpensive formulation additives that enhance activity of at least *T. harzianum*. However, an additional requirement also exists. Any such formulation must be capable of stimulating growth of *T. harzianum* without a commensurate enhancement of growth of pathogens and competitive microflora. To meet this criterion, knowledge of stimulants for germination of pathogen propagules is essential; such materials either must be avoided or be formulated in such a way that they are available only, or primarily to, the biocontrol agent. The example of the relative efficacy of simple treatments on snap beans and cucumbers, plus the demonstration that the fungus is indeed effective on these and many more crops if treatments permit sufficient spermosphere competence, demonstrate that solutions exist. The challenge will be to gather the appropriate data and then, through extensive empirical testing, devise economically reasonable solutions.

REFERENCES

Ahmad J S; Baker R (1988) Implications of rhizosphere competence of *Trichoderma harzianum*. *Canadian Journal of Microbiology* **34**, 229-234.

Baker, R. (1986) Biological Control: An Overview. *Canadian Journal of Plant Pathology* **8**, 218-221.

Chet, I. (1987) Trichoderma-Application, mode of action, and potential as a biocontrol agent of soilborne plant pathogenic fungi. In: *Innovative Approaches to Plant Disease Control*. Chet , I. (ed.), J. Wiley and Sons. New York, pp. 137-160.

de Lorenzo V; Herrero M; Jakubzik U; Timmis K N (1990) Mini-Tn*5* transposon derivatives for insertion mutagenesis, promoter probing, and chromosomal insertion of cloned DNA in gram-negative eubacteria. *Journal of Bacteriology* **172**, 6568-6572.

Di Pietro, A.; Lorito, M.; Hayes, C. K.; Broadway, R. M.; Harman, G. E. (1993) Endochitinase from *Gliocladium virens*: isolation, characterization and synergistic antifungal activity in combination with gliotoxin. *Phytopathology* **83**, 308-313.

Duguid J P; Old D C (1980) Adhesive properties of Enterobacteriaceae. In: Bacterial Adherence. E.H. Beachey (ed.). Chapman & Hall, London. pp 187-217.

Duijff B J; Meijer J W; Bakker P A H M; Schippers B (1991) Suppression of *Fusarium* wilt of carnation by *Pseudomonas* in soil; mode-of-action. In: *Plant Growth-Promoting Rhizobacteria: Progress and Prospects*. Keel, C. Koller, B. and Defago, G.. (eds.). pp 152-157.

Elad Y; Chet I (1987) Possible role of competition for nutrients in biocontrol of *Pythium* damping-off by bacteria. *Phytopathology* **77**, 190-195.

Farrand S K; Slota J E; Shim J S; Kerr A (1985) Tn5 insertions in the agrocin 84 plasmid: the conjugal nature of pAgK84 and the locations of determinants for transfer and agrocin 84 production. *Plasmid* **13**, 106-117.

Gilbert R A; Linderman R G (1971) Increased activity of soil microorganisms near sclerotia of *Sclerotium rolfsii* in soil. *Canadian Journal of Microbiology* **17**, 557-562.

Gilbert R G; Griebel G E (1969) The influence of volatile substances from alfalfa on verticillium dahliae in soil. *Phytopathology* **59**:, 1400-1403.

Gorecki R J.; Harman G E; Mattick L R (1985) The volatile exudates from germinating pea seeds of different viability and vigor. *Canadian. Journal of Botany* **63**, 1035-1039.

Gutterson N I; Layton T J; Ziegle J S; Warren G J (1986) Molecular cloning of genetic determinants for inhibition of fungal growth by a fluorescent pseudomonad. *Journal of Bacteriology* **165**, 696-703.

Hamdan H; Weller D M; Thomashow L S (1991) Relative importance of fluorescent siderophores and other factors in biological control of *Gaeumannomyces graminis* var. *tritici* by *Pseudomonas fluorescens* 2-79 and M4-80R. *Applied and Environmental Microbiology* **57**, 3270-3277.

Haran, S., Schickler, H., Pe'er, S., Logeman, S., Oppenheim, A.; Chet, I. (1993) Increased constitutive chitinase activity in tranformed *Trichoderma harzianum*. *Biological Control* **3**, 101-108.

Harman, G. E.; Taylor, A. G.; Stasz, T. E. (1989) Combining effective strains of *Trichoderma harzianum* and solid matrix priming to improve biological seed treatments. *Plant Disease* **73**, 631-637.

Harman, G. E. (1990) Deployment tactics for biocontrol agents in plant pathology. *New Directions in Biological Control: Alternatives for Suppressing Agricultural Pests and Diseases*. Baker, R. R. and Dunn, P. E. (eds.), Alan R. Liss, Inc. New York. pp. 779-792.

Harman, G. E. (1991) Seed treatments for biological control of plant disease. *Crop Protection*. **10**, 166-171.

Harman, G. E; Hayes, C. K. (1993) The genome of biocontrol fungi: Modification and genetic components for plant disease management. In: *Pest Management: Biologically Based Technologies*. Lumsden, R. D. and Vaughn, J. L.(eds.), American Chemical Soc. Washington, D. C.

Harman, G. E.; Hayes, C. K.; Lorito, M.; Broadway, R. M.; Di Pietro, A; Peterbauer, C; Tronsmo, A. (1993) Chitinolytic enzymes of *Trichoderma harzianum*: purification of chitobiase and endochitinase. *Phytopathology* **83**, 313-318.

Harman, G. E. and Stasz, T. E. (1986) Influence of seed quality on soil microbes and seed rots. In: *Physiological-Pathological Interactions affecting Seed Deterioration.* West, S. (ed,), Crop Sci. Soc. Am. CSSA Spec. Pub. 12. Madison, WI. pp. 11-37.

Harman G E; Mattick L R; Nash G; Nedrow B L (1980) Stimulation of fungal spore germination and inhibition of sporulation in fungal vegetative thalli by fatty acids and their volatile peroxidation products. *Canadian Journal of Botany* **58,** 1541-1547.

Harman G E; Nedrow B; Nash G (1978) Stimulation of fungal spore germination by volatiles from aged seeds. *Canadian Journal of Botany* **56,** 2124-2127.

Hayes, C K.; Klemsdal, S; Lorito, M; Di Pietro, A; Peterbauer, C; Nakas, J. P; Tronsmo, A; Harman, G. E. (1994) Isolation and sequence of an endochitinase gene from a cDNA library of *Trichoderma harzianum. Gene* (In Press).

Howell, C R (1987) Relevance of mycoparasitism in the biological control of *Rhizoctonia solani* by *Gliocladium virens. Phytopathology* **77**, 992-994.

Howell, C R; Stipanovic, R D (1983). Gliovirin, a new antibiotic from *Gliocladium virens*, and its role in the biological control of *Pythium ultimum. Canadian Journal of Microbiology* **29**, 321-324.

Howell C R; Beier R C; Stipanovic R D (1988) Production of ammonia by *Enterobacter cloacae* and its possible role in the biological control of *Pythium* preemergence damping-off by the bacterium. *Phytopathology* **78**, 1075-1078.

Hubbard, J P; Harman, G E; Hadar, Y (1983) Effect of soilborne *Pseudomonas* sp. on the biological control agent, *Trichoderma hamatum*, on pea seeds. *Phytopathology* **73**, 655-659.

Jin, X; Hayes, C. K; Harman, G. E. (1992) Principles in the development of biological control systems employing *Trichoderma* species against soil-borne plant pathogenic fungi. In: *Frontiers in Industrial Mycology.* Letham, G. F. (ed.), 174-195. Chapman and Hall. New York.pp. 174-195.

Jones J D G,; Grady K L; Suslow T V; Bedbrook J R (1986) Isolation and characterization of genes encoding two chitinase enzymes from *Serratia marcescens. European Molecular Biology Organization Journal* **5**, 467-473.

Keel C; Schnider U; Maurhofer M; Voisard C,; Laville J; Burger U; Wirthner P; Haas D; Defago G (1992) Suppression of root diseases by *Pseudomonas fluorescens* CHA0: Importance of the bacterial secondary metabolite 2,4-diacetylphloroglucinol. *Molecular Plant-Microbe Interactions* **5**, 4-13.

Keel C; Voisard C; Berling C H; Kahr G; Defago G (1989) Iron sufficiency, a prerequisite for the suppression of tobacco black root rot by *Pseudomonas fluorescens* strain CHA0 under gnotobiotic conditions. *Phytopathology* **79**, 584-589.

Laville J; Voisard C; Keel C; Maurhofer M; Defago G; Haas D (1992) Global control in *Pseudomonas fluorescens* mediating antibiotic synthesis and suppression of black root rot of tobacco. *Proceedings of the National Academy of Science, USA* **89**, 1562-1566.

Lifshitz, R; Windham, M T; Baker, R (1986) Mechanism of biological control of preemergence damping-off of pea by seed treatment with *Trichoderma* spp. *Phytopathology* **76,** 720-725.

Linderman R G;Gilbert R G (1969) Stimulation of *Sclerotium rolfsii* in soil by volatile components of alfalfa hay. *Phytopathology* **59**, 1366-1372.

Linderman R G; Gilbert R G (1975) Influence of volatiles of plant origin on soil-borne plant pathogens. In: *Biology and Control of Soilborne Plant Pathogens.* G.W. Bruehl (ed.). The American Phytopathological Society, St Paul. pp. 90-99.

Loper J E (1988) Role of fluorescent siderophore production in biological control of *Pythium ultimum* by a *Pseudomonas fluorescens* strain. *Phytopathology* **78**, 166-172.

Lorito, M; Harman, G E; Hayes, C K; Broadway, R M; Tronsmo, A.; Woo, S L; Di Pietro, A.; (1993a) Chitinolytic enzymes produced by *Trichoderma harzianum*: antifungal activity of purified endochitinase and chitobiosidase. *Phytopathology* **83**, 302-307.

Lorito, M; Di Pietro, A; Hayes, C K; Woo, S L; Di Pietro, A; Harman, G E (1993b) Antifungal synergistic interaction between chitinolytic enzymes from *Trichoderma harzianum* and *Enterobacter cloacae. Phytopathology* **83**: 721-728.

Lorito, M; Peterbauer, C; Hayes, C K; Di Pietro, A; Harman, G. E. (1993c) Synergistic interaction between fungal cell wall degrading enzymes and different antifungal compounds. *Journal of General Microbiology* (in press),

Maloney A P; Nelson E B; Ishamaru C A; Loper J E (1994) Identification of a gene, *psp1*, from *Enterobacter cloacae*, with a major role in suppression of *Pythium* seed rot. *Applied and Environmental Microbiology* (in preparation).

Nelson E B (1987) Rapid germination of sporangia of *Pythium* species in response to volatiles from germinating seeds. *Phytopathology* **77**, 1108-1112.

Nelson E B (1990) Exudate molecules initiating fungal responses to seeds and roots. *Plant and Soil* **129**, 61-73.

Nelson E B (1992) Bacterial metabolism of propagule germination stimulants as an important trait in the biocontrol of *Pythium* seed infections. In: *Biological Control of Plant Diseases: Progress and Challenges for the Future*. Tjamos, E. C. , Papavizas, G. C. and Cook, R. J (eds). Plenum Publishing Company, New York. pp. 353-357.

Nelson E B; Chao W L; Norton J M; Nash G T; Harman G E (1986) Attachment of *Enterobacter cloacae* to hyphae of *Pythium ultimum*: Possible role in the biological control of *Pythium* preemergence damping-off. *Phytopathology* **76**, 327-335.

Nelson E B; Craft C M (1989) Comparative germination of culture-produced and plant-produced sporangia of *Pythium ultimum* in response to soluble seed exudates and exudate components. *Phytopathology* **79**, 1009-1013.

Norton J M; Harman G E (1985) Responses of soil microorganisms to volatile exudates from germinating pea seeds. *Canadian Journal of Botany* **63**, 1040-1045.

Ownley, B H; Weller, D M; Thomashow, L S (1992) Influence of in situ and in vitro pH on suppression of *Gaeumannomyces graminis* var. *tritici* by *Pseudomonas* fluorescens 2-79. *Phytopathology* **82**, 178-184.

Papavizas G C; Adams P B (1969) Survival of root-infecting fungi in soil XII. Germination and survival of endoconidia and chlamydospores of thielaviopsis basicola in fallow soil and in soil adjacent to germinating seeds. *Phytopathology* **59**:, 371-378.

Paulitz T C (1990) The stimulation of *Pythium ultimum* by seed volatiles and the interaction of *Pseudomonas putida*. *Phytopathology* **80**, 994-995.

*Paulitz T C (1991) Effect of *Pseudomonas putida* on the stimulation of *Pythium ultimum* by seed volatiles of pea and soybean. *Phytopathology* **81**, 1282-1287.

Paulitz T C; Loper J E (1991) Lack of a role for fluorescent siderophore production in the biological control of cucumber by a strain of *Pseudomonas putida*. Phytopathology **81**, 930-935.

Shapira R; Ordentlich A; Chet I; Oppenheim A B (1989) Control of plant diseases by chitinase expressed from cloned DNA in *Escherichia coli*. *Phytopathology* **79**, 1246-1249.

Slota J E; Farrand S K (1982) Genetic isolation and physical characterization of pAgK84, the plasmid responsible for agrocin 84 production. *Plasmid* **8**, 175-186.

Taylor, A G.; Harman, G E (1990) Concepts and technologies of selected seed treatments. *Annual Review of Phytopathology* **28**, 321-339.

Taylor, A G; Min, T-G; Harman, G E; Jin, X. (1991) Liquid coating formulation for the application of biological seed treatments of *Trichoderma harzianum*. *Biological Control* **1**, 16-22.

Thomashow L S and Weller D M 1988 Role of a phenazine antibiotic from *Pseudomonas fluorescens* in biological control of *Gaeumannomyces graminis* var. *tritici*. *Journal of Bacteriology* **170**, 3499-3508.

Vannacci, G;   Harman, G E (1987) Biocontrol of seedborne *Alternaria raphani* and *A. brassicicola*. *Canadian Journal of Microbiology* **33**, 850-856.

Vincent M N; Harrison L A; Brackin J M; Kovacevich P A; Mukerji P; Weller D M; Pierson E A (1991) Genetic analysis of the antifungal activity of a soilborne *Pseudomonas aureofaciens* strain. *Applied and Environmental Microbiology* **57**, 2928-2934.

Voisard C, Keel C, Haas D and Defago G (1989) Cyanide production by *Pseudomonas fluorescens* helps suppress black root rot of tobacco under gnotobiotic conditions. *European Molecular Biology Organization Journal* **8**, 351-358.

BACTERIZATION TO PROTECT SEED AND RHIZOSPHERE AGAINST DISEASE

B.J.J. Lugtenberg, L.A. de Weger

Leiden University, Institute of Molecular Plant Sciences,
Clusius Laboratory, Wassenaarseweg 64, 2333 AL Leiden, The Netherlands

B. Schippers

University of Utrecht, Department of Plant Ecology and Evolutionary
Biology, Sorbonnelaan 16, 3584 CA Utrecht, The Netherlands

ABSTRACT

Microbiological protection against plant diseases is gaining momentum, partly because of the negative attitude towards chemical pesticides. In order to protect the plant against a pathogen, a biocontrol
agent should not only produce a protective factor, it should also
deliver it by colonization at the right site at the right time in
optimal levels. The present knowledge of mechanisms of protection and
colonization are discussed in this paper. Major factors limiting
further application of microbiological control agents that will be
discussed are colonization, survival on seed, microbiological activity during seed germination, and public acceptance.

INTRODUCTION

The interaction with microbes is of utmost importance in a plant's
life. Therefore microbe-plant interactions are also economically important.
Pathogenic microbes are responsible for the loss of approximately one third
of the world's crop. Fortunately, many microbe-plant interactions are
beneficial for the plant, e.g. the interaction of leguminous plants with
Rhizobium and Bradyrhizobium can result in biological nitrogen fertilizati-
on, whereas microbial antagonists can cause a reduction of plant diseases.
Finally, some microbe-plant interactions are neither harmful nor beneficial
to the plant. Campbell & Greaves (1990) reported on experiments in which
roots of wheat seedlings were inoculated with fluorescent Pseudomonas spp.
from the rhizosphere of field-grown wheat. Of 150 isolates studied,
approximately 40 % inhibited root growth, 40 % stimulated it, whereas the
remaining 20 % had no effect.

Benefical effects of naturally-occurring plant-associated bacteria on
plant growth can be direct of indirect, i.e. in the absence of pathogens or
through limitation of the growth of the pathogen, respectively. Indirect
microbial plant growth stimulation is often referred to as biocontrol in
order to stress the contrast with chemical control.

Direct microbial plant growth promotion is due to the fact that a large number of soil bacteria produce plant growth promoting hormones, e.g. _Azospirillum_, _Azotobacter_ , _Pseudomonas_ and _Bacillus spp_. Other microbes which directly promote plant growth are for example nitrogen-fixing bacteria (e.g. _Rhizobium_, _Bradyrhizobium_, _Frankia spp._).

Although most studied biocontrol microbes attack microbial pathogens, it is interesting to note that other biocontrol microbes act by attacking insects or weeds. Interest in biocontrol has gained momentum, partly because of the negative attitude towards chemical pesticides. This review will focus on the _bacteria_ that can be used in biological protection against disease.

PLANT GROWTH - PROMOTING RHIZOBACTERIA (PGPR's)

Monoculture of crops can result in decrease of crop yield due to an increase of pathogens in soil. After several years of monoculture plant yield can increase again as a result of changes in the microbial population. It is said that disease-conducive soil changes to disease-suppressive soil. Microbial biocontrol agents can be isolated from plant roots present in the latter soil. They belong to the PGPR's. PGPR's often are fluorescent _Pseudomonas_ spp. (e.g. Weller & Cook, 1983) but strains of _Agrobacterium_, _Alcaligenes_, _Arthrobacter_, _Bacillus_, _Enterobacter_, _Erwinia_, _Flavobacterium_, _Serratia_, _Streptomyces_ and _Xanthomonas_ have also been reported to promote plant growth.

MECHANISMS OF INDIRECT PLANT GROWTH STIMULATION

In order to protect the plant against a pathogen, a biocontrol agent usually must fullfil two criteria. Firstly, it must produce a protective factor that somehow selectively harms the pathogen, e.g. an antibiotic or cell wall-degrading enzyme. Secondly, it must be present at the right site in the rhizosphere at the right time to deliver optimal levels of the protective factor. Substantial knowledge has been collected, at least at the level of laboratory know-how, of the protective factors and their biosynthesis. However, knowledge of the molecular basis of rhizosphere colonization is virtually lacking as is know-how on spatial and temporal colonization strategies of pathogens and biocontrol agents.

PROTECTIVE FACTORS PRODUCED BY PGPR's

Antibiosis is the best-known strategy of PGPR's (Tomashow, 1991). The _P. fluorescens_ strain Pf-5 produces pyrrolnitrin and protects cotton against a variety of fungi (Howell, 1990). Similarly, _P.fluorescens_ 2-79 which protects wheat against _Gaeumannomyces graminis_ var. _tritici_, produces phenazine-1-carboxylic acid (Gurusiddaiab _et al._, 1986). _P. fluorescens_ strain CHAO, which can protect tobacco against _Thielaviopsis basicola_, produces hydrogen cyanide (HCN), 2,4-diacetyl-phloroglucinol, monoacetylphloroglucinol, pyoluteorin, salicylic acid (Défago *et al.*,1990) as well as indole-3-acetic acid (Oberhänsli *et al.*, 1991). Agrocin 84 is produced by <u>Agrobacterium</u> <u>radiobacter</u> var. <u>radiobacter</u> strain 84 which is applied to protect stone fruit and rose cuttings against <u>A. tumefaciens</u> (Ryder & Jones, 1990).

The ability of biocontrol strains to produce antibiotics does not necessarily mean that the antibiotic is involved in biocontrol. For example, work using mutants deficient in antibiotic production has in one case shown lack of evidence (Kraus & Loper, 1992), and in another case strong evidence (Tomashow & Weller, 1988), for a major role of the antibiotic in biocontrol. Interestingly, Laville *et al.* (1992) reported that a <u>gacA</u> mutant of <u>P. fluorescens</u> strain CHAO is impaired in the production of the two antibiotics 2,4 diacetylphloroglucinol and pyoluteorin as well as in the production of HCN. The authors provide evidence for a role of GacA protein as a global regulator of secondary metabolism whose expression is increased in the stationary phase.

Siderophores, i.e. compounds with an extremely high affinity for Fe^{3+}, are secreted by most bacteria upon sensing Fe^{3+}-deficiency. Fe^{3+}-concentrations in neutral and alkaline soil are so low that under those conditions soil micro-organisms are subject to Fe^{3+}-limitation. They can produce a range of different siderophores. The resulting siderophore-Fe^{3+} complexes are taken up through often specific receptors in the bacterial outer membrane (Neilands, 1982). Certain fluorescent biocontrol pseudomonads are very efficient in the competition for Fe^{3+}-ions, thereby further limiting the amount of Fe^{3+} available for the pathogen (Kloepper *et al.*, 1980; Geels & Schippers, 1983; Weller & Cook, 1983). The plant apparently can cope with this situation since it grows even better in the presence of the biocontrol bacterium.
The production of siderophores appeared to be a prerequisite for the biocontrol activity of these pseudomonads, since siderophore-negative mutants were unable to increase potato yields in contrast to the wildtype strain (Bakker *et al.*, 1986; Kloepper *et al.*, 1980).
Interestingly one of these plant growth-promoting strains, <u>P. putida</u> strain WCS358, possesses multiple siderophore-uptake systems which are induced by the presence of heterologous siderophores (Koster *et al.*, 1993) These multiple uptake systems enable this strain to utilize the siderophores of many other microorganisms which presumably is an advantage in the rhizosphere.

A role of volatile substances like ammonia and cyanide in biocontrol has been implicated. Voisard *et al.* (1989) provide evidence for the role of HCN in controlling black root rot of tomato by <u>P.fluorescens</u> strain CHAO. Circumstancial evidence for quite another role of cyanide, namely in inhibiting plant development by affecting root energy metabolism, has been proposed by Schippers *et al.* (1990).

Chitinase secreted by the soil bacterium <u>Serratia</u> <u>marcescens</u> plays a role inprotecting beans against certain fungi (Jones *et al.*, 1986; Ordentlich *et al.*, 1987).

Competition for niches is best-known from the use of non-pathogenic "ice-minus" mutants of <u>P. syringae</u> to protect plant leaves against the wild type bacterium which can cause frost damage (Lindow, 1990). Similarly, avirulent mutants of <u>P. solanacearum</u> are effective as antagonists for the control of bacterial wilt of tomato (Trigalet & Trigalet-Demery, 1990).

Competition for nutrients is considered as both a directly protective factor as well as a colonization factor. A non-pathogenic <u>Fusarium oxysporum</u> strain was able to reduce the occurence of Fusarium wilt in carnation. When <u>P. putida</u> strain WCS358 was combined with this non-pathogenic <u>F. oxysporum</u> strain, an increased suppression of the disease was observed. For this cooperative plant-disease control the production of siderophores by the <u>P. putida</u> strain was essential, since siderophore-negative mutants did not increase the effects of the <u>F. oxysporum</u> strain (Lemanceau *et al.*, 1992). The non-pathogenic <u>F. oxysporum</u> was shown to be less sensitive for competition for Fe $^{3+}$ than the pathogen and is therefore supposed to more succesfully compete for carton with the weakened pathogen. (Lemanceau *et al.*, 1992)

Co-operation of (serveral) microbial biocontrol agents is supposed to be responsible for the prevention of particular soil-borne diseases in disease-suppressive soils (Schippers, 1992). Preinoculation of carnation roots with <u>Pseudomonas fluorescens</u> WCS 417r protects the plant against a vascular disease caused by <u>F. oxysporum f. sp. dianthi</u> wilt (van Peer *et al.*, 1991). The protection was related to accelerated and increased phytoalexin accumulation at the site of infection, which occurs only after infection with the pathogen. It was hypothesized that signals triggered by the <u>Pseudomonas</u> bacterium in the root system induce sensitization of defence responses against the fungus in the stem, including increased accumulation of phytoalexins. This protection seems to be a form of systemic acquired resistance, which can also be brought about in plants by exogenously applied chemicals like 2,6-dichloroisonicotinic acid and salicylic acid (Uknes *et al.*, 1992).

RHIZOSPHERE COLONIZATION

Colonization is of crucial importance to deliver the protective biocontrol factor at the right site in the rhizosphere at the right time. Colonization will be influenced by the root's surface components, the exudate composition, the local bacterial growth conditions in the various niches of the rhizosphere, the soil type, and various other biotic and abiotic factors. Our poor knowledge of colonization is directly related to our poor knowledge of these factors, both individually as well as in the way they mutually influence each other.

We have approached the question of which bacterial traits are important for colonization by genetic studies. In the first approach we tried to list putative bacterial colonization traits, e.g. mobility. We subsequently isolated mutants deficient in such a trait and compared their colonizing ability to that of the wild type strain. Using this approach we have shown that the presence of flagella, and perhaps mobility or even chemotaxis (de Weger *et al.*, 1987), as well as the ability to produce O-antigen side chains of lipopolysaccharide (de Weger *et al.*, 1989) are crucial factors for colonization of potato roots by the tested <u>Pseudomonas</u> strains. Similarly, we intend to indentify quantitatively the major components of root exudates in order to test whether their utilization as a carbon source is involved in colonization. The second genetic approach assumes that we cannot predict all colonization traits. After screening random mutants for their colonizing ability, we subsequently plan to isolate colonization (<u>col</u>) genes. Nucleotide sequence analysis of these genes will presumably lead to novel traits involved in colonization.

BACTERIAL BIOCONTROL PRODUCTS AND THEIR FUTURE

Quite a few bacterial products for plant growth stimulation are commercially available (Kloepper, 1991; Lewis, 1991).
For seed companies it is interesting to sell their products with a coating containing one or more biocontrol strains, thereby using the know-how that the microbial population in a plant's rhizosphere can be determined by that present in the spermosphere (Lynch, 1990).
A limiting factor for the efficacy of the bacterial biocontrol products, as well as for the number of products, is our poor insight in the molecular basis of interactions in the spermosphere and in the rhizosphere. Often the bacterial factors which cause biocontrol are unknown. If they are known, knowledge of the effect of environmental factors on the levels of productive factor produced <u>in situ</u> are important. A description of spacial-temporal colonization patterns as given by Bahme and Schroth (1987) should be carried out for both biocontrol organisms as well as for pathogens under a range of relevant conditions. In this respect it is interesting to note that bacteria isolated from different sites along the wheat root differ in their physiological properties (Liljeroth *et al.*, 1991) Finally, mechanisms used for colonization are unknown, despite the fact that in several field trials colonization is the limiting step for biocontrol (Schippers *et al.*, 1987; Weller, 1988).

Whereas survival is hardly a problem with spore-formers, survival during coating on seed and during storage on the shelf is a problem when Gram-negative bacteria are used as biocontrol organisms. This problem may be overcome by increasing our knowledge of survival strategies of Gram-negative bacteria. Similarly, knowledge of bacterial physiology could be important to ensure a fast and correct response of the biocontrol strain upon germination of the seed.

The composition of root exudate is of prime importance for interactions between biocontrol strain, pathogen and resident microflora. Quantitatively the major exudate compounds are neutral sugars, organic acids and amino acids (Vancura, 1964). However, for many crop plants this knowledge, if available, is not public. Moveover, exudates can contain a number of other compounds that can have profound effects on interactions in the rhizosphere. Various flavonoids have been reported to activate the NodD protein as an early step in nodulation of leguminous plants by <u>Rhizobium</u>

bacteria (see chapters of Long *et al.*, Redmond *et al.*, Wijffelman *et al.* and Firmin *et al.* in Lugtenberg, 1989). Flavonoids have also been reported to increase the growth rate of specific soil bacteria (Hartwig *et al.*, 1991), to promote development of spores of the vesicular-arbuscular mycorrhizal fungus <u>Glomus etunicatum</u> (Tsai & Phillips, 1991), and to function as chemoattractants for zoo-spores of the pathogenic fungus <u>Phytophtora sojae</u> (Morris & Ward, 1992).
Microbial genes which have a function in the interaction with a plant are often induced by secreted plant components. Also biotic and antibiotic conditions proposed to be valid at the site of the interaction may stimulate the induction (e.g. Lugtenberg & de Maagd, 1991; Stachel *et al.*, 1985; Mantis & Winans, 1992; Schulte & Bonas, 1992). Minor exudate components which can play major roles in such interactions can be purified by using reporter genes behind the promoter of the induced gene (e.g. Zaat *et al.*, 1987).

Although some information is available about concentrations of various components in soil and in exudate, only local conditions in their particular niche are relevant for micro-organisms in the rhizosphere. In our group we have started to study local rhizosphere conditions with respect to phosphate-availability (de Weger *et al.*, in press) and amino acid availability (Simons *et al.*, in preparation). Another technique, which uses magnetic beads containing specific antibodies, provides the possibility of isolating cells of a specific bacterial strain from a majority of other bacteria (de Weger *et al.*, in preparation). This technique, therefore, seems to offer the possibility of introducing a specific bacterium (e.g. with an inducible promoter in front of a reporter gene), allowing it to grow in the rhizosphere, and subsequently reisolating and analysing specifically the cells of the introduced strain (e.g. after activation of the studied promoter <u>in situ</u>).

Increased know-how about rhizosphere events will allow us to construct genetically improved strains for biocontrol purposes. Indeed, Maurhofer *et al.* (1992) have shown that increased antibiotic production in <u>P. fluorescens</u> strain CHAO resulted in better protection of cucumber against <u>Pythium ultimum</u>. In the absence of a pathogen the increased antibiotic production had a deleterious effect on cress and sweet corn. The authors conclude that, depending on the host-pathogen system, enhanced antibiotic production by <u>P. fluorescens</u> may result in improved disease suppression or, in contrast, in a toxic effect on the plant.

As far as we know, resistance of pathogens towards protective factors has not been reported, but can certainly not be excluded. Therefore strategies for using microbial control agents should take this possibility into account. A combination of at least two protective factors with different mechanism of action can overcome the putative problem of resistance. Co-operation between several microbial biocontrol agents seems to be responsible for the efficiency of suppression of particular soil-borne diseases in natural disease-suppressive soils. Co-inoculation of biocontrol agents therefore may also improve protection against disease by seed bacterization (Schippers, 1992).

We agree that before application genetically modified strains will have to go through a risk assessment procedure. We disagree with certain environmental activists who oppose more or less on principle the application of genetically modified bacteria in the environment. Public acceptance is negatively influenced by their opposition and sometimes also their ignorance as well as by lack of public knowledge of these matters. In this respect it is important to note that <u>Rhizobium</u> an <u>Bradyrhizobium</u> inoculants have been used in massive amounts for almost a century without a single reported accident. Considering the poor sterilisation facilities in the earlier decades, it is likely that the inoculants not only contained the namedbacteria but also a broad range of other organisms. We see a role for scientists to educate the public on what is really going on and for companies to co-operate in an attempt to explain the balance of advantages and disadvantages of their future products.

REFERENCES

Bahme, J.B.; Schroth, M.N. (1987) Spatial-temporal colonization patterns of a rhizobacterium on underground organs of potato, *Phytopathology*, **77**, 1093-1100.

Bakker, P.A.H.M.; Lamers, J.G.; Bakker, A.W.; Marugg, J.D.; Weisbeek, P.J.; Schippers, B. (1986) The role of siderophores in potato tuber yield increase by *Pseudomonas putida* in a short rotation of potato. *Netherlands Journal of Plant Pathology*, **92**, 249-256.

Campbell, R.; Greaves, M.P. (1990) Anatomy and community structure of the rhizosphere. In: *The rhizosphere*. J.M. Lynch (Ed.) Wiley, New York, pp. 11-34

Défago, G.; Beiling, C.H.; Burger, U.; Haas, D.; Kahr, G.; Keel, C.;Voisard, C.; Wirthner, P.; Wüthrich, B. (1990) Suppression of black root rot of tobacco and other root diseases by strains of *Pseudomonas fluorescens*. Potential applications and mechanisms. In: *Biological Control of Soil Borne Plant Pathogens*, D. Hornby, R.J. Cook, Y. Henis, W.H. Ko, B. Schippers, and P.R. Scott (Eds), Wallingford: CAB International, pp. 93-108.

Geels, F.P.; Schippers, B. (1983) Reduction of yield depressions in high frequency potato cropping soil after seed tuber treatments with antagonistic fluorescent *Pseudomonas* spp. *Phytopathologische Zeitschrift*, **108**, 207-214.

Gurusiddaiah, S.; Weller, D.M.; Sarkar, A.; Cook, R.J. (1986) Characterization of an antibiotic produced by a strain of *Pseudomonas fluorescens* inhibitory to *Gaeumannomyces graminis* var. tritici and Pythium spp. *Antimicrobial Agents and Chemotherapy*, **29**, 488-495.

Hartwig, U.A.; Joseph, C.M.; Phillips, D.A. (1991) Flavonoids released naturally from alfalfer seeds enhance growth rate of Rhizobium meliloti, *Plant Physiology*, **95**, 797-803.

Howell, C.R. (1990) Fungi as biological control agents. In: *Biotechnology of Plant-Microbe Interactions*, J.P. Nakas and C. Hagedom (Eds), New York: McGraw-Hill, pp. 257-286.

Jones, J.D.G.; Grady, K.L.; Suslow, T.V.; Bedbrook, J.R. (1986) Isolation and characterization of genes encoding two chitinase enzymes from *Serratia marcescens*, *EMBO Journal*, **5**, 467-473.

Kloepper, J.W.; Leong, J.; Teintze, M.; Schroth, M.N. (1980) Enhanced plant growth by siderophores produced by plant growth-promoting rhizobacteria. *Nature*, **286**, 885-886.

Kloepper, J.W. (1991) Plant growth-promoting rhizobacteria as biological control agents of soilborne diseases. In: Biological control of diseases, J. Bay-Petersen (Ed.), *FFTC Book Series No 42. Food and Fertilizer Technology Center*, Taipeh, Taiwan, pp. 142-152.

Koster, M.; van de Vossenberg, J.; Leong, J.; Weisbeek, P.J. (1993) Identification and characterization of the pupB gene encoding an inferric-pseudobactin receptor of *Pseudomonas putida* WCS358. *Molecular Microbiology*, **8**, 591-601.

Kraus, J.; Loper, J.E. (1992) Lack of evidence for a role of antifungal metabolite production by *Pseudomonas fluorescens* Pf-5 in biological control of pythium damping-off of cucumber, *Molecular Plant Pathology*, **82**, 264-271.

Laville, J.; Voisard, C.; Keel, C.; Maurhofer, M.; Défago, G.; Haas, D. (1992) ,Global control in *Pseudomonas fluorescens* mediating antibiotic synthesis and suppression of black root rot of tobacco, *Proceedings of the National Academy of Sciences of the United States of America*, **89**, 1562-1566.

Leeman, M.; Van Pelt, J.A.; Bakker, P.A.H.M.; Schippers, B. (*submitted*). Involvement of lipopolysaccharides in <u>Pseudomonas</u>-mediated induced resistance in radish against Fuserium wilt.

Lemanceau, P.; Bakker, P.A.H.M.; de Kogel, W.J.; Alabouvette, C.; Schippers, B. (1992) Effect of pseudobactin 358 production by *pseudomonas putida* WCS358 on suppression of fusarium wilt of carnations by nonpathogenic fusarium oxysporum Fo47. *Applied and Environmental Microbiology*, 2987-2982.

Lewis, J.A. (1991) Formulation and delivery systems of biocontrol agents with emphasis on fungi. In: *The rhizosphere and plant growth*, D.L. Keister and P.B. Cregan (Eds), Netherlands: Kluwer academic publishers, pp. 279-287.

Liljeroth, E.; Burgers, S.L.G.E.; van Veen, J.A. (1991) Changes in bacterial populations along roots of wheat (*Triticum aestivum* L.) seedlings. *Biology and Fertility of Soils*, **10**, 276-280.

Lindow, S.E. (1990) Use of genetically altered bacteria to achieve plant frost control. In: *Biotechnology of Plant-Microbe Interactions*, J.P. Nakas and C. Hagendorn (Eds), New York: McGraw-Hill, pp. 85-110.

Lynch, J.M. (1990) Introduction: Some consequences of microbial rhizosphere competence for plant and soil. In: *The rhizosphere*. J.M. Lynch (Ed.). Wiley, New York, pp. 1-10

Lugtenberg, B.J.J. (1989) Recognition in microbe-plant symbiotic and pathogenic interactions, *NATO ASI Series H4*, Springer Verslag, Berlin, Heidelberg.

Lugtenberg, B.J.J.; de Maagd, R.A. (1991) Recent advances in plant-microbe interactions.In: *Cell to cell signals in plants and animals*. NATO ASI Series H51, V. Neuhoff and J. Friend (Eds), Springer Verlag ,Heidelberg, pp. 15-26.

Mantis, N.S.; Winans, S.C. (1992) The *Agrobacterium tumefaciens vir* gene transcriptional activator VirG is transcriptionally induced by acid pH and other stress stimuli,*Journal of Bacteriology.*, **174**, 1189-1196.

Maurhofer, M.; Keel, C.; Schnider, U.; Voisard, C.; Haas, D.; Défago, G. (1992) Influence of enhanced antibiotic production in *Pseudomonas fluorescens* strain CHAO on its disease suppressive capacity, *Phytopathology*, **82**, 190-195.

Morris,P.F.; Ward, E.W.B. (1992) Chemoattraction of zoospores of the soybean pathogen, *Phytophtora sojae*, by isoflavones, Physiological Plant Pathology, **40**, 17-22

Neilands, J.B. (1982) Microbial envelope proteins related to iron, *Annual Review Microbiology*, **36**, 285-309.

Oberhänsli, T.; Défago, G.; Haas, D. (1991) Indole-3-acetic acid (IAA) synthesis in the biocontrol strain CHAO of *Pseudomonas fluorescens*: role of tryptophan side chain oxidase, *Journal of General Microbiology*, **137**, 2273-2279.

Ordentlich, A.; Elad, Y.; Chet, I. (1987) Rhizosphere colonization by *Serratia marcescens* for the control of *Sclerotium rolfsii*, *Soil Biology and Biochemistry*, **49**, 747-751.

Peer, R. van; Niemann, G.J.; Schippers, B. (1991) Induced resistance and phytoalexin accumulation in biological control of fusarium wilt of carnation by *Pseudomonas* sp. strain WCS417r, *Phytopathology*, pp. 728-734.

Peer,R. van; Schippers, B. (1992) Lipopolysaccharides of plant-growth promoting *Pseudomonas* sp. strain WCS417r induce resistance in carnation to Fusarium wilt. *Netherlands Journal of Plant Pathology*, **98**, 129-199.

Ryder, M.H.; Jones, D.A. (1990) Biological control of crown gall. In: *Biological control of soil-borne plant pathogens*, D. Hornby (Ed.), Wallingford: CAB International, pp. 123-130.

Schippers, B.; Bakker, A.W.; Bakker, P.H.A.M. (1987) Interactions of delete rious and beneficial rhizosphere microorganisms and the effect of cropping practices, *Annual Review of Phytopathology*, **25**, 339-358.

Schippers, B.; Lugtenberg, B.J.J.; Weisbeek, P.J. (1987) Plant growth control by fluorescent pseudomonads. In: *Innovative approaches to plant disease control*. I. Chet (Ed.). New York: Wiley, pp. 19-39.

Schippers, B.; Bakker, A.W.; Bakker, P.A.H.M.; van Peer, R. (1990) Beneficial and deleterious effects of HCN-producing *Pseudomonads* on rhizosphere interactions. *Plant and Soil*, **129**, 75-83.

Schippers, B. (1992) Prospects for managements of natural suppressiveness to control soilborne pathogens. In: *Biological control of plant diseases*. Progress and challenges for the future, E.C. Tjamos, G.C. Papavizas and R.J. Cook (Eds), Plenum Press, New York, pp. 21-34.

Schulte, R.; Bonas, U. (1992) A *Xanthomonas* pathogenicity locus is induced by sucrose and sulfur-containing amino acids, *The Plant Cell*, **4**, 79-86.

Stachel, S.E.; Messens, E.; van Montagu, M.; Zambryski, P. (1985) Identification of the signal molecules produced by wounded plant cells that activate T-DNA transfer in agrobacterium tumefaciens. *Nature*, **318**, 624-629.

Tomashow, L.W.; Weller, D.M. (1988) Role of a phenazine antibiotic from *Pseudomonas fluorescens* in biological control of *Gaeumannomyces graminis* var. tritici, *Journal of Bacteriology.*, **170**, 3499-3508.

Tomashow, L.S. (1991) Molecular basis of antibiosis mediated by rhizosphere pseudomonads. In: *Plant growth-promoting rhizobacteria progress and prospect*, C. Keel, B. Knoller and G. Défago (Eds), International organization for biological and integrated control of noxious animals and plants, pp. 109-114.

Trigalet, A.; Trigalet-Demery, D. (1989) Use of avirulent mutants of *Pseudomonas solanacearum* for the biological control of bacterial wilt of tomato plants, *Physiological and Molecular Plant Pathology*, **36**, 27-38.

Tsai, S.M.; Phillips, D.A. (1991) Flavonoids released naturally from alfalfa promote development of symbiotic Glomus spores in vitro, *Applied and Environmental Microbiology*, **57**, 1485-1488.

Uknes, S.; Mauch-Mani, B.; Moyer, M.; Potter, S.; Williams, S.; Dincher, S.; Chandler, D.; Slusarenko, A.; Ward, E.; Ryals, J. (1992) Acquired resistance in arabidopsis. *The Plant Cell*, **4**, 645-656.

Vancura, V. (1964) Root exudates of plants. I. Analysis of root exudates of barley and wheat in their initial phases of growth, *Plant and Soil*, **2**, 231-248.

Voisard, C.; Keel, C.; Haas, D.; Défago, G. (1989) Cyanide production by *Pseudomonas fluorescens* helps suppress black root rot of tobacco under gnotobiotic conditions, *EMBO Journal.*, **8**, 351-358.

Weger, L.A. de; Vlugt, C.I.M. van der; Wijfjes, A.H.M.; Bakker, P.A.H.M.; Schippers, B.; Lugtenberg, B.J.J. (1987) Flagella of a plant growth-stimulating *Pseudomonas fluorescens* strain are required for colonization of potato roots, *Journal of Bacteriology*, **169**, 2769-2773.

Weger, L.A. de; Bakker, P.A.H.M.; Schippers, B.; Loosdrecht, M.C.M. van; Lugtenberg, B.J.J. (1989) *Pseudomonas spp.* with mutational changes in the Oantigenic side chain of their lipopolysaccharide are affected in their ability to colonize potato roots. In: *Signal molecules in plants and plant-microbe interactions*, B.J.J. Lugtenberg (Ed.), NATO ASI series **H36**, pp. 197-202.

Weger, L.A. de; Dekkers, L.C.; Van der Bij, A.J.; Lugtenberg, B.J.J. (1993) Use of phosphate reporter bacteria to study phosphate limitation in the rhizosphere and in bulk soil. *Applied and Environmental Microbiology*, in press.

Weller, D.M.; Cook, R.J. (1983) Disease control and pest management. suppression of takeall of wheat by seed treatments with fluorescent pseudomonads, *Phytopathology*, **73**, 463-469.

Weller, D.M. (1988) Biological control of soilborne plant pathogens in the rhizosphere with bacteria. *Annual Review of Phytopathology.* **26**, 379-407.

Zaat, S.A.S.; Wijffelman, C.A.; Spaink, H.P.; Brussel, A.A.N. van; Okker, R.J.H.; Lugtenberg, B.J.J. (1987) Induction of the nodA promoter of *Rhizobium leguminosarum* Sym plasmid pRLIJI by plant flavanones and flavones, *Journal of Bacteriology*, **169**, 198-204.

BIOLOGICAL SEED TREATMENTS - THE DEVELOPMENT PROCESS

D.J. RHODES AND K.A. POWELL

ZENECA Agrochemicals, Jealott's Hill Research Station, Bracknell, Berks,
RG12 6EY

ABSTRACT

Biological seed treatments for plant pest and disease control have been
the focus of significant research effort for the past 15 years, in the
hope of replacing chemical seed treatments with biologically based
products. Commercial products are beginning to reach the market place,
but growth of the field has been slowed by formulation difficulties,
limited understanding of rhizosphere ecology, poor reproducibility of
field effects, and uncertainties over the size of the commercial
opportunity, the strength of patents, and the cost of registration.

INTRODUCTION

Biological treatments may be applied to seed in order to introduce
rhizobia or other beneficial symbionts, or to achieve biological control of
pests or diseases. While production of legume symbionts is a well-established
industry, it is only during the past 15 years that application of biological
control agents to seed has received serious attention. The purpose of this
paper is to discuss the potential for biocontrol agents and plant growth-
promoting rhizobacteria (PGPR) applied to seed, the development process
through which such products must pass before reaching the market place, and
the effect of development costs on the growth prospects for seed-applied
biocontrol agents.

Market for biopesticides

The market for biopesticides has grown considerably in recent years from
negligible levels in the 1970s to at least $100 million today, and is expected
to grow to $400-500 million by the end of the decade. This growth has been
driven by environmental concerns, the high cost of re-registering existing
chemical products, resistance, and reduced tolerance for chemical residues on
food. In contrast, the crop protection industry as a whole, after several
decades of growth, has begun to decline in real terms (Anon., 1993a). The
growth prospects for biological products have prompted considerable interest
in biopesticide research on the part of industry. The majority of this
research is not directed towards seed treatment, however, since the vast
majority of biocontrol products are bioinsecticides applied to foliage, water
or soil. Products based on *Bacillus thuringiensis* accounted for 92% of
biopesticide sales in 1990 (Anon., 1991), although nematode-based products are
growing in importance.

Biological seed treatments

Although the research effort devoted to biological seed treatments has
increased sharply in recent years, the concept is not new. Seed treatments
with bacterial inoculants such as *Azotobacter chroococcum* and *Bacillus
megaterium* has been practised for many decades in the former Soviet Union with
the objective of enhancing crop growth and yield, while other organisms such

as *Pseudomonas* and *Streptomyces* species have been investigated as potential seed treatment additives (Cooper, 1959; Brown, 1974; Filippov, 1989).

Western interest in biological seed treatments has increased greatly during the past 20 years, beginning with the demonstration by Merriman *et al.* (1974) that applications of bacteria to seed resulted in increased plant growth in the field. Kloepper *et al.* (1980) demonstrated that such growth increases were associated with aggressive colonization of the rhizosphere by the. applied bacteria, and it was subsequently demonstrated by Weller & Cook (1983) that a major disease, take-all of wheat, could be controlled by application of fluorescent pseudomonads to seed. Harman et al. (1980) demonstrated that such affects could also be obtained by application of fungi to seed. The past decade has been characterised by attempts to understand the ecology and mode of action of organisms with potential as seed treatments, and also to exploit such organisms industrially.

AVAILABLE PRODUCTS

<u>Chemical products</u>

A number of chemical products are applied to seed in order to control various pests and diseases. This method of application is increasing in popularity since it makes efficient use of active ingredient, is relatively inexpensive, avoids the need for specialised equipment where seed is treated off-farm, and minimises the volume of chemical applied per unit area. Although the range of active ingredients which can be applied by seed treatment is increasing, as is the range of pests and diseases which can be controlled in this way, seed treatment still represents a very small proportion of total agrochemical use. Some of the most important targets for seed treatment products, both registered and in development, are listed in Table 1.

<u>Biological products</u>

The majority of microbial seed treatment products currently available (Table 2) have been developed for control of soilborne damping-off diseases, caused mainly by *Pythium*, *Rhizoctonia*, and *Fusarium* species, or for stimulation of early plant growth. Biological seed treatments have made essentially no impact in the substantial market for cereal seed treatments, which are mostly used for control of seedborne pathogens, or on the market for seed-applied insecticides. A novel seed-applied biological insecticide is, however, being investigated by Crop Genetics International Inc., consisting of a strain of the endophyte *Clavibacter xyli* pv. *cynodontis* which expresses a Bt endotoxin. It is hoped eventually to introduce the bacterium into seed, possibly by vacuum infiltration, for control of European Corn Borer (*Ostrinia nubilalis*). The biological seed treatment products which are currently available, or in late stages of development, are listed in Table 2.

TABLE 1. Major pests and pathogens controlled by seed treatment

Crop	Pest or pathogen
Sugar beet, wheat, maize, vegetables, legumes	Soil pests, aphids
Cereals	*Tilletia caries* *Ustilago* spp. *Pyrenophora* spp. *Fusarium* spp. *Septoria* spp.
Cotton	*Pythium* spp., *Rhizoctonia solani* *Xanthomonas campestris*
Vegetables	*Pythium* spp., *Rhizoctonia solani*
Maize	*Pythium, Fusarium* spp.
Sugar beet	*Pythium, Aphanomyces* spp., *Phoma betae*
Potatoes	*Fusarium* spp., *Rhizoctonia solani*
Peas, beans, soya	*Pythium* spp., *Rhizoctonia solani* *Peronospora* spp.

TABLE 2. Biological seed treatment products

Trade name	Manufacturer	Active ingredient	Crops treated
Quantum 4000	Gustafson	*Bacillus subtilis*	Various
Kodiak	Gustafson	*Bacillus subtilis*	Cotton, peanuts, beans
Mycostop	Kemira	*Streptomyces griseoviridis*	Vegetables, ornamentals
F-Stop	Under negotiation	*Trichoderma harzianum*	Various
Blue Circle	Stine Microbial Products	*Pseudomonas cepacia*	Various
Nogall	Bio Care Technology	*Agrobacterium radiobacter*	Tree seedlings

STRAIN SELECTION STRATEGIES

Two basic strategies are available for selecting the strain which comprises the active ingredient in a biological product; the strain may be deliberately manipulated in order to confer the required characteristics or may be selected from the natural microflora according to the ability to fulfil various criteria. To date, only one of the commercially-available seed treatment products contains a strain which was genetically modified using recombinant DNA technology, the *Agrobacterium radiobacter* strain sold in Australia as Nogall, in which the transfer region has been deleted from the plasmid coding for agrocin production (Ryder, 1991). There are several reasons for this. Firstly, although considerable progress has been made in recent years in understanding the molecular basis of the interactions between biocontrol agents and plant pathogens, the ability to achieve significant improvements in strain performance through genetic manipulation has yet to be proved in the field. Achieving this is, in turn, hampered by the reluctance of regulatory authorities to permit release of genuinely "improved" microbial strains, although a number of genetically marked strains have been released into the environment for tracking studies. Given this situation, commercial interest in genetically engineered strains for seed treatment is likely to be limited, since the market opportunity is too small to allow the substantial research and development costs to be recouped. However, authorities in both Europe and the USA have sought to remedy this situation by funding basic research on ecology and genetic stability of recombinant microorganisms which should allow rational decisions on field release and registration to be made in future. Despite the present difficulty of developing products based on recombinant strains, other strain improvement strategies are possible, such as the protoplast fusion approach described by Harman (1991).

Selection of strains from nature presents a number of challenges, given the essentially infinite number of strains available in nature, and the multiple factors which must be taken into account during the screening process. These factors are summarised in Table 3. Strain selection is, in general, heavily dependent on biological efficacy data. Efficacy data may have been over-emphasised, since a number of promising strains have not progressed beyond the research phase due to difficulties with formulation, production, or inconsistency of effect across a range of environmental conditions.

In general, as the complexity of the screening process increases, the throughput (i.e. the number of strains which can be tested per unit time) decreases. The challenge in designing a screening cascade is to ensure that the assays reflect, as closely as possible, the situation which is encountered in the field, without compromising throughput unduly. The throughput which is required in order to select an effective strain depends on the strictness of the criteria which are imposed during the screening process (for example, biocontrol activity against more than one pathogen), and the abundance of the necessary traits in nature. It is not possible to calculate the necessary rate of throughput until a screening cascade has been running for some time. Screens may consist of mechanistic assays, in which strains are tested for some pre-defined trait, such as hyperparasitism, antibiotic or siderophore production, which is believed to contribute to the biological effect. An example is the *in vitro* screen used by Geels & Schippers (1983) to select *Pseudomonas* strains which counteract yield depression in potato. Non-mechanistic *in planta* assays, such as those developed by Kloepper (1991) incorporate plant-surface interactions, and do not require the mode of action to be defined at the outset. A rapid throughput can usually be achieved using assays of this type, although at some cost in terms of the environmental factors which can be incorporated into the assay. The third approach which has been employed for selecting microbial strain is the use of microcosms or mesocosms, ecosystems which incorporate as many features as possible of the field situation. Microcosms are simulated ecosystems, constructed in the laboratory or glasshouse, for example those of Nelson & Craft (1992), while mesocosms consist of field plots, as described by Hagedorn *et al.* (1993). Our experience has been that *in planta* screens are preferable to *in vitro* assays, but that miniaturisation is required in order to screen an adequate number of strains (typically several thousand). This strategy was outlined by Renwick *et al.* (1990).

TABLE 3.   Strain selection strategies for microbial seed treatments

Component of the screening cascade	Available strategies
Source of isolates	Seeds Random soil samples Suppressive soils Hyperparasites from pathogen biomass Isolation under selection pressure e.g. for rhizosphere competence
Nature of isolates	Random Genera selected for ease of production or formulation
Biological efficacy	Mechanistic *in vivo* screening Non-mechanistic *in planta* screening Screening in microcosms or mesocosms Spectrum of biocontrol activity
Ecology	Host specificity Colonisation of plant surfaces Effect of temperature, moisture, soil type
Secondary evaluation	Production parameters (growth rate, stability) Evaluation of strains in seed-applied formulation

PATENTING

Microorganisms, unlike chemicals, can easily be isolated and propagated by individuals other than the original producer. It is therefore even more essential to protect microbial products from unlicensed manufacture than is the case with chemicals. In the past, patenting of microbial strains has been a controversial issue, but this has now largely been resolved in the major industrialised nations (Anon, 1993b). The fact that a microbial product consists of a living organism is no longer considered to be a barrier to patentability. Large numbers of patents covering both microorganisms and microbial genes have now been granted in Europe, the USA and Japan. However,

as with any other patent, the criteria of novelty, non-obviousness and utility
must be fulfilled. While the number of challenges to granted patents covering
wild-type organisms has been limited, and while novelty and non-obviousness
have become perhaps more difficult to prove as the prior art has expanded,
there is no evidence to suggest that patents of this type are intrinsically
weak. Protection of intellectual property has also been greatly strengthened
by availability of DNA profiling and other diagnostic tools which allow
precise identification of microbial strains.

PRODUCTION AND FORMULATION

Fortunately, the majority of bacterial agents can be produced by liquid
(submerged) fermentation, which is the standard method for producing microbial
biomass in Europe and North America. Although more technically difficult, it
is also possible to produce fungal biomass, for example that of *Trichoderma*,
in this way (Harman, 1991). Following production, sufficient numbers of
viable propagules, usually in a stabilised form, must be applied to seed to
achieve the required biological effect. This has clearly been achieved in the
case of *Rhizobium* and related inoculants using peat-based and liquid inocula,
but is proving more difficult where numbers of cells per seed is critical as
with biocontrol agents. The storage stability requirements for biological
agents used to treat seed vary widely. Seed is sown immediately after hopper
box treatments, but may also be treated soon after harvest and stored until
the following growing season. The technology available for formulation and
application of biological products to seed has been discussed in a number of
reviews (Connick, 1990, Harman, 1991; Lewis, 1991; Rhodes, 1993) and is
summarised in Table 4.

TABLE 4. Formulation technology for application of biological
agents to seed

Dry formulations	Aqueous fomulations	Coatings
Gum/talc powders	Sprays	Pellets
Dusts	Dips	Liquid
Dry spores	Fluid drilling gels	coatings
	Solid matrix priming	

REGISTRATION

Microbial products are regulated, not only according to pesticide
regulations, but by regulations governing the handling and release of
microorganisms. Although the cost of registering microbial products may still
be considerably lower, in most countries, than that of chemical products,
registration costs and delays are still a major factor in determining the
commercial viability of products in development. It is necessary to interact
with regulatory authorities at a number of points throughout the development
process. Significant differences exist between countries; indeed most
countries have not yet formulated specific guidelines for the registration of
microbial products, but deal with microbial products on a case-by-case basis
according to chemical legislation. Specific legislation covering microbial
products is probably best advanced in the USA, under the guidelines produced

by Subdivision M of the Environmental Protection Agency. Specific legislation also applies in Canada and France, and will soon cover all the member states of the European Community as Directive 91/414/EEC Part B comes into force.

It is impossible, therefore, to describe a universal scheme for registration of microbial crop protection products. However, a typical example of the evidence required by regulatory authorities in the course of development of a microbial product is presented in Table 5. Requirements may be waived where the organism concerned is thoroughly documented in the scientific literature, or has a long history of safe use.

TABLE 5. Typical registration requirements for a biological control agent.

Development phase	Evidence required
Field release	Identity with indigenous organism Genetic stability Ecotoxicology
Registration	Identity and biological properties of the microorganism Description of the formulation Manufacturing process and product analysis Residues (frequently not required) Efficacy Toxicology, pathogenicity and infectivity Effects on non-target organisms Environmental fate

CONCLUSIONS

A large number of potential biological seed treatments have been described in the literature, while others are in the process of commercial development. Given the inherently limited commercial opportunity for such products, it is clear that not all will be commercially viable. Success will depend not only on possessing strains which are clearly differentiated from those of competitors, but on strong patents, formulation technology, and the ability to minimise delays during registration.

REFERENCES

Anon. (1991) *Wood Mackenzie Agrochemical Monitor*, September.
Anon. (1993a) *Wood Mackenzie Agrochemical Monitor*, September.
Anon. (1993b) European Federation of Biotechnology Task Group on Public Perception of Biotechnology, Briefing Paper 1, Patenting Life.
Brown, M.E. (1974) Seed and root bacterization. *Annual Review of Phytopathology*, 12, 181-197.
Connick, W.J. Jr., Lewis, J.A., Quimby, P.C. Jr. (1990) Formulation of biocontrol agents for use in plant pathology. In: *New Directions in Biological Control*, R.R. Baker and P.E. Dunn (Eds), New York: Liss, pp. 345-372.
Cooper, R. (1959) Bacterial fertilizers in the Soviet Union. *Soils and Fertilizers*, 22, 327-333.

Filippov, N.A. (1989) The present state and future outlook of biological control in the USSR. *Acta Entomologica Fennica*, **53**, 11-18.

Geels, F.P.; Schippers, B. (1983) Selection of antagonistic fluorescent *Pseudomonas* spp. and their root colonization and persistence following treatment of potatoes. *Phytopathologische Zeitschrift*, **108**, 193-206.

Hagedorn, C.; Gould, W.D.; Bardinelli, T.R. (1993) Field evaluations of bacterial inoculants to control seedling disease pathogens on cotton. *Plant Disease*, **77**, 278-282.

Harman, G.E. (1991) Seed treatments for biological control of plant disease. *Crop Protection*, **10**, 166-171.

Harman, G.E.; Chet, I.; Baker, R. (1981) Factors affecting *Trichoderma hamatum* applied to seeds as a biocontrol agent. *Phytopathology*, **71**, 569-572.

Kloepper, J.W. (1991) Development of in vivo assays for prescreening antagonists of *Rhizoctonia solani* on cotton. *Phytopathology*, **81**, 1006-1013.

Kloepper, J.W.; Schroth, M.N.; Miller, T.D. (1980) Effects of rhizosphere colonization by plant growth-promoting rhizobacteria on potato plant development and yield. *Phytopathology*, **70**, 1078-1082.

Lewis, J.A. (1991) Formulation and delivery systems of biocontrol agents with emphasis on fungi. In: *The Rhizosphere and Plant Growth*, D.L. Keister and P.B. Cregan (Eds), Netherlands: Kluwer, pp. 279-287..

Merriman, P.R.; Price, R.D.; Kollmorgen, J.F.; Piggott, T.; Ridge, E.H. (1974) Effects of seed inoculation with *Bacillus subtilis* and *Streptomyces griseus* on the growth of cereals and carrots. *Australian Journal of Agricultural Research*, **25**, 219-226.

Nelson, E.B.; Craft, C.M. (1992) A miniaturized and rapid bioassay for the selection of soil bacteria suppressive to Pythium blight of turfgrasses. *Phytopathology*, **82**, 206-210.

Renwick, A.; Powell,K.; DeBruyne, E.; Fattori, M.; Iriarte, V. (1990) Biological control of soil-borne plant pathogens. *Brighton Crop Protection Conference - Pests and Diseases 1990*, **1**, 227-231.

Rhodes, D.J. (1993) Formulation of biological control agents. In: *Exploitation of Microorganisms*, D. G. Jones (Ed), London: Chapman & Hall, pp. 411-439.

Ryder, M. (1991) Biological control of crown gall using *Agrobacterium* strains K84 and K1026. In: *Plant Growth-Promoting Rhizobacteria: Progress and Prospects*, C. Keel, B. Koller and G. Defago (Eds.), Interlaken: IOBC WPRS, p. 96.

Weller, D.M.; Cook, R.J. (1983) Suppression of take-all of wheat by seed treatments with fluorescent pseudomonads. *Phytopathology*, **73**, 463-469.

THE SEED INDUSTRY'S VIEW ON BIOLOGICAL SEED TREATMENTS

RUDY J. SCHEFFER

S&G Seeds B.V., P.O.Box 26, 1600 AA Enkhuizen, the Netherlands

Abstract

Biological seed treatments are attractive because of the combination of their specific effect and limited environmental impact in comparison with agrochemicals or fertilizers. Inoculation of legume crops with rhizobia already is widely employed. Other plant growth-promoting microorganisms have been tested, as well as bacterial and fungal strains that deter a number of soil-borne diseases. The first products are on the market. Limitations for biological seed treatments do exist in the following fields: their efficacy, in an absolute sense against the disease or pest, and in comparison to resistance in the host or to chemical control measures, cost, and legislative restrictions on use. Biological seed treatments will find their place where an acceptable effectiveness, reliability and competitive pricing can be provided.

INTRODUCTION

Biological seed treatments have many potential uses, but currently only the first applications are emerging, both for growth/yield improvement and for control of soil-borne diseases. The use of microbial inocula has a longer history of practical use, with nitrogen-fixing bacteria employed in legume fields and mycorrhiza in tree nurseries being the prime examples.

Commercialization of nitrogen-fixing rhizobia already had begun last century after a U.S. patent on the use of pure cultures was issued to Nobbe and Hiltner in 1896. Inoculation of legume crops became widespread in the first quarter of this century. Currently inoculation is standard for legume crops such as soybeans and alfalfa grown under relatively low nitrogen levels and, as the application technology improves, new products also emerge (Smith, 1992).

Practical use of mycorrhiza is restricted to environmental extremes, such as nutrient deficiency (especially phosphorus) or prolonged drought. Under such conditions, vesicular-arbuscular mycorrhizae and ectomycorrhizae have been shown to benefit plants (Linderman, 1988).

Although biocontrol of insects employing *Bacillus thuringiensis*, baculoviruses or predators beautifully demonstrates the possibilities that exist for successful biocontrol practices, these successes are almost exclusively with foliar applications, and for that reason are not discussed   here (Jutsum, 1988; Reinecke, 1990).

SEED TREATMENTS

Using the seed as a carrier can improve the efficacy of microbial inocula. Examples are biocontrol of *Pythium* and *Rhizoctonia* damping-off by several fungi and bacteria. Less obvious are the significant effects of seed treatments on diseases that affect the plant after its seedling stage; for instance the biocontrol of *Aphanomyces* root rot of pea caused by *A. euteichis* f.sp. *pisi* (Parke et al., 1991), control of take-all in wheat (Thomashow & Weller, 1990), and also in our work with a fluorescent *Pseudomonas* isolate we repeatedly found that a seed treatment was more effective against *Fusarium* than a soil drench. Apparently the seed treatment positioned the inoculum where it could most effectively colonize the emerging root and therefore deter the pathogen.

Seed treatments have been attempted in various forms: commercial or semi-commercial are only simple dustings such as with a *Streptomyces griseoviridis* strain now marketed as Mycostop (Tahvonen, 1988; White et al., 1990), or with a *Bacillus subtilis* marketed as Kodiak and Quantum 4000. Much work has also been done on biocontrol with *Gliocladium virens* as the active ingredient, some of it on seed coatings (Howell, 1991), more on soil or substrate applications for which one product, Gliogard is now on the U.S. market (Fravel et al., 1985; Lumsden & Locke, 1989; Lynch et al, 1991). Incorporating a biocontrol agent in seed pellets was done for *Pythium oligandrum* (Lutchmeah & Cooke, 1985), but the same was also coated onto seeds (McQuilken, 1990). The most subtle seed treatments tested were based on combinations of priming (also known as osmoconditioning) of seeds in the presence of the biocontrol agent, which is thought to establish itself on the seed during the process (Harman et al., 1989; Callan et al. 1990). The examples of seed treatments outperforming soil or substrate applications are not rare, but apparently in some cases the amount of inoculum needed cannot be applied as a seed treatment. An example of this may be control of *Fusarium* by a non-pathogenic *Fusarium* isolate, which was shown to be effective if the antagonist could be applied at a much higher level than the pathogen (Alabouvette, 1990; Lemanceau & Alabouvette, 1991).

BIOLOGICAL SEED TREATMENTS: LIMITATIONS AND PROSPECTS

If very low inoculum densities in the soil cause serious crop losses already, or if relatively mobile organisms such as nematodes or insect larvae have to be controlled, the inoculum dose feasible with a seed treatment may be insufficient, although seedling protection still can be a valid reason for development of a such a product.

An example is white rot of onion, caused by the fungus *Sclerotium cepivorum*. After failure of an onion crop due to the pathogen hundreds of sclerotia can be present per kg of soil which will remain viable for many years. However, less than one sclerotium per kg soil already precludes growing onions. Any control has to be extremely effective to reduce the inoculum potential in the soil far enough to keep the infection level acceptable (Entwistle, 1990).

The efficacy of biological seed treatments in comparison with alternatives such as resistance in the host or chemical control is an important issue. The outcome of a comparison of the various options will depend on the individual crop/parasite combination.

Resistance to a pest or disease in the host plant is attractive because of its often absolute character. Obviously, sources of resistance have to be available before breeding for resistance becomes an option. The long time frame for a breeding success and therefore the high costs, the specificity of breeding for resistance (only the newly bred varieties carry the desired gene) and the negative correlation with yield can be arguments against resistance.

Novel seed treatments with fungicides and insecticides are being developed that combine improved efficacy with the use of lower amounts of the active ingredients. Large reductions in the quantities of pesticides can be realized by employing such seed treatments. An example of this is the seed treatment we developed to control the cabbage root fly (*Delia radicum*) in cauliflower and Brussels sprouts. Use of seeds coated with chlorpyrifos reduced the use of insecticide by over 99% and combines efficacy, a very acceptable environmental impact, and safe use for the grower (Kosters et al., 1993).

Biological seed treatments will find their place where alternatives are not cost effective or not available.

The economics of biological seed treatments are most complicated. If a biological seed treatment is extremely crop, parasite and probably even environment specific, that minimizes the overall environmental impact of the 'biological'. However, such a specificity also restricts use so much that for a company it may be impossible to do the research needed to develop a practical application based on an already repeatedly demonstrated effect. Therefore, a certain generality will probably be needed; the few current commercial products such as Mycostop are indeed not extremely crop and pathogen specific.

A final limitation in the application of microbial inocula is in the legislative restrictions on use. Despite the low intrinsic risks in comparison with agrochemicals, government regulations are complex, quickly changing and very different from country to country. Standards on acceptable environmental impacts associated with the introduction of beneficial microorganisms are badly needed.

Outlined above are a number of limitations and opportunities that do exist for microbial seed treatments. Combinations of efficacy, economic constraints and legislative restrictions will determine the products that will be developed. If an acceptable efficacy, reliability and competitive pricing can be provided to the grower, a product will find its place in the market. Environmental benefits will convince both the grower and the consumer.

REFERENCES

Alabouvette, C. 1990. Biological control of fusarium wilt pathogens in suppressive soils. In: *Biological control of soil-borne plant pathogens*. D. Hornby (Ed.), Wallingford, Oxon, UK; CAB International, pp. 27-43.

Callan N.W.; Mathre D.E.; Miller J.B. 1990. Bio-priming seed treatment for biological control of *Pythium ultimum* preemergence damping-off in sh2 sweet corn. *Plant Disease* **74**, 368-372.

Entwistle A.R. 1990. Allium white rot and its control. *Soil Use and Management* **6**, 201-209.

Fravel D.R.; Marois J.J.; Lumsden R.D.; Connick W.J. 1985. Encapsulation of potential biocontrol agents in an alginate-clay matrix. *Phytopathology* **75**, 774-777.

Harman, G.E.; Taylor, A.G.; Stasz, T.E. 1989. Combining effective strains of *Trichoderma harzianum* and solid matrix priming to improve biological seed treatments. *Plant Disease* **73**, 631-637.

Howell, C.R. 1991. Biological control of *Pythium* damping-off of cotton with seed-coating preparations of *Gliocladium virens*. *Phytopathology* **81**, 738-741.

Jutsum, A.R. 1988. Commercial application of biological control: status and prospects. *Philosophical Transactions of the Royal Society of London B 318*, 357-373.

Kosters, P.S.R.; Hofstede, S.B.; Ester, A. 1993. Effects of film-coating Brussels sprouts seeds with various insecticides on germination and on the control of cabbage root fly (*Delia radicum*). In: *Fourth international workshop on seeds: basic and applied aspects of seed biology, Vol. 3*. D. Côme and F. Corbineau (Eds.), Université Pierre et Marie Curie, Paris, pp. 1081-1086.

Lemanceau, P.; Alabouvette, C. 1991. Biological control of *Fusarium* diseases by the association of fluorescent *Pseudomonas* and non-pathogenic *Fusarium*. *Bulletin SROP* **14**, 45-50.

Linderman R.G. 1988. VA (Vesicular-Arbuscular) Mycorrhizal Symbioses. *ISI Atlas of science: animal and plant sciences*, 183-188.

Lumsden R.D.; Locke J.C. 1989. Biological control of damping-off caused by *Pythium ultimum* and *Rhizoctonia solani* with *Gliocladium virens* in soilless mix. *Phytopathology* **79**, 361-366.

Lutchmeah, R.S.; Cooke, R.C. 1985. Pelleting of seed with the antagonist *Pythium oligandrum* for biological control of damping off. *Plant Pathology* **34**, 528-531.

Lynch, J.M.; Lumsden, R.D.; Atkey, P.T.; Ousley, M.A. 1991. Prospects for control of *Pythium* damping-off of lettuce with *Trichoderma*, *Gliocladium*, and *Enterobacter spp*. *Biology and Fertility of Soils* **12**, 95-99.

Mcquilken, M.P.; Whipps, J.M.; Cooke, R.C. 1990. Control of damping-off in cress and sugar-beet by commercial seed coating with *Pythium oligandrum*. *Plant Pathology* **39**, 452-462.

Reinecke P. 1990. Biological control products: Demands of industry for succesful development. In: *Pesticides and alternatives: Innovative chemical and biological approaches to pest control*. J.E. Casida (Ed.), Elsevier science publishers, pp. 99-108.

Smith, R.S. 1992. Legume inoculant formulation and application. *Canadian Journal of Microbiology* **38**, 485-492.

Tahvonen, R.T. 1988. Microbial control of plant disease with *Streptomyces* spp. *Bulletin OEPP* **18**, 55-59.

Thomashow, L.S.; Weller, D.M. 1990. Application of fluorescent pseudomonads to control root diseases of wheat and some mechanisms of disease suppression. In: *Biological control of soil-borne plant pathogens*. D. Hornby (Ed.), Wallingford. Oxon, UK; CAB International, pp. 109-122.

White, J.G.; Linfield, C.A.; Lahdenpera, M.L.; Uoti, J. 1990. Mycostop - a novel biofungicide based on *Streptomyces griseoviridis*. *Brighton Crop Protection Conference, Pests and Diseases - 1990*, **1**, 221-226.

RHIZOSPHERE CONSTRAINTS AFFECTING BIOCONTROL ORGANISMS APPLIED TO SEEDS

J. W. DEACON

Institute of Cell and Molecular Biology, University of Edinburgh, Rutherford Building, Mayfield Road, Edinburgh EH9 3JH, UK

ABSTRACT

The ecological constraints affecting success of seed-applied biocontrol agents of soil-borne pathogens are discussed and new approaches for overcoming them are reviewed. Both evidence and theory suggest that progress has been limited by the use of single clonal strains of biocontrol agents, which inevitably are ecologically constrained in competitive soil environments. They achieve transitory, localised dominance of the rhizosphere, and in only some soils and seasons, leading to inconsistency and poor efficacy of applied biocontrol, compared with natural biocontrol which is multifactorial. The more promising new approaches attempt to exploit the diversity of biocontrol agents by identifying ecotypes of individual control agents, for use singly or combined, and by combining different biocontrol agents for synergism or more persistent control.

INTRODUCTION

There is a truism in plant pathology: most plants are resistant to most pathogens; disease is the exception rather than the rule. It is equally true that biological control is omnipresent in natural and agricultural environments, contributing to the suppression of pathogens and moderation of disease. This natural biocontrol is particularly important in soil. However, it is complex and multifactorial. It operates not only by protection of infection courts but also through competition and antagonism in the soil-borne phase of pathogens; and it involves not just one biocontrol agent but several, against a background of other soil properties (physical, chemical) that favour biocontrol agents but limit pathogen activities.

Specific examples of this stable, multifactorial biocontrol are seen in suppressive soils. Suppression of the cyst nematode *Heterodera avenae* in cereal monoculture involves at least two fungi and requires suitable moisture conditions for one of them - the zoosporic fungus *Nematophthora gynophila* (Kerry *et al.*, 1980). Natural soil suppressiveness to fusarium wilts probably involves both non-pathogenic *Fusarium* spp. and fluorescent pseudomonads (e.g. Lemanceau & Alabouvette, 1991); also, it is associated with specific soil properties such as the type of clay constituents (Stotzky & Martin, 1963). Take-all decline or equivalent take-all suppression in cereal monoculture is a complex phenomenon (Hornby, 1979) even where fluorescent pseudomonads have been implicated as the primary biological cause (Weller, 1983). One crucial line of evidence shows the multifactorial nature of biocontrol: we can seldom, if ever, reproduce it by introducing a single biocontrol agent into <u>untreated</u>, natural soil.

Most current research on biocontrol of soil-borne plant pathogens is far removed from these natural situations, being aimed at the development of single biocontrol agents - usually single clonal strains - for use as commercial inocula. It has succeeded in a few cases, notably with *Phlebia gigantea* for protection of pine stumps against *Heterobasidion annosum* and with strain K84 of *Agrobacterium radiobacter* for control of crown gall. But these successes can be attributed to special factors (Deacon, 1991). Usually the result has been inconsistent control across sites and seasons (Table 1) or early-season control that is not carried through to harvest (e.g. Geels & Schippers, 1983) or a degree of control sufficient only for low pathogen levels (e.g. Whipps *et al.*, 1993). So, biocontrol now has the reputation of being inefficient and inconsistent compared with control by chemicals or plant breeding (Becker & Schwinn, 1993).

BIOCONTROL BY INOCULATION: HOW SHOULD WE PROCEED?

It remains to be seen whether biocontrol of soil-borne pathogens with inoculant organisms must remain a restricted activity or will flourish. But it is clear that the approaches used of late have not produced the hoped-for success and so must be reappraised. It is also clear that this reappraisal must involve a return to first principles of microbial ecology, which tell us this: that individual organisms (and especially individual strains) can occupy only a restricted niche when in competition with a multitude of other organisms, and especially when those others are resident in a site and thus adapted to the site factors. The niche that an introduced organism can occupy is likely to be much narrower than the range of conditions that it can exploit in laboratory culture or even in experimental work in a glasshouse.

Should we expect a single clonal strain to protect all of the root system? No, it may flourish for a while by virtue of its high initial inoculum level, but will progressively be restricted to specific sites or microenvironments that suit it more than they suit any resident organism. Should we expect that strain to function in all soils, on all crops or in all seasons? No, for the same types of reason. And should we expect it to confer protection against a range of pathogens as it does in laboratory culture or in glasshouse trials? No, because each pathogen also has specific ecological attributes - a different site of infection, time of infection, or set of conditions in which it is active. The single clonal biocontrol strain cannot be expected to cover all these situations <u>in a competitive environment</u> where its own activities are constrained.

There is nothing inherently complex, or new, in these statements. They are obvious. But it has taken many years of failure for us to recognise them, and still the research literature (and attempted commercialisation) is dominated by single clonal strains. The rest of this paper is devoted to expansion of these points and discussion of how they are being, or could be, overcome. But first it is necessary to consider (1) a special case in which single strains might prove effective and (2) a case that seems special at first sight but is not so.

Biocontrol in glasshouses

Glasshouses and similar protected cropping environments are highly artificial, so that many of the ecological constraints on biocontrol are removed. Control agents can be selected for activity in the specific environmental conditions of the crop. Moreover, in soil-less rooting media or pasteurised soil the biocontrol agent should experience little competition during its initial establishment. These highly favourable conditions may not persist through the life of the crop, but at least the control of seedling diseases or damage to cuttings should be achievable with relative ease. There are several impending products for these circumstances, including 'Mycostop' (*Streptomyces griseoviridis*; Mohammadi, 1992) and 'Gliogard' (*Gliocladium virens*). It is envisaged that *G. virens* will be added to rooting media and formulated in a prill with nutrients to support its production of antibiotics, and that cuttings of ornamentals will be introduced later to coincide with the high antibiotic levels (R.D. Lumsden, pers. comm.). Even so, there might be an additional benefit in using mixtures of strains because Howell *et al.* (1993) have identified two groups of strains of *G. virens* with different antibiotic profiles. One group produces an antibiotic active against *Pythium* but not *Rhizoctonia solani*; the other group has the opposite attributes.

Control of seedling diseases in field sites

Control of seedling diseases requires only short-term activity of a biocontrol agent and has been demonstrated repeatedly in glasshouse experiments with seed-applied inocula. Seed-derived nutrients often promote the activities of biocontrol agents so that it should be possible to achieve an overwhelmingly high, localized population of the control agent, sufficient to overcome any ecological constraints, even in field conditions. However, this apparently special case has two principal limitations in practice. First, the seedling pathogens - especially *Pythium* spp. - can initiate infection extremely quickly, before a seed-applied biocontrol agent can exert its effects. Second, the seedling pathogens are seldom confined to infection of the seed but also attack the young root and shoot tissues. Protection of these requires that the biocontrol agent spreads rapidly from its site of application, and this is often the stage at which biocontrol breaks down. G.E Harman and E.B. Nelson (this volume) discuss the future options for this form of control.

MICROSITES ON ROOTS

In this section it will be argued that roots and other plant surfaces provide many microsites (or microenvironments or niches) that favour specific microorganisms, that single biocontrol agents thus cannot give generalized protection, and that future work must focus on mixtures of compatible biocontrol agents to achieve effective crop protection.

Application to seeds would be the favoured means of delivering biocontrol agents to the root zone for control of root diseases in field sites. Many laboratory, glasshouse and field experiments show that seed-applied microbes can spread down roots, either by carriage in percolating water (e.g. Bahme & Schroth, 1987; Lascaris & Deacon, 1991; U. Krauss, this volume) or in some cases perhaps by physical carriage by the root tip. Gammack *et al.* (1992) reviewed the subject. Some bacterial biocontrol agents, especially pseudomonads, can even maintain a partial or complete dominance of the microflora at the extending root tip. This ability to proliferate on roots in competitive conditions is termed rhizosphere competence (e.g. Ahmad & Baker, 1987). But the term is potentially misleading because it can obscure the fact that different organisms exploit different regions of roots - at least in competitive conditions. There are rhizosphere competences, not just rhizosphere competence.

The consequence is that spread of a seed-applied biocontrol agent down roots does not necessarily confer protection of the whole root system, only where the control agent can proliferate. Such localized proliferation is reflected in the fact that populations of individual microorganisms on plant surfaces typically show a lognormal distribution (Hirano *et al.*, 1982; Loper *et al.*, 1984), i.e. exponential growth where the conditions especially suit them but only poor growth elsewhere.

Weller (1983) reported on a specific case in which the spatial (and temporal) variation in population of a biocontrol agent influenced the efficiency of disease control. In field studies a take-all suppressive fluorescent pseudomonad initially grew well on the seedling roots of autumn-sown wheat, keeping pace with the rate of root tip extension. It also gave effective early-season control of take-all. But its population level decline markedly over winter, and in spring it was found mainly on the younger root regions, with only low levels on the more mature root regions. The roots subsequently developed take-all lesions in their older regions, and the bacterial strain was then found to proliferate on the lesions. The likely explanation is that this strain, like most rhizosphere pseudomonads, is specialized to exploit the relatively large amounts of simple soluble nutrients released from root tips or disease lesions but competes poorly for other types of nutrient elsewhere on the roots.

In terms of the distribution of microsites (niches) for microbial growth on roots, we know perhaps more for cereals and grasses (Deacon, 1987) than for any other type of plant. Fig. 1 depicts a cereal root in a simplified way, considering only the major sources of root-derived nutrients and not other features such as differences in root surface properties or physico-chemical factors that can influence microbial activities.

Figure 1. Simplified succession of zones for microbial growth along cereal roots, relating to the supply of root - derived nutrients.

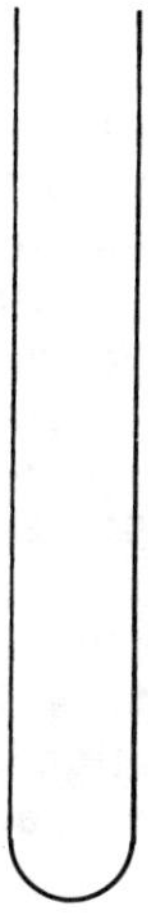

4. Zone of secondary invaders, including mycoparasites, that exploit tissues occupied by primary invaders.

3. Zone of progressive, natural root cortical senescence, favouring weak parasites that exploit incipiently senescing cells.

2. Progressive limitation of simple, soluble nutrients as bacterial population reaches carrying capacity. Deleterious microorganisms are now favoured.

1. Root tip; abundant supply of soluble nutrients, including amino acids, favouring pseudomonads.

Even from such a simplified picture we see that there is a progression of zones that selectively favour particular types of microbe along the root. The root tip and youngest root region favour at least some types of pseudomonad, which multiply rapidly in a nutrient pool. Behind this zone the microbial population level stabilizes because it uses all the continuing supply of soluble nutrients just to maintain itself (Newman & Watson, 1977). Deleterious microorganisms will be favoured in this zone because only they can gain access to further nutrients, by damaging the plant cells in their vicinity. These organisms will include those that invade the tissues, but also those that grow on the root surface and produce phytotoxic metabolites such as hydrogen cyanide (e.g. Bakker & Schippers, 1987). Further back, the cortex senesces progressively - a natural, programmed phenomenon. This favours weakly parasitic fungi that exploit incipiently senescing cells and that are known to control take-all in glasshouse conditions. They include *Phialophora graminicola* (principally in grass turf) and *Idriella bolleyi* which is common on cereals in agricultural field conditions. Still further back, the conditions become favourable for secondary invaders, including mycoparasites. The mycoparasites are of different types, with different modes of action (Deacon & Berry, 1992), but have one feature in common and it may be more important than the parasitism of other fungi *per se*: the ability to invade substrata that have already been colonized by other fungi. This seems to be a rare ability among fungi in general, and it forms the basis of a selective isolation technique for presumptive mycoparasites: soil or root pieces (preferably air-dried) are placed directly on agar precolonized by other fungi. There is even evidence of fungus-specific effects - at least *in vitro* - because the type of mycoparasite that grows on the agar plates depends on the fungus used as the primary colonizer (Mulligan & Deacon, 1992). The implication is that zone 4 (Fig. 1) might select for different secondary colonizers, depending on the microbes that became established in the earlier zones - a further dimension to the microsites along roots.

Microsite-based biocontrol

Understanding of the ecology of the rhizosphere - even at the rather gross level represented by zones of activity in Fig. 1 - should enable us to improve the prospects for biocontrol. It is notable that different potential biocontrol agents are favoured by the different zones on cereal roots: pseudomonads near the tip (or on take-all lesions), *Phialophora* and *Idriella* further back, and mycoparasites still further back. Even the deleterious agents of zone 2 (Fig. 1) might function as biocontrol agents if they induce host resistance to more aggressive pathogens (reviewed by Kuc & Strobel ,1992, and an example in agricultural practice was described by Komada , 1990). This zonation of biocontrol agents might explain why natural take-all suppression is relatively stable and long-lasting. An obvious future approach would be to explore the use of mixtures of inoculant organisms, selected for occupation of the different niches. For example, provided that they are compatible as seed-applied inocula, we could expect a combination of a pseudomonad and *Idriella bolleyi* to give better and more durable protection against take-all than would either alone. They would distribute themselves naturally on the developing roots according to their niches.

This kind of approach has already been reported for control of fusarium wilt diseases, where combinations of fluorescent pseudomonads and non-pathogenic fusaria were found to be highly successful in experimental conditions, even when the pseudomonads did not show biocontrol activity alone (e.g. Park *et al.*, 1988; Lemanceau & Alabouvette, 1991).

SOIL AND SEASONAL FACTORS IN BIOCONTROL

Becker & Schwinn (1993) pointed to the 'actual or perceived' inconsistency of biocontrol as a major factor that limits the acceptance of this technology. To overcome it they advocated technical solutions such as better quality control for inoculum preparations and attention to the genetic stability of biocontrol strains. These are important considerations, but are not the solution. Inconsistency is inevitable if single clonal biocontrol strains are expected to compete with the panoply of indigenous microbes in different sites and seasons. This is illustrated in Table 1, with data from the field trials of Suslow & Schroth (1982) on plant growth-promoting bacteria, which act at least partly by controlling soil-borne pathogens. The strains in these trials were the best of numerous bacteria screened in glasshouse and laboratory tests and they were carefully maintained. Yet none gave a consistently significant growth response across a range of sites and (or) growing seasons. Cook (1992) described similar variations in control of take-all by selected fluorescent pseudomonads in north-western U.S.A. Soil factors seemed to be the important variables in this case.

The solution that Cook and his colleagues are exploring is a 'customised' approach, in which fluorescent pseudomonads for take-all control are isolated from soils/regions where they are intended for use as inoculants. As an extension of this approach in general, it could be useful to identify strains that perform well in seasons where others have failed (e.g. from studies such as those in Table 1). Then it might be possible to develop mixed-strain inocula that give consistent control across sites and seasons. Schroth & Becker (1991) identified the use of strain mixtures as an important area for future research. It will require the answers to many questions. Would a mixture of seed-applied strains be mutually cancelling so that none proliferates enough to give control? How rapidly would one strain be selected from the mixture according to environmental conditions? And how many strains would be necessary in order to cover most eventualities?

TABLE 1. Per cent increase in yield of sucrose from sugar beet crops when seeds were treated with individual plant growth-promoting *Pseudomonas* strains and grown in different field sites in California (CA) or Idaho; data abstracted from Suslow & Schroth (1982)

Bacterial strain[+]	Field location and year of trial					
	Westside, CA, 1977	Davis, CA 1977	Idaho 1977	Davis, CA 1978	Imperial, CA 1979	Tracy, CA 1979
B4	-5	5	12[*]	10[*]	10	nd
SH5	15[*]	nd	0	13[*]	19	8
RV3	nd	16[*]	12[*]	10[*]	8	-13
A1	nd	nd	nd	12[*]	nd	5

+ All strains were isolated from sugar beet roots. * Significant yield increase compared with untreated control. nd, no data.

HOW MUCH DIVERSITY HAS GONE UNNOTICED IN BIOCONTROL AGENTS?

The questions in the section above take us into largely uncharted territories of biocontrol: we have little knowledge of the ecological variation that exists between strains of biocontrol agents. Meanwhile, population geneticists have made quite rapid progress in charting the variation within plant pathogens, with evidence that new, host-adapted species of *Phytophthora*, for example, have arisen within the relatively recent period of developed agriculture in North America (Hansen, 1989). For successful biocontrol we need as much information about our friends as about our enemies.

There is preliminary evidence that rhizosphere bacteria (though not necessarily biocontrol strains) can be 'host-adapted' and that this is partly associated with their ability to be agglutinated by root-surface polysaccharide (e.g. Chao *et al.*, 1988; but see Glandorf, 1992). This could be important in the selection of effective biocontrol strains (or mixtures) of either broad or narrow host range. For leaf surface (phylloplane) bacteria there is evidence of inter-strain differences in microsite selection (resource utilisation patterns). Lindow (1992) reviewed this in relation to biocontrol of ice-nucleation active bacteria. These bacteria, such as *Pseudomonas syringae*, have a surface protein that acts as a nucleus for ice crystal formation, leading to frost damage of the leaves. They can be controlled by mutants (Ice$^-$) that lack the protein, and the basis of control is simple competitive exclusion: if the Ice$^-$ mutants are applied early then they can use the leaf surface resources and exclude the natural Ice$^+$ population. But Lindow and his colleagues found that, in experimental conditions, the best control was obtained by Ice$^-$ mutants derived from individual Ice$^+$ strains: the progeny were better at controlling their parents than at controlling non-parental strains, even though there was no apparent difference between the wild-type strains. Thus it was suggested, first that wild-type populations of *P. syringae* consist of a mixture of strains with different resource utilisation patterns and, second, that the Ice$^-$ mutants retain these patterns so that they can better compete with their parents. The question (Lindow, 1992) is: how many such patterns are there in a wild-type population, and how many Ice$^-$ strains might be needed in an inoculum mixture to give the best biocontrol? Identical questions face us in the rhizosphere.

Recent work in this laboratory suggests that there is substantial inter-strain variation within *Idriella bolleyi*. This fungus is common on both roots and stem bases of cereals, and is implicated in control of several cereal pathogens including take-all (Kirk & Deacon, 1987a), eyespot (Reinecke & Fokkema, 1981), *Fusarium* spp. (Reinecke *et al.*, 1979) and *Pythium graminicola* (Waller, 1979). Such a broad range of activities might be expected to be associated with ecologically different strains, especially as the basis of control seems to be competitive exclusion (Kirk & Deacon, 1987b). With this in mind, L. I. Douglas (unpubl.) has compared the tolerance of different strains to water stress *in vitro*, because this correlates with the conditions in which different cereal root-rot and foot-rot pathogens occur in field sites (Cook & Baker, 1983). Of three strains of *I. bolleyi* taken at random from a culture collection, one showed low stress-tolerance, equivalent to that of the take-all fungus; one showed extreme stress-tolerance like that of *Fusarium culmorum*, and one was intermediate in behaviour. These differences were evident on osmotically (KCl) adjusted agar and in liquid media supplemented with polyethylene glycol, which exerts a matric water stress. They were also evident when assessed by different criteria - spore germination, mycelial growth or ability to sporulate.

These *in vitro* findings must be extended to soil conditions, and to comparisons of strains isolated from cereal roots as opposed to stem bases in field sites. But the variation seen to date is astonishing because it spans almost the whole range of *in vitro* water-stress responses of the pathogens that *I. bolleyi* might be used to control. Equivalent work is needed on the potential ecotypic variation in other biocontrol agents.

CONCLUSION

As we look back over the history of biocontrol of plant disease (Garrett, 1965) and especially the more recent history (Deacon, 1991) there is a clear dichotomy. On the one hand, real progress has been made in understanding the processes and mechanisms that limit pathogen activities in nature. And the focused application of molecular tools and mathematical modelling (e.g. Lindow, 1992) will yield even more rapid progress in this area. On the other hand, the achievements in applying this understanding to practical cropping systems have been modest. In fact, we have moved backwards because the few notable successes in the use of microbial inocula for biocontrol do not compensate for the abandonment of traditional cropping practices that fostered natural biocontrol.

The approaches advocated here could provide the bridge between natural and inoculant-based biocontrol. It is the responsibility of the research community to build this bridge and to ask the relevant questions. For example, how many types of control agent would be needed for effective control of a pathogen, and how many strains of each would ensure that this control is consistent across sites and seasons. This information could then provide the basis for rational decisions by companies that are interested in producing biocontrol inocula. Of course, single control agents, and single strains, would be preferred, especially under the current regulatory conditions. But there should be no inherent barrier to the development of mixed inocula, because these are already available in commercial silage additives. The overriding criterion must be efficacy in practice.

Truman Capote is said to have recommended taking an instant dislike to Simone de Beauvoir because 'it saves time'. For the same reason perhaps we should take an instant dislike to single clonal biocontrol agents.

REFERENCES

Ahmad, J.S.; Baker, R. (1987) Rhizosphere competence of *Trichoderma harzianum*. *Phytopathology*, **77**, 182-189.

Bahme, J.B.; Schroth, M.N. (1987) Spatial-temporal colonisation patterns of a rhizobacterium on underground organs of potato. *Phytopathology*, **77**, 1093-1100.

Bakker, A.W.; Schippers, B. (1987) Microbial cyanide production in the rhizosphere in relation to potato yield reduction and *Pseudomonas* spp.-mediated plant growth stimulation. *Soil Biology and Biochemistry*, **19**, 451-457.

Becker, J.O.; Schwinn, F.J. (1993) Control of soil-borne pathogens with living bacteria and fungi: status and outlook. *Pesticide Science*, **37**, 355-363.

Chao, W.L.; Li, R.K.; Chang, W.T. (1988) Effect of root agglutinin on microbial activities in the rhizosphere. *Applied and Environmental Microbiology*, **54**, 1838-1841.

Cook, R.J. (1992) A customised approach to biological control of wheat root diseases. In: *Biological Control of Plant Diseases*, E.C. Tjamos, G. C. Papavizas and R.J. Cook (Eds), New York: Plenum, pp. 211-222.

Cook. R.J.; Baker, K.F. (1983) *The Nature and Practice of Biological Control of Plant Pathogens*. St. Paul: American Phytopathological Society.

Deacon, J.W. (1987) Programmed cortical senescence: a basis for understanding root infection. In: *Fungal Infection of Plants*, G.F. Pegg and P.G. Ayres (Eds), Cambridge: Cambridge University Press, pp. 285-297.

Deacon, J.W. (1991) Significance of ecology in the development of biocontrol agents against soil-borne plant pathogens. *Biocontrol Science & Technology*, **1**, 5-20.

Deacon, J.W.; Berry, L.A. (1992) Modes of action of mycoparasites in relation to biocontrol of soilborne plant pathogens. In: *Biological Control of Plant Diseases*, E.C. Tjamos, G.C. Papavizas and R.J. Cook (Eds),New York: Plenum, pp. 157-167.

Gammack, S.M.; Paterson, E.; Kemp, J.S.; Cresser, M.S.; Killham, K. (1992) Factors affecting the movement of microorganisms in soil. In: *Soil Biochemistry*, Vol. 7, G. Stotzky and J.-M. Bollag (Eds), New York: Marcel Dekker, pp. 263-305.

Garrett, S.D. (1965) Toward biological control of soil-borne plant pathogens. In: *Ecology and Management of Soil-borne Plant Pathogens*, K.F. Baker and W.C. Snyder (Eds), London: John Murray, pp. 4-17.

Geels, F.P.; Schippers, B. (1983) Reduction of yield depression in high frequency potato cropping soil after seed tuber treatments with antagonistic fluorescent *Pseudomonas* spp. *Phytopathologische Zeitschrift*, **108**, 207-214.

Glandorf, B. (1992) Root Colonization by Fluorescent Pseudomonads. *PhD Thesis, University of Utrecht*, Netherlands.

Hansen, E.M. (1989) Speciation in plant pathogenic fungi: the influence of agricultural practice. *Canadian Journal of Plant Pathology*, **9**, 403-410.

Hirano, S.S.; Nordheim, E.V.; Arny, D.C.; Upper, C.D. (1982) Lognormal distribution of epiphytic bacterial populations on leaf surfaces. *Applied and Environmental Microbiology*, **44**, 695-700.

Hornby, D. (1979) Take-all decline: a theorist's paradise. In: *Soil-borne Plant Pathogens*, B. Schippers and W. Gams (Eds), London: Academic Press, pp. 133-156.

Howell, C.R.; Stipanovic, R.D.; Lumsden, R.D. (1993 in press) Antibiotic production by strains of *Gliocladium virens* and its relation to the biocontrol of cotton seedling diseases. *Biocontrol Science and Technology*, **3**.

Kerry, B.R.; Crump, D.H.; Mullen, L.A. (1980) Parasitic fungi, soil moisture and multiplication of the cereal cyst nematode, *Heterodera avenae*. *Nematologica*, **26**, 57-68.

Kirk, J.J.; Deacon, J.W. (1987a) Control of the take-all fungus by *Microdochium bolleyi*, and interactions involving *M. bolleyi*, *Phialophora graminicola* and *Periconia macrospinosa* on cereal roots. *Plant and Soil*, **98**, 231-237.

Kirk, J.J.; Deacon, J.W. (1987b) Invasion of naturally senescing root cortices of cereal and grass seedlings by *Microdochium bolleyi*. *Plant and Soil*, **98**, 239-246.

Komada, H. (1990) Biological control of fusarium wilts in Japan. In: *Biological Control of Soil-borne Plant Pathogens*, D. Hornby (Ed.), Wallingford: C.A.B. International, pp. 65-75.

Kuc, J.; Strobel, N.E. (1992) Induced resistance using pathogens and nonpathogens. In: *Biological Control of Plant Diseases*, E.C. Tjamos, G.C. Papavizas and R.J. Cook (Eds), New York: Plenum, pp.295-303.

Lascaris, D.; Deacon, J.W. (1991) Colonization of wheat roots from seed-applied spores of *Idriella (Microdochium) bolleyi*: a biocontrol agent of take-all. *Biocontrol Science and Technology*, **1**, 229-240.

Lemanceau, P.; Alabouvette,C. (1991) Biological control of fusarium diseases by fluorescent *Pseudomonas* and nonpathogenic *Fusarium*. *Crop Protection*, **10**, 279-286.

Lindow, S.E. (1992) Ice⁻ strains of *Pseudomonas syringae* introduced to control ice nucleation active strains on potato. In: *Biological Control of Plant Diseases*, E.C. Tjamos, G.C. Papavizas and R.J. Cook (Eds), New York: Plenum, pp. 169-174.

Loper, J.E.; Suslow, T.V.; Schroth, M.N. (1984) Lognormal distribution of bacterial populations in the rhizosphere. *Phytopathology*, **74**, 1454-1460.

Mohammadi, O. (1992) Mycostop biofungicide - present status. In:*Biological Control of Plant Diseases*, E.C. Tjamos, G.C. Papavizas and R. J. Cook (Eds), New York: Plenum, pp. 207-210.

Mulligan, D.F.C.; Deacon, J.W. (1992) Detection of presumptive mycoparasites in soil placed on host-colonized agar plates. *Mycological Research*, **96**, 605-608.

Newman, E.I.; Watson, A. (1977) Microbial abundance in the rhizosphere: a computer model. *Plant and Soil*, **48**, 17-56.

Park, C.S.; Paulitz, T.C.; Baker, R. (1988) Biocontrol of fusarium wilt of cucumber resulting from interaction between *Pseudomonas putida* and non-pathogenic isolates of *Fusarium.Phytopathology*, **78**, 190-194.

Reinecke, P.; Fokkema, N.J. (1981) An evaluation of methods of screening fungi from the haulm base of cereals for antagonism to *Pseudocercosporella herpotrichoides* in wheat. *Transactions of the British Mycological Society*, **77**, 343-350.

Reinecke, P.; Duben, J.; Fehrmann, H. (1979) Antagonism between fungi of the foot rot complex of cereals. In:*Soil-borne Plant Pathogens*, B. Schippers and W. Gams (Eds), London: Academic Press, pp. 327-337.

Schroth, M.N.; Becker, O. (1991) Concepts of ecological and physiological activities of rhizobacteria related to biological control and plant growth promotion. In: *Biological Control of Soil-borne Plant Pathogens*, D. Hornby (Ed.), Wallingford: C.A.B.International, pp. 389-414.

Stotzky, G.; Martin, R.T. (1963) Soil mineralogy in relation to the spread of *Fusarium* wilt of banana in Central America. *Plant and Soil*, **18**, 317-337.

Suslow, T.V.; Schroth, M.N. (1982) Rhizobacteria of sugar beets: effects of seed application and root colonization on yield. *Phytopathology*, **72**, 199-206.

Waller, J.M. (1979) Observations on *Pythium* root rot of wheat and barley. *Plant Pathology*, **28**, 17-24.

Weller, D.M. (1983) Colonization of wheat roots by a fluorescent pseudomonad suppressive to take-all. *Phytopathology*, **73**, 1548-1553.

Whipps, J.M.; McQuilken, M.P.; Budge, S.P. (1993) Use of fungal antagonists for biocontrol of damping-off and *Sclerotinia* diseases. *Pesticide Science*, **37**, 309-313.

INCORPORATION OF MICROENCAPSULATED BENEFICIAL ORGANISMS INTO ENVIRONMENTALLY ACCEPTABLE SEED COATINGS TO ENHANCE CROP PERFORMANCE IN SOYBEANS

J. S. BURRIS

Seed Science Center, Iowa State University, Ames,IA

Y. Chen

Ball Seed Co., West Chicago, IL

ABSTRACT

The effectiveness of microencapsulation technology to deliver beneficial organisms on seeds was evaluated using *Rhizobia/Glycine* as a model system. The survival of microencapsulated *Rhizobia* applied to the seed was evaluated by standard bacteriological techniques and by nodulation tests. The effects of coating material on seed quality and *Rhizobia* survival was also determined. A total of 8 seed lots with various coating material combinations were evaluated. A multi-coating delivery system was developed for microencapsulation which included both fluidized-bed and pan coating processes. Both warm and cold germination test results indicate no significant effects on germination due to the microencapsulation. Microencapsulated seed lots showed a significant amount of live *Rhizobia* on the seed at 7 and 30 days after microencapsulation. This result was further verified by the nodulation test in the greenhouse. These results indicate that the 'sandwich' seed microencapsulation technology provides a suitable environment for survival of *Rhizobia* on seeds which can then be stored and handled safely.

INTRODUCTION

Microencapsulation technology had its origin in the initial work of Barry Green of the National Cash Register Company who developed the first process of microencapsulation by coacervation, Green 1956. This process provided the world with the first carbon-less copying paper, and, more importantly, described a technology which would have far reaching applications beyond the initial objectives in office systems. Green's successful use of microencapsulation stimulated researchers to adapt microencapsulation techniques to other areas such as agriculture, advertising, pharmaceuticals and to other industries. Seed is a suitable and effective delivery system for many biotechnically altered organisms designed to enhance crop performance or replace pesticides. However, survival of these organisms on the seed is often very low. Microencapsulation of these organisms in a suitable buffered substrate may increase their survival. In 1978, Lim patented a process by which viable cells, tissues, and other labile biological substances could be microencapsulated within a semipermeable membrane (Lim 1984). To date, however, this technology has not been applied to living organisms which are routinely or appropriately applied to seed. These microcapsules could then be incorporated into biodegradable polymeric seed coatings which are being developed, to provide an environmentally safe alternative to fungicide seed treatment of agricultural crops. To assure successful microencapsulation of living cells and/or

tissues, several conditions are necessary including the selection of coating materials and conditions which are totally nontoxic to living organisms (Lim 1984). For seed microencapsulation, an additional concern relates to the effect of the coating on seed performance. The current research attempts to develop the technology to incorporate beneficial organisms into environmentally acceptable seed coatings to enhance crop performance. *Rhizobial* bacteria are an important component of soybean production and this technique was used as a model system.

MATERIALS AND METHODS

Soybean seed (*Glycine max* L.) variety Wells provided by the Committee for Agricultural Development, Iowa State University, and *Rhizobia japonicium* carried in peat (manufactured by Titre Inc., Ryegate, MT) was used for seed microencapsulation. *Rhizobia* was applied by Pharmaceutical Services (UI) in a fluidized bed and by the Seed Science Center (ISU) using both fluidized bed and pan coating equipment. Three biodegradable polymeric coating materials, Sepiret (Seppic, Paris, France), Polyvinylpyrrolidone (PVP) and Aquacoat (FMC Pharmaceuticals, Philadelphia,PA), were used. The formulations are listed on Table 1.

TABLE 1. Coating materials evaluated as delivery systems for application of *Rhizobium j.* to Soybean *Glycine max.* L

10% Sepiret + *Rhizobia*/Peat + 5% PVP, applied by ISU
15% Sepiret + *Rhizobia*/Peat + 5% PVP, applied by ISU
20% Sepiret + *Rhizobia*, applied by UI
5% Polyvinylpyrrolidone + Talc + *Rhizobia*,applied by UI
7% Polyvinylpyrrolidone + Talc + *Rhizobia*, applied by UI
10% Polyvinylpyrrolidone + Talc + *Rhizobia*, applied by UI
‹5% Aquacoat + *Rhizobia*, Cured 48hrs at 40C, applied by UI
5% Aquacoat + *Rhizobia*, Not cured, applied by UI

Coatings were applied using both conventional pan and fluidized bed coating equipment designed and constructed by ISU for laboratory research and having a coating capacity of upto 500 gm of seed. Seeds were initially coated with 5% PVP (% seed weight) as a base coat, and then followed with a peat-based *Rhizobia* coat evenly over the PVP base coat. The *Rhizobia* coat was then coated with a 10 to 15% (% seed weight) coat of Sepiret to form a sandwich coating(Figure 1). The effectiveness of *Rhizobia* on microencapsulated seed was evaluated either 3 days after coating or stored at room temperature for later evaluation. Sixty seeds were soaked in 50 ml of autoclaved water for 24 hours with shaking. A yeast mannitol agar medium was used to count the viable colonies of *Rhizobia* (Vincent, 1970). *Rhizobium* colonies were characterized by morphological, staining and cultural tests, and finally verified by the invasiveness test, i.e. nodulation of soybean under bacteriologically controlled conditions. All steps and equipment were under bacteriological control. Soybean were grown in a lighted and temperature controlled greenhouse. Seeds were planted in autoclaved sand. Root nodules were then counted at 21, 28, and 35 days after planting to evaluate the effectiveness of the microencapsulation. The standard warm germination and a sand/soil (4:1) cold test were conducted. Warm germination was evaluated after 7 days at 25°C, and the cold

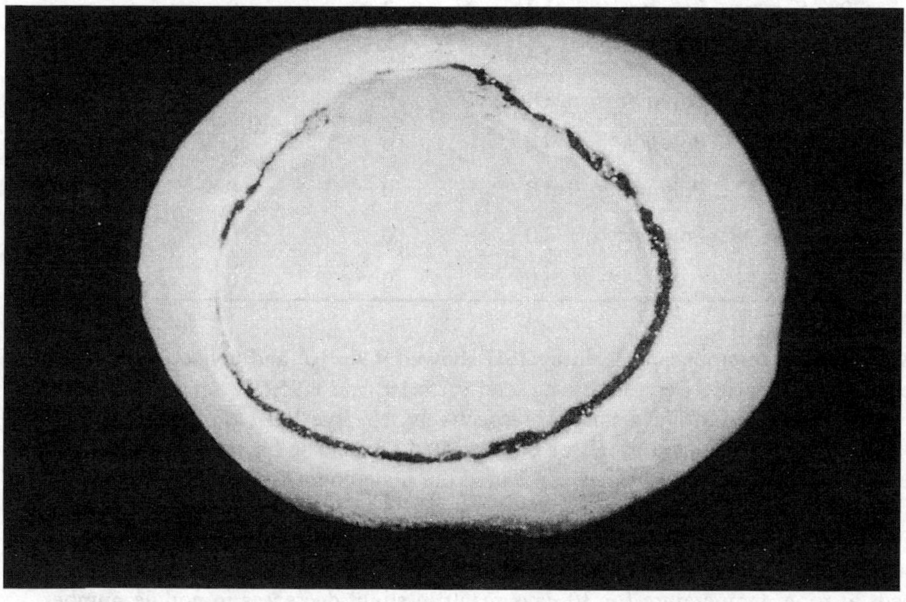

Figure 1. Soybean seed microencapsulated by the ISU technique showing the "sandwich" structure.

germination test was evaluated after 14 days, the first 7 days at 10°C, followed by 7 days at 25°C. The AOSA (Association of Official Seed Analysts) seed testing rules were used as a guide for seedling interpretation (Anon. 1986).

RESULTS AND DISCUSSION

The effectiveness of seed microencapsulated *Rhizobia* was evaluated by a bacterial plate count test after 7 and 30 days of storage at room temperature (Table 2).

TABLE 2. Viable number of *Rhizobium* cells on microencapsulated soybean seed

Coating Materials	Rhizobia Carrier	Method	Days After Microencapsulation	
			7	30
			MPN (Rhizobia/seed)	
10% Sepiret	Peat	ISU	9.96×10^6	6.97×10^6
15% Sepiret	Peat	ISU	9.25×10^6	8.95×10^6
20% Sepiret	Polymer	UI	0	0
5% PVP	Polymer	UI	0	0
7% PVP	Polymer	UI	0	0
10% PVP	Polymer	UI	0	0
Aquacoat-C	Polymer	UI	0	0
Aquacoat-NC	Polymer	UI	0	0
Untreated		UI	0	0

Two seed lots microencapsulated by ISU showed 9.96×10^6 and 6.97×10^6 live *Rhizobia* per seed for the 10% Sepiret coating, and 9.25×10^6 and 8.95×10^6 for 15% Sepiret coating after 7 and 30 days storage, respectively. No live *Rhizobia* were found on the 6 lots of microencapsulated by UI or on untreated seeds. The bacteriologically controlled nodulation test (Table 3) further confirmed the plate count results. Peat-base *Rhizobia* microencapsulated with 10 and 15% Sepiret began to form nodules at about 3 weeks after planting and the nodule number per plant was similar to peat-base *Rhizobia* inoculated on the seed immediately before planting. Storage of microencapsulated seeds at room temperature for 30 days result in slight decreases in nodule number (Table 3). The live *Rhizobia* number by plate count tests also exhibited a slight decline (Table 2). However, both live *Rhizobia* number on seed and nodule number per plant of 7 day-old peat-base *Rhizobia* inoculated seed were much less than that of 30 day-old microencapsulated seed. These results indicate that the 'sandwich' microencapsulation technology employed by ISU provided a suitable microenvironment for the survival of *Rhizobia* on seed. In this technology, *Rhizobia* were buffered against the stress of the coating conditions by the peat carrier. Preliminary storage studies showed that the 'sandwich' microencapsulated seeds can be stored and handled safely (Table 2, 3). When the coating polymer was directly used as the *Rhizobia* carrier, the bacteria did not survive the coating conditions. In addition, the coating materials also may directly effect the survival of the *Rhizobia*.

TABLE 3. Nodulation of microencapsulated soybean with *Rhizobium japonicum*

Storage Days	Coating Materials	*Rhizobia* Carrier	Method	Days After Planting 21	28	35
				Nodules / Plant		
3	10% Sepiret	Peat	ISU	9	14	13
3	15% Sepiret	Peat	ISU	0	21	24
7	20% Sepiret	Polymer	UI	0	0	0
7	5% PVP	Polymer	UI	0	0	0
7	7% PVP	Polymer	UI	0	0	0
7	10% PVP	Polymer	UI	0	0	0
7	Aquacoat-C	Polymer	UI	0	0	0
7	Aquacoat-NC	Polymer	UI	0	0	0
0		Peat	Inoculate	10	17	19
0	Untreated			0	0	0
30	10% Sepiret	Peat	ISU	4	9	8
30	15% Sepiret	Peat	ISU	5	15	20
30	20% Sepiret	Polymer	UI	0	0	0
30	5% PVP	Polymer	UI	0	0	0
30	7% PVP	Polymer	UI	0	0	0
30	10% PVP	Polymer	UI	0	0	0
30	Aquacoat-C	Polymer	UI	0	0	0
30	Aquacoat-NC	Polymer	UI	0	0	0
30	Untreated			0	0	0
7		Peat	Inoculate	2	5	7

The interaction of two living organisms, seed and microorganisms, make seed microencapsulation complicated. Therefore, the effects of coating materials on seed and seedling performance were determined (Table 4). Both warm and cold germination results indicate no significant effect on germination by microencapsulation as compared to control seed, except for the cold germination of the 10% PVP coated seed. In fact, microencapsulated seed exhibited an increased germination in most cases. Aquacoat coated seed exhibited significantly higher incidence of abnormal seedling in the warm germination test, however, it also exhibited higher seed vigor in the cold test. Aquacoat coated seed also appear to be free of bacteria in the *Rhizobium* plate count test. This coating material may have some toxic effect on both seed and microorganisms.

CONCLUSIONS

These results indicate the need for caution when coating materials are used directly as microorganism carriers. The use of a safe carrier as a buffered medium for microorganism could be a key factor in the success of microencapsulation to incorporate the beneficial microorganisms into seed coatings. The application of the microencapsulation technology to *Rhizobia/Glycine* system will greatly improve the soybean seed inoculation with superior *Rhizobia* and the microencapsulated seed can be handled safely. Once this technology is developed it should be applicable to other microorganisms and other seed types.

TABLE 4. The effect of microencapsulation materials on seed and seedling performance.

Coating Material	*Rhizobia* Carrier	Method	Warm Nor.	Warm Abnor	Cold Nor.	Cold Abnor.
				Germination %		
10% Sepiret	Peat	ISU	82.3	12.0	41.5	13.3
15% Sepiret	Peat	ISU	83.3	13.0	45.3	12.0
20% Sepiret	Polymer	UI	84.3	9.8	50.5	6.3
5% PVP	Polymer	UI	87.0	9.8	59.8	7.0
7% PVP	Polymer	UI	86.7	8.8	50.5	6.3
10% PVP	Polymer	UI	83.8	9.5	13.8	3.8
Aquacoat-C	Polymer	UI	8.8	90.3	84.3	2.5
Aquacoat-NC	Polymer	UI	13.8	84.3	83.0	2.8
Untreated			75.3	15.5	36.3	14.5

ACKNOWLEDGMENTS

Journal Paper No. J-15566 of the Iowa Agriculture and Home Economics Experiment Station, Ames, Iowa. Project No. 2526.

REFERENCES

Anon. 1986. Rules for testing seed. Journal of Seed Technology, 6:1-126.

Green, B.K. and L. Schleicher. U.S. Patent 2,730,456.

Lim, F. 1984. Biomedical application of microencapsulation. Boca Raton, FL. CRC Press. 160p.

Vincent, J. M. 1970. A manual for the practical study of root-nodule bacteria, Oxford:Blackwell Scientific Publications, 164p.

TESTS FOR BIOLOGICAL CONTROL OF SEED AND SEEDLING DAMPING-OFF DISEASES OF PEAS AND BEANS USING *BACILLUS* SPECIES.

R. WALKER, A.A. POWELL, B. SEDDON.

Department of Agriculture, University of Aberdeen, Aberdeen AB9 1UD, UK.

ABSTRACT

Potential bacterial antagonists to the damping-off fungi *Pythium mamillatum* and *Botrytis cinerea* were isolated from the testae and cotyledons of peas and dwarf French beans. The isolates were screened for antagonism by *in vitro* dual culture analysis of pathogen against potential antagonist. Five *Bacillus* isolates showed distinct antagonism to the two pathogens *in vitro* and were selected for further screening as seed treatments together with *Bacillus brevis* strain Nagano and a *Bacillus licheniformis* isolate previously shown to be antagonistic to *B. cinerea*. When applied as seed treatments to peas and beans sown on agar inoculated with the pathogens, six of these antagonists reduced *Pythium* and *Botrytis* infection levels compared to controls. When pea seeds were sown in infested compost a reduction of *Pythium* infection levels was again observed with these seed treatments.

INTRODUCTION

Damping-off diseases cause considerable yield losses in many important agricultural crops worldwide. These losses may occur before seed germination or seedling emergence (pre-emergence damping-off) or the juvenile tissues of older seedlings may be attacked producing brown, watery lesions and eventual tissue collapse (post-emergence damping-off). A number of soilborne pathogenic fungi have been implicated in damping-off diseases including species of *Botrytis* and *Pythium* (Lucas et al, 1985).

Elimination of soilborne fungi by control measures including chemicals is unrealistic and only limited success has been achieved with the use of single control measures to reduce crop disease caused by such pathogens (Martin et al). Therefore control of these pathogens by the utilisation of integrated strategies including biological control should be considered in order to reduce the general overuse of fungicides in the environment (Rishbeth, 1988; Nelson, 1989). The use of microbial antagonists already present in the spermosphere as biocontrol agents applied as seed treatments against these fungi is one possible option. Antagonists must remain viable and withstand desiccation, heat treatment, formulation and field application (possibly in association with a reduced level of fungicide). Bacterial spores are considered to be ideal candidates for such applications (Roberts & Hitchins, 1969; Rhodes, 1990).

The aims of this study were; to isolate *Bacillus* species from the testae of peas and dwarf French beans, to screen these isolates *in vitro* for antagonism to the damping-off pathogens *P. mamillatum* and *B. cinerea* and to apply these biocontrol agents (BCA's) as seed treatments on peas and beans to achieve successful control of damping-off.

METHODS

Plant material

The plant material used in this work consisted of two seed lots of peas cv. Puget and two different cultivars of dwarf French beans; a black-seeded cv. Sunray and a white-seeded cv. Processor.

Isolation and characterisation of isolates

In an attempt to optimise the variety and number of *Bacillus* species obtained a combination of techniques were used in the bacterial isolations . The potential microbial antagonists were isolated from the testae of single seeds (either whole seeds or with the testa removed from the embryo) of peas and dwarf French beans by the techniques of washing, sonication, incubation in nutrient broth and heat treatment.

Four sets of three replicate seeds of the four seed types were each subjected to one of the following treatments. The first set of replicate seeds were individually washed in Ringer's solution using a whirlimix followed by plating out the washing medium. The second set of replicates were individually sonicated in Ringer's solution using a MSE bench sonicator followed by plating out the sonication medium. The third set of replicates were incubated in nutrient broth for 48h at three different incubation temperatures (10°C, 25°C and 37°C) followed by plating out dilutions of the broth. Heat treatment was also used to bias the selection of isolates towards spores of *Bacillus* species. This was done by incubating a fourth set of replicate seeds in nutrient broth for 5d to encourage the *Bacillus* species present to produce spores. Three different incubation temperatures (10°C, 25°C and 37°C) were used to optimise the number and variety of *Bacillus* species present. Heat treatment at 80°C for 10 min was then used to kill vegetative cells in the samples allowing only spores to survive. Dilutions of the broth were then plated out as before.

All isolations were plated out onto nutrient agar supplemented with cycloheximide antibiotic to inhibit fungal growth. All isolates obtained were characterised using standard bacteriological methods (Bradbury, 1988).

Screening for antagonism *in vitro*

The isolates were screened for antagonism *in vitro* against *Pythium mamillatum* and *Botrytis cinerea* using dual culture analysis of pathogen against potential antagonist on nutrient agar (NA), potato dextrose agar (PDA) and malt extract agar (MEA). This was achieved by inoculating the same culture plate with both the pathogen and the potential BCA and observing the interaction of the two microorganisms against control plates inoculated with the pathogen only. The plates were assessed for antagonism when the control plates were fully developed. Any zones of inhibition of fungal growth produced by the BCA were measured and recorded.

Screening for antagonism *in vivo*

Bacillus isolates which showed distinct antagonism to the two pathogens *in vitro* were selected for further screening as seed treatments together with *Bacillus brevis* strain Nagano and a *Bacillus licheniformis* isolate, both previously shown to be antagonistic to *B. cinerea* (Edwards & Seddon, 1992).

Spores of these *Bacillus* isolates were applied as seed treatments to all four seed types. The seed treatments were applied by soaking the seeds in a suspension of spores in sterile distilled water for 1h followed by air-drying in aseptic conditions. The concentrations of the *Bacillus* spore suspensions used were in the magnitude of 10^7-10^9 spores ml^{-1}. The seeds were then sown in magenta pots, three replicate seeds per treatment against each pathogen, on water agar which had been supplemented with Dowson's plant growth medium and inoculated with the pathogens. As a control seeds were surface sterilised and soaked in sterile distilled water (SDW) for 1 h before air-drying. A second set of control seeds remained unchallenged by the pathogens. The pots were incubated in a plant growth chamber at 20°C. *Pythium* infection levels of the seeds were assessed after 7 d by surface sterilisation of the seeds followed by removal of the testa and plating out the cotyledons onto water agar. Seeds from which *P. mamillatum* were re-isolated were deemed to be infected. *Botrytis* infection levels of seed on agar were assessed visually after 14 d. After this time period seeds infected with *B. cinerea* were covered in mycelia and conidia.

Pea seeds treated with these *Bacillus* isolates (fifteen replicate seeds per isolate) were also sown in *Pythium* infested compost. *Pythium* infection levels were assessed using the surface sterilisation method outlined previously. A further fifteen seeds were surface sterilised, soaked in SDW and sown as controls.

RESULTS

Isolation and characterisation of isolates

Ninety two bacterial isolates were obtained by the isolation methods outlined above. Sixty six of these isolates were identified as members of the genus *Bacillus*. The remaining isolates belonged to the genera *Erwinia*, *Lactobacillus*, *Pseudomonas* and *Rhodococcus* although some isolates could not be assigned to any single genus on the basis of this characterisation.

<u>Screening</u> <u>for</u> <u>antagonism</u> *in vitro*

Five *Bacillus* isolates showed distinct antagonism to the two pathogens *in vitro* on all three types of media tested. *B. brevis* strain Nagano and *B. licheniformis* showed antagonism to *B. cinerea* but limited antagonism to *P. mamillatum*. The results are presented in Table 1.

TABLE 1   Antagonism of isolates to *Botrytis cinerea* and *Pythium mamillatum in vitro*

Isolate	*Botrytis cinerea*			*Pythium mamillatum*		
	NA	MEA	PDA	NA	MEA	PDA
B3	Z1	7	11	*	*	3
C1	Z1	9	9	5	*	*
D4	Z1	6	*	10	4	5
J7	Z1	4	*	3	*	*
M10	Z1	5	1	10	3	6
Bacillus brevis (strain Nagano)	Z1	-	*	Z1	-	-
Bacillus licheniformis	Z1	*	*	Z1	*	-

Figures represent zone of inhibition in mm.
Z1, zone of inhibition present (unmeasured);
*, some antagonism <u>cf</u> control; -, no antagonism.

<u>Screening</u> <u>for</u> <u>antagonism</u> *in vivo*

Pythium infection levels were reduced most effectively on white dwarf French beans and pea lot 1 where six out of the seven potential BCA's were effective on both seed types. Only *B. brevis* and *B. licheniformis* were effective on black dwarf French beans and only *B. brevis* and treatments C1 and D4 were effective on pea lot 2 (Table 2). *Botrytis* infection levels were reduced by all treatments on all seed types except on black dwarf French beans which appeared to be very susceptible to *B. cinerea* (Table 2).

TABLE 2. Effect of *Bacillus* seed treatments on infection levels of damping-off fungi in peas and beans sown on agar

		Bacillus seed treatments						
Pathogen	Seed Type	B3	C1	D4	J7	M10	*Bacillus brevis*	*Bacillus licheni-formis*
Botrytis	DFB	0	0	0	0	0	0	0
cinerea	DFW	3	3	3	3	2	3	3
	P1	3	2	3	3	2	2	2
	P2	3	3	2	3	1	3	3
Pythium	DFB	0	0	0	0	0	1	1
mamillatum	DFW	3	3	3	0	3	2	3
	P1	1	2	2	1	1	0	2
	P2	0	2	2	0	0	2	0

Figures represent number of seeds protected from infection c̲f̲ controls.
0, No reduction of infection compared to controls; DFB, black dwarf French bean; DFW, white dwarf French bean; P1, pea lot 1; P2, pea lot 2.

Six of the isolates tested consistently reduced pre-emergence infection by *Pythium* in infested compost (Figure 1). However this work needs to be repeated on a greater number of replicate seeds to produce statistical comparisons.

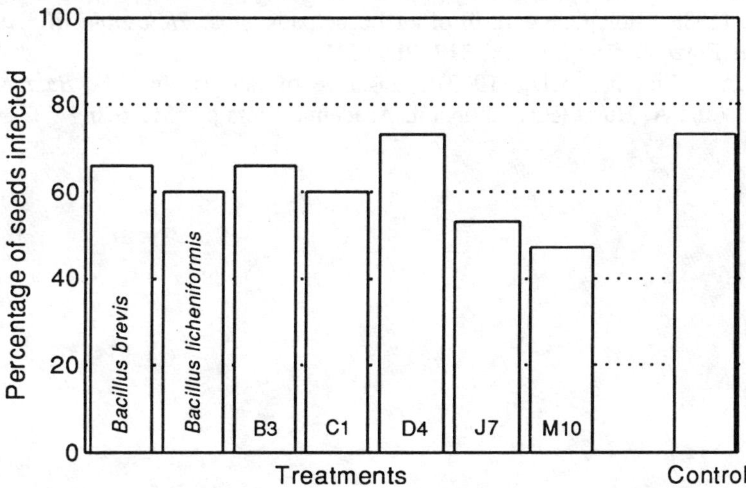

Figure 1. Effect of *Bacillus* seed treatments on *Pythium* infection of peas sown in infested compost.

DISCUSSION

Ninety two potential antagonists were isolated from the testae of peas and dwarf French beans. Five isolates of *Bacillus* showed distinct antagonism to both *B. cinerea* and *P. mamillatum in vitro* on three types of media. These five isolates together with *Bacillus licheniformis* and *Bacillus brevis* strain Nagano reduced *Botrytis* infection levels on peas and white French beans sown on agar. Six of the seven isolates appear to consistently reduce *Pythium* infection levels in pea seeds sown in infested compost. Further work is underway to investigate the mechanism of antagonism and to determine if these treatments can yield significant reductions in disease incidence.

REFERENCES

Bradbury, J.F. (1988) Identification of cultivable bacteria from plants and plant tissue cultures by use of simple classical methods. *Acta Horticulturae.* **225**, 27-37.

Edwards, S.G.; Seddon, B. (1992) *Bacillus brevis* as a biocontrol agent against *Botrytis cinerea* on protected Chinese cabbage. *Proceedings of the Xth Botrytis Symposium, Heraklion, Crete.* Wageningen: Pudoc pp. 272-276.

Lucas, G.B.; Campbell, C.L.; Lucas, L.T. (1985) *Introduction to Plant Diseases - Identification and Management.* Van Nostrand Reinhold, pp. 151-154.

Martin, S.B.; Abawi, G.S.; Hoch, H.C. (1985) Biological control of soilborne pathogens with antagonists. In: *Biological Control in Agricultural IPM Systems*, M.A. Hoy and D.A. Herzog (Eds). London: Academic Press pp. 433-434.

Nelson, M.R. (1989) Biological Control; The second century. *Plant Diseases.* **73**, 616.

Rhodes, D.J. (1990) Formulation requirements for biological control agents. In: *The exploitation of microorganisms in applied biology. Aspects of Applied Biology No. 24* Association of Applied Biologists., Warwick, pp. 145-153.

Rishbeth, J. (1988) Biological control of air-borne pathogens. *Philosophical Transactions of the Royal Society London.* **318**, 265- 281.

Roberts, T.A.; Hitchins, A.D. (1969) Resistance of Spores. In: *The Bacterial Spore* G.W. Gould and A. Hurst (Eds). London: Academic Press pp. 611-670.

SPORE MOVEMENT OF *MUCOR HIEMALIS* IN THE RHIZOSPHERE OF GROUNDNUT IN NATURAL FIELD CONDITIONS

ULRIKE KRAUSS

Institute of Cell and Molecular Biology, University of Edinburgh, EH9 3JH, UK

ABSTRACT

A pimaricin-resistant strain of *Mucor hiemalis* was applied as spores to the hypocotyl of groundnut seedlings and to the soil surface of unplanted soil. Its movement in the soil profile was studied in natural field conditions in Malawi. The procedures for monitoring spread of the inoculant organism in field sites are described and their general applicability is discussed.

Fungal spread was greatest along the groundnut tap root, exceeding 40 cm after 51 days. Soil below inoculated plants was colonized to a depth of 22 cm, but non-planted soil only down to 7 cm. Although transport along the tap roots was high, their colonization was patchy, as the fungus was diluted out with increasing depth in the profile. Fungal establishment on lateral roots was poor.

It is concluded that <u>soil</u> application of BCAs is unlikely to yield success. In contrast, inocula applied to <u>seed</u> or <u>stem bases</u> have the potential to keep up with the rapidly expanding root system. The development of biological seed inocula is particularly promising for tropical climatic conditions. A novel method for the selective isolation of rhizosphere-competent fungi is suggested.

INTRODUCTION

The use of biocontrol agents (BCAs) against soil-borne diseases requires an efficient delivery system of the inoculum into the root zone of the crop. Population dynamics of seed-applied bacteria in the rhizosphere have received appreciable attention (recently reviewed by Elsas & Heijnen, 1990; Gammack *et al.*, 1992). Knowledge of fungal spore movement in natural soil is scarce.

In a podsol profile in Cheshire, U. K., spores of *Mucor ramannianus* freely passed through the sandy A horizon, but were unable to penetrate the more compacted B_1 horizon (Hepple, 1960). Tropical soils lack the structural horizons typical for soils in temperate areas (Lowole & Banda, 1986). Additionally, the rainfalls during the cropping season typically exceed precipitations in the UK.

Little information is available on the influence of a root system on fungal spore movement. Spores in percolating water, rather than relatively slow-growing hyphae, are believed to be responsible for colonization of root apices which can grow several centimetres per day (Huisman, 1982). Secondary sporulation on seeds and roots of wheat has been shown for seed-applied spores of the fungus *Idriella bolleyi* in artificial conditions (Lascaris & Deacon, 1991). Bahme & Schroth (1987) demonstrated the importance of irrigation water in movement of *Pseudomonas fluorescens* down potato roots; they also reported a log-normal distribution of the inoculant bacterium on roots, indicating that population levels fall off rapidly with distance from the inoculum source.

The present study investigates the movement of *Mucor hiemalis* spores in a ferric luvisol in Malawi and the influence of the root-soil interface on the transport process. The work was conducted in a natural, agricultural soil profile, because sieving and repacking disrupts the continuity of natural channels and increases the surface area available to entrap organisms (White, 1985).

MATERIALS AND METHODS

Groundnut plants (cv. JL 24) were planted on one half of either side of a trench. Twelve days after sowing, hypocotyls and non-planted soil were inoculated with 3 ml of a spore suspension (total 0.5×10^7 spores) of a pimaricin resistance-marked *M. hiemalis* strain. The wild type originated from healthy groundnut roots in the same field.

During sampling the soil from the trench wall was carefully removed. When the root system was approached, soil cores (diameter 8 mm) were taken at 2 cm intervals down the profile. Afterwards the plants were excavated and the tap roots were cut into 2 cm segments. Laterals were severed 2 cm away from the tap root and the following 2 cm pieces were collected (Fig. 1).

Air-dried samples (0.25 g soil or 2 cm root) were enriched for *M. hiemalis* in submerged liquid culture in potato-sucrose broth containing 40 μg ml^{-1} penicillin and 50 μg ml^{-1} streptomycin. After three days incubation, the cultures were streaked onto potato-dextrose agar containing 40 μg ml^{-1} penicillin, 50 μg ml^{-1} streptomycin and 20 μg ml^{-1} pimaricin.

Rainfall was measured in a gauge adjacent to the trench, and evaporation by Chitedze Meteorological Station. The minimum depth D of the wetting front in non-saturated soil was calculated according to the equation $D = R_e/\theta_{FC}$, assuming dry soil, where R_e is the excess rainfall (rainfall minus evaporation) and θ_{FC} the volumetric water content at field capacity, which was 20.1 %. A detailed description of procedures will be published elsewhere.

RESULTS

Pimaricin-resistant *M. hiemalis* was never recovered from non-inoculated soil, nor from the trench floor. In all enriched rhizosphere samples examined by microscopy, *Mucor*-type spores were observed, but only occasionally in bulk soil. Sensitive detection of the inoculant strain was possible by simultaneous use of (1) enrichment by employing the fungus's ability to proliferate rapidly in liquid culture and (2) the genetic marker by exploiting the fungicide resistance.

Fig. 2 shows the average depth in the profile from which the marked strain was recovered. In bare soil movement increased over time but did not exceed 7 cm. In soil below inoculated plants, *M. hiemalis* was gradually transported to a depth of 22 cm. Apart from day 4, when soil and root colonisation coincided, the fungus moved approximately twice as far along the root than it did in adjacent soil, exceeding 40 cm at 51 d. Downward transport was significantly correlated with cumulative rainfall in bare soil (r=0.874, P<0.05), planted soil (r=0.880, P<0.01) and along the tap root (r=0.957, P<0.001). The calculated wetting front soon exceeded the length of the tap root and surpassed spore movement in bulk soil.

With increasing depth, tap root colonization became patchy as the fungus was diluted out (Fig. 1). Colonisation of lateral roots was poor in comparison to the tap root. The greatest recorded depth of occurrence on laterals was 9 cm at 18 d. At later

Fig. 1: A. Summary representation of occurrence (solid symbol) of inoculant strain at different sampling depths (open symbol = no detection) (a) non-planted soil, (b) planted soil, (c) tap root and (d) lateral roots. B. Exposed root system in trench; cork-borer (for sampling) arrowed.

Fig. 2: Tap root length and maximum distance of pimaricin-resistant *Mucor hiemalis* down the tap root, root laterals or in soil, in planted or non-planted inoculation sites. Data points are means of two replicates for roots and root-free soil. Cumulative rainfall and calculated wetting front are also indicated.

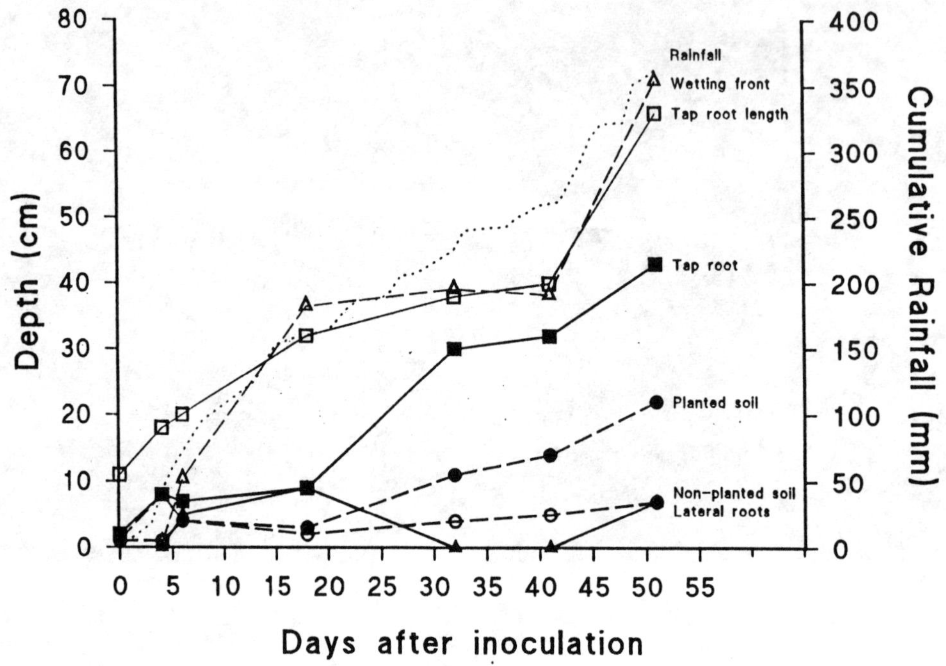

sampling dates, the first-order laterals in the upper regions of the tap root had largely senesced and decomposed. Younger laterals were formed further down the profile, but *M. hiemalis* was not found on them at 32 and 41 d, and by 51 d, the fungus was only detected down to 7 cm depth on laterals, the same depth it had reached in non-planted soil.

DISCUSSION

Despite the lack of structural horizons and the relatively high precipitation of 363 mm during the 51 days, movement in non-planted soil was low. The wetting front calculated from excess rainfall in non-saturated soils reached far beyond the detection limit of the spores. A similar lag behind the water front has been reported for zoospores and cysts of *Phytophthora megasperma* and for the bacterium *Serratia marcescens* (Wilkinson *et al.*, 1981). Although spores of the Mucorales are water-dispersed (Dobbs, 1942), water movement was not efficient in carrying the spores through the soil matrix. Establishment of a biocontrol agent applied to bulk soil thus seems unpromising.

In contrast to bulk soil, fungal movement rates in the rhizosphere exceeded those previously reported by an order of magnitude. Water films surrounding the root cortex are believed to be of primary importance for the mobility of microbes (Gammack, 1992; Parke *et al.*, 1986). Chao *et al.* (1986) observed *Trichoderma harzianum* spores on pea roots 5 cm below the inoculated seed if percolating water was present, but only at 2 cm below the seed in a humid chamber in the absence of water flow. It has been suggested that, under field conditions, the rate of root growth often exceeds the movement of water films in unsaturated soils (Huck & Hoogenboom, 1990). The groundnut tap root can grow at rates greater than 3 cm day-1 . Under conditions of tropical rainfalls in this experiment, however, the calculated minimum water front soon caught up with the advancing apex. Even under drier soil moisture regimes, the formation of water films on roots at night time has been reported (Huck *et al.*, 1970). High evapotranspiration rates during hot and dry spells may cause the roots to contract. Changes in root diameter during high transpirational demands have been demonstrated for sunflower (Faiz & Weatherley, 1982) and cotton (Huck *et al.*, 1970), two crops of the semi-arid tropics. Root contraction and thus gap formation at the root-soil interface is quite conceivable, especially since groundnuts are virtually free from root hairs (Chandler, 1978; Yarbrough, 1949) which could bridge gaps (Tinker, 1976). Such voids can readily be flushed with the next rain. It seems likely that water flow within these root channels mediated transport along the tap root and in rhizosphere soil. Thus, under tropical conditions, water-dispersed spores can effectively be transported along root channels.

M. hiemalis was only detected on a proportion of tap root segments and the colonisation of laterals was poor. Competition has been held responsible for the lack of root (Chao *et al.*, 1986) and soil (Wilkinson *et al.*, 1981) colonisation by fungi and bacteria in soil columns packed with non-sterile soil. Transport rates in autoclaved soil were consistently higher. However, this might have reflected a failure to establish measurable population levels, rather than a lack of transport through soil. The same might have been true for tap root segments and their laterals on which *M. hiemalis* was not detected, although it must have passed them. However, *M. hiemalis* is an ubiquitous and highly competitive fungus. In an agricultural soil, even at inoculum ratios as unfavourable as 1:60 it could pre-empt colonization of soil and substrates by, and replace, *T. harzianum*, a fungus often cited for its biocontrol potential (Wardle *et al.*, 1993). The ability of *M. hiemalis* to become established in soil when supplied with a food base merits further study. Spores of *M. hiemalis* dried onto seeds remained viable for over three weeks at room temperature (unpublished). Coating of spores onto seeds could provide a nutrient boost to the fungus during seed germination. Furrow application of the inoculant into the planting hole, however, has the advantage that the agent is not removed from soil when the cotyledons emerge (Windels *et al.*, 1983) or the testa is shed.

Using surface-sterilized seeds, Lascaris & Deacon (1991) provided evidence that seed-derived nutrients and early establishment in root regions close to the seed facilitate root colonisation by the biocontrol fungus *Idriella bolleyi*. In contrast to their investigation, the inoculant in this study was exposed to the natural soil flora and, by inoculating the hypocotyl at twelve days after sowing, was deprived of seed exudates which are highest after 48 h (Subrahmanyam *et al.*, 1983). Nevertheless, *M. hiemalis* was able to colonize proximal regions of the root system and spread downwards along the tap root. This is likely to be at least partly due to secondary sporulation (Windels *et al.*, 1983; Lascaris & Deacon, 1991). Thus, rhizosphere competence might be linked to the ability to sporulate rapidly in water films surrounding roots. The chosen isolation technique - enrichment in liquid culture - mimics this and could be a promising tool for the selective isolation of biocontrol candidates. It marries ecological significance with the advantages of fermentation technology for the mass production of biological crop inoculants.

ACKNOWLEDGEMENTS

This work was done during the tenure of a scholarship by the Carnegie Trust for the Universities of Scotland. Field and laboratory facilities were generously provided by SADC/ICRISAT, Chitedze, Malawi. Special thanks to J. W. Deacon, G. Russell, P. Subrahmanyam and C. J. Matabwa for useful discussion and support.

REFERENCES

Bahme, J. B. & Schroth, M. N. (1987) Spatial-temporal colonisation patterns of a rhizobacterium on underground organs of potato. *Phytopathology*, **77**, 1093-1100.

Chandler, M. R. (1978) Some observations on infection of *Arachis hypogaea* L. by *Rhizobium*. *Journal of Experimental Botany*, **29**, 749-755.

Chao, W. L.; Nelson, E. B.; Harman, G. E.; Hoch, H. C. (1986) Colonization of rhizosphere by biological control agent aplied to seeds. *Phytopathology*, **76**, 60-65.

Dobbs, C. G. (1942) Spore dispersal in the Mucorales. *Transactions of the British Mycological Society,* **25**, 441.

Elsas, J. D. Van; Heijnen, C. E. (1990) Methods for the introduction of bacteria into soil: a review. *Biology and Fertility of Soils*, **10**, 127-133.

Faiz, S. M. A.; Weatherley, P. E. (1982) Root contraction in transpiring plants. *New Phytologist*, **92**, 333-343.

Gammack, S. M.; Paterson, E.; Kemp, J. S.; Cresser, M. S.; Killham, K. (1992) Factors affecting the movement of microorganisms in soil. In: *Soil Biochemistry*, G. Stotzky & J.-M. Bollag (Eds), Marcel Dekker, New York, Vol 7, pp. 263-305.

Hepple, S. (1960) The movement of fungal spores in soil. *Transactions of the British Mycological Society*, **43**, 73-79.

Huck, M. G.; Hoogenboom, G. (1990) Soil and plant water flux in the rhizosphere. In: *Rhizosphere Dynamics*, J. E. Box & L. C. Hammond (Eds), AAAS, Westview Press, Boulder, pp. 268-312.

Huck, M. G.; Klepper, B.; Taylor, H. M. (1970) Diurnal variations in root diameter. *Plant Physiology*, **45**, 529-530.

Huisman, O. C. (1982) Interrelation of root growth dynamics to epidemiology of root-invading fungi. *Annual Review of Phytopathology*, **20**, 303-327.

Lascaris, D.; Deacon, J. W. (1991) Colonisation of wheat roots from seed-applied spores of *Idriella (Microdochium) bolleyi*: a biocontrol agent of take-all. *Biocontrol Science and Technology*, **1**, 229-240.

Lowole, M. W.; Banda, P. (1986) Extent, distribution and properties of red soils in Malawi. In: *The Red Soils of East and Southern Africa*, K. Nyamapfene, J. Hussein and K. Asumadu (Eds), Proceedings of an International Symposium, Harare, Zimbabwe, International Development Research Centre, Canada, 1988, pp. 22-27.

Parke, J. L.; Moen, R.; Rovira, A. D.; Bowen, G. D. (1986) Soil water flow affects the rhizosphere distribution of a seed-borne biological control agent, *Pseudomonas fluorescens*. *Soil Biology and Biochemistry*, **18**, 583-588.

Subrahmanyam, P.; Reddy, M. N.; Rao, A. S. (1983) Exudation of certain organic compounds from seeds of groundnut. *Seed Science and Technology*, **11**, 267-272.

Tinker, P. B. (1976) Transport of water to plant roots in soil. *Philosophical Transactions of the Royal Society of London*, **B 273**, 445-462.

Wardle, D. A.; Parkinson, D.; Waller, J. E. (1993) Interspecific competitive interactions between pairs of fungal species in natural substrates. *Oecologia*, **94**, 165-172.

White, R. E. (1985) The transport of chloride and non-diffusible solutes through soil. *Irrigation Science*, **6**, 3-10.

Wilkinson, H. T.; Miller, R. D.; Millar, R. L. (1981) Infiltration of fungal and bacterial propagules into soil. *Soil Science Scociety of America Journal*, **45**, 1034-1039.

Windels, C. E.; Kommedahl, T.; Sarbini, G.; Wiley, H. B. (1983) The role of seeds in the delivery of antagonists into the rhizosphere. In: *Ecology and Management of Soilborne Plant Pathogens*, C. A. Parker, A. D. Rovira, K. J. Moore, P. T. W. Wong and J. F. Kollmorgen (Eds), American Phytopathological Society, St Paul, Minnesota.

Yarbrough, J. A. (1949) *Arachis hypogaea*. The seedling, its cotyledons, hypocotyl and roots. *American Journal of Botany* **36**, 758-772.

BIOCONTROL OF SEED-BORNE <u>BOTRYTIS ALLII</u> USING AN ANTAGONISTIC BACTERIUM

LINDSEY PEACH, R. B. MAUDE, G. M. PETCH

Horticulture Research International, Wellesbourne, Warwick, CV35 9EF, UK

ABSTRACT

A bacterium, <u>Enterobacter agglomerans</u>, isolated from neck rot infected onions, was antagonistic to the causal fungus (<u>Botrytis allii</u>) <u>in vitro</u>. Film coat applications of the bacterium and also of benomyl, applied at a standard rate, to naturally infected seeds gave similar levels of control of the fungus in laboratory tests and of the disease (neck rot) in stored bulbs produced from field-grown crops of onions.

INTRODUCTION

In 1987 and 1988 while isolating fungal pathogens from commercial consignments of imported onions, some of which were infected with neck rot (<u>Botrytis allii</u>), an unknown bacterium was observed which appeared to be inhibiting the growth of that fungus on some bulbs.

The bacterium was isolated into pure culture and identified as a member of the Enterobacteriaceae using the methods of Lelliot & Stead (1987). It was identified to specific level as <u>Enterobacter agglomerans</u> using an API 20E Bacterial Identification Strip supplied by API System, La Balme Les Grotte, 38390 Montalieu Vercieu, France.

The bacterium was tested for its inhibitory effect on the growth of <u>B allii</u> in culture and for control of onion neck rot when applied to naturally infected seeds.

<u>IN VITRO</u> TESTS

The bacterium produced flat and domed colonies when grown on King's medium B (KB). The flat form was stable when sub-cultured but sub-cultures from domed colonies yielded both types. Tests for antibiosis were made on KB agar and on prune lactose yeast agar (PLYSE).

Bacterial streak test

Materials and methods
Domed and flat colony forms of the bacterium were tested independently. Bacterial cells from agar cultures were spread using a sterile loop across the upper third of a 90mm plate of KB agar. The plates were incubated for 24h at 25°C to establish the bacteria. Each of 4 replicate plates per bacterial colony form was then inoculated with a 6mm disk of mycelium taken from the edge of a colony of <u>B. allii</u> (isolate B4037 which had been obtained from an onion bulb which bore the bacterium). The disk was placed 50mm from the bacterial streak (centre to centre). Plates were incubated for 8 days at 20°C when the radial growth across two diameters of each fungal disk was measured in mm.

The test was repeated on PLYSE agar with five isolates of <u>B. allii</u> (isolate B4037, two isolates from different seed samples, B4072, B3922, one Australian isolate, B4017 and one isolate from shallots, B4089).

<u>Results</u>
By comparison with the control (25.4mm mean radial growth inclusive of disk) bacteria of the domed colony form reduced growth of <u>B. allii</u> on KB agar to a mean of 9.5mm (a 63% reduction in growth). Bacteria of the flat colony form reduced growth to a mean of 13.4mm (a 47% reduction).

Both forms of the bacterium reduced the growth of all five isolates of the fungus from different sources (Table 1).

TABLE 1   Antagonistic effects of <u>Enterobacter agglomerans</u> to isolates of <u>Botrytis allii</u> grown on PLYSE agar

| | Growth of <u>B. allii</u> from disks in mm | | | | |
| | | Bacterium (1975D) | | % reduction in growth | |
Source of isolate	No bacterium	flat	domed	flat	domed
Onion seed (B4072)	18.6	10.8	8.3	42	55
Shallots (B4089)	19.8	8.6	8.6	57	57
Onion (Australia)(B4017)	14.5	13.8	11.0	5	24
Onion seed (B3922)	20.0	7.9	11.8	61	41
Onion (B4037)	19.5	16.6	14.3	15	27

(   ) = isolate numbers

Reduction in growth of <u>B. allii</u> isolate B4037 by both forms of the bacterium was less on PLYSE than KB. Reduction in mycelial growth of isolates of different origin ranged from 24 to 57% for the domed colony form of the bacterium and from 5 to 61% for the flat colony form. The predominantly domed colony form was, in general, more antagonistic to the growth of different isolates of the fungus and it was used in all subsequent tests.

SEED TREATMENT TESTS AND FIELD STUDIES

To ensure maximum inhibition of the fungus the bacterium was cultured in nutrient broth and was applied to seed at the highest concentration achievable.

<u>Seed treatment test 1</u>

<u>Method and Materials</u>

Using fluidised bed film coating equipment (Maude & Suett, 1986) <u>E. agglomerans</u> bacteria were applied to seeds in a polyvinyl acetate film. Thirty ml of a mixture of 5ml of cells from a 2-day old culture + 25ml nutrient broth + sterile distilled water (1:1) + 0.25% of seed weight of polyvinyl acetate (PVA), as Vinamul R18160 (Vinamul Ltd., Carshalton, London) was sprayed per 50g infected onion seeds. Colony counts based on serial dilution plates indicated that 2-day old cultures contained 3.4×10^8 viable cells per ml.

Two hundred seeds per treatment (8 replicates of 25 seeds each) of a single stock of naturally infected onion seeds obtained from a commercial source were placed on moist filter paper and on PLYSE agar plates and assessed for colonies typical of the fungus and of the bacterium after 7 days.

Result
There was a significant reduction (P<0.05) in recovery of the fungus on filter paper compared with PLYSE (Table 2). The application of PVA caused a further significant reduction (P<0.05)in the recovery of seed-borne <u>B. allii</u> on PLYSE. The fungus was not recovered on either substrate from seeds to which bacteria had been applied by film coating.

TABLE 2 Laboratory test of the effect of film coating naturally infected onion seeds with <u>Enterobacter agglomerans</u> cells on the control of seed-borne <u>Botrytis allii</u>

Treatment	Test substrate	% seeds from which <u>B. allii</u> was recovered	% seeds from which bacteria were recovered
Untreated seeds	Filter paper	6	0
" "	PLYSE agar	28	0
PVA sprayed	Filter paper	3	0
" "	PLYSE agar	14	0
Bacteria + PVA	Filter paper	0	100
" "	PLYSE	0	100
	LSD (<0.05)	3.5	-

All seeds treated with the bacterium were enveloped in pools of the organism on both test substrates.

Seed treatment test 2 and field trial

Method and materials
Benomyl and two rates (x and 2x) of the bacterium were applied in a film

coat or in suspension to onion seeds. The x rate of the bacterium was prepared as described above; in this case 6.6ml of the bacterial cell concentrate in nutrient broth was applied to 66g seeds and 13.2ml were used to obtain the 2x rate. Bacteria in suspension were applied to seeds in a rotating bowl and talcum powder was added to dry the bacteria onto the seeds. Benomyl was applied as Benlate (50% wp - Du Pont Ltd.) either as a dust (1g a.i./kg seed) or sprayed in water plus sticker at the same rate of active ingredient to form a film coat. PVA sticker alone was applied to seeds at the rate specified in the previous test.

Two hundred seeds per treatment were plated out on PLYSE as described above. The remaining seeds were drilled in the field on 12 April 1989 in individual plots of 5 rows each 15.25m long and 1.3m apart. There were 4 replicate plots per treatment in a randomised block layout from which plots grown from untreated seeds were excluded. Paired plots of untreated seeds and of the treatments were also sown isolated from each other on the farm at Wellesbourne. Records of emergence of onions per m row were made and the mature bulbs were harvested on 5 October 1989. Onions were stored in nets (about 270 bulbs per net) in an onion store until 9 January 1990 when samples of bulbs were split vertically and recorded for the presence or absence of neck rot (Maude & Presly, 1977). Three nets from each of the three isolation plots per treatment were removed and 50 bulbs taken at random from each net and assessed. This provided a total of 450 bulbs per treatment. Six hundred

TABLE 3 Effect of bacterial seed treatments on the control of <u>B. alli</u> on the seeds in the laboratory and on stored bulb onions

| | | | % neck rot in bulbs (stored for 3 months) from | | | |
| | % <u>B. allii</u> on agar | | Isolation plots | | Randomised plots | |
Biocide and rate	D	FC	D	FC	D	FC
Benomyl 1g a.i./kg	0	0	1.1	1.1	0.2[a]	1.4[f]
Bacteria x	7.5	0.5	8.1	2.5	3.5[bd]	2.3[f]
Bacteria 2x	6.5	2.0	5.4	1.8	1.7[ce]	1.3[f]
Nil + FC	14.0		8.0		-	-
Nil	27.0		12.9		-	-
LSD[1] (<0.05)	2.09		-			
LSD[2] (<0.05)	4.18		-			

D = dusted or dried on to seeds; FC = film coated onto seeds; LSD[1] = significant differences between application methods and LSD[2] between treatments; different letters indicate significant differences between treatments based on χ^2 analysis bulbs per treatment comprising 150 bulbs per replicate in 3 sub-samples of 50 bulbs per net were taken for assessment from the randomised block experiment.

Results

In the laboratory and field test, film coat applications of bacteria and benomyl and dust application of benomyl were more effective than suspensions of bacteria dried on to the surface of seeds with talcum powder (Table 3).

Film coating with bacteria (either rate) reduced but did not completely eradicate the fungus in the agar test as it had done in Seed Test 1. The reduction obtained by application of a PVA film only was similar to that obtained in the previous test. Benomyl either dusted or film coated onto seeds eliminated the fungus in the agar test.

None of the seed treatments affected emergence which ranged from 11.7 to 13.7 plants per m row in the randomised part of the experiment. Very similar numbers of onions per m row were obtained from untreated seeds in the isolation plots.

The bacteria applied in a film coat to seeds gave similar levels of control as benomyl applications in bulbs assessed for neck rot after 3 months in store. The bacteria applied as a suspension in nutrient broth and dried on to the seeds with talcum powder were not as effective in reducing neck rot.

In comparison with the laboratory test there was a reduction in transmission of fungus from the seed in field sowings of untreated seeds and seeds treated with PVA alone and this resulted in a lower incidence in neck rot in store.

DISCUSSION

Biological control of plant pathogens by seed treatment with micro-organisms has been directed mainly against soil-borne organisms (Harman, 1991). Fungi and bacteria have been applied to seeds for this purpose (Taylor and Harman, 1990). Generally treatments shown to be effective under artificial conditions in inoculated sterile soil are less effective when exposed to the variability of the soil rhizosphere. However, there are examples where micro-organisms applied as seed treatments have been as effective as manufactured pesticides in achieving control of soil-borne diseases (Parke, 1990; McQuilken et al., 1990).

Micro-organisms have also been applied successfully to control the seed-borne phases of certain fungal and bacterial diseases, for example, black leg of sugar beet (Phoma betae) (Walther and Gindrat, 1987; Gordon-Lennox et al., 1987) and black arm of cotton (Xanthomonas campestris pv. malvacearum) (Randhawa et al., 1987).

In this study a bacterium Enterobacter agglomerans, isolated from imported onion bulbs, was shown to be antagonistic in vitro to the growth of Botrytis allii (onion neck rot) the main source of which in the UK is imported infected onion seeds. Eradication of the seed-borne phase virtually eliminates the disease from the stored bulb crop (Maude, 1983) so the possibility of achieving this by means of a bacterial seed treatment was tested. The film coating method has been used to achieve accurate dosing of agrochemicals onto seeds (Maude & Suett, 1986; Maude,1990) and in research on biocontrol methodology to apply oospores of Pythium oligandrum to seeds

(McQuilken _et al._, 1990).

The method of application of the bacteria to seeds affected the efficacy of seed treatment. Bacteria applied to seeds in nutrient broth which was then dried on to the seeds by the addition of talcum powder were only partly effective. The reason for this is not known. However, bacteria applied by film coating were as effective as benomyl in controlling the seed-borne phase of the fungus on agar. This reduced its transmission in the crop and ultimately resulted in control of neck rot in the stored bulbs. In this respect the microbial treatment appears to have been as effective as the standard rate of benomyl previously used to control the disease in commerce.

The use of sticker alone also reduced the incidence of the fungus and the disease, a phenomenon also reported for seed-borne _Alternaria brassicicola_ when the effect was considered to have been due to toxicity to the more superficial seed-borne inoculum (Maude and Suett, 1986).

The bacterial seed treatments did not increase or reduce crop emergence.

REFERENCES

Gordon-Lennox, G.; Walther, D.; Gindrat, D. (1987) Utilisation d' antagonistes pour l'enrobage des semences: efficacité et mode d'action contre les agents de la fonte des semis. _EPPO Bulletin_ 17, 631-637.

Harman, G. E. (1991) Seed treatments for biological control of plant disease. _Crop Protection_ 10, 166-171.

Lelliot, R. A.; Stead, D. E. (1987) _Methods for the Diagnosis of Bacterial Diseases of Plants._ London : BSPP, Blackwell Scientific Publications,216 pp.

Maude, R. B. (1983) The correlation between seed-borne infection by _Botrytis allii_ and neck rot development in store. _Seed Science and Technology_ 11, 829-834.

Maude, R. B. (1990) Seed Treatment. _Pesticide Outlook_ 1,16-22.

Maude, R. B.; Presly, A. H. (1977) Neck rot (_Botrytis allii_) of bulb onions. 11. Seed-borne infection and its relationship to the disease in store and the effect of seed treatment. _Annals of Applied Biology_ 86, 181-188.

Maude, R. B.; Suett, D. L. (1986) Application of fungicide to brassica seeds using a film-coating technique. _Proceedings 1986 British Crop Protection Conference_ - _Pests and Diseases_, pp. 237-242.

McQuilken, M. P.; Whipps, J. M.; Cooke, R. C. (1990) Control of damping-off in cress and sugar beet by commercial seed-coating with _Pythium oligandrum_. _Plant Pathology_ 39, 452-462.

Parke, J. L. (1990) Population dynamics of _Pseudomonas cepacea_ in the pea spermosphere in relation to biocontrol of _Pythium_. _Phytopathology_ 80, 1307-1311.

Randhawa, P. S.; Singh, N. J.; Schaad, N. W. (1987) Bacterial flora of cotton seeds and biocontrol of seedling blight cause by _Xanthomonas campestris_ pv. _malvacearum_. _Seed Science and Technology_ 15, 65-71.

Taylor, A. G.; Harman, G. E. (1990) Concepts and technologies of selected seed treatments. _Annual Review of Phytopathology_ 28, 321-339.

Walther, D.; Gindrat, D. (1987) Biological control of _Phoma_ and _Pythium_ damping-off of sugar-beet with _Pythium oligandrum_. _Journal of Phytopathology_ 119, 167-174.

Session 7

Application of Seed Treatments, Coatings, Pelleting and Other Techniques

Chairman and
Session Organiser   D L SUETT

LARGE-SCALE SEED PRIMING TECHNIQUES AND THEIR INTEGRATION WITH CROP PROTECTION
TREATMENTS

D GRAY

Horticulture Research International, Wellesbourne, Warwick   CV35 9EF, UK

ABSTRACT

Priming, a treatment of seeds in which they are hydrated
sufficiently to allow the preparative events for germination to
take place but insufficiently hydrated to allow the radicles to
emerge, followed by drying before sowing advances and
synchronises germination. In turn seedling emergence is more
predictable; it is advanced and more synchronised, giving earlier
growth. Three main priming techniques have been developed to
successfully treat both large and small-seeded species; solid
matrix priming (SMP) using materials such as calcined clay,
polyethylene glycol (PEG) priming in bioreactors and drum
priming. Provided comparisons are made for seeds achieving the
same water potential during treatment, the effects on germination
and seedling emergence are identical. It is possible to
integrate these 'physiological' treatments successfully with the
application of crop protectant chemicals to seeds and with seed
pelleting techniques. Priming provides a means to introduce
biocontrol agents on to seeds though to date the potential of
this has been demonstrated largely only for SMP.

INTRODUCTION

Seed treatment is an ancient technology and the soaking of seed in dung
for short periods prior to sowing, and 'pelleting' in dung, have been reported
at regular intervals by writers down the ages from Roman times. Such
treatment has been claimed to increase establishment and provide nutrients,
particularly trace elements, so increasing yields on poorly manured soils.
Despite this, physiological treatments of seed have not been widely exploited
in commerce in contrast to the pervasive uptake of crop protectants applied
to seeds. The commercial use of physiological seed treatments has been
restricted largely to the alleviation of dormancy, a problem of limited
significance amongst cultivated agronomic and vegetable crop species.

However, in the early 1970s Heydecker *et al*. (1973) showed that a
treatment of seeds, called priming, in polyethylene glycol (PEG) advanced
germination and emergence and reduced the time over which the seedling
population emerged in the field. The potential of this treatment to improve
crop establishment and reduce variation in plant size was widely appreciated
and the benefits have been confirmed (see Bradford, 1986 for a review) using
seeds primed on a small scale in Petri dishes (Brocklehurst & Dearman, 1983)
or in small bubble columns (Darby & Salter, 1976). As a result of this and
the interest of seed companies in the technique, substantial effort has been
directed recently into large-scale priming using a variety of methods.

The purpose of this paper is to outline the mechanism of priming, the
basis for improved crop response, the systems that are currently or have been
developed for large-scale priming and how these processes can be integrated

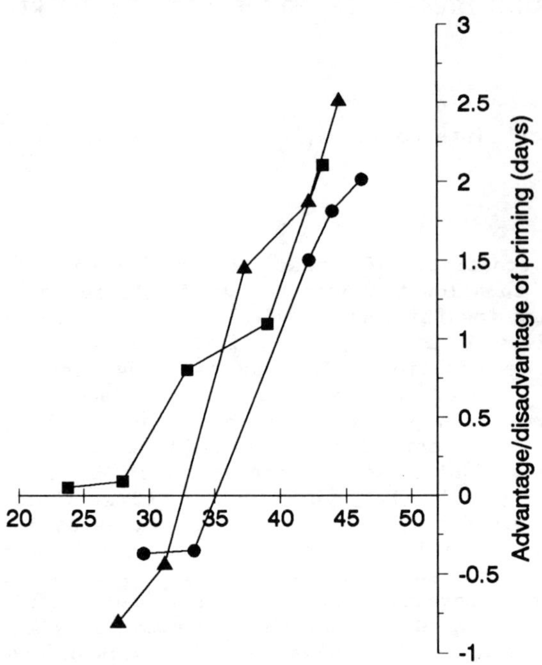

Seed moisture content (Fresh weight basis)

FIGURE 1 Relationship between response to priming and seed moisture content for leek (•), onion (▲) and carrot (■) (after Gray *et al.*, 1990).

with current technology for applying crop protectants to seeds.

MECHANISM OF PRIMING

Essentially, in a seed priming regime seed water potential (ψ) is maintained at a level sufficient to initiate metabolic events in phase II of the germination process (Bewley & Black, 1978) but which prevents radicle emergence. Measurable germination responses to priming are obtained at seed moisture contents of approximately 30% and the response increases linearly over the range 30% to 45-50%, the precise upper limit depending upon species (Fig.1; Gray *et al.*, 1990). A detailed analysis of the water relations of priming seeds is given in Bradford (1986, 1990). The maintenance of the seed water potential below the threshold necessary for radicle expansion allows the slower germinating seeds in the population to 'catch up' with the faster germinating ones giving, once the 'restraint' to germination is removed, more synchronous and more rapid germination. There is much debate about the biochemical changes such a treatment might produce and whether 'repair' mechanisms are stimulated but there is little evidence that they are intrinsically different from those occurring during normal imbibition in water. Even if differences do exist there is little evidence that these are reflected in differences in seedling relative growth rates between populations of primed and unprimed seeds (Brocklehurst *et al.*, 1984).

354

RESPONSES TO PRIMING

The main benefits of priming are to advance and synchronise the germination of seeds in a population. The precise effects depend upon the seed moisture status and the temperature of priming (Fig.2). As the temperature at which seeds are primed falls, the response to priming increases for the same number of 'day degrees' of priming but, accompanying this, there is an increase in the proportion of abnormal seedlings. The 'crossover' point where the maximum response to priming and minimum level of abnormal seedlings occurs, in leek, for example, is 10 d priming at 15°C.

Whilst reports in the literature invariably record earlier seedling emergence from priming it is not invariably more synchronous even though germination is. This is due to a number of factors principally that, in the field, priming often secures the emergence, after a time, of weaker seedlings from poorly germinating seeds which usually would not survive from untreated seeds. This extends the period over which all seedlings emerge. In addition, some species, eg Umbelliferae, exhibit a wide range of variation in embryo size which will contribute to asynchronicity of germination and seedling emergence. However, priming does increase the predictability of percentage seedling emergence over a wide range of conditions and sowing occasions (Finch-Savage, 1990) and in this way contributes to uniformity of the crop through its effects on plant density and, in turn, on mean plant size. A summary of the effects of priming on seed germination and seedling emergence is given by Bradford (1986) and Khan (1992). The earlier emergence from primed seed advances growth and, provided harvests are made during the exponential phase of growth, 'yields' are higher from primed than untreated seeds (Brocklehurst & Dearman, 1983). However, at maturity, effects of priming on total plant weight, after eliminating the effects of the treatment on plant density, are small (see Bradford, 1986; Brewster *et al.*, 1991). An unexpected consequence of priming has been the finding that the rate of loss of viability of the seeds is increased relative to untreated seeds (see Tarquis & Bradford, 1992). However, the benefits of priming can be maintained over a period of two years in onion, leek and carrot with suitable storage temperatures and scheduling of seed treatment and sowing times (Gray, unpublished data).

LARGE SCALE PRIMING TECHNIQUES

The maintenance of a specific ψ during phase II is the basis for all the techniques currently being developed for large-scale priming. Three main techniques are used in commerce. These are solid matrix priming (SMP) developed in the USA (Taylor *et al.*, 1988; Eastin, 1990; Khan, 1992), PEG priming (Nienow & Brocklehurst, 1987), and drum priming (Rowse, 1991, 1992) both developed in the UK. Provided similar levels of seed ψ are achieved in each system the subsequent effects of priming by different methods on seedling emergence are virtually identical (Gray *et al.*, 1992).

<u>Solid matrix priming</u>

The use of PEG, with its low oxygen solubility (dO_2) and high viscosity has not proved to be suitable for large seeds and it is not possible to obtain high seed: PEG ratios to make the system economic. As a result, efforts have been made to use solid materials. The solid materials used for priming, ideally, should be capable of generating a low water potential, produce few solutes and have high 'flowability'. Materials made from leonardite shale,

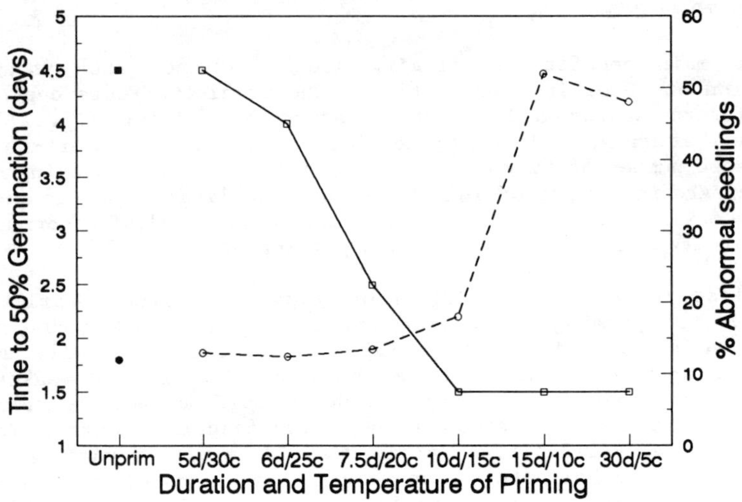

FIGURE 2 Relationship between advancement in germination from priming and the number of abnormals produced in leek. (□), time to 50% germination; (○), percentage of abnormal seedlings.

exfoliated vermiculite, diatomaceous silica and expanded calcined clay are commonly used for various methods of solid matrix priming (Taylor *et al.*, 1988; Khan, 1992). These materials are capable of generating water potentials ranging from -0.4 to -1.5 MPa. The ratio of solid material to seed in solid matrix priming described by Taylor *et al.* (1988) is typically 1.5:1 whereas in the system described by Khan (1992) it is typically 0.2 to 0.4:1. In some procedures the priming material is removed before sowing and in others it is left on the seeds. A disadvantage of the method is the lack of precision in control of ψ in the material and the seed and hence in 'repeatability' of the priming effect.

<u>PEG priming</u>

Two types of vessel are used for PEG priming, a bubble column and a stirred bioreactor. Details of the systems used are given in Nienow and Brocklehurst (1987), and Gray *et al.* (1992). With bubble columns it is possible to achieve greater gas liquid mass transfer coefficients (k_La) than with stirred bioreactors at equal energy input (Fig.3) and there is less risk of damage to seeds by abrasion in such systems compared with stirred bioreactors where high impeller speeds are needed to maintain dO_2 above 80% (Nienow & Brocklehurst, 1987). However, if oxygen enriched air is used (Bujalski *et al.*, 1989) then stirred bioreactors are more cost effective, since lower rates of enriched air sparging are possible, 0.02 to 0.05 vvm (volume of gas per volume of liquid per minute) as compared with 1.2 vvm for bubble columns. Nevertheless, both systems can be operated to give reliable responses to priming. However, in some species e.g. onion, even under these conditions variable responses to priming have been obtained and indicate that k_La from PEG to seeds was limiting even when dO_2 was between 80 and 100%. By using oxygen enriched air, dO_2 values can be increased to 150% and improved, more reliable responses to priming have been achieved with onion (Bujalski and Nienow, 1991), with leek (Bujalski *et al.*, 1991a) and with

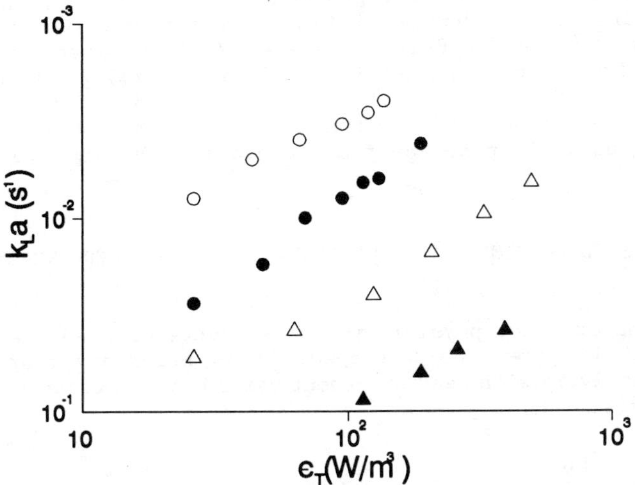

FIGURE 3 Relationship between k_La/s and specific energy dissipation rate (e_T) for water (open symbols) and PEG 6000 (closed symbols) (-1.5 MPa) in bubble columns (circles) and stirred bioreactors (triangles).

carrot (Bujalski *et al.*, 1991b) but not with some flower species (Finch-Savage, 1991). At present, PEG 6000 is widely used but there would be engineering advantages in reduced power input and in obtaining higher dO_2 values by using lower molecular weight PEGs. Priming with PEG 600 and 6000 (see Gray *et al.*, 1992) give similar responses and there are no adverse effects on plant growth from the lower molecular weight PEG. As PEG 600 is a liquid, there could be additional advantages in enabling simple control systems to be devised for maintaining prescribed ψ of the PEG in bioreactors. A major problem with PEG priming arises from the need for and cost of disposal of PEG. It is possible, despite massive increases in microflora, to re-use PEG (Petch *et al.*, 1991) as no adverse effects on germination or seedling emergence responses have been noted with two cycles of use, though after three a slight reduction in seed viability was detected. Whilst this may reduce costs of treatment it may not fully compensate for the costs of final disposal of PEG.

Drum priming

The drum priming technique, developed at Wellesbourne (Rowse, 1991, 1992) involves the hydration of seeds over a period of 24 to 48 h in a drum revolving at 1-2 cm/sec. Mixing of seeds is very uniform and seeds at the end of this period are plump but surface dry. The degree of hydration is determined by a simple calibration test for each lot and a computer is used to calculate the rate of wetting and control of water supply. The seeds are then tumbled for between 5 and 15 days. The technique is now operated under licence.

Drying techniques

Drying of the seed after priming is an essential pre-requisite for storage and handling. Several systems are used based on forced air drying through a thin layer of seeds or in fluidized beds (Maude *et al.*, 1988). The latter has the advantage that drying can be combined with the application of pesticides. Using these techniques and with temperatures within the range 15-20°C and airflow rates in fluidized beds of 0.34 m/sec to 0.05 m/sec, repeatable results without loss of the benefits of priming have been obtained (Bujalski *et al.*, 1991b). However, drying the seeds after treatment does reduce the advantage in earliness resulting from priming though this amounts to 0.5 to 1 d, equivalent to the time needed for the seed to imbibe water fully.

INTEGRATION OF SEED PRIMING WITH CROP PROTECTION AND OTHER TREATMENTS APPLIED TO SEEDS

If priming or other physiological treatments of seed are to be widely taken up and their potential developed, it is necessary that they can be integrated effectively with seed treatment using crop protectant chemicals and biological agents.

Solid matrix priming

In the SMP priming system described by Khan (1992), prior or concurrent treatments with fungicides have advanced seedling emergence and proved effective in controlling disease (Khan *et al.*, 1992) and Khan & Ptasnik (1992) have effectively primed seeds after treatment with insecticides. In a few cases the prolonged seed treatment period with priming has been exploited effectively to introduce biological agents to control disease, eg 'damping off' from *Pythium ultimum* using *Trichoderma* (Harman & Taylor, 1988; Harman *et al.*, 1989) and *Pseudomonas fluorescens* (Callan *et al.*, 1990).

PEG and drum priming

Seeds routinely treated with Benlate-T have been primed in PEG successfully (Maude *et al.*, 1992; Bujalski & Nienow, 1991; Finch-Savage *et al.*, 1991) but for *Alternaria dauci* infection such treatment only partially reduced infection as did the addition of iprodione or thiram to the PEG during the priming process (Maude *et al.*, 1992). In both situations further dusting with iprodione after treatment and drying or alternatively application of it in a polymer film coat as the final stage in treatment was necessary to control the disease fully. Such combined treatment increased plant stand and yield. The feasibility of following PEG priming by both drying and the coating of the seeds with pesticide in a single operation yet retaining the benefits of priming have been demonstrated (Table 1: Bujalski *et al.* 1992). It has also been shown that, for tomato, onion, leek, carrot, parsnip and lettuce, priming can be fully integrated with modern seed pelleting techniques without loss of the benefits of PEG and drum priming (Gray, unpublished data; Valdes & Bradford, 1987). Khan & Taylor (1986) have reported that PEG can be directly incorporated into the pelleting material, improving the rate and final percentage germination, possibly as a result of a priming action.

CONCLUDING REMARKS

The benefits of priming in terms of earliness of emergence and more reliable seedling establishment (Finch-Savage, 1990) are well appreciated and, despite the increased cost of the seed compared with untreated lots, primed seeds are in widespread use for a number of species. With flower seeds, which often exhibit erratic seed germination as a result of dormancy, the benefits of priming are substantial and several companies in the USA and EC offer primed seeds for sale. With vegetables, primed celery and lettuce seeds, which give better germination at thermodormancy-inducing temperatures, have been available for some years from at least two companies. Details of the techniques used have not been published though details of the Quick Pill technique are available (Jacob, 1982). More recently, the introduction of the HRI drum priming technology, through licensing arrangements with the British Technology Group, has led to the commercial sale on an annual basis of several hundred kilograms of primed seed of onion, leek, carrot and parsnip in the UK.

TABLE 1. Effects of priming, drying and coating treatments on seedling emergence of leek (Bujalski *et al.*, 1992)

Priming method	Drying method	Coated with	Mean emergence time (days)	Spread of emergence 5-95% (days)	Percentage emerged seedlings
Bioreactor	Thin-layer	-	2.0	4.7	84
Filter paper	Thin-layer	-	2.3	4.9	89
Bioreactor	Spouted-bed	-	2.3	5.2	76
Bioreactor	Spouted-bed	binder	2.5	4.7	83
Bioreactor	Spouted-bed	thiram+binder	2.3	5.7	78
Untreated		binder	4.9	7.1	85
Untreated		thiram+binder	4.8	7.5	79
Untreated			5.0	7.3	81
LSD (p=0.01)			0.45	2.0	14.1

A major outlet for primed seeds is in the plant raising industry where control of the post sowing environment can be readily achieved so maximising the benefits. For vegetable transplants and bedding plants raised in modules the improvement in 'cell fill' and seedling uniformity is clearly evident and improves the appearance and performance of the product. There is increasing evidence of the use of primed seed sown directly in the field but it is necessary to sound a note of caution for the success of such treatments. Finch-Savage (1990) has shown that although, on average, over a number of successional sowings primed seeds performed better than untreated seeds this was not always the case for every sowing. This is because changes in the seed bed, notably the availability of soil moisture at the time of radicle emergence do not always coincide for both primed seeds and the untreated seeds which germinate later. To maximise the benefits from primed seed in the field it is necessary also to optimise the post-sowing environmental conditions. This can be achieved readily in outdoor seed beds and there could be therefore considerable scope to examine the possibility of priming for hardy ornamental and tree species.

Good progress has been made in integrating priming technology with existing seed treatment technology for applying crop protectant pesticides (Talavera-Williams *et al.*, 1991) and with pelleting, without loss of seed viability or the benefits of priming. It is also possible from simple tests to predict in advance the likely response to a given priming treatment or, alternatively, the particular priming conditions can be adjusted to maximise

the response for each seed lot (Rowse, pers. comm.). Currently, several
techniques are used to prime seeds and each has its advantages. PEG-priming
can be now routinely carried out on 10 kg batches of seeds for a wide range
of species and it would be possible to contemplate priming much larger batches
since the scale-up relationships for large bioreactors are well known
(Talavera-Williams *et al.*,1991). However, operations on such a large-scale
pose problems for the disposal of PEG and this may limit the commercial growth
of such a system compared with drum priming which, theoretically, can be
carried out using seed lots ranging from 1g up to 100kg or more. Furthermore,
it can be used for treating large seeds, currently treated using SMP
techniques. For some species which have seed coat inhibitors, eg celery, and
which require a pre-treatment, priming of seeds can be accomplished more
quickly by PEG than drum priming.

The preference for one particular method may also be influenced by the
development of other seed treatment technologies, for example the application
of micro-organisms to control disease. It is claimed that one of the
advantages of solid matrix priming is that the material used can provide also
a base for the beneficial activity of protective micro-organisms and some
progress is being made using both antagonistic fungi and bacteria. No
comprehensive studies have been carried out using biological agents in PEG or
drum priming but it has been shown that *Enterobacter agglomerans* can be film-
coated onto onion seeds and is effective in reducing the incidence of neck rot
(Maude, pers. comm.). This suggests that the particular conditions offered
by SMP priming are not necessarily essential to the development of successful
biological disease control systems for seeds and that successful systems could
also be associated with PEG or drum priming.

ACKNOWLEGEMENT

The Agricultural & Food Research Council, Ministry of Agriculture,
Fisheries and Food, the Horticultural Development Council and the Seed
Industry supported the work carried out at HRI.

REFERENCES

Bewley, J. D.; Black, M. (1978) Physiology and biochemistry of seeds in
    relation to germination. I. Development, germination and growth.
    Springer Verlag, Berlin.
Bradford, K. J. (1986) Manipulation of seed water relations via osmotic
    priming to improve germination under stress conditions. *HortScience*,
    21, 1105-1112.
Bradford, K. J. (1990) A water relations analysis of seed germination rates.
    Plant Physiology, **94**, 840-849.
Brewster, J. L.; Rowse, H.R.; Bosch, A.D. (1991) The effects of sub-seed
    placement of liquid N and P fertiliser on the growth and development of
    bulb onions over a range of plant densities using primed and non-primed
    seeds. *Journal of Horticultural Science*, **66**, 551-557.
Brocklehurst, P. A.; Dearman, J. (1983) Interactions between seed priming
    treatments and nine seed lots of carrot, celery and onion. II. Seedling
    emergence and plant growth. *Annals of Applied Biology*, **102**, 585-593.
Brocklehurst, P. A.; Dearman, J.; Drew, R. L. K. (1984) Effects of osmotic
    priming on seed germination and seedling growth in leek. *Scientia
    Horticulturae*, **24**, 201-210.
Bujalski, W.; Nienow, A. W.; Gray, D. (1989) Establishing the large-scale

osmotic priming of onion seeds by using enriched air. *Annals of Applied Biology*, **115**, 171-176.

Bujalski, W.; Nienow, A. W. (1991) Large-scale osmotic priming of onion seeds: a comparison of different strategies for oxygenation. *Scientia Horticulturae* **46**, 13-24.

Bujalski, W.; Nienow, A. W.; Petch, G. M. (1991a) The bulk priming of leek seeds - the influence of oxygen enriched air. *Process Biochemistry*, **26**, 281-286.

Bujalski, W.; Nienow, A. W.; Petch, G. M.; Gray, D. (1991b) Scale-up studies for osmotic priming and drying of carrot seeds. *Journal of Agricultural Engineering Research*, **48**, 287-302.

Bujalski, W.; Nienow, A. W.; Petch, G. M.; Drew, R. L. K.; Maude, R.B. (1992) The process engineering of leek seeds: a feasibility study. *Seed Science and Technology*, **20**, 129-139.

Callan, N. W.; Mathre, D. E.; Miller, J. B. (1990) Biopriming seed treatment for biological control of *Pythium ultimum* pre-emergence damping off in sweet corn. *Plant Disease* **74**, 368-372.

Darby, R. J.; Salter, P. J. (1976) A technique for osmotically pre-treating and germinating small quantities of seeds. *Annals of Applied Biology*, **83**, 313-315.

Eastin, J. A. (1990) Solid matrix priming of seeds. US Patent No. 4912874.

Finch-Savage, W. E. (1990) The effects of osmotic seed priming and the timing of water availability in the seedbed on the predictability of carrot seedling establishment in the field. *Acta Horticulturae*, **267**, 209-216.

Finch-Savage, W. E. (1991) Development of bulk priming/plant growth regulator seed treatments and their effect on the seedling establishment of four bedding plant species. *Seed Science and Technology*, **19**, 477-485.

Finch-Savage, W. E.; Gray, D.; Dickson, G. M. (1991) The combined effects of osmotic priming with plant growth regulator and fungicide soaks on the seed quality of five bedding plant species. *Seed Science and Technology*, **19**, 495-503.

Gray, D.; Steckel, J. R. A.; Hands, L. J. (1990) Controlled hydration of vegetable seeds. *Journal of Experimental Botany*, **66**, 227-235.

Gray, D.; Rowse, H. R.; Finch-Savage, W. E.; Bujalski, W.; Nienow, A. W.; (1992) Priming of seeds - scaling up for commercial use. Fourth International Workshop on Seeds.Basic & Applied Aspects of Seed Biology.Angers, France 1992. Vol. 3.

Harman, G. E.; Taylor, A.G. (1988) Improved seedling performance by integration of biological control agents at favourable pH levels with solid matrix priming. *Phytopathology*, **77**, 520-525.

Harman, G. E.; Taylor, A. G.; Stasz, T. E. (1989) Combining effective strains of *Trichoderma harzianum* and solid matrix priming to improve biological seed treatments. *Plant Disease* **73**, 631-637.

Heydecker, W.; Higgins, J.; Gulliver, R. L. (1973) Accelerated germination by osmotic seed treatment. *Nature*, **246**, 42-44.

Jacob, S. (1982) Process for bringing pre-germinated seed in a sowable form and some time storage form as well as pilled pre-germinated seeds. European Patent Application No.00623832 (EP 8220 0 400).

Khan, A. A. (1992) Preplant physiological seed conditioning. *Horticultural Reviews*, **14**, 131-181.

Khan, A. A.; Abawi, G. S.; Maguire, J. D. (1992) Integrating matriconditioning and fungicidal treatment of table beet seed to improve stand establishment and yield. *Crop Science*, **32**, 231-237.

Khan, A. A.; Ptasnik, W. (1992) Integrating matriconditioning of snap bean seeds with pesticides, hormones and drying treatments. *National Symposium for Stand Establishment in Horticultural Crops*. Fort Myers, Florida: 16-20 November, 1992.

Khan, A. A.; Taylor, A. G. (1986) Polyethylene glycol incorporation in table beet seed pellets to immprove emergence and yield in wet soil. *HortScience* **21**, 987-989.

Maude, R. B.; Suett, D. L.; Springer, P. H.; Nienow, A. W.; Bujalski, W.; Maroglou, A.; Petch, G. M.; (1988) A prototype fluidised bed film-coating seed treater. Application to Seeds and Soil 1988. *BCPC Monograph*, **39**, 403.

Maude, R. B.; Drew, R. L. K.; Gray, D.; Petch, G. M.; Bujalski, W.; Nienow, A. W. (1992) Strategies for control of seed-borne *Alternaria dauci* (leaf blight) of carrots in priming and process engineering systems. *Plant Pathology*, **41**, 204-214.

Nienow, A. W.; Brocklehurst, P. A. (1987) Seed preparation for rapid germination - engineering studies. *Paper 5: International Conference on Bioreactors and Biotransformations*. Gleneagles, Scotland, UK: 9-12 November 1987.

Petch, G. M.; Maude, R. B.; Bujalski, W.; Nienow, A. W. (1991) The effects of re-use of polyethylene glycol priming osmotica upon the development of microbial populations and germination of leeks and carrots. *Annals of Applied Biology*, **119**, 365-372.

Rowse, H. R. (1991) Methods of priming seeds. UK Patent No. 2192781.

Rowse, H. R. (1992) Methods of priming seeds. US Patent No. 5119589.

Talavera-Williams, C. G.; Pacek, A. W.; Bujalski, W.; Nienow, A. W. (1991) A feasibility study of the bulk priming and drying of tomato seeds. *Transactions of the Institute of Chemical Engineers, part C*, **69**, 134-144.

Tarquis, A. M.; Bradford, K. J. (1992) Prehydration and priming treatments that advance germination also increase the rate of deterioration of lettuce seeds. *Journal of Experimental Botany*, **43**, 307-317.

Taylor, A. G.; Klein, D. E.; Whitlow, T. H. (1988) SMP: Solid matrix priming of seeds. *Scientia Horticulturae*, **37**, 1-11.

Valdes, V. M.; Bradford, K. J. (1987) Effects of seed coating and osmotic priming on the germination of lettuce seed. *Journal of the American Society for Horticultural Science* **112**, 153-156.

THE DEVELOPMENT OF QUALITY SEED TREATMENTS IN COMMERCIAL PRACTICE - OBJECTIVES
AND ACHIEVEMENTS

P. HALMER

Germain's (UK) Ltd., Hansa Road, Hardwick Industrial Estate, King's Lynn,
Norfolk, PE30 4LG

ABSTRACT

Recent developments in seed treatment procedures are reviewed and
the implications of treatment quality fluctuations are discussed.

INTRODUCTION

At the time of the first BCPC Symposium on Application to Seeds and Soil
(Martin, 1988) there was much discussion of developments in seed treatment
application technology, such as film-coating, and the ways these techniques
might be used to deliver a wider range of chemicals and biologically active
materials to crops (Clayton, 1988a,b; Halmer, 1988; Suett, 1988). The aim of
this paper is to provide an update review, from the perspective of the seed
and seed treatment industries, of recent technical developments and trends,
with particular reference to issues of the quality of application. Because
these industries and their markets have fragmented and complex structures,
which differ in virtually every country and crop, it is only possible in this
paper to discuss themes of progress and change in a general way.

Seed treatments have historically tended to be regarded within the seed
industry as a necessary but secondary production step in the process of the
breeding, production and sale of varieties. In practice, the treatment of
seed presents the industry with special challenges in production logistics and
inventory management. Often the time available is short. In some crops, the
seed itself may be only recently produced - in Europe, for example, autumn
sown cereals and oilseed rape are cleaned, processed and treated within about
a 10 week period following harvest. Even where processed seed is available
in good time to the seed company, sales orders from the farmer may not be
finalised until a relatively late date. In some cases a range of varieties
must be considered in possible combination with alternative seed treatments,
perhaps with different application rates or even formulations. Considerable
problems would arise if the wrong quantities of treatments were prepared. In
general, therefore, treatments - especially those with higher value - tend to
be produced to order, rather than speculatively, to minimise the amount of
unsold surplus carried over at the end of the season.

It is common to consider seed production in terms of the high-volume or
agricultural crops (small grain cereals, maize, pulses, oil seeds, cotton,
grasses and sugar beet), and the low-volume or minor crops (chiefly
horticultural vegetable and flower species). Continuous throughput equipment
is used to treat high volume crops, whereas in general batch treaters are used
for the low volume crops. Arguably, in practical terms, cereals merit a
category of their own, because of the large quantities involved: at least
15t/h is desirable for cereal treatment in fixed installations, and some
facilities achieve 40t/h, though lower rates are acceptable for other crops.
Sugar beet also is a special case, in Europe and the USA, because it is mostly
treated, on a batch basis, through pelleting or encrusting processes.

A substantial proportion of agricultural crop seed treatment worldwide is carried out on farmer-saved seed (with the notable exception of the hybrid crops, maize and sugar beet) - to a greater extent where yields are low, such as Southern Europe and the American Midwest - either using the farmer's own equipment, or by specialist mobile treating operations, with throughputs much slower than static equipment. In many countries, farmers' own machines used are of a basic design in which liquids or powders are added to seed metered directly into the auger - "planter box application".

As a generalisation, the most advanced seed treatment technology is employed in Europe and North America, though standards are variable. Technology is less developed in Eastern European countries and South America. In the majority of Africa and Asia (excepting Japan) and the Middle East (with certain exceptions), where there is neither the economic strength or infrastructure to support seed treatment technology, the majority of seed is farmer-saved, and treatments are applied using the simplest techniques.

DEVELOPMENTS IN APPLICATION TECHNOLOGY

Application standards

The requirements for application of seed treatments are often stated along the following lines.

Loading efficiency
The correct target dose should be reliably applied to the seed bulk, and maintained on the seed up to the point of sowing.

Drillability
Materials should be securely attached to the seed surface, and non-tacky, so as not to impair flow in the hopper and coulter tubes or the action of the seed drill itself.

Uniformity
Materials should be distributed uniformly from seed to seed. The meaning and implications of this principle are more complex than may first appear, and this point will be discussed in greater depth later in the paper.

Effect on seed quality
It is important that seed treatments have well defined and minimal effects on the physiological quality of seed. Application methods must be safe, causing minimal mechanical damage and maintaining seed moisture at acceptable levels for storage. The germination and vigour of varieties are of particular importance for module- or space-planted crops like vegetables and sugar beet, where rapid and uniform crop establishment is important in ensuring yield and harvest quality. In these crops too, seed treatments may now need to be integrated with commercial priming techniques, so as not to lose the germination enhancement benefits (Gray, 1994).

Application to pelleted and encrusted seed
Most pelleted and encrusted seeds are treated with a fungicide, and in some species an insecticide also - notably sugar beet and some vegetables in Western Europe. The main purpose of encrusting and pelleting is to alter the dimensions of seed to add weight or permit space planting through a variety of seed drills and sowing mechanisms. Coated seed must therefore meet integrity, weight, size and shape standards, usually specified by grading

through round hole and slotted sieves.

Handling safety
Aside from directly indicating poor application, dustiness may be of particular concern to those who handle seed treated with insecticides, or with materials that have irritant effects.

Appearance
Cosmetic appearance is an important factor in the perception of quality, especially when the pesticide or coating adds substantially to the cost. Visible dustiness apart, appearance is also the most obvious indicator (though not always a true reflection) of uniformity of distribution of the applied material between and on individual seeds. An attractive features of film-coating is the characteristically more or less continuous covering of the seed surface, and pelleted and film-coated seed is commercially produced in a range of appealing colours, with polished or pearlised finishes.

The impact of Operating Regulations and Quality Standards

The seed industry has to operate under increasingly rigorous regulatory standards, governing both the environment as well as health and safety in the workplace. Regulatory changes are part of a general trend affecting industrial production which has been identifiable over the past two decades, particularly in developed countries. The rate at which these changes are occurring, and the form they take, varies between countries, with "green pressures" perhaps strongest in Scandinavia, Germany, Holland, the UK and California. In the UK, for example, operators of seed treatment equipment now need to pass through a certification procedure, as will in the near future those who sell treated seed. Companies treating seed must also handle the increasingly expensive and difficult disposal of treatment waste products, such as contaminated dust and water from cleaning equipment, and the safe disposal of empty pesticide containers.

At the same time, there are the beginnings of a trend within the seed and agrochemical industries to seek voluntary accreditation under Quality Management System standards (ISO 9000, or BS 5750 in the UK) which, through external audit of operational procedures, are aimed at assuring the customer of product and service quality. Also in view for industry in general is the creation of standards for Environmental Management Systems (e.g. BS 7750), through which companies can establish and minimise the environmental impact of their operations, on a voluntary basis.

FORMULATIONS AND BINDERS

The trend continues towards liquid formulations, whose advantages include ease of mixing, relatively low sedimentation rates, decreased loss in loading, and safety to the environment. Liquid formulations, usually water-based true solutions or flowable suspensions or slurries, now are preferred in all European countries and by commercial treaters in North America and Australia. However, powders are still predominantly used for on-farm treatments in the US. Differences in practices exist. In France, Belgium, Holland, and Denmark, for example, because of the equipment installed in seed treatment plants there is a preference to use formulations in a dilute form, sometimes with film-coating polymers, whereas in the UK concentrates containing at least 50% solids are used for direct application to seed.

Film-coating was first developed for vegetables, and now has become a standard method in all major markets for applying treatments, typically fungicides, to these species. The most readily apparent characteristics of film-coating, in comparison with "conventionally" treated seed, are the absence of dust, and uniform coverage between and on individual seeds, and improved flow through the drill.

There has also been a concerted move in recent years in the French market to film-coat low-value treatments onto high-volume crop seeds, notably oilseed rape, sunflower and maize. Similarly, film-coated cereals, oilseed rape and grasses are now available in the UK. In large part this seems to have been done for reasons of brand-image, rather than improved biological efficacy. (Indeed it is worth mentioning here that, in general, there seems to be no evidence that an even coverage of active ingredients over the surface of each seed is of any particular advantage to the biological efficacy of a pesticide treatment.) The high aesthetic standards adopted involve the use of relatively high amounts of film-coating binders, which are applied both through modified conventional continuous-throughput machines, or specialised equipment (Bazin *et al.*, 1992a). With the recent advent of treatments at high rates that need to be applied with relatively large volumes of liquid - notably the registration in France of imidacloprid for maize and sunflower seed, applied at 700 g and 1150 g formulation/100 kg seed respectively - film-coating methods involving drying become technically necessary to achieve the best application standards.

Film-coating is also widely used in western European markets to apply treatments onto previously pelleted sugar beet seed. Though the range of products used in each country is relatively small, there are a large number of possible permutations of pellet size, with fungicides (thiram or iprodione, and hymexazol) and insecticides (carbofuran, methiocarb, furathiocarb, carbosulfan, imidacloprid, or tefluthrin) at various rates (Fauchere, 1992), throughout the market as a whole. In particular, some of the insecticides are applied at high levels, to replace soil-applied granules (Dewar, 1992), and add substantially to the cost of the seed. However, because of the nature of the pelleting process, and the relatively late ordering of varieties and treatments by growers, there is not enough time in practice to produce all the required combinations on each variety of seed, without resorting to costly overproduction. Film-coating, in addition to providing an efficient, uniform and safe application method, gives pelleters a high degree of flexibility in producing to order the exact required amounts of different treatments.

Polymer binders for seed treatment, with film-forming abilities, are available from specialist suppliers, co-formulated with colorants (dyes, pigments and opacifiers) and plasticizers to improve film coverage, in liquid or powder forms, suited for different crops. Filler or absorbent "drier" powder materials to add bulk or reduce stickiness are also available. Though film-coating tends to be equated with "polymer coating", a similar coating quality can be achieved for some seeds - like onion, carrot or sugar beet - at moderate weight increases using encrusting techniques.

SEED TREATMENT EQUIPMENT

<u>The design of equipment</u>

Most present day treatment machinery is designed to meet the several operational criteria discussed above. These can be summarised (Jeffs &

Tuppen, 1986; Clayton, 1993) as follows:
        accuracy and uniformity of application
        ease of operation
        the ability to handle a range of formulations and seed species absence
        of seed damage
        easy clean-down to prevent cross contamination between seed bulks
        high throughput.

Enclosure of treatment chambers and packing units to prevent contamination of the workplace with dust and sprays is now standard. Most modern equipment is automated to some degree, often with sophisticated - and expensive - features to prevent operator mistakes, including safety interlocks and alarms designed to ensure operations occur in the correct sequence, and only when all components are present. In large processing plants, controls may also regulate the automatic delivery of seed from holding bins, or to gear treatment throughput to the rate of packing.

High-throughput conventional seed treatment equipment
A review of machinery was conducted by Jeffs & Tuppen (1986), and Clayton (1993) reviewed the range of equipment currently available for treating agricultural seeds, listing the major machinery models, and their design features and capabilities, of which the following is a summary.

Almost all recent developments in treatment equipment have been on continuous-throughput systems, designed for use with liquid formulations. In these designs, the accuracy and uniformity of treatment application depend on the principles and accuracy of seed and liquid metering used, on how liquid is distributed on to the seed, and on the degree of secondary re-mixing.

Liquid metering
In newer machines, liquid flows are controlled through forms of metering pump or volumetric dosing jar systems. Application to the seed itself is typically by rotary atomisers, often cup-shaped spinning discs, or less commonly though air or airless nozzles.

Seed metering
The counterbalanced weigher bucket principle is common in older continuous throughput designs, but in newer machines higher and more even seed throughputs are achieved using the variable regulating collar system, in which seed flow over a cone is restricted by a collar seated around it.

Secondary Mixing
Almost all machines incorporate a remixing stage, most commonly using horizontal or inclined augers, with various diameters and configurations of internal mixing baffles/flights depending on seed type, or with brushes for fragile seed like pulses or maize.

Horizontal drums of different designs with a gentle mixing action, incorporating direct liquid spray systems, are used to treat species such as cotton, soya and maize that are susceptible to damage.

Film-coating equipment for low volume crops
The film-coating process uses specialised equipment to spray seeds with relatively large volumes of formulations and polymer binders, with simultaneous drying, so that the seed moisture content remains virtually unchanged throughout the treatment. Typically each seed goes through several cycles of spraying and drying within the coating equipment, and is effectively

treated with several layers of the coating materials. It is therefore possible to apply different, perhaps chemically incompatible, solutions sequentially. Depending on seed size, a critical minimum amount of time and fluid is required to film-coat satisfactorily. The key to high quality coating lies in the means and efficiency of seed mixing.

Low volume crops, and pelleted sugar beet, tend mostly to be treated on a single- or continuous- batch basis - though seed bulks may amount to several tonnes each. For film-coating vegetable seed, companies consider it appropriate to have batch equipment capable of treating a range of quantities, from 100 g or less to several hundred kg.

Spouted Beds
The earliest seed film-coating equipment was based on stirred fluidised bed principles (Halmer, 1988; Clarke, 1988). Even coverage depends on mixing seed at greater than the fluidisation point and, because of the irregular shape of many seeds, flow in the bed tends to be turbulent and erratic. Capacities are typically about 5-20 kg seed, though some commercial systems achieve continuous throughput by using fluidised beds operating in series (Horner, 1988). Scaling up beyond this point in a single batch requires a disproportionate increase in the amount of air required, which is costly. Indeed, the air flow necessary to circulate seed can cause abrasion in some species (Drew, 1989) leading, as quantities increase, to potential reduction of seed quality, or cosmetic imperfections in the coating.

Drum coaters
Fluidised bed coaters are in widespread commercial use, but the trend within the vegetable seed industry in recent years has been to use batch drum coaters, which achieve comparable results. This process is carried out in a rotating cylindrical pan equipped with spraying units, with dry warmed air drawn in through perforations in the drum wall (Halmer, 1988). The design and configuration of stirrer blades is important to ensure even gentle mixing of seed of different shapes and size, so that all seeds are equally presented to the spraying system. Because of the flow behaviour of seed in the mass, it is necessary to build a range of coaters to handle different batch capacities. In practice, individual batch drums seem to have an upper size limit of about 350 kg seed, depending on seed size. However, larger capacity batch coaters using longer drums can be produced, using a design that joins shorter drums together, so that seed is more or less constrained within each successive section, and lateral movement through the drum is minimal.

Film-coating equipment for high volume crops
Because the quantities of liquid involved in the conventional treatment of seed are relatively small, it is not necessary to dry off the applied water: either seed "self-dries" or absorbent "drying" powders may sometimes be used. Film-forming binders are often incorporated in the liquids used in conventional treatment - which is in the technical sense film-coating, though the cosmetic results may or may not be a conspicuous coated layer. However, where the amounts of material and/or binder - and hence the volume of liquid - to be applied reaches a certain point, it becomes necessary to remove some water by drying. Commercially this can be achieved by modifying existing equipment or by adding an extra drying stage (Currey & Clayton, 1987).

In recent years, continuous throughput "true" film-coaters have begun to be developed commercially (e.g. Bazin *et al.*, 1992b), based on the principle of spraying with simultaneous drying. At present, only a limited assessment of the equipment is possible. Most designs are based on perforated drums with integral spray systems, thus resembling batch drum coaters in some respects.   To   deliver   coatings   with   good   loading   and   uniformity characteristics, these machines must ensure that seed is mixed thoroughly during its short dwell-time in the drum. As the volumes to be applied to a given seed throughput increase, the faster spray rates require higher flows of drying air, at high temperatures in some designs, which is liable to cause turbulence and may distort spraying patterns and loading efficiency.

MEASUREMENT OF LOADING EFFICIENCY AND UNIFORMITY

<u>Analytical procedures</u>

     The standard of application is directly measured by the analytical recovery of active ingredients from representative samples of commercial lots both in terms of bulk chemical loading - on a number or a mass of seed basis - or on individual seeds to determine the seed-to-seed distribution of material. Chromatographic techniques, such as tlc, glc and now most commonly hplc, are widely used as analytical tools by the agrochemical industry, by seed treatment companies and by Institutes, both for research purposes and quality monitoring of seed treatment. In the UK, for example, because of their close historical link in the provision and maintenance of much of the cereal treatment machinery, there is a well established system in place through which the major agrochemical companies receive samples of treated bulks for recovery analysis.

     Chromatographic analysis is costly and time consuming, and therefore impractical for direct quality control purposes in most treatment facilities. There is thus an advantage in developing methods that give quick, if somewhat approximate, results for on-site use. Koch & Spieles (1988) proposed using measurements of fluorescence from the dye present in cereal seed treatment formulations as an indirect method of treatment distribution. Dyes have been incorporated routinely in German seed treatment formulations for some time, and this allows such checks to be carried out simply, such as the surveys made by Koch & Spieles (1989) and Rietz (1989). Ciba have developed a validated seed loading analysis kit (SLAK), based on the dye-formulation principle, for use in on-site quality control by seed treaters (C. Martin *pers. comm.*): this system is now in commercial use in the UK for the recently approved fenpiclonil cereal seed treatment.

     Pesticide recovery analysis from treated seed can be complicated by the interaction of active ingredients with each other, with coating techniques and with storage duration. Certain active ingredients can be partly degraded during extraction to their component moieties, such as benomyl to carbendazim, or some carbamate insecticides (e.g. carbosulfan, furathiocarb) to carbofuran. Much analytical investigation has been carried out on treatments applied to pelleted sugar beet seed within that industry. Huijbrechts & Gijssel (1989) showed that the actual concentrations of active ingredients recovered from pellets differed from the applied doses, as a result of decomposition, irreversible absorption or conversion to other active ingredients. Interactions with certain carbamate insecticides were found to affect the analytical recovery of the fungicide hymexazol, although no clear correlation was seen with differences in biological efficacy of the latter (Pussemier *et al.*, 1994). For some materials, such as thiram and hymexazol, there is also a decrease in recovery after periods of storage in air-dry conditions, and the degree of decrease can vary between different pelleting processes (Heijbroek *et al.*, 1993).

<u>Sampling Considerations</u>

Product quality analysis relies on the general principle that the sample being examined is truly representative of the whole. Commercial seed bulks are not always completely homogeneously mixed after processing and conditioning, and accordingly the seed industry operates under internationally-agreed procedures for the representative sampling of bulks for Seed Certification purposes. Bulk homogeneity can also be affected by differential settling of seeds of different size and shape, during packing and distribution to the farmer or in the seed drill itself, or by storage conditions such as humidity and temperature. It therefore becomes more difficult to obtain representative samples at points further on in the distribution chain. Agreed sampling protocols will need to be developed as measurements of the recovery of applied active ingredients at the buyer level become standard practice. Agreement will also be needed on the appropriate numbers of seeds to be used for the determination of seed-to-seed uniformity. A CIPAC method for the determination of uniformity of application of liquid seed treatment formulations will indeed soon be published.

<u>Practical performance of seed treatment equipment</u>

There is relatively little published on the performance of conventional seed treatment equipment. Early studies (Griffiths, 1986) revealed that overall loadings and seed-to-seed distributions of cereal grain treatments were very variable, and on average achieved less than 50% of target. The practical situation in the UK now is much improved according to a recent survey (Suett *et al.*, 1994). These authors have reported that, in analyses of 21 batches of barley seed commercially treated with ethirimol (as Ferrax), 18 batches had mean loadings greater than 75% of target, and there was also a high degree of consistency between samples taken during the processing of 14 of them. In all batches, approximately 90% of individual seeds held 50-150% of the target, expressed on a weight basis. Distribution appeared to be similarly variable in commercially treated grain in Germany where, in two surveys made using the dye-fluorescence analytical method, (1) 80% of seed lots treated with various seed dressings were found to have individual seed loadings within 50% of the mean dose (Rietz, 1989), whereas (2) half the lots had CVs >50% higher than 50% (Koch & Spieles, 1989).

Film-coating through batch coaters is capable of a greater degree of uniformity. In commercial systems with film-coated pellets or vegetable seed, it is reported that the almost all seeds have individual loadings within 30% of the mean dose (Halmer, 1988; Horner, 1988).

As already mentioned, much attention is now being paid by the buyers of treated seed to recovery analysis from pelleted sugar beet sown in certain Western European countries. Sugar companies and growers' representatives in Holland, Belgium, France and the UK now routinely analyse pesticides applied to commercial seed bulks. This is done to meet concerns that required doses have in fact been applied - both to ensure optimal biological protection for the crop, and to assure the end-user that the often expensive treatments have been applied in full measure. Analytical methods have been developed, and their repeatability has been established by inter-laboratory ring tests involving the pelleters (Heijbroek *et al.*, 1993). Indeed, over the past few years, sugar beet pelleters have moved to the point of routine in-house recovery analysis, for quality control and assurance purposes.

Batch analytical recovery is not at present a prominent quality

assurance factor in other crops. However, increasing attention to treatment
quality by farmers is foreseeable as high-performance high-cost treatments
become available, and perhaps as food-processors and supermarket chains seek
to assure themselves about all aspects of produce quality. The far-reaching
consequence of this trend could be a situation where pesticide loadings on
seeds, determined by recovery analysis, become an additional condition of
trade between the seed industry and its customers.

PRACTICAL LIMITATIONS ON ACHIEVING UNIFORMITY

How important is uniformity?

Theoretical considerations

It is generally accepted on theoretical grounds that all individual
seeds in a treated bulk should as far as possible carry the same loadings of
active ingredients. A *priori*, an equal distribution would be thought
necessary to give the same degree of protection to each plant, especially with
AIs that have a systemic mode of action and where crops are sown at wide
spacings so that each seed is isolated from its neighbour. It should also
minimise the risk of phytotoxicity. Erratic loading and seed-to-seed
distribution of pesticides could have complex effects on pest and disease
control, depending upon the nature of the AI, its mode of action and the soil
type (Griffiths, 1986).

In practice however, seed size naturally varies to some degree within
every bulk of seed, even after size-grading, and it is therefore inevitable
that treatment doses applied to individual seeds will differ. Suett *et al.*
(1994) report up to 20% CV in individual barley seed weights, and up to 45%
CV in individual ethirimol loadings, within commercial batches. Even with
commercial pelleting, though the range of seed size tends to be narrowed, seed
drills can satisfactorily accommodate fairly wide size ranges: currently in
Europe for example, diameter tolerances for sugar beet pellets are 1.00 or
1.25 mm above a minimum of 3.50-3.75 mm, depending on the country (Fauchere,
1992). This means that, at the extreme, seed or pellet surface area can
differ by 50% or more from the smallest to the largest seed in the batch.
Westwood *et al.* (1994) have shown that, in a batch of pelleted seed in which
63 out of 100 individuals had loadings within 22% of the mean, individual seed
loading was highly significantly covariant with weight and surface area.

Most registered seed treatment rates are expressed on the basis of a
weight or volume of formulation per weight of seed. However, the number of
seeds for a given weight can differ greatly from batch to batch, or variety
to variety. This makes it illogical to argue on the grounds of optimal
efficacy that all seeds in a given batch must receive the same dose, if that
individual-seed dose will be different from one batch to the next. This
situation does not arise when treatment doses are expressed on the basis of
a weight per number of seeds. Such "unit" rates are used in most European
countries for the treatment of sugar beet seed, but this would not necessarily
be a likely or practical proposition for other high-volume crops.

There is little doubt that it should be of concern if distribution is
grossly uneven or skewed, but the situation is not so clear where %CVs are
about 50% or less. Koch & Spieles (1989) propose that the proportion of
cereal grains carrying between 75% and 125% of the correct amount of dressing
is the most important parameter in assessing the quality of seed treatment.

<u>Relationship to biological efficacy</u>

There is a surprising lack of published evidence about the relationship between the degree of uniformity and biological efficacy, no doubt partly because it is a difficult matter to investigate experimentally. Suett & Maude (1988) found marked differences in the uniformity of uptake of chlorfenvinphos by individual seedlings of film-coated cabbage, carrot and onion seed. Along similar lines, Westwood *et al.* (1994) report marked differences in the concentration of imidacloprid in sugar beet cotyledons during early seedling development from seeds of different sizes.

Suett & Maude (1988) and Jukes *et al.* (1994) have developed an experimental approach that involves mixing together, in different proportions, film-coated samples that have been prepared on a batch of seed: this is done in such a way as to make up sets of experimental treatments, in which the biological efficacy of any given bulk-loading level of active ingredient can be tested at different degrees of seed-to-seed uniformity. These authors have shown that radish film-coated with chlorpyrifos or chlorfenvinphos and sown at various seed spacings, exhibited the same degree of control in conditions of severe infestation of cabbage root fly in the field, despite having seed-to-seed variabilities ranging from about 15% to 65% (as %CV). The effect was evident over a wide dose range: though the overall degree of protection was reduced at lower doses, performance of the seed treatment was unaffected by the uniformity of dosing within each mean dose level. The authors speculate that this can be best explained in terms of the tolerance distribution of the pest population in response to the insecticide dose on the seed population taken together. Minimising seed-to-seed distribution in order to optimise biological efficacy therefore may not always be necessary beyond the point already achieved in commercial practice, and at least not in the case of the control of cabbage root fly on radish.

Clearly, more research is needed about the degree of uptake of active ingredients from pesticide-treated seed, and the interaction between a pest or disease organism and the developing crop, before the importance of a high degree of uniformity can be properly assessed.

THE PROSPECTS FOR QUALITY SEED TREATMENTS

It is possible to see a long term trend in the quality of seed treatment application technology - from powder to liquid formulations, and from conventional application to film-coating. It is technically possible now to deliver as much quality in the application of a particular seed treatment as the farmer or the seed treater is prepared to pay for. What this amounts to varies between crops, and from country to country. Where the seed itself has a high value, the seed treater may be able to readily absorb the costs of applying even low value treatments using quality coating techniques.

The use of seed treatments, and the methods to apply them, are always vulnerable to changes and fluctuations in the agricultural climate. The effect of the recent reforms in the CAP, for example, has been to switch financial support for farmers from crop prices to farm incomes, on the basis of rotational set-aside of arable land. Crop economic inputs have thus been placed under greater pressure. One consequence has been to further increase the proportion of farmer-saved seed in crops like cereals and oilseed rape. Another is that it becomes more difficult for the farmer to justify using relatively expensive seed treatments when buying seed, especially where yield improvements are likely to be small. In this market situation the seed

merchant may not be able to justify the cost of using the highest quality
coating techniques to apply treatments. A degree of technical compromise may
then become necessary, to apply treatments in conventional rather than more
expensive specialised equipment - a trend seen recently in the UK with coated
cereal and oilseed rape products.

New high-performing seed treatments must offer a sufficient added value
to both the farmer and the agrochemical and seed industries, in order to
develop as an economical proposition, and they are likely to be costly. As
such seed treatment products are introduced, agrochemical companies will need
to ensure product efficacy, and control product liability, by safeguarding the
quality of application. This is a situation that is easiest to manage in
distribution systems where seed is treated by producers, merchants or
specialist coating service companies - where application quality can be
assured by formal agreements, such as Bayer's "Quality Charter" for the recent
introduction of imidacloprid seed treatment. The trend in high value seed
treatment technology is clearly towards efficient application of expensive AIs
in factory situations, using techniques such as film-coating.

ACKNOWLEDGEMENTS

I am grateful for discussions with many colleagues in the seed and
agrochemical industry, whose help was invaluable in writing this paper.

REFERENCES

Bazin, M.; Bourreau, F.-Y.; Souche, J.-L.; de Vergnes, B. (1992a) CERPRO -
    Une nouvelle technique de traitment des semences de mais. In: *4th
    International Workshop on Seeds. Basic and Applied Aspects of Seed
    Biology*, D. Come, F. Corbineau (Eds), Paris:Universite Pierre et Marie
    Curie, **3**, 1087-1094.
Bazin, M.; Denise-Lecat, S.; Souche, J.-L.; de Vergnes, B. (1992b) Le
    pelliculage multicouche CERTOP. Un moyen d'ameliorer la qualite des
    semences traitees. In: *4th International Workshop on Seeds. Basic and
    Applied Aspects of Seed Biology*, D. Come, F. Corbineau (Eds),
    Paris:Universite Pierre et Marie Curie, **3**, 935-942.

Clarke, B. (1988) Seed coating techniques. In: *Application to Seeds and
    Soil*, T.J. Martin (Ed.), *BCPC Monograph No. 39*, Thornton Heath: BCPC
    Publications, pp. 205-211.
Clayton, P.B. (1993) Seed treatment. In: *Application Technology for Crop
    Protection*, G.A. Matthews and E.C. Hislop (Eds), Wallingford: CAB
    International, pp. 329-349.
Clayton, P.B. (1988a) Seed treatment technology - the challenge ahead for he
    agricultural chemicals industry. In: *Application to Seeds and Soil*,
    T.J. Martin (Ed.), *BCPC Monograph No. 39*, Thornton Heath: BCPC
    Publications, pp. 247-256.
Clayton, P.B. (1988b) The implications for industry of technological
    advances in seed treatment. *Brighton Crop Protection Conference -
    Pests and Diseases*, **3**, 823-832.
Currey, R.P.; Clayton, P.B. (1987) UK Patent 2182864A.
Dewar, A.M. (1992) The effects of pellet type and insecticides applied to the
    pellet on plant establishment and pest incidence in sugar beet in some
    European countries. *Proceedings of the I.I.R.B. Winter Congress No.*

55, 89-112.

Drew, R.L.K. (1989) The effects of duration and temperature of treatment in a prototype fluidised-bed seed-treater on the subsequent germination of oilseed rape (*Brassica napus* L.). *Seed Science and Technology*, **17**, 7-13.

Fauchere, J. (1992) Les semences de betteraves: des qualites equilibrees. *Proceedings of the I.I.R.B. Winter Congress No. 55*, 121-126.

Gray, D. (1994) Large-scale seed priming techniques and their integration with crop protection treatments. *These proceedings*.

Griffiths, D.C. (1986) Insecticidal treatment of cereal seeds. In: *Seed Treatment*, K.A. Jeffs (Ed.), Thornton Heath: BCPC Publications, pp. 113-137.

Halmer, P. (1988) Technical and commercial aspects of seed pelleting and film-coating. In: *Application to Seeds and Soil*, T.J. Martin (Ed.), *BCPC Monograph No. 39*, Thornton Heath: BCPC Publications, pp. 191-204.

Heijbroek, W.; Huijbrechts, A.W.M.; Gijssel, P.D.; Vanstallen, M. (1993) Influence of storage on treated sugar beet seeds. *Proceedings of the I.I.R.B. Winter Congress No. 56*, 449-470.

Horner, E.L. (1988) SHR: coating technology for application of pesticides to seeds. In: *Application to Seeds and Soil*, T.J. Martin (Ed.), *BCPC Monograph No. 39*, Thornton Heath: BCPC Publications, pp. 271-276.

Huijbrechts, A.W.M.; Gijssel, P.D. (1989) Dosage and release patterns of insecticides and fungicides in pelleted seed. *Proceedings of the I.I.R.B. Winter Congress No. 52*, 131-143.

Jeffs, K.A.; Tuppen, R.J. (1986) Application of pesticides to seeds. Requirements for efficient treatment of seeds. In: *Seed Treatment*, K.A. Jeffs (Ed.), Thornton Heath: BCPC Publications, pp. 17-45.

Jukes, A.A.; Suett, D.L.; Phelps, K.; Sime, S.; Cooke, S.J. (1994) The influence of seed treatment uniformity on the biological performance of chlorfenvinphos and chlorpyrifos. *These proceedings*.

Koch, H.; Spieles, M. (1988) Determination of the quality of seed treatment by single seed analysis. In: *Application to Seeds and Soil*, T.J. Martin (Ed.), *BCPC Monograph No. 39*, Thornton Heath: BCPC Publications, pp. 181-188.

Koch, H.; Spieles, M. (1989) [New assessment of the quality of dressing of cereal seed by means of single seed analysis.] *Gesunde Pflanzen*, **41**, 18-24.

Martin, T.J. (Ed.) (1988) *Application to Seeds and Soil, BCPC Monograph No. 39*, Thornton Heath: BCPC Publications, 404pp.

Pussemier, L.; Debongnie, P.; van Elsen, Y. (1994) Interactions between hymexazol, furathiocarb and some clay materials used for seed treatment. *These proceedings*.

Rietz, S. (1989) [What eveness of distribution on single grains is produced by seed dressing equipment?] *Gesunde Pflanzen*, **41**, 24-28.

Suett, D.L. (1988) Application to seeds and soil: recent developments, future prospects and potential limitations. *Brighton Crop Protection Conference - Pests and Diseases*, **2**, 823-832.

Suett, D.L.; Hewett, P.J.; Jukes, A.A.; Morgan, L.; Fox, V. (1994) The ethimirol content of commercially-treated cereal seeds. *These proceedings*.

Suett, D.L.; Maude, R.B. (1988) Some factors influencing the uniformity of film-coated seed treatments and their implication for biological performance. In: *Application to Seeds and Soil*, T.J. Martin (Ed.), *BCPC Monograph No. 39*, Thornton Heath: BCPC Publications, pp. 25-32.

Westwood, F.; Bean, K.M.; Dewar, A.M. (1994) The effect of pellet weight on the distribution of imidacloprid applied to sugar-beet pellets. *These proceedings*.

THE INFLUENCE OF SEED TREATMENT UNIFORMITY ON THE BIOLOGICAL PERFORMANCE OF CHLORFENVINPHOS AND CHLORPYRIFOS

A.A. JUKES, D.L. SUETT, K. PHELPS, S. SIME & S.J. COOKE

Horticulture Research International, Wellesbourne, Warwick, CV35 9EF, UK

ABSTRACT

Radish seeds were treated with insecticides at 3 dose levels and 4 levels of between-seed uniformity and were field-sown at 3 different spacings. Damage by cabbage root fly larvae was influenced by dose and seed spacing but not by the level of treatment uniformity.

INTRODUCTION

To optimise the biological efficacy of seed treatments, dosing of individual seeds should be as uniform as possible. However, in a field study with insecticide treated radish seed, similar high levels of protection were achieved with treatments having wide-ranging between-seed uniformities (Suett & Maude, 1988). That study, with a single insecticide and single crop spacing, had limited implications. The results of a more extensive investigation, using reduced doses, different seed spacings and two insecticides, are presented in this paper.

MATERIALS AND METHODS

Radish seeds were treated with chlorfenvinphos and chlorpyrifos at various doses. Seed batches were then mixed in different proportions in order to achieve mean target doses of 3, 6 and 12 g AI/kg and target between-seed uniformities of 15, 25, 38 and 50 %CV. Each of these 24 treatments was field sown at spacings of 2.5, 5 and 10 cm.

Seed treatment

Radish seeds, cv French Breakfast (2.75 - 3.00 mm diameter), were film coated at Horticulture Research International, Wellesbourne (HRI-W) in a fluidised-bed unit (Maude & Suett, 1986). Chlorfenvinphos (50% AI wettable powder, Shell Agrochemicals UK Ltd.) and chlorpyrifos (25% AI wettable powder, Dow Agriculture UK Ltd.) were applied at target doses of 1, 2, 3, 4, 5, 6, 8, 10, 12, 16 and 20 g AI/kg. A polymer sticker (polyvinylacetate, Vinamul formulation R18160) was used at a rate of 25 g/kg for all applications. Each treatment was applied over 15 min at a bed temperature of 30°C.

Seed Mixing

In order to achieve a range of between-seed uniformities at the different mean dose levels required, batches of seeds at different doses were mixed in the fluidised-bed unit at 25°C for 1 minute. The compositions of the required mixtures were calculated using a model which assumed a normal distribution of seed doses within each batch and, on the evidence of previous studies, an inherent level of between-seed treatment uniformity of CV = 15%. Table 1 shows the composition of mixtures required to give target levels of variation. The doses a, b, c, d, and e were 1, 2, 3, 4 and 5 g AI/kg to give a mean

TABLE 1. Composition of seed mixtures required to achieve different levels of dose uniformity, expressed as percentage coefficient of variation (%CV).

CV (%)	a	b	c	d	e
		Seed dose (% in mixture)			
15	0	0	100	0	0
25	0	15	65	20	0
38	5	20	25	25	25
50	20	20	20	20	20

mixture dose of 3 g AI/kg and were increased proportionally to give mean doses of 6 and 12 g AI/kg.

<u>Residue Analysis</u>

The mean doses of the seed batches treated initially were determined by extracting 3 replicate samples of 100 seeds per treatment with methanol (100 ml). The mean doses and between-seed variabilities of each mixture were assessed by extracting 50 single seeds per treatment with methanol (1 ml per seed).

Solution concentrations were determined by hplc using a Spectra Physics model 8810 pump with a 25 cm Spherisorb ODS II cartridge column. For chlorpyrifos the mobile phase was water + methanol (5 + 95, 1.3 ml/min) and for chlorfenvinphos the mobile phase was water + methanol (15 + 85, 1.5 ml/min). Retention times were 3.4 min and 3.3 min respectively. Detection was by Cecil 2112 variable wavelength uv detector at 220nm. All samples were injected automatically using a Spectra Physics 8775 autosampler. Peak areas were measured by a Spectra Physics 4270 computing integrator and were quantified by comparison with external standards.

<u>Field performance</u>

The experiment was established on a light sandy-loam soil at HRI-W in July 1988. Separate areas were allocated to each of the 3 seed spacings so

TABLE 2. Insecticide doses achieved in initial batches.

Target dose (g AI/kg)	Achieved dose g AI/kg (± sd)	
	Chlorfenvinphos	Chlorpyrifos
1	1.27 (0.04)	0.90 (0.01)
2	2.14 (0.22)	1.90 (0.07)
3	3.01 (0.11)	2.88 (0.02)
4	4.49 (0.12)	3.79 (0.02)
5	5.42 (0.15)	4.50 (0.02)
6	6.60 (0.15)	5.91 (0.01)
8	9.50 (0.33)	7.78 (0.00)
10	11.24 (0.35)	9.36 (0.10)
12	12.60 (0.39)	11.42 (0.04)
16	14.32 (0.25)	14.80 (0.06)
20	23.10 (0.47)	18.89 (0.08)

that comparison of treatments would not be affected by possible differences
in attractiveness of different plant densities to cabbage root fly. Each area
contained 3 replicate blocks of 6 randomised insecticide/dose combinations.
Within each of these plots, 4 treated and 1 untreated sub-plot were
randomised, each sub-plot comprising a single row of 130 seeds. Treatments
were arranged in a split plot design with uniformity variates as sub-plots.
Each plot therefore contained all 4 variabilities of a single dose of one
insecticide.

The seeds were hand-sown to a depth of 2 cm on 11-14 July and the mature
plants, excluding the first 20 and last 20 in each row, were harvested on 11-
12 August, 5 weeks after sowing. Roots were washed and assessed for the
presence or absence of damage caused by cabbage root fly (*Delia radicum*)
larvae. Crop stand was assessed at harvest by comparison of the numbers of
roots in the samples taken for damage assessment.

TABLE 3. Mean doses (g AI/kg) and between-seed uniformities (% CV)
achieved at the different target doses.

| | Dose (g AI/kg) | | | % CV | |
Target	Chlorfenvinphos	Chlorpyrifos	Target	Chlorfenvinphos	Chlorpyrifos
3	2.44	2.94	15	14.4	15.8
3	2.50	3.06	25	36.1	29.0
3	3.27	4.42	38	42.5	41.9
3	2.96	3.67	50	49.9	60.0
6	6.24	6.47	15	15.3	17.2
6	6.31	6.28	25	21.6	25.0
6	7.91	6.33	38	34.8	42.0
6	6.56	5.92	50	47.2	65.3
12	9.95	12.5	15	21.1	11.5
12	10.7	13.6	25	28.3	27.6
12	14.6	16.0	38	47.3	30.3
12	13.4	12.8	50	50.7	42.7

RESULTS

Seed analyses

The mean doses of the initial batches of chlorfenvinphos and chlorpyrifos
treated seeds were within 30% and 11% respectively of target doses (Table 2).
The mean doses of the subsequent mixtures are shown in Table 3. Between-seed
variabilities of the mixed seeds ranged from 14.4% to 50.7% for the
chlorfenvinphos treated seeds and 15.5% to 65.3% for the chlorpyrifos treated
seeds. Mean insecticide loadings and between-seed variabilities were within
25% of target in 21 out of the 24 mixtures. Results are therefore presented
on the basis of the initial target doses and %CV.

Field performance

Analysis of variance of the numbers of roots harvested from each
treatment (Table 4) confirmed that there were no significant differences

TABLE 4. The mean numbers of radish roots at harvest.

Target dose (g AI/kg)	Target CV (%)	Mean number of roots Chlorfenvinphos	Chlorpyrifos	Untreated
Untreated				91
3	15	89	89	
3	25	94	98	
3	38	87	93	
3	50	94	95	
6	15	86	92	
6	25	88	92	
6	38	87	92	
6	50	84	93	
12	15	86	92	
12	25	91	89	
12	38	89	92	
12	50	86	92	

p = 0.05) in crop-stand between any of the treatments and the untreated radish. There was therefore no evidence of phytotoxic effects even at the largest dose of the more phytotoxic chlorfenvinphos (Thompson *et al.*, 1980).

There was a high level of insect infestation with the proportion of attacked roots in the untreated row in each plot ranging from 62-90% at 2.5 cm spacing up to 69-95% at 10 cm spacing and the random nature of this attack between untreated subplots suggested that they may not be a reliable estimation of the level of attack within a plot. Data from untreated subplots was therefore not used as a baseline when comparing data from the treated subplots. Nevertheless, the split plot experimental design meant that comparisons between levels of treatment uniformity were the least likely to be affected by different levels of insect attack across the experiment. The percentage of unattacked roots at the different doses, uniformities and spacing are shown in Figure 1. Analyses of variance using an angular trans-

TABLE 5. Comparison of insecticide performance at 3 seed spacings by the estimated percentage reduction in the numbers of cabbage root fly larvae.

	Estimated % reduction in numbers of larvae (± sd)		
Seed spacing (cm)	2.5	5.0	10
Insecticide g AI/kg			
chlorfenvinphos 3	81.9 (6.8)	78.6 (8.7)	70.4 (10.2)
chlorfenvinphos 6	92.3 (5.3)	90.6 (6.0)	87.5 (9.5)
chlorfenvinphos 12	92.9 (5.1)	95.8 (6.0)	96.6 (2.8)
chlorpyrifos 3	75.2 (14.2)	74.8 (15.1)	58.9 (7.5)
chlorpyrifos 6	80.3 (11.0)	80.6 (12.9)	71.5 (10.2)
chlorpyrifos 12	92.9 (5.5)	85.4 (6.7)	79.0 (10.4)

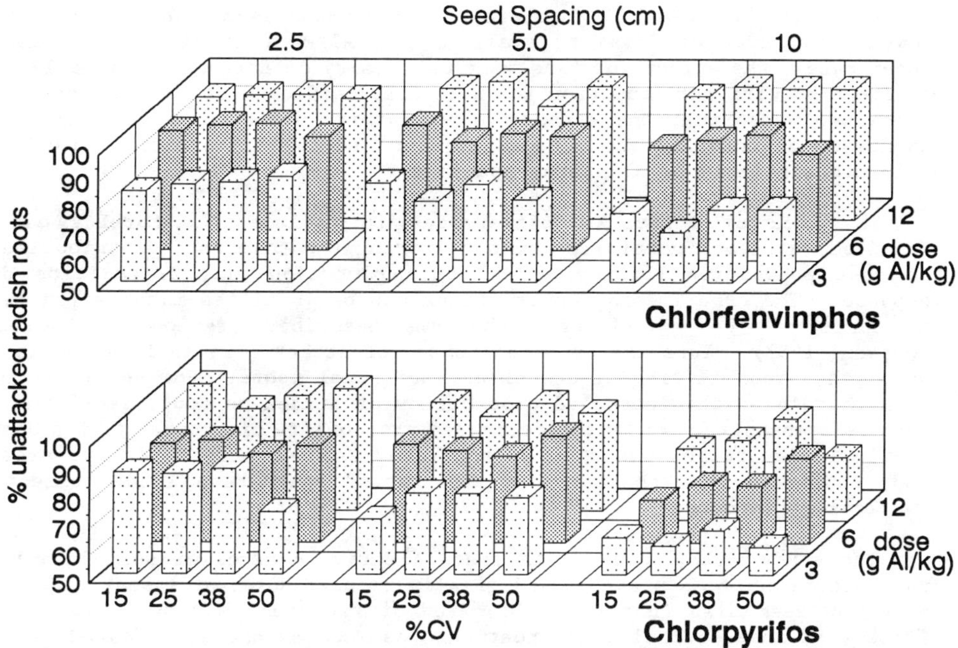

Figure 1.   The % of unattacked radish roots at 3 seed spacings, 3 doses and 4 levels of seed treatment uniformity.

formation on the percentage of attacked roots in each subplot showed that, within each spacing, chlorfenvinphos performed significantly better than chlorpyrifos ($p < 0.05$) and that larval attack decreased with increasing dose. However, there was no effect of, nor interaction with, treatment uniformity ($p = 0.05$).

For each plant spacing, the mean of all the untreated plots was used as a baseline in the estimation of the percentage reduction in the numbers of cabbage root fly larvae (Wheatley & Freeman, 1982) by the various treatments measured over all the uniformity treatments and the results are presented in Table 5.   All treatments (except the smallest dose of chlorpyrifos at 10 cm spacing) reduced the numbers of larvae by more than 70% and chlorfenvinphos at 6 and 12 g AI/kg and chlorpyrifos at 12 g AI/kg (except at 10 cm spacing) reduced numbers of larvae by more than 85%.

DISCUSSION

The study supported the earlier suggestion (Suett and Maude, 1988) that large variations in insecticide loadings on individual seeds have relatively little effect on the mean efficacy of a treatment against cabbage root fly. Although the performance of both insecticides was influenced significantly by the mean dose level, as well as by plant spacing, performances within each of these variables were unaffected by the uniformity of dosing within each mean dose level.   It had been suggested that the failure of the initial study (Suett & Maude, 1988) to reveal differences in performance resulting from treatment uniformity differences might be ascribed to the similarly high efficacies of the 15 and 30 g chlorpyrifos/kg doses used. The present study showed that, even at doses as small as 3 g AI/kg (<1% of the recommended linear rate), the effects of treatment uniformity differences were negligible despite a significant reduction in the overall efficacy of the mean dose. It was evident also that the effects of uniformity differences were not being suppressed by interactions between adjacent seeds, since mean uniformities performed similarly at all three spacings.

The lack of effect of differences in treatment uniformity can be explained, at least partially, by considering the tolerance distribution of the pest population (Finney, 1971) in conjunction with insecticide dose response. The experimental observations can be justified mathematically on the basis of the underlying mechanisms described previously (Phelps & Thompson, 1983). Thus the expected number of attacked roots in a row is the sum of the probabilities of attack of individual roots. For any given mean dose, therefore, the increased probability of attack on roots receiving low doses is compensated for by the decreased probability of attack at high doses. Numerical simulations based on appropriate response curves for the present study confirmed that the range of CVs tested would have little effect on practical performance.

The study showed once again that effective control of soil-inhabiting insect pests can be achieved using greatly-reduced doses of insecticides. It would now seem also that non-uniformity of treatment may not threaten the efficacy of insecticidal seed treatments as much as had been feared.

ACKNOWLEDGEMENTS

The authors are grateful to Mr G. Petch for seed treatment. The study was supported by the Ministry of Agriculture, Fisheries and Food. Mr S.J. Cooke was funded by a Link Scheme established between the Agricultural & Food Research Council and the University of Birmingham.

REFERENCES

Finney, D.J. (1971) Probit Analysis (3rd Edition) Cambridge University Press.
Maude, R.B.; Suett, D.L. (1986) Application of pesticide to brassica seeds using a film-coating technique. *1986 British Crop Protection Conference - Pests and Diseases* 1, 237-242
Phelps, K.; Thompson, A.R. (1983) Methods of analysing data to compare effects on populations of carrot fly (*Psila rosae*) larvae on carrots of treatments with exponentially-increasing doses of granular insecticide formulations. *Annals of Applied Biology*, 103, 191-200.
Suett, D.L.; Maude, R.B. (1988) Some factors influencing the uniformity of film-coated seed treatments and their implications for biological performance. In: *Application to Seeds and Soil, BCPC Monograph No. 39.* T.J. Martin (Ed.), Thornton Heath: British Crop Protection Council, pp. 25-32.
Thompson, A.R.; Suett, D.L.; Percivall, A.L. (1980) Protection of radish from cabbage root fly damage by seed treatments with organophosphorus and carbamate insecticides. *Annals of Applied Biology*, 94, 1-10
Wheatley, G.A.; Freeman, G.H. (1982) A method of using proportions of undamaged carrots or parsnips to estimate the relative population densities of carrot fly (*Psila rosae*) larvae, and its practical application. *Annals of Applied Biology*, 100, 229-244.

THE LIFE CYCLE OF A SEED TREATMENT FORMULATION DEVELOPMENT

L. R. Frank

Ciba-Geigy Ltd., Product Development, Seed Treatment Formulations,
CH-4333 Münchwilen; Switzerland

ABSTRACT

In addition to a short introduction to the development of seed treatment formulations, the life cycle of an active ingredient and its formulations are presented. The development cycle, starting from an early stage of screening up to market introduction, is discussed using a new seed treatment product as an example. Interactions between departments involved in this development, such as process development, analytical development, toxicology, registration, sales, product development and relationships to third parties, is illustrated.

INTRODUCTION

The history of formulation development for seed treatment is strongly influenced by the development of new technologies and attempts to increase user safety. In addition to presenting the benefits of seed treatment, and the special characteristics of seed treatment formulations, the paper will put special emphasis on the factors during the life-cycle, which could have an unpredictable influence on the progress of the development of a seed treatment formulation. Furthermore, an estimation of the time schedules for the development of an active ingredient for seed treatment, from discovery up to first sales, will be illustrated.

HISTORY AND BENEFITS OF SEED TREATMENT

Since ancient times, plant protection agents have been used to control pests which attack crops, stored foods, domestic livestock, arable lands, domiciles, and the human body. Homer (c.1000 B.C.) refers in the Odyssey to the "pest-averting sulfur" and Cato (c.200 B.C.) describes the boiling of bitumen (asphalt) to produce insecticidal fumes which controlled pests in vineyards. The ancient Egyptians employed fumigation to reduce infestation in stored grains. From the beginning of agriculture, farmers were always looking for methods to protect the seeds for the crop.

For many years, it was common practice to use mercury compounds, but this has since been banned, due to the ecological and toxicological risks. In the last few years, seed treatment has become an important segment of the plant protection market, and many agrochemical companies started to set up development groups, with special emphasis on the development of formulations and application technologies for seed treatment uses.

In comparison with conventional chemical treatments, seed treatment offers many advantages. Whereas in other methods of controlling seedborne diseases, large amounts of products had to be sprayed or otherwise applied, seed treatment makes it possible to reduce the amount of active ingredient which is applied. With new methods, the required amount of active ingredient can be reduced by up to 90%, compared with conventional application.

Being in close contact to the seed kernel itself, the active ingredient is located directly at the place where it is needed and forms a type of 'survival bag' for the young seedling. This can control early season diseases and soil pests, as well as seedborne infections, and may allow the farmer to reduce the number of foliar sprays. Last but not least, due to sophisticated new formulation technologies and new methods of application, the risk of the farmer being exposed to products is reduced, as well as the risks for contamination of the environment. Seed treatment can clearly be considered as a very safe method for man, as well as for nature.

<u>Seed Treatment Formulation Types</u>

With the increased need for safe formulations, formulation types and the methods for application for seed treatment, have become more and more sophisticated.

The first and oldest types of formulations for seed treatment were all powder based formulations. Dust formulations for seed treatment (DS) are the most common types of classical formulations. They consist of an active ingredient and some inorganic carriers. The application of these dusts is very inconvenient for the user and the quality of the treatment is poor, because much of the product is lost during the process of application. This type of formulation is still of importance in some less developed countries, where modern application machinery is too expensive, or not available. The first step towards safer formulation types was the development of wettable powders for seed treatment (WS), where the addition of special additives made the preparation of water based slurries possible. The dust is reduced and the application properties improved, resulting in less abrasion, and significantly less risk of contamination for the operator.

Modern formulations are all liquid - mainly water based - formulations, although the oldest types are true solutions (LS). In this case, the active ingredient is dissolved, for example in water. Whenever possible, organic solvents are avoided due to the potential risk of phytotoxicity to the seeds. In cool climates, where cold stability could be a problem, solvents are chosen very carefully, in line with ecological safety aspects.

Emulsions (ES), which contain respectively small liquid droplets, flowables (FS), which contain finely ground solid particles, and capsule suspensions (CS), which contain small polymer balls, filled with active ingredient, are all stabilised by the same special surfactant systems. Emulsions are prepared by using special mixing equipment, where knives, rotating at a very high r.p.m., produce small particles. Capsule Suspensions are formed by a chemical reaction at the interface of small emulsion droplets of liquid or dissolved active ingredient, which are formed in a preliminary step, where a fine membrane, forming the wall of the polymer balls, is produced by interfacial-polymerisation with the surrounding media. By adjusting the ratio of the polymerisation components, the thickness of the polymer walls, and thus the release characteristics, can be varied.
By encapsulation of furathiocarb for example, the toxicity of the formulation PROMET® 400 CS can be reduced, in comparison to that of the active ingredient, from an LD_{50} of 30 mg/kg to an LD_{50} of 3000 mg/kg (rat) - significantly increasing the safety for the user.

The formulation to be chosen for a particular active ingredient is a function of the physical properties of the active ingredient, as well as a function of its uses. Different application machinery used for various types of seeds may neccesarily require different types of formulations, or at least different formulation properties.

THE BOTTLENECK IN DEVELOPMENT

The screening for potential active ingredients for fungicides and insecticides for seed treatment starts at a very early stage in research, where thousands of new compounds are screened, with the aid of several standard tests, for their activities against the main diseases and pests. Only a small proportion show sufficient potential activity to proceed to the next stage, where a closer study is made of the new compounds. After a second evaluation, active ingredients are selected for testing in the field. Only a few dozen are left at this stage, about which time, chemical research is looking for the first routes of active ingredient production, and special research formulations are prepared for the first time. At the next stage of development, chemical production processes are developed and optimised.

Only a handful of active ingredients a year reach the final development stage, and even fewer of these are released eventually for sales. At this time, a formulation and packaging concept is established. After collecting enough data, registration is initiated. Complete product dossiers are submitted to the competent authorities, which include studies on toxicity, environmental behaviour, safety and ecological aspects. To save development time, different investigations have to be carried out simultaneously. For example, formulation development may have to commence before the manufacturing procedure for the chemical itself has been optimised fully.

TIME SCHEDULES

The complete development of a new product takes about ten years. During this period, work is carried out simultaneously in research and development, analytics, chemistry and product development. Toxicological studies and registration procedures are also ongoing. These studies and registration are largely independent of product development <u>but</u> tox and registration could have a profound influence on the life-time of the product.

<u>Factors which affect formulation</u>

Adverse ecological behaviour, toxicological results and sometimes formulation issues may influence, or even cause a sudden halt in development. But development is not always stopped - sometimes formulations may have an effect on this decision. There are possibilities - as mentioned before in the LD_{50}-reduction of the PROMET® formulation, where, by using adequate formulation technology, the toxicological effects could be minimised. This entails experiments with special formulation technologies, such as encapsulation.

If changes in manufacturing procedures of the active ingredient become necessary, due to environmental or cost aspects, these might cause problems in formulation development. Even slight changes in the specification, such as the spectra of by-products, might cause incompatibilities with special surfactant systems, making the development of new formulation variants necessary.

Last but not least, changes in the quality of the components used in the different formulations, might be an additional source of problems in the life-cycle of a formulation. If, for example, a manufacturer of a special additive decides to change the specification of one of his products, this might lead to difficulties in formulation development. It is even more complicated: some suppliers will not disclose the composition of their surfactants. If the supplier then subsequently decides to change, either the composition or the production process, difficulties in product development can occur.

<u>The fenpiclonil example</u>

An example of the interactions between chemical and product development, as well as the relationship to suppliers of auxiliaries, has been the development of fenpiclonil - BERET® - flowables. A new chemical class of fungicides, the phenylpyrroles, was introduced for seed treatment, using new and sophisticated chemical processes, as well as modern formulation types.

Continuous improvements of the chemical process, resulting from the learning progress in handling that new class of fungicides, led to an active ingredient with very high purity. In the meantime, product development planned to produce several tons of formulation. A scale-up production in the pilot plant using the new AI quality, led to severe thickening of the flowable. An intensive testing program revealed that the active ingredient, which had unexpectedly high purity, interacted with the antifoam agent, and this caused the thickening. By changing the antifoam agent, re-testing the physico/chemical stability and re-testing the application properties of the product, as well as supplying the necessary documents, the product could be produced on time. Only application properties had changed slightly, due to the use of a minor amount of additive (<0.5 %).
Finally, the manufacturer of the new antifoam agent guaranteed a shelf life of only 6 months for his product. When out-of-date antifoam-agent was used in manufacturing, separate small additive droplets appeared on the surface of the formulation. This was not well received by the customers. This problem was solved by changes in the specifications of the additive and careful checking-procedures of the new antifoam batch.
A second adaptation in the manufacturing process of the active ingredient, which allowed recycling of the organic solvent used as medium of reaction, had no negative side effects in formulation development.

CONCLUSIONS

The above examples clearly demonstrate factors for success and the necessity for total communication of all involved parties, during the whole development life-cycle of a product, in order to achieve a fast and successful market introduction.

ACKNOWLEDGEMENTS

I would like to thank my co-workers, Mr. René Bircher, Robert Wullschleger and Miss Christina Bühlmann for their enthusiasm during the past few years.
Without their efforts, many developments would not have been possible.

TEMPERATURE EFFECTS ON OSMOTIC PRIMING OF LEEK SEEDS

W. BUJALSKI, A.W. NIENOW

The SERC Centre for Biochemical Engineering, School of Chemical Engineering, The University of Birmingham, Edgbaston, Birmingham B15 2TT, UK

R.B. MAUDE, D. GRAY

Horticulture Research International, Wellesbourne, Warwick CV35 9EF, UK

ABSTRACT

Leek seeds were osmotically primed in polyethylene glycol solutions in priming bioreactors sparged with air. Combinations of a range of temperatures from 5 to 30°C in steps of 5°C and priming durations from 5 to 30 days were used to give 6 treatments each of thermal time of 150 day-degrees. During the treatment samples of seeds were removed at about 30 day-degrees intervals including a final sample of fully primed seeds. These samples were surface dried only and were assessed by a germination test. The reminder of the fully primed seeds were dried in thin layers for 2 days in an air flow at 15°C and then tested for germination. These seeds were retested after 3 months in store.

There were no differences in percentage germination between the different temperature/time treatments or compared with untreated seeds. However, only seeds treated at temperatures lower than 20°C showed significant improvements in terms of reduced germination time; the range of temperatures from 5 to 15°C gave the best priming performance which was almost equal for both surface dried and dried seeds, with the 15°C treatment being the only one to reduce the spread of germination times. These effects were retained in the dried seeds during a storage period of 3 months but the proportion of abnormal seedlings increased significantly, especially for the longer treatments.

INTRODUCTION

Osmotic priming of seeds in polyethylene glycol (PEG 6000; -1.5 MPa) solutions increases the rate and synchrony of germination (Heydecker, 1977; Heydecker and Coolbear,1977) and may also improve emergence characteristics, plant fresh weight and final yield for a number of species (Bradford, 1986). The literature on priming (Heydecker and Coolbear,1977; Bradford, 1986) mainly reports the effects of priming and of drying seeds using small scale, laboratory methods. These include priming in Petri dishes, on filter paper moistened with suitable PEG (or salt) solution, with thin-layer drying in near-ambient air temperatures, either on the bench in an open environment or in temperature-controlled drying cabinets. However, recently there has been significant development of methods for priming and drying large quantities of seeds (Nienow and Brocklehurst, 1987; Nienow et al, 1991;

Bujalski and Nienow, 1991; 1992; Bujalski et al., 1991b; 1992). The effects of priming in different bioreactors and drying in various ways on the storage potential of leek seeds following conventional storage have also been assessed (Maude et al., 1993a, b).

This paper describes the priming of leek (*Allium porrum* L.) seeds in stirred bioreactors and the influence of the temperature and duration of priming on the intermediate and final treatment effects. All treatments were subjected to priming temperature and duration equivalents of 150 degree-days (0°C base). This approach was adopted on the basis of analogy with the response of seed germination to constant heat sum (also called thermal time) (Bierhuizen and Wagenvoort, 1974) and to constant temperature and osmotic potential i.e. the hydrothermal time to germination concepts (Gummerson, 1986). 150 day-degrees was selected because of the success of priming carried out for 10 days at 15°C (Gray et al., 1991) and because such heat sum constitutes only about 70% of the amount of a thermal time found to be necessary for leek seed germination at constant temperature (Bierhuizen and Wagenvoort, 1974), thus preventing seed from undesired germination during the priming process.

MATERIALS AND METHODS

Seeds and Storage

Leek seed cv Verina raised at Luddington EHS in 1988 was used. Seeds were kept in a seed store at 10°C and 40% r.h. prior to use and after completion of the treatments.

Germination tests and seedling assessments

In all experiments, four replicates, each of 50 seeds from each treatment, were laid out as a randomised block on a Copenhagen tank at 15°C under 16 h light/8 h dark regime. Germination was counted daily for 14 days then seedlings were removed and assessed under a magnifying lens for normal and abnormal development (Beckendam and Grob, 1979).

Statistical analysis of the data

Germination and presence of abnormal seedlings were calculated as a percentage of the number of seeds sown. Spread of germination times was expressed as log variance of time to germination. The data were subjected to analysis of variance and, where appropriate, the angular transformation of the data was used for the analysis.

Priming

Seeds (140 g) were osmotically primed in 1.4 l PEG 6000 solutions (at concentration of 100 g seed/l) in bioreactors sparged with 700 ml air/min. The PEG concentration was adjusted to give solutions of -1.5 MPa osmotic potentials (Michel, 1983) the respective treatment temperature. A range of temperatures (5 to 30°C in steps of 5°C) and priming durations (5 to 30 days), giving the thermal time of 150 day-degrees, was tested. Seeds were kept in suspension by the action of a double impeller system (bottom impeller pumping upwards and the upper one downwards) at 400 rpm. This geometry (baffled vessel, two impellers of diameter equal to half of the vessel diameter, pumping in the opposite directions, variable speed motor capable of running in either direction) gives the greatest flexibility for seed suspension (Chapman et al., 1983) and also good results for air dispersion (Kuboi and Nienow, 1982). The dissolved oxygen level in the priming liquid was kept above 70% saturation throughout the treatment periods, thus ensuring consistent priming responses (Bujalski et al., 1991a; 1993). Samples of about 300 seeds were removed at approximately 30 day-degree intervals throughout the treatment period with a final sample taken when the seeds were fully primed. All of these were surface dried only and were assessed immediately for germination. The reminder of fully primed samples (120 g each) were dried in thin layers for 2 days in an air flow at 15°C to their original moisture content. These seeds were then tested for germination and the test was repeated after the seeds had been kept for three months in a seed store.

RESULTS AND DISCUSSION

There were no differences in percentage germination between comparable temperature-time treatments at either the intermediate (data not shown) or final sample assessments. Only seeds treated at 30°C for five days and dried showed a significant (P<0.05) decrease in germination and the three months storage had no further effect (Table 1).

TABLE 1. Effects of thermal time of 150 day-degrees during priming treatments at different temperature/time combinations on dried seeds germinated immediately and after three months storage.

Priming treatments	Tested					
	Immediately after priming and drying			After 3 months storage		
	%G	Spread (hrs)	%A	%G	Spread (hrs)	%A
Untreated seeds	97(82)	6.65	11	97(82)	6.65	11
30°C for 5 days	91(72)	6.47	13	91(73)	6.51	13
25°C for 6 days	95(77)	6.79	12	97(81)	6.32	9
20°C for 7.5 days	95(78)	6.67	13	94(76)	6.42	24
Untreated seeds	96(78)	6.34	15	96(78)	6.34	15
15°C for 10 days	95(77)	5.35	20	97(82)	5.69	27
10°C for 15 days	95(79)	6.24	21	95(77)	5.95	27
5°C for 30 days	94(76)	6.93	24	94(75)	5.87	33
LSD (P<0.05)	3.9(6.3)	0.532	5.7	3.9(6.3)	0.532	5.7

%G = % germination; Spread = log (variance of germination times); %A = % abnormals; (angularly transformed data in brackets).

The development of the response in terms of reducing the mean germination time due to priming at different temperatures as a function of thermal time is shown in Figure 1.

Treatments carried out at 25 and 30°C produced almost no priming effect, the reduction in the mean germination time being accounted for by the fact that these samples which were surface dried only had the advantage over the controls of being almost fully imbibed when set for the germination test.

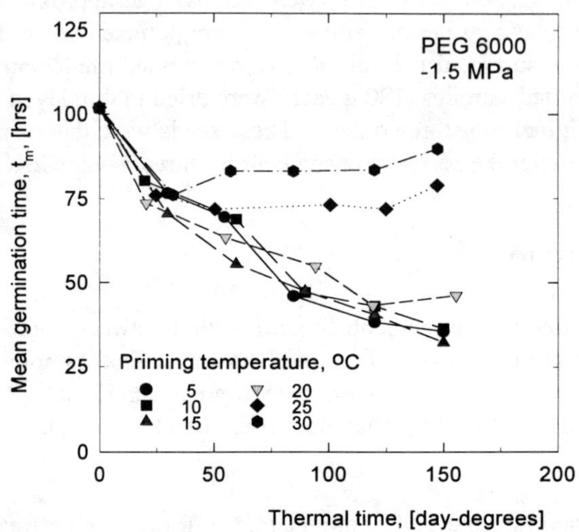

FIGURE 1. Priming responses of leek seeds vs. thermal time.

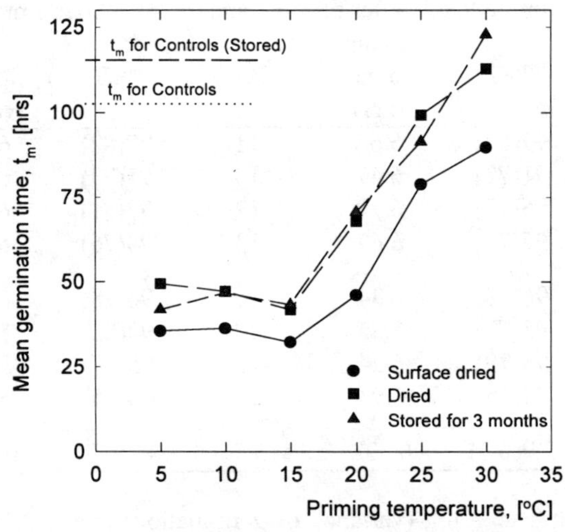

FIGURE 2. Responses for primed seeds either surface dried, dried
and dried and stored.

There were significant (P<0.001) differences between treatments in mean germination times for fully primed seeds, ranging from about 34 hrs for seeds treated at 5 to 15°C and surface dried to 112 hrs for those primed at 30°C and dried (Figure 2). Similar effects were obtained after three months storage (Figure 2). The spread of germination times (log variance values) was significantly reduced only for seeds fully primed at 15°C which was 4.66 when surface dried (data not shown) and this effect was retained after drying and storage (Table 1). The long treatments at 10°C and 5°C showed significant increases in abnormal seedlings after drying and also further increases after the storage period. There was also a small increase after storage for the 15°C and 20°C treatments (Table 1).

CONCLUSIONS

The reduction in mean germination time by priming was constant and greatest for equivalent thermal time treatments at temperatures less than 20°C, with the overall optimum response at a temperature of 15°C due to the reduction in the spread of the germination times. However, all these treatments also resulted in an increased proportion of abnormal seedlings, especially after storage. The longer the treatment period, the more severe was the deterioration. The effect of storage accords well with earlier studies (Maude et al., 1993a, b). However, the problem here appeared more severe and occured without storage for the longer treatments, possibly because of the prolonged shear and mechanical forces of the impellers acting on the seeds.

Clearly, though seeds can be successfully primed in bulk, prolonged low temperature treatments can be deleterious and the effect also magnifies the reduction in storage life. On the other hand, high temperatures do not lead to effective priming. This work suggests that there is an optimum working temperature for bulk seed priming.

ACKNOWLEDGEMENTS

This work was supported by an AFRC Grant for the Linked Scheme between The School of Chemical Engineering, The University of Birmingham and Horticulture Research International, Wellesbourne. We also thank Mr D. Ward for technical assistance and Mrs K. Phelps for statistical advice.

REFERENCES

Beckendam, J.; Grob, R. (1979) Handbook for seedling evaluation. *The International Seed Testing Association*, Zurich, Switzerland, 2nd Edition, pp. 130.

Bierhuizen, J. F.; Wagenvoort, W. A. (1974) Some aspects of field germination in vegetables. 1. The determination and application of heat sums and minimum temperature for germination. *Seed Science and Technology*, **2**, 213-219.

Bradford, K. J. (1986) Manipulation of seed water relations via osmotic priming to improve germination under stress conditions. *HortScience*, **21**, 1105-1112.

Bujalski, W.; Nienow, A. W. (1991) The large-scale osmotic priming of onion seeds: A comparison of different strategies for oxygenation. *Scientia Horticulturae*, **46**, 13-24.

Bujalski, W.; Nienow, A. W. (1992) An evaluation of the design and operation of bioreactors for seed priming. *The 1992 IChemE Research Event, 9-10 January 1992, UMIST, IChemE Rugby*, UK, 184-186.

Bujalski, W.; Nienow, A. W.; Maude, R. B.; Gray, D. (1993) Priming responses of leek (*Allium porrum* L.) seeds to different dissolved oxygen levels in the osmoticum. *Annals of Applied Biology*, **122**, 569-577.

Bujalski, W.; Nienow, A. W.; Petch, G. M. (1991a) The bulk priming of leek seeds - the influence of oxygen enriched air. *Process Biochemistry*, **26**(2), 281-286.

Bujalski, W.; Nienow, A. W.; Petch, G. M.; Drew, R. L. K.; Maude, R. B. (1992) The process engineering of leek seeds: a feasibility study. *Seed Science and Technology*, **20**, 129-139.

Bujalski, W.; Nienow, A. W.; Petch, G. M.; Gray, D. (1991b) Scale-up studies for osmotic priming and drying of carrot seeds. *Journal of Agricultural Engineering Research*, **48**, 287-302.

Chapman, C. M.; Nienow, A. W.; Cooke, M.; Middleton, J. C. (1983) Particle-gas-liquid mixing in stirred vessels. Part III: Three phase mixing. *Chemical Engineering Research and Design*, **61**, 167-181.

Gray, D.; Drew, R. L. K.; Bujalski, W.; Nienow, A. W. (1991) Comparison of polyethylene glycol polymers, betaine and L-proline for priming vegetable seed. *Seed Science and Technology*, **19**, 581-590.

Gummerson, R. J. (1986) The effect of constant temperature and osmotic potential on the germination of sugar beet. *Journal of Experimental Botany*, **37**, 729-741.

Heydecker, W. (1977) Stress and seed germination: an agronomic view. In: *The Physiology and Biochemistry of Seed Dormancy and Germination.* Khan, A. A. (Ed.), Amsterdam, Elsevier, , pp. 237-282.

Heydecker W.; Coolbear, P. (1977) Seed treatments for improved performance - survey and attempted prognosis. *Seed Science and Technology*, **5**, 353-425.

Kuboi, R.; Nienow, A. W. (1982) The power drawn by dual impeller systems under gassed and ungassed conditions. *Proceedings of the 4th European Conference on Mixing*, 27-29 April 1982, Delft University of Technology, Holland, BHRA Fluid Engineering, Cranfield, Paper G2, 247-261.

Maude, R. B.; Drew, R. L. K.; Gray, D.; Bujalski, W.; Nienow, A. W. (1993a) The effect of storage on the performance of leek seeds primed and dried by different methods. I. Development of special techniques for the assessment of abnormal seedlings. *Seed Science and Technology*, (submitted).

Maude, R. B.; Drew, R. L. K.; Gray, D.; Bujalski, W.; Nienow, A. W. (1993b) The effect of storage on the performance of leek seeds primed and dried by different methods. II. Application of special assessment technique for comparison of treatments effects. *Seed Science and Technology*, (submitted).

Michel, B. E. (1983) Evaluation of the water potential of solutions of polyethylene glycol 8000 in the absence and presence of other solutes. *Plant Physiology*, **72**, 66-70.

Nienow, A. W.; Brocklehurst, P. A. (1987) Seed preparation for rapid germination - engineering studies. In: Moody, G. W. and Baker, P. B. (eds) *Bioreactors and Biotransformations*, London, Elsevier, pp. 52-63.

Nienow, A. W.; Bujalski, W.; Petch, G. M.; Gray, D.; Drew, R. L. K. (1991) Bulk priming and drying of leek seeds: the effects of two polymers of polyethylene glycol and fluidised bed drying. *Seed Science and Technology*, **19**(1), 107-116.

THE PHYSIOLOGICAL ADVANCEMENT OF SUGAR-BEET SEEDS

T.H. THOMAS, K.W. JAGGARD, M.J. DURRANT, S.J. MASH

AFRC Institute of Arable Crops Research, Broom's Barn Experimental Station, Higham, Bury St Edmunds, Suffolk IP28 6NP

ABSTRACT

Since sugar-beet yields are limited by the amount of solar radiation intercepted, particularly during May and June, treatments to stimulate earlier seedling emergence should be advantageous. However, earliness of sowing is limited by exposure to cool weather which vernalizes the plants and induces 'bolting'. A treatment which both advances and devernalizes seeds has been developed which allows sowing to be brought forward safely by about 10 days. Annual experiments since 1988 show that sugar yield was increased by 0.19 t/ha as a consequence of seed advancement giving more rapid emergence and by 0.042 t/ha/day as a result of earlier sowing, In practice, the improvement which was produced by advancing and devernalizing beet seeds was c 0.35 t/ha, about 5% of the average national yield. A 3-year study indicated that advanced seeds can be stored for at least 3 years without any major reduction in their performance.

INTRODUCTION

Two main factors determine the earliest date at which the sugar-beet seeds should be sown in the UK. Firstly, if seeds are sown too early the developing seedlings can be exposed to cool temperatures which induce bolting in the growing crop. This creates a weed beet problem and can result in considerable yield loss (Longden, 1991). Data from the 1970s indicated that most bolting resistant varieties should not be sown before 20 March (Jaggard *et al*, 1983) and although current varieties are slightly more bolting-tolerant, this continues to be a guideline for sugar-beet farmers. Secondly, cold, wet conditions in early spring can reduce the proportion of seeds which will produce established plants. These factors combine to give crops in which the ability to intercept solar radiation in May and June is restricted, thus limiting yield. Consequently, seed advancement treatments have been sought which will accelerate seed germination and seedling development without introducing the risk of bolting and flowering in the growing crop. Early work by Longden (1971), based on hardening techniques used on carrot seeds, demonstrated the potential to shorten the period between sowing and germination. Although the effects on yield were not examined, positive responses of developing seedlings indicated the potential benefit for the UK crop if bolting could be controlled. Between 1984 and 1989 seed treatments were developed at Broom's Barn Experimental Station which both devernalized seeds, thereby reducing the number of bolters, and improved seedling establishment (Durrant & Jaggard, 1988).

DEVELOPMENT OF THE SEED ADVANCEMENT TREATMENT

Initial methods of advancement, which involved steeping seeds in water followed by incubation with NaCl solution, caused germination and hypocotyl extension to occur sooner than with either untreated or water-steeped seeds in laboratory tests. Seeds given a 6-day advancement treatment and sown in early March 1984 emerged about 12 days earlier in the field than untreated seeds, plant establishment was higher and bolting was reduced from 33 to 27% (Durrant & Jaggard, 1988). Subsequent experiments testing the effect of seed water content, duration of treatment and incubation temperature showed that, under restricted moisture conditions in the absence of NaCl, seeds could be advanced for at least 5 days at 25°C without deleterious effects on germination or hypocotyl growth (Durrant & Mash, 1990).

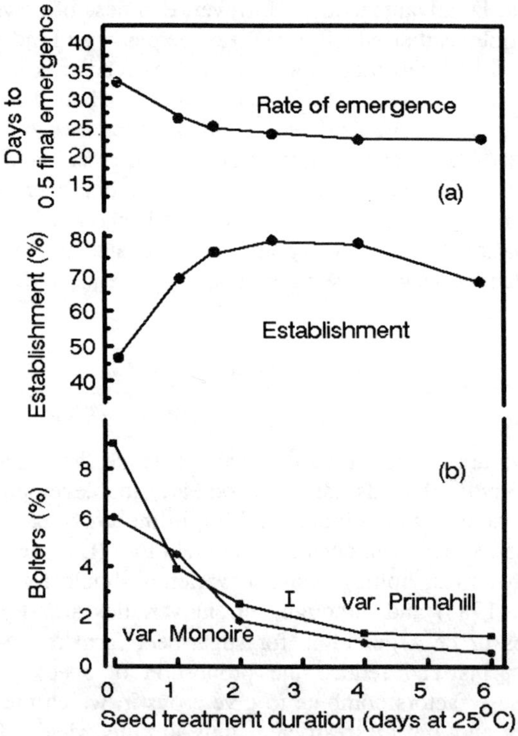

FIGURE 1. Relationship between seed treatment duration at 25°C on a) plant establishment and time required for half of the final number of seedlings to emerge from a 13 March sowing in 1987 and b) the number of bolters on 15 October from a 17 March sowing in 1986. (Durrant & Mash, 1990).

Based on the results of experiments carried out in 1986 and 1987 (Figure 1), an advancement treatment of 4 days was identified as being necessary for maximum benefit in terms of improved emergence and establishment and decreased bolting. The prolonged, high temperature devernalization treatment makes use of the principles of hydro-thermal time (Gummerson, 1986). The seeds are physiologically active during the high temperature treatment but do not germinate because their water content is close to the base water potential for germination.

EXPERIMENTS WITH ADVANCED SEEDS

The advancement treatment used consisted of steeping sugar-beet seeds in an agitated 0.2% (w/v) aqueous suspension of thiram at 25°C for 8 h, partially drying to 124% of the original weight, incubating at 25°C for 88 h and then air-drying (ADV). Control seeds were steeped in thiram for 8 h then air dried (TS). The ratio of seeds to thiram suspension was 1:4 (w/v). The TS treatment is analogous to the current commercial treatment and the ADV treatment has now been slightly modified for commercial testing.

Field experiments were carried out between 1988 and 1991 with seeds of the varieties Amethyst (1988/90) and Celt (1991), pelleted by Germain's (UK) Limited. Over the 4 years, 11 sowings were made (Durrant *et al*, 1993). Overall, seed treatment had little effect on the proportion of seeds giving established plants but establishment increased consistently as sowing was delayed in March. Advancement consistently reduced the time between sowing and emergence; this was about 9 days from early March sowings and between 1 and 4 days when seeds were sown in early April The benefits from advancement showed considerable seasonal variation but, on average, ADV seeds needed 4 fewer days for half the seedlings to emerge than TS seeds. Results from a similar experiment in 1993 are presented in Figure 2 in which, although the advancement treatment provided a maximum benefit of about 3 days earlier emergence from March sowings, the consistency of the advancement effect overall was well demonstrated.

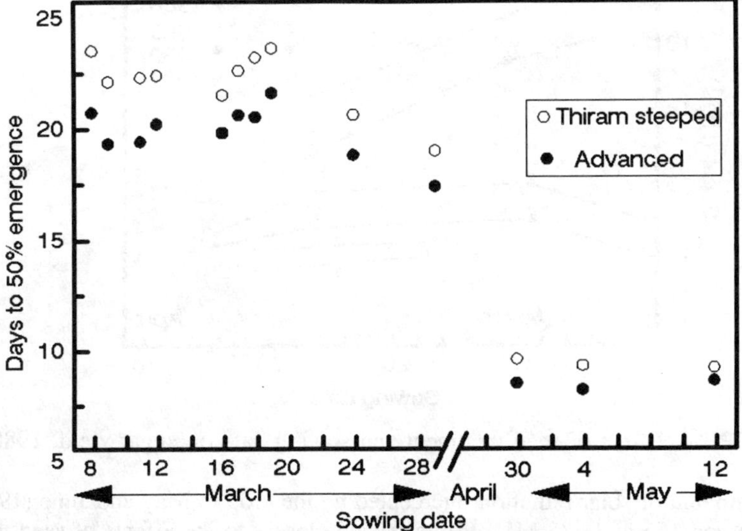

FIGURE 2. The effect of seed advancement on the number of days to 50% emergence from a range of sowing dates in 1993 (data courtesy of Germain's UK Ltd.).

The range of years, seed treatments and sowing dates had large effects on the proportion of land covered by leaves. Seed advancement increased ground cover by between 1% and 6% and the ability to sow advanced seeds earlier without the risk of bolting provided an extra benefit in this respect, as shown for 1990 (Figure 3).

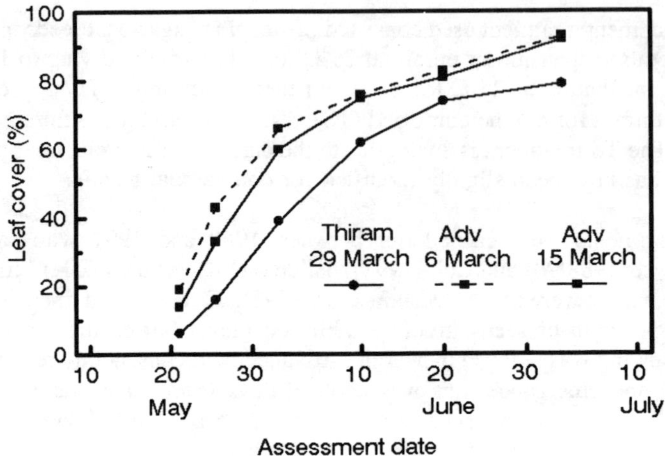

FIGURE 3. Effect of combination of early sowing and seed advancement on leaf cover, 1990

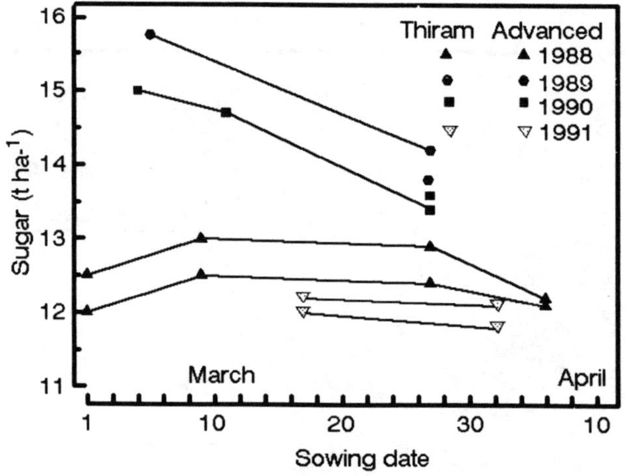

FIGURE 4. Effect of seed treatment and sowing date on sugar yield, 1988-1991.

The amount of solar radiation intercepted by the crop in May and June (1988-1991) ranged between 155 and 485 MJ/m^2; this related closely to the effects of seed treatments and sowing dates on canopy development by the end of May. In 1988, when TS seeds were sown on 2 March or 9 March, 20% and 9% of the plants respectively bolted by late October; this figure was halved by the advancement treatments, In other years, very few plants bolted from any of the seed treatments, presumably because of mild conditions during the spring.

Different relationships between sowing rate and sugar yield occurred (Figure 4). In 1989 and 1990, yield was increased substantially by progressively earlier sowing of advanced seed whereas this achieved little effect in 1988 and 1991. In 7 of the 8 comparisons, yield was larger from ADV than from TS seeds, although in individual comparisons the effect was never large enough to be significant at the 5% probability level.

COMMERCIAL TRIALS

In 1991, seeds of 5 commercial varieties were given scaled-up TS and ADV treatments, pelleted and tested in 8 trials drilled between 4 March and 2 April; the majority were drilled earlier than the UK national crop. As in the small scale trials, ADV seed of all varieties emerged more rapidly than TS seeds at the first two assessment dates, with no major adverse effects being observed at final establishment (Thomas *et al*, 1993). Three replicated trials in 1992 compared advanced and standard seeds drilled earlier than currently recommended as against a normal drilling date. No significant differences in yield were found between ADV and TS seeds at either drilling date but at 2 of the sites sugar yield was increased significantly by earlier drilling. When averaged over the 3 trials there was an increase of approximately one tonne of sugar/ha for crops drilled at least 10 days earlier than normal

STORAGE OF ADVANCED SEED

TS and ADV seeds kept in sealed polythene bags at room temperature were tested for laboratory germination and field emergence after 9 and 36 months of storage (Thomas *et al*, 1993). As an average over the 4 varieties tested, germination remained consistent at 97-98% throughout the study. Seed vigour, as measured by the number of day degrees above 3°C required to produce hypocotyls 2 cm high, was not impaired during the first 2 years of storage. After 9 months of storage ADV seeds emerged faster and seedling establishment was greater than from TS seeds sown in the field on 16 March. However, after 3 years of storage the ADV advantage was smaller, mainly due to decreased performance of 1 of the 4 varieties. In general, ADV seeds required 7 to 15 d°C less than TS seeds for half of the seedling hypocotyls to grow to 2 cm in standard laboratory tests. Results from late spring sowings during the third year of storage showed no significant difference (p = 0.5) in emergence percentage between ADV and TS seeds and all seedlings were normal.

CONCLUSIONS

The potential advantages to be gained from sugar-beet seed advancement are two-fold. Firstly, advancing the seeds to complete part of the germination process before sowing reduces the time for seedlings to emerge as compared with thiram-treated seeds. On average, this effect increased sugar yield by 0.19 t/ha (0.048 t/ha/day). Secondly, 'devernalization' of the seeds allows earlier sowing, so increasing yield for every day gained during March by about 0.035 t/ha/day. The number of days likely to be gained by advancing the earliest safe sowing date from 20 March to about 11 March will depend on whether or not the soil is friable and dry enough to be worked by machinery.

The average sowing for sugar beet in the UK during the 1980s was between 19 March and 4 April. Bringing forward the current recommendation for the earliest safe date to sow to 11 March would, on average, bring the mean date of sowing forward by 9 days, each worth 0.035 t/ha of sugar, a total of 0.315 t/ha. Thus the likely sugar yield improvement due to the advancement treatment is 0.315 t/ha, plus the benefit from advancement *per se* of 0.19 t/ha, an overall increase of 0.5 t ha^{-1}.

Although the combination of the advancement treatment and the improved opportunity to sow the crop early are, on the basis of experimental evidence, likely to be worth an extra 0.5 t/ha of sugar, it is unlikely that all of this will be achieved in practice. In similar situations the yield which arrives at the sugar factory gate is approximately 70% of what would be predicted as present in the middle of the fields (Scott & Jaggard, 1992). Using this correction, the treatment should be worth an extra 0.35 t/ha of sugar to the grower and processor, which is about 5% of the average national yield.

ACKNOWLEDGEMENTS

We thank John Webb and Ray Bugg for help with the experiments, Chris Clark for the crop cover measurements, and Germain's for pelleting the seed. The study was financed from the Sugar Beet Research and Education Fund.

REFERENCES

Durrant, M.J.; Jaggard, K.W. (1988) Sugar-beet seed advancement to increase establishment and decrease bolting. *Journal of Agricultural Science, Cambridge*, **110**, 367-374.

Durrant, M.J.; Mash, S.J. (1990) Sugar-beet seed treatments and early sowing. *Seed Science and Technology*, **18**, 839-850.

Durrant, M.J.; Mash, S.J.; Jaggard, K.W. (1993) Effect of seed advancement and sowing date on establishment, bolting and yield of sugar beet. *Journal of Agricultural Science, Cambridge*. In press.

Gummerson, R.J. (1986) The effect of constant temperatures and osmotic potentials on the germination of sugar beet. *Journal of Experimental Botany*, **37**, 729-741.

Jaggard, K.W.; Wickens, R.; Webb, D.J.; Scott, R.K. (1983) Effects of sowing date on plant establishment and bolting and the influence of these factors on yield of sugar beet. *Journal of Agricultural Science, Cambridge*, **101**, 147-161.

Longden, P.C. (1971) Advanced sugar-beet seed. *Journal of Agricultural Science, Cambridge*, **77**, 43-46.

Longden, P.C. (1991) Weed beet on the wane. *British Sugar Beet Review*, **59**(2), 18-19.

Scott, R.K.; Jaggard, K.W. (1992) Crop growth and weather: can yield forecasts be reliable? *Proceedings of the 55th Winter Congress of the Institut International de Recherches Betteravières*, Brussels, 169-187.

Thomas, T.H.; Jaggard, K.W.; Durrant, M.J.;Mash, S.J.; Armstrong, M.J. (1993) Development of the sugar-beet seed advancement treatment in England. *Proceedings of the 56th Winter Congress of the Institut International de Recherches Betteravières*, Brussels, 437-448.

THE ETHIRIMOL CONTENT OF COMMERCIALLY-TREATED CEREAL SEEDS

D.L. SUETT

Horticulture Research International, Wellesbourne, Warwick CV35 9EF

P.J. HEWETT

126 Thornton Road, Cambridge CB3 0ND

A.A. JUKES, L. MORGAN, V. FOX

Horticulture Research International, Wellesbourne, Warwick CV35 9EF

ABSTRACT

Analysis of >750 samples of treated seed showed that most mean doses of ethirimol were within 20% of target and that, within batches, >90% of samples contained the mean dose ±10%. The results of between-seed variability and loading retention studies are also presented.

INTRODUCTION

More than 3 million ha of wheat and barley are grown currently in the UK and it is estimated that at least 85% of these crops receive a seed treatment prior to sowing (Davis *et al*., 1990). A comprehensive survey of commercial seed treatments more than 20 years ago revealed considerable variations in loading accuracy and uniformity (Lord *et al*., 1971) and prompted the industry to examine treatment efficiencies. However, despite many subsequent modifications to the range of available pesticides and formulations, as well as to application machinery, there has been no major assessment of the impact of these major developments on treatment efficiency.

This deficiency was recognised by the BCPC Seed Treatment Working Party, who initiated recently two 6-month studies of commercial seed treatment quality. Some of the results of these studies are presented in this paper.

METHODS

The studies were done at Horticulture Research International, Wellesbourne, during July - December 1990 and April - September 1992. The treatment selected for study was Ferrax (Zeneca Agrochemicals), a liquid formulation applied to winter and spring barley seeds at a rate of 500 ml product/100 kg seeds and which contains ethirimol (400 g/l), flutriafol (30 g/l) and thiabendazole (10 g/l). Samples were taken from 5 types of treatment machine at 13 plants in different regions of the UK. In 1990, 367 samples were taken from 14 treatment runs. In 1992, 292 samples were obtained from 7 runs and a further 120 samples were taken at a single site from 1-tonne sacks in which treated seeds were being stored prior to bagging. Between-seed treatment variability was also assessed and the physical stabilities of treatments were measured in drop tests and by passing treated seeds through a drill unit.

<u>Seed sampling</u>

Samples from treatment machines were taken either from the stream of seed as it left the treatment chamber or from the tops of filled sacks prior to sealing. Individual sacks on pallets were sampled centrally using a spear. The 1-tonne sacks were sampled by taking four replicate samples from the top, middle and bottom of each sack. The samples, approximately 10 g, were taken and retained in plastic tubes, 80 x 20 mm, which were sealed immediately after sampling. Between-seed variability was determined by analysing 25 individual, weighed seeds from 2 treatment runs in 1990 and from 2 runs in 1992. Drop tests were done in 1990 and 1992 and were based on the procedure of Jeffs (1973); a sample of seeds (200 g) was dropped 100 times down a 40 cm-long plastic tube on to a wire mesh and a 10 g subsample was removed after every 10 drops. The influence of the drilling procedure on loading retention was assessed in 1992 by transferring the contents of a 50 kg sack of treated seed into the hopper of a 25-outlet Ransomes Nordsten Liftomatic Model 300 seed drill and activating the drill. Samples were taken from the sack before opening, from the hopper and from all 25 outlets, 6 of which were selected for analysis.

<u>Analytical methods</u>

It was established at the outset that the objectives of this study would be achieved by determination of only ethirimol, which was the principal component of the formulated product. Ethirimol was analysed by hplc using a Hypersil ODS column. The mobile phase was phosphate buffer (pH 8.0): methanol: acetonitrile (4:3:3) and the eluant was monitored at 220 nm. The sample extractant comprised buffer: methanol: tetrahydrofuran (3:4:3). Samples were weighed then extracted by transferring the contents of each sample tube, with rinsing, to a 200 ml Duran bottle and tumbling with 50 ml extractant for 30 min. Extracts were diluted x 5 with mobile phase solution prior to analysis by hplc. The stock standard solution (1.00 mg ethirimol/ml) was checked against a freshly-prepared stock solution every 4 wk and working standard solutions were shown to be stable for at least 1 wk after dilution. The analytical procedure was monitored by incorporating a control sample into the study, one portion of which was analysed for every 20 samples received. By the end of the 1990 study, 30 control subsamples had been analysed.

RESULTS

Results of the analyses of the 30 control samples analysed during July - December 1990 were highly consistent. Ethirimol levels ranged from 1520 - 1740 mg/kg and, with a CV of 3.5%, were all within ± 8% of the mean loading of 1650 mg/kg.

<u>Samples from machines and sacks</u>

Table 1 summarises the analyses of the batches of treated seeds sampled in 1990 and 1992. In most of the batches the ethirimol contents of the individual samples were highly uniform. Thus, in 14 of the 21 batches, at least 90% of the samples contained the mean dose ± 10% and, in 6 of these, all of the samples were within this dose range. The overall uniformity of treatment was reflected by the coefficients of variation which, with the exception of batches P and V, were all <8%. Batch P was particularly anomalous, with 8 of the 20 samples containing 990 - 1070 mg/kg and the remainder containing 1600 - 1810 mg/kg. The overall mean loadings of the

TABLE 1. Ethirimol loadings on seeds taken from machines or 50 kg sacks

Year + sample code		No of samples	Mean load		Range mg/kg	% CV	Samples containing ± 10% of mean load, %
			mg/kg	% of target			
1990	A	30	1870	93.5	1730-1970	2.7	100
	B	54	1830	91.5	1600-2060	6.0	89
	C	23	1640	82	1480-1750	3.7	100
	D	24	1520	76	1260-1620	5.7	92
	E	29	1780	89	1620-2000	5.7	97
	F	28	1230	61.5	1160-1380	3.7	96
	H	21	1570	78.5	1400-1700	6.2	100
	L	37	1840	92	1700-1950	3.5	100
	M	14	1580	79	1490-1690	4.1	100
	N	36	1420	71	1240-1620	6.4	94
	P	20	1430	71.5	990-1810	24.5	0
	S	22	1550	77.5	1350-1790	7.2	73
	T	12	1610	80.5	1520-1690	2.9	100
	V	17	1550	77.5	1300-1950	10.2	73
1992	BB	27	2040	102	1710-2230	5.8	97
	CC	60	1630	81.5	1340-1860	7.5	91
	DD	27	1930	96.5	1350-2130	7.8	85
	EE	31	1770	88.5	1390-2070	7.9	83
	FF	83	1720	86	1550-1920	6.1	96
	GG	29	1770	88.5	1400-2080	7.4	82
	HH	35	1950	97.5	1770-2240	6.8	90

individual batches ranged from 1230 - 2040 mg/kg. BB was the only batch in which the mean dose exceeded the target dose for ethirimol of 2000 mg/kg and a further 5 batches had mean doses within 90% of target. Twelve of the remaining 15 batches contained 75 - 90% of the target dose.

Samples from 1-tonne sacks

The overall mean ethirimol loading on the 120 samples from the ten 1-tonne sacks was 1970 mg/kg, equivalent to 98.5% of target. In individual sacks, mean doses ranged from 94.5 - 113% of target. The mean loading of 2030 mg/kg on samples from the tops of sacks were significantly greater (p <0.01) than those in the middle and bottom of the sacks (1940 and 1930 mg/kg respectively).

Between-seed variability

The results of the analyses of single seeds (Table 2) are expressed as mg/kg and are based on the mean 100-seed weight determined for each batch. The weights of individual seeds ranged from 28 - 67 mg and there was a 2.0 - 2.4-fold weight range within each batch, the coefficient of variation ranging from 16 - 21%. Ethirimol loadings on individual seeds were much more variable, with coefficients of variation ranging from 29 - 45%. There were 3 - 7-fold ranges of dose within batches, with some seeds coated with the equivalent of almost 3 times the target dose. Nevertheless, in all batches approximately 90% of the seeds held 50 - 150% of target.

TABLE 2. Mass and ethirimol content of individual seeds analysed in 1990 and 1992. Each sample comprised 25 single seeds

Year	Sample	Seed mass, mg			Ethirimol content, mg/kg		
		Mean	Range	% CV	Mean	Range	% CV
1990	1	51.3	33.2-66.4	16.5	1670	860-6120	45.6
	2	50.9	33.8-66.6	16.0	1840	1040-5400	41.8
1992	3	51.3	28.3-67.4	18.9	2080	1100-3790	28.7
	4	43.7	24.8-60.9	21.1	2470	1080-5950	39.6

Physical stability

Data from the drop tests done in 1990 and 1992 are shown in Figure 1 as proportions of the initial loadings of 1420 and 2170 mg/kg (1990 and 1992 respectively). The results from the two tests were similar, with both of the initial loadings declining by 10, 20 and 30% after 10, 30 and 60 drops respectively. Subsequent losses were small and, after 100 drops, approximately 65% of the initial doses remained.

Changes in seed loadings at different stages of the drilling procedure are shown in Table 3. There were no significant differences (p=0.05) in the mean loadings of seeds taken from the top, middle and bottom of a single 50 kg sack. Although the overall mean dose declined by 3% following transfer from the sack to the seed hopper and by a further 4.8% between the hopper and the outlets of the drill unit, neither of these single-step differences were significant at the 95% confidence level. However, the overall decline of the initial mean dose from 1950 mg/kg to a mean of 1800 mg/kg on the seeds emerging from the drill outlets was significant at this level of confidence.

TABLE 3. Ethirimol loadings (mg/kg) on seeds before, during and after passage through a seed drill

Sack			Hopper			Drill outlet		
Position	mg/kg	CV %	Position	mg/kg	CV %	Outlet	mg/kg	CV %
Top	1920	5.6	1	1780		1	1870	
Middle	1920	4.1	2	1910		5	1780	
Bottom	2000	3.6	3	1980		10	1820	
Mean	1950	4.4	4	1970		15	1770	
			5	1800		20	1770	
			Mean	1890	4.9	25	1780	
						Mean	1800	2.4

DISCUSSION

This survey showed that there had been a marked improvement in the quality of commercial seed treatments since the earlier extensive studies of Lord et al (1971). However, it was still evident that the majority of seed treatment plants were failing to achieve the target dose of fungicide. In some cases, eg samples F, N and P, mean doses of ethirimol were so much below target that they clearly reflected serious deficiencies in the treatment

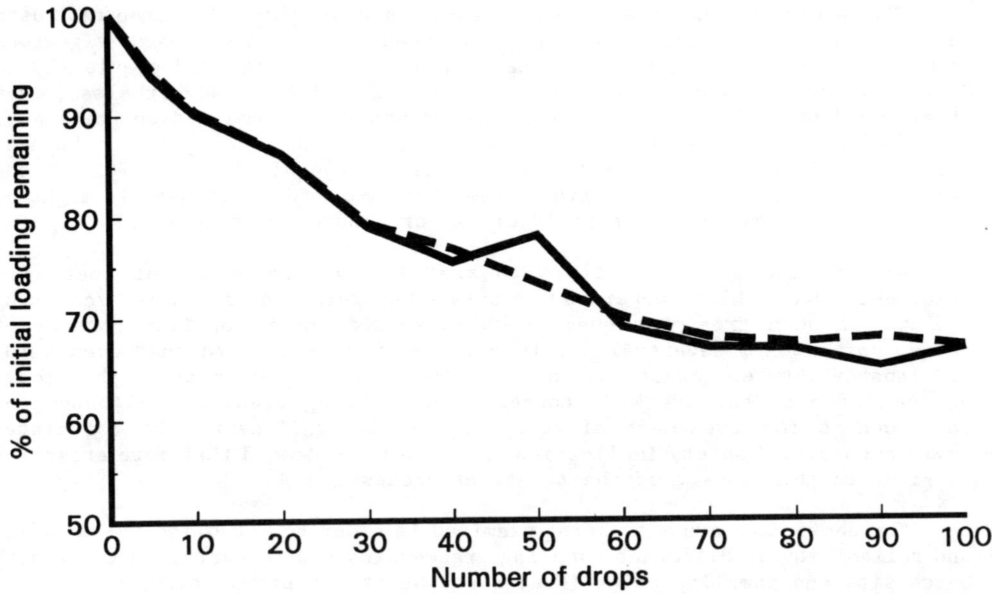

FIGURE 1:   Decline of ethirimol loading on seeds in retention tests in
- - - - - - - - 1990 and in ___________ 1992

procedure. Nevertheless, despite the overall low dosing, the mean doses of
the majority of treatments were well within recommended guidelines of 70%
(Slawson, 1994) or 75% (Koch and Spieles, 1989) of target.

    This generally satisfactory achievement of treatment accuracy was
associated with a much improved level of mean dosing uniformity between
individual samples in each batch, with the great majority of samples
containing within ± 10% of their respective mean doses. Although measurements
of the loadings on individual seeds (Table 2) revealed CVs of up to 45%, there
was already a marked inherent variability in the sizes of individual treated
seeds. It seems unlikely, therefore, that even the most precise application
could achieve variabilities significantly smaller than 25% CV. Under such
circumstances, the production of seeds with between-seed loading variabilities
of only 29 - 45% CV is no small achievement, especially at typical rates of
throughput of 15 - 20 tonnes/hr. Furthermore, the proportion of each batch
containing ± 50% of the target dose was always >90%. This compared favourably
with the 80% of single grains from 67 seed lots which Reitz (1989) found were
within this range. Precise definition of specifications for between-seed
uniformity have yet to be established in the the UK. It is possible that
different criteria may have to be established for different purposes. For
example, uniformity requirements for effective eradication of a seed-borne
pathogen are likely to be more stringent than for systems in which the seed
is used simply to transport active ingredients into the soil. Furthermore,
recent studies have shown that the efficiencies of insecticide seed treatments
against field populations of cabbage root fly were influenced little by the
level of between-seed treatment uniformity (Jukes *et al*, 1994). Clearly there
is still much to be learned about this aspect of dosing variability before
realistic guidelines can be established.

The results of the retention studies are encouraging. The drop test used in the present studies subjected the seeds to a much more rigourous examination than that imposed by the 3 or 6 drops recommended previously in Germany and the UK respectively (Kohsiek and Jeffs, 1986). Nevertheless, even after 100 drops, 65% of the initial dose of ethirimol remained on the seeds, little short of the 70% minimum proposed for the UK (Slawson, 1994). The mean dose on seeds emerging from the seed drill was equivalent to 90% of the target, suggesting that, with liquid seed treatment formulations, it might be realistic to recommend at least 10 drops for studies of retention.

Despite the general failure to meet the specified target dose, the treatment plants which cooperated in this study were, on the whole, achieving a level of seed treatment quality which should ensure optimum control of target pathogens. Nevertheless, it remains of some concern that even a 10% discrepancy between target and achieved doses would have entailed the "loss" of >4kg of a.i. for every 20 tonnes of seed being treated. Although the consequences for treatment efficacy may be insignificant, the associated environmental and safety implications are likely to demand that more attention is given to this aspect of the treatment process.

The above data are also being examined in order to establish the accuracy and reliability of different sampling frequencies and to correlate these with batch size and sampling point (Hewett and Suett, in preparation).

ACKNOWLEDGEMENTS

The cooperation of Zeneca Agrochemicals and of staff at seed treatment plants is acknowledged gratefully. We are especially grateful to the BCPC and the Home Grown Cereals Association for their financial support.

REFERENCES

Davis, R.P., Garthwaite, D.G.; Thomas, M.R. (1990) *Pesticide Usage Survey Report 78: Arable Farm Crops 1988*, London: MAFF, 61pp.

Jeffs, K.A. (1973) Tests for retention of powders on seeds. *Rothamsted Experimental Station Report for 1973*, 177.

Jukes, A.A.; Suett, D.L.; Phelps, K.; Sime, S.; Cooke, S.J. (1994) The influence of seed treatment uniformity on the biological performance of chlorfenvinphos and chlorpyrifos. *These proceedings*.

Koch, H.; Spieles, M. (1989) New assessment of the quality of dressing of cereal seed by means of single seed analysis. *Gesunde Pflanzen* **41**, 18-24.

Kohsiek, H.; Jeffs, K.A. (1986) Application of pesticides to seeds: assessment of application methods and prevention of loss. In: *Seed Treatment*, K.A. Jeffs (Ed.), 2nd edition, Thornton Heath: British Crop Protection Council, pp. 46-50.

Lord, K.A.; Jeffs, K.A.; Tuppen, R.J. (1971) Retention and distribution of dry powder and liquid formulations of insecticides and fungicides on commercially dressed cereal seed. *Pesticide Science* **2**, 49-55.

Rietz, S. (1989) What evenness of distribution on single grains is produced by seed dressing equipment? *Gesunde Pflanzen* **41**, 24-28.

Slawson, D.D. (1994) Efficacy and physical/mechanical data requirements for approval of seed treatments in the UK. *These proceedings*.

THE EFFECT OF PELLET WEIGHT ON THE DISTRIBUTION OF IMIDACLOPRID APPLIED TO SUGAR-BEET PELLETS

F. WESTWOOD, K.M. BEAN, A.M. DEWAR

AFRC Institute of Arable Crops Research, Broom's Barn Experimental Station, Higham, Bury St Edmunds, Suffolk, IP28 6NP

ABSTRACT

The quantity of imidacloprid applied to individual pelleted seed of sugar beet was variable, ranging from 0.6 to 1.8 mg per pellet. This variation was partially due to small differences in pellet size, which led to proportionally larger differences in surface area. The higher doses of imidacloprid on larger pellets caused significant slowing of emergence in the laboratory, but did not reduce final emergence. The presence of imidacloprid in the leaves was greater from larger seeds 19 days after sowing at the 2-leaf stage, but not before or after that stage.

INTRODUCTION

The technology for applying small quantities of pesticide to individual seeds has improved immensely in recent years (Maude, 1990), largely by adopting techniques used in the pharmaceutical industry. However, distribution on individual seeds can still be variable, especially when pesticides are applied to the surface of the pelleted seed. Some of this variation is caused by small differences in seed pellet size leading to proportionally larger differences in surface area. This in turn can lead to large difference in the amount of pesticide applied to individual pellets (Suett & Maude, 1988), which can result in under or over dosing. The consequences of the former may be insufficient activity of the pesticide, leading to pest attack, and the consequences of the latter may be phytotoxicity.

Imidacloprid (1-[6-chloro-3-pyridinyl] methyl 4-5 dihydro N-nitro(1-imidazol-2-amine), commercially known as Gaucho (Bayer plc), is a new systemic insecticide which gives good control of soil and foliar pests of sugar beet when applied to the pelleting material surrounding sugar-beet seed (Altmann, 1991). The rate of application to sugar beet which has been approved in France, Belgium, the Netherlands, Finland and, this year, in the UK is 90 g AI per unit (1 unit = 100,000 seeds). This constitutes a relatively high proportion (approximately 4%) of the total pelleting material surrounding the seed, and may be responsible for the slowing of emergence (Heatherington & Meredith, 1992) and reduction in plant number (Dewar, 1992) observed in some trials.

The present study was carried out to determine the distribution of imidacloprid on commercially-pelleted seeds, and whether variation in this distribution caused adverse effects on the germination, emergence and survival of plants growing from them.

MATERIALS AND METHODS

Variability in seed and pellet size

Commercial raw monogerm sugar-beet seeds are botanically fruits, each containing a true seed. The raw seeds, which are discus-shaped at harvest, are rubbed to remove excess cortical tissue, and then graded to remove too small or too large seed. The minimum and maximum diameters of seeds stipulated by British Sugar plc for use in the UK are 3.00 to 4.25 mm (J.W.F. Prince, personal communication). Pelleting the seed renders it more spherical and so much easier to sow mechanically. However, pellet size still varies between 3.5 and 4.75 mm in diameter, which means that the largest pellets are 36% bigger than the smallest. Only 5% of pelleted seed outside that range is allowed in commercial batches.

The sugar-beet seed, cv. Saxon, used in this study was either pelleted in the normal way, but without the standard methiocarb at 2 g AI per unit, or treated with imidacloprid at 90 g AI per unit. The imidacloprid was applied as a wettable powder to the outside of previously-formed pellets which had also received basal applications of thiram, partially in a steep process and partially added to the pellet (a total of 5 g AI per unit), and hymexazol at 10.5 g AI per unit. These two fungicides were applied to control the fungal diseases *Phoma*, *Aphanomyces* and *Pythium* (Asher & Dewar, 1994). The imidacloprid was bound to the outer surface of the pellet by a coloured polymer to prevent loss of active ingredient (Halmer, 1988); untreated seed received neither imidacloprid nor polymer.

To determine the variability of seed used in this study, 500 pelleted seeds from each of the untreated and imidacloprid-treated batches were weighed individually on a high-precision microbalance (LeCo-250), and their length and breadth measured using a very accurate digital calliper. The approximate surface area of each seed was calculated by using the equation for a prolate spheroid: $2\pi b^2 + 2\pi$ (ab/e) arcsin e where a = length (mm), b = width (mm) and e = $\sqrt{(1-[b^2/a^2])}$ (Korn & Korn, 1961).

<u>Variability in application of imidacloprid</u>

To determine the variability in application of imidacloprid to individual pellets, one hundred pelleted seeds were selected at random and weighed. Imidacloprid was extracted by placing each seed in 10 ml of acetonitrile : water (80:20), homogenising gently to break up the pellet, and left overnight on an orbital shaker. One ml aliquots from each sample were filtered through a DynaGuard filter (0.2μm) prior to analysis by high performance liquid chromatography (HPLC). After extraction the, by then, depelleted seeds were recovered from the solvent, air-dried at room temperature for 10 hrs, followed by a further 8 hrs at 30°C in a drying cabinet before being weighed. The weight of pelleting material surrounding each seed was assumed to be the difference between the two weights.

<u>The effect of pellet size on rate of emergence</u>

Fifty seeds from each of three different size grades (20-30 mg, 30-40 mg, 40-50 mg) from each of the untreated and imidacloprid-treated batches were sown individually in compost in 9 cm pots and placed in a controlled environment room at 20°C. The rate and extent of emergence was recorded daily for 13 days after sowing. Subsequently, 10 plants from the imidacloprid-treated batch were bulk-harvested 12, 19, 24, 31 and 41 days after sowing when plants were at the cotyledon, 1-2 leaf, 3-4 leaf, 5-6 and 7-8 leaf stages respectively. Imidacloprid was extracted from the plant tissue by placing 1 g of fresh leaf tissue from each bulk in 25 ml of acetonitrile : water (80:20), macerating for 10 min, and centrifuging (10 000 g) for 10 min at 10°C. The supernatant was evaporated down to 1 ml and filtered through a DynaGuard filter (0.2 μm) before analysis by HPLC.

Analytical Procedure

Imidacloprid was analysed by HPLC (Westwood & Dewar, 1993) using a Hypersil ODS 150 x 4.6 mm (ID) column. The mobile phase was water : acetonitrile (75:25) at pH 3.5, and the eluant was monitored at 265 nm and a flow rate of 1 ml per minute.

RESULTS

Variability in pellet size

Pelleted seed from both untreated and imidacloprid-treated batches varied in weight from 25 mg to almost 60 mg. The mean weight of untreated seed was 34.4 ± 5.5 mg, almost 2 mg less than the treated seed (36.3 ± 5.2 mg). The difference is attributable to the quantity of active ingredient and polymer added to the treated pellet. For both batches of seed there were highly significant positive correlations between weight and surface area (P >0.001) (Fig. 1), and these relationships were of similar slope in each case, but the treated seed had a consistently larger surface area of about 1 mm^2 for the same weight of seed. Again this must be attributable to the application of the treatment and polymer. In both batches the surface area of large pellets (*ca* 50 mg) was almost double that of small pellets (*ca* 25 mg) but the majority (approximately 70%) were within ± 10% of the means.

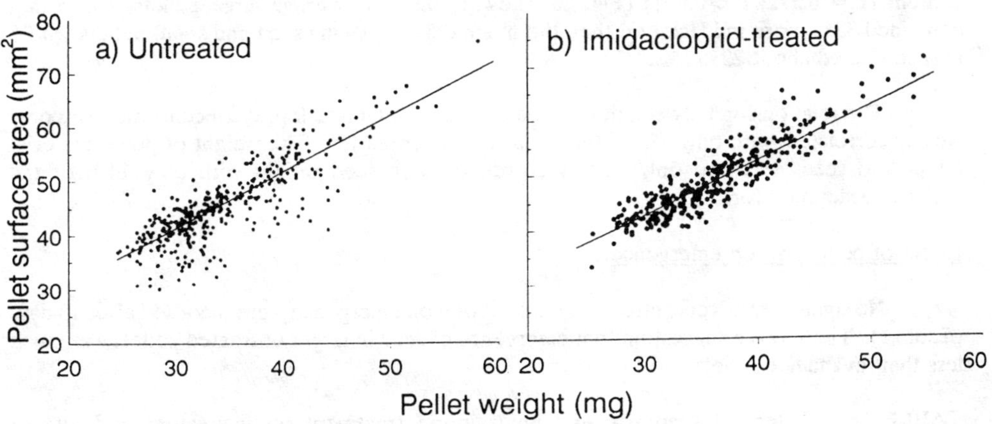

FIGURE 1   Comparison between pellet weight and surface area for a) untreated seed; y = 11.058 + 1.014x and for b) imidacloprid-treated seed; y = 12.012 + 1.014x, where x = pellet weight (mg) and y = surface area (mm^2).

Variability in imidacloprid concentration

The weights of imidacloprid-treated pelleted seeds which were subsequently extracted to determine the amount of imidacloprid varied from 24 mg to almost 50 mg. Imidacloprid concentration also varied, from 0.56 mg/seed to over 1.7 mg/seed, a 300% difference, although the majority of observations (63%) lay within 0.2 mg/seed of the mean dose of 0.9 mg/seed (equivalent to 90 g AI/unit) (Fig. 2b).

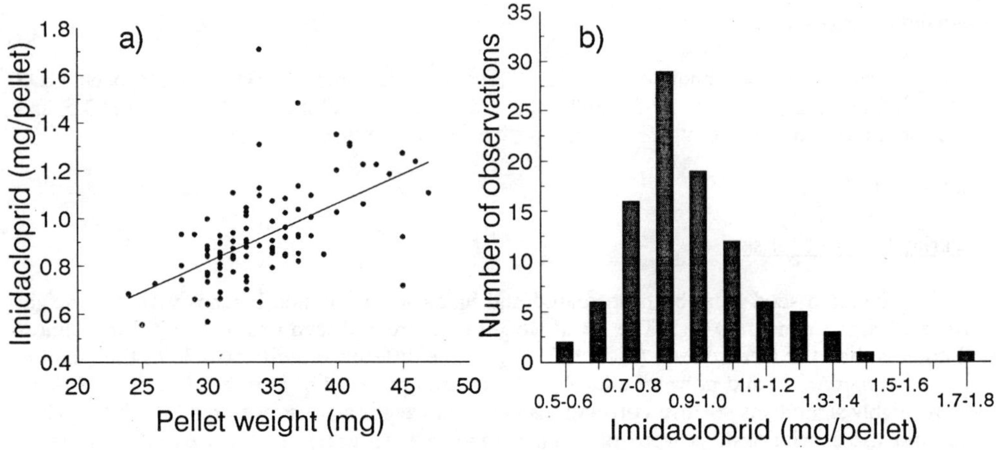

FIGURE 2   a) Comparison of imidacloprid concentration with pellet weight:
y = 10.7 + 2.412x when y is the amount of imidacloprid (mg/seed) and x = pellet weight (mg):
b) frequency distribution of imidacloprid on sugar beet pellets.

There was a significant positive relationship between pellet weight and imidacloprid content (r = 0.572; P >0.001) (Fig. 2a) showing that, on average large pellets (*ca* 45 mg) contained 33% more imidacloprid than the mean dose of 0.9 mg/seed and small pellets (*ca* 25 mg) contained about 22% less.

The relationship between depelleted seed size and imidacloprid concentration was only weakly correlated, with only 7% of the variance accounted for. The weight of pellet material applied to seeds was also only weakly correlated with seed weight with only 14% of the variance accounted for.

<u>Effect of pellet size on emergence</u>

No significant adverse effects of imidacloprid on emergence were recorded after 13 days (Table 1). There was a suggestion that emergence of seed in larger untreated pellets was much less than in smaller pellets.

TABLE 1.   Effect of seed size and imidacloprid treatment on the extent and rate of emergence of sugar beet seedlings.

Seed weight (mg)	% Emergence after 13 days		Days to 50% emergence	
	Untreated	Imidacloprid	Untreated	Imidacloprid
20-30	100	96	6.1	6.9
30-40	84	94	6.3	7.4
40-50	86	94	6.7	7.4

Plants emerged consistently later from imidacloprid treated pellets than untreated over all seed sizes (X^2 = 51.85; P >0.01) (Table 1). Within each batch emergence was significantly faster from smaller pellets than larger ones (X^2 = 28.13, 8 d.f. and 47.92, 8 d.f. for untreated and imidacloprid-treated respectively).

Uptake of imidacloprid

There were no differences between pellet weights in the uptake of imidacloprid 12, 24, 31 and 41 days after sowing, but 19 days after sowing (2-leaf stage) at least twice as much imidacloprid was detected in the leaves of plants from large pellets compared to those from smaller pellets (Fig. 3). The apparent decline in concentration of imidacloprid with time is due to dilution within plant tissue as the plants grow larger, not to a decrease in total uptake within the plant.

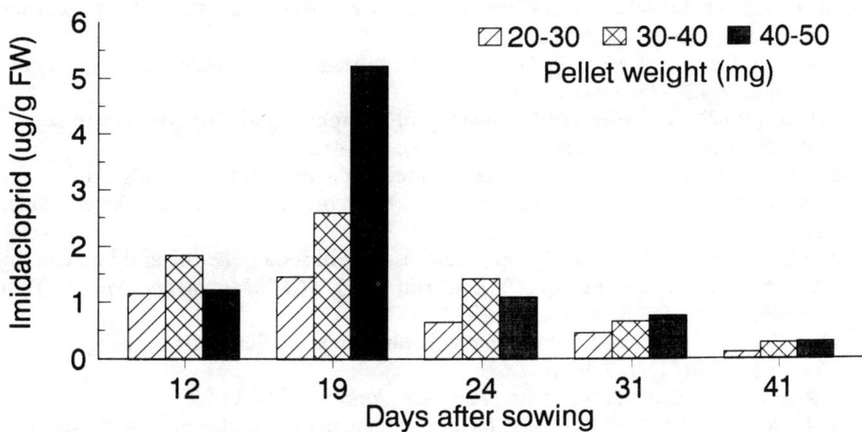

FIGURE 3.   The effect of pellet weight on the concentration of imidacloprid in the leaves of sugar-beet seedlings.

DISCUSSION

Sugar-beet pelleted seed as supplied to growers in the UK varied in size from 3.50 mm to 4.75 mm in diameter, and in weight from about 25 mg to 50 mg. This resulted in a range of surface areas from about 35 mm^2 to 68 mm^2 - almost a 100% difference. It was not surprising therefore that the amount of chemicals such as imidacloprid, applied to this surface area, vary by similar proportions. This study confirms that the amount of imidacloprid on individual pellets varied by as much as 300%, although the mean value of the 100 seeds tested here was 89.1 g AI, only 1% less than the target.

The consequences of this variation, which is partially related to pellet size, are that larger pellets have larger doses of imidacloprid and that this can slow emergence significantly, although final emergence was not affected. However, larger untreated pellets were also slower to emerge and the proportion of these seeds not germinating was noticeable. Therefore these results remain inconclusive and the problem needs further investigation.

The concentration of imidacloprid in plants growing from larger pellets was only greater than that from smaller pellets on one sample occasion (the 2-leaf stage), but this could have important consequences, especially if herbicides are applied at that growth stage. Exposure to high doses of more than one chemical at vulnerable growth stages may cause death of some plants.

ACKNOWLEDGEMENTS

Thanks are due to Alan Todd of IACR Rothamsted Experimental Station's Statistics Department for analysis of data.

REFERENCES

Altmann, R. (1991) Gaucho - a new insecticide for controlling beet pests. *Pflanzenschutz - Nachrichten Bayer*, **44(2)**, 159-174.

Asher, M.J.C.; Dewar, A.M. (1994) Control of pests and diseases in sugar beet seed treatments. *These proceedings*.

Dewar, A.M. (1992) The effects of imidacloprid on aphids and virus yellows in sugar beet. *Pflanzenschutz - Nachrichten Bayer*, **45(3)**, 423-442.

Heatherington, P.J.; Meredith, R.H. (1992) United Kingdom field trials with Gaucho for pest and virus control in sugar beet, 1989-91. *Pflanzenchutz-Nachrichten Bayer*, **45(3)**, 491-523.

Halmer, P.H. (1988) Technical and commercial aspects of seed pelleting and film coating. In: *Application to Seeds and Soil*, T.J. Martin (Ed.), *BCPC Monograph No. 39*, Thornton Health: BCPC Publications, pp 191-204.

Korn, G.A.; Korn, T.M. (1961) Mathematical Handbook for Scientists and Engineers. New York: McCraw Hill, 749 pp.

Maude, R.B. (1990) Seed Treatments. *Pesticide Outlook*, **1(4)**, 16-22.

Suett, D.L.; Maude, R.B. (1988) Some factors influencing the uniformity of film-coated seed treatments and their implications for biological performance. In: *Application to Seeds and Soil*, T.J. Martin (Ed.), *BCPC Monograph No. 39*, Thornton Heath: BCPC Publications, pp 25-32.

Westwood, F.; Dewar, A.M. (1993) Sugar beet under analysis: the search for pesticide residues. *British Sugar Beet Review*, **61(2)**, 38-40.

THE COATING OF CARROTS WITH CHLORFENVINPHOS

H. WEBER, O. FUSS

SUET Saat- und Erntetechnik GmbH, D-3440 Eschwege, Germany

ABSTRACT

Considerable yield losses in the cultivation of carrots are caused by the carrot fly (*Psila rosae*). It is possible to limit damage by treating the crop in the field with chlorfenvinphos. Similar efficacy is achieved with a chlorfenvinphos seed treatment, which uses only 2% of the amount of AI required for a soil or crop treatment. On the basis of the physical properties of chlorfenvinphos, and for hygienic, technological and physiological reasons, it was formulated as a wettable powder.

INTRODUCTION

From the end of May up to autumn, carrots are liable to attack by two generations of the carrot fly (*Psila rosae*). One week after laying eggs the larvae hatch out. They migrate into the main root and develop there over a period of 4 to 7 weeks. After pupation, the second generation hatches out from the middle of August. The damage is caused by the larvae which cause the carrots to taste bitter, smell unpleasant and rot. It is therefore almost impossible to grow carrots commercially without controlling this pest.

Insect control

The insecticidal effectiveness of chlorfenvinphos is well-known (Beynon *et al.*, 1968). Chlorfenvinphos (Birlane, Sapecron, Haptarax, Haptasol: Shell) blocks the enzyme acetylcholinesterase and acts as a contact and breathing poison. The pure substance is a colourless oil with a boiling point of 110°C at 0.0013 mbar and a vapour pressure of 2.2×10^{-7} mbar at 25°C. It has a water solubility of 145 mg/l at 23°C. To date, its application has been limited to field treatments with an emulsifiable concentrate or to row treatments with granules at doses up to 5 kg AI/ha. These doses may need to be even greater if prolonged control of second generation larvae is required. This paper describes studies undertaken to control first generation carrot fly larvae with a seed treatment. An essential advantage of this method is the reduction of the amount of active ingredient to only 0.1 kg per ha.

METHODS

Preparation of a wettable powder

Since the usual trade formulations of chlorfenvinphos were not suitable for treating seeds, the active substance was absorbed in a nearly pure state (96% AI) on an inert, inorganic carrier and converted into a dry powder that contained 60% AI. This powder did not contain any formulation aids such as organic solvents, emulsifiers, adhesives or antifreeze substances that might have adversely influenced the germination of seeds. It was readily stirred into an aqueous suspension together with fungicides, germination aids and binders and was sprayed on to the seeds in a technical process. After drying,

the carrier prevented the migration and penetration of the active substance
on or into the seeds and also suppressed possible effects on germination.
Despite the oily characteristics of chlorfenvinphos, the seeds did not stick
together, thus avoiding potential difficulties during sowing.

<u>Treatment of carrots</u>

The treatment of the carrots was carried out in a technical installation
for seed coating (Hoerner, 1985; Halmer, 1988). In this process, the aqueous
suspension was sprayed on to a bed of seeds that was fluidised by warm air.
This was performed in a container suitable to the amount and variety of the
seeds. As they passed the spraying zone, the seeds picked up the suspension,
while at the same time the warm air was drying the seeds. After dropping down
the container wall the cycle was completed. Treatment continued until the
calculated amount of suspension had been applied. At an air temperature of
40°C, 10 kg seed could be treated in 15 min with 4 kg of a suspension
containing 25% w/w solids (fungicide, chlorfenvinphos, pigment and binder).

<u>Analytical investigations</u>

The analytical determination of chlorfenvinphos in a representative
sample and measurements of a sufficient number of single seeds give
information about the quality of the coated seeds. In numerous assessments,
the mean achieved dose was found to be within 90-100% of the target dose of
25 g AI/kg. At this dose level, and with a mean seed weight of 2 g/1000
seeds, the target dose was approximately 50 μg chlorfenvinphos per seed. The
distribution of AI on 100 individual seeds is shown in Figure 1.

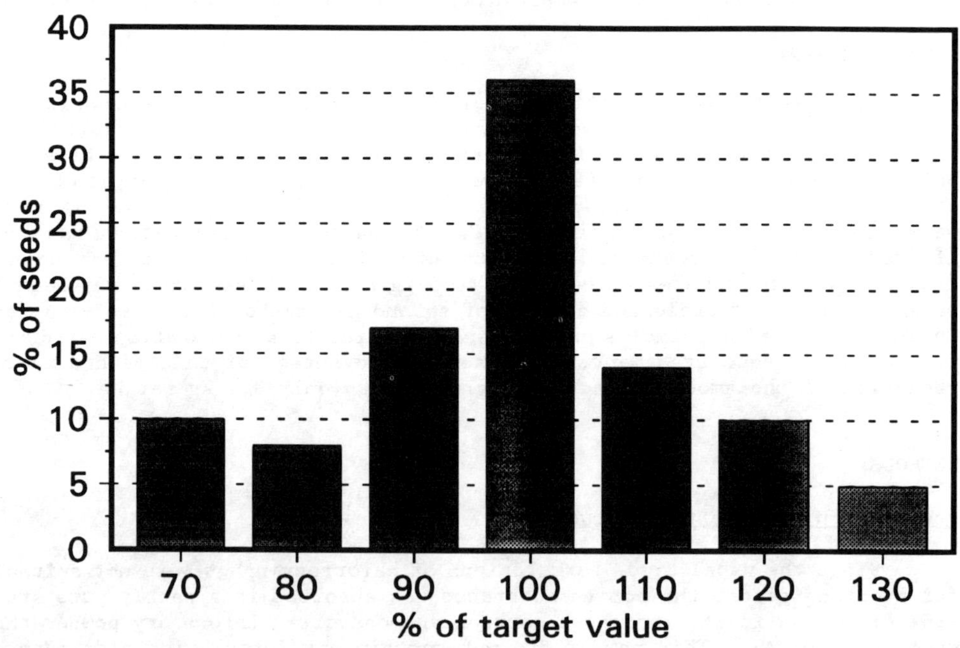

FIGURE 1. Distribution of chlorfenvinphos on 100 carrot seeds. Target value
= 50 μg/seed)

All the seeds contained 70-130% of the target dose of chlorfenvinphos. The coefficient of variation (CV) was ± 17%.

Storage stability

The seeds were stored under usual conditions at 18°C for 3 years to examine the effects on germination and on the stability of the chlorfenvinphos treatment. Results showed that there was no significant decomposition of chlorfenvinphos.

RESULTS

Field trials under practical conditions indicated that the chlorfenvinphos seed treatment was very effective in reducing damage by 1st generation carrot fly larvae (Table 1).

TABLE 1. Effect of different treatments of chlorfenvinphos against 1st generation carrot fly larvae (cv. Napoli)

	% of plants infested		
Treatment	1989 sown 20 March evaluation 27 June		1990 sown 15 March evaluation 6 June
	field a	field b	field c
1. Control (untreated)	10.5	17.0	50.0
2. Surface band treatment with granules at sowing	13.0	12.5	-
3. Field treatment with EC before sowing	-	-	9.5
4. Coated seed	4.3	12.3	12.1

With a seed density of approximately 2 million seeds/ha, the amount of chlorfenvinphos applied via the seed treatment was equivalent to 100 g AI/ha, compared with 5 kg/ha with the granule or EC treatment. Analyses of carrots at harvest showed that residues of chlorfenvinphos were below the limit of detection (< 0.01 mg/kg). The seed treatment with chlorfenvinphos thus appears to be an effective and economic alternative to more conventional methods of applying the insecticide to control 1st generation carrot fly on carrots.

ACKNOWLEDGEMENTS

The analytical determinations of chlorfenvinphos were performed in the laboratory of SUET, the residue investigations were done in the Institute Fresenius, Taunusstein. We thank Prof. Kretschmer, Geisenheim for carrying out and evaluating the field trials and colleagues of the Dutch Shell Company, with whom we coordinated the work.

REFERENCES

Beynon, K.I.; Edwards, M.J.; Elgar, K.; Wright, A.N. (1968) Analysis of crops and soils for residues of chlorfenvinphos insecticide and its breakdown products. *Journal of the Science of Food and Agriculture*, **19**, 302-307.
Halmer, P. (1988) Technical and commercial aspects of seed pelleting and film coating. In: *Application to Seed and Soil*, T.J. Martin (Ed.), *BCPC Monograph No. 39*, Thornton Heath: BCPC Publications, pp. 191-204.
Hoerner, E.L. (1985) SHR - a new technology for seed processing. *International Agrophysics*, **1**, 141-145.

INTERACTIONS BETWEEN HYMEXAZOL, FURATHIOCARB AND SOME CLAY MATERIALS USED FOR SEED TREATMENT

L. PUSSEMIER, Ph. DEBONGNIE, Y. VAN ELSEN

Institute for Chemical Research, Ministry of Agriculture, Leuvensesteenweg 17, B-3080 Tervuren, Belgium

ABSTRACT

The transformation of furathiocarb was studied after deposition of this pro-pesticide onto clay films in the presence or absence of hymexazol as well as other model compounds. It appeared that, in the absence of hymexazol, there was very little transformation (<2.5%) of furathiocarb into carbofuran. In the presence of hymexazol the transformation into carbofuran was much greater and increased progressively with time (up to 27%), especially with some specific clay materials such as bentonite, hectorite and vermiculite. It was also shown that model compounds structurally related to hymexazol were unable to induce this acceleration of the transformation of furathiocarb. Similar trends were observed when the clay materials were used in seed treatments, and the instability of furathiocarb was again greater when hymexazol and bentonite were used together in the same treatment.

INTRODUCTION

One of the main benefits from pesticide application on seeds is the marked reduction of the application rate and, therefore, of the potential hazard for the environment. Significant progress has been achieved with the sugar beet crop where a fungicide such as hymexazol can be applied in the seed pellets, either alone or together with an insecticide such as furathiocarb. Some problems, however, may occur with this combination of pesticides. Furathiocarb, for instance, lacks stability when applied together with hymexazol on seed pellets prepared with clay materials (Huijbrechts & Gyssel, 1989 ; Pussemier *et al.*, 1990). The aim of the present study was to follow the transformation of furathiocarb into carbofuran which, although not being the active ingredient *sensu stricto*, seems to be, under certain circumstances, the major biocidal derivative. Thus the transformation of furathiocarb, on clay films and in experimental seed treatments based on the same materials, has been studied with special emphasis on the role of the clay material and of the accompanying organic compound (hymexazol or other model compounds).

MATERIALS AND METHODS

Materials

Furathiocarb, technical grade, was obtained from Ciba-Geigy, Basle. Carbofuran and hymexazol were obtained by extracting the active ingredients from their commercial formulations (Curater (Bayer Ltd) and Tachigaren (Sankyo Ltd), respectively). The identity and purity of the chemicals obtained after crystallization were confirmed by determination of their melting points.

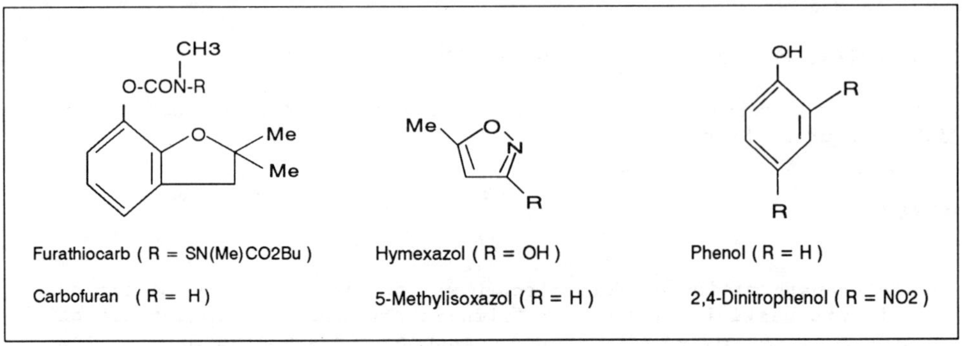

FIGURE 1.   Chemical structure of the selected pesticides and model organic chemicals

5-Methylisoxazol, phenol and 2,4-dinitrophenol, all analytical grade, were purchased from Janssen Pharmaceutica, Beerse. The chemical structures of all the compounds are detailed in Figure 1.

The clay materials selected were : kaolin from Eire, bentonite from Wyoming, vermiculite from Texas, and hectorite from California. Methylcellulose (Tylose from Merck), a binder widely used in the seed treatment technology, was taken as a reference material and, because of its polysaccharidic composition, was assumed to be chemically inert with respect to the pesticides.

<u>Clay films</u>

For the studies dealing with the transformation of furathiocarb on solid films, 2.5 g of the selected clay and reference materials were suspended in 50 ml distilled water. Two ml of the suspensions were poured into glass vials (3 cm diameter) and allowed to dry for 16 to 20 hours at 60°C. One ml of a 20 mM furathiocarb solution in methanol:water (1:1) was then poured onto the films, together, when appropriate, with hymexazol or with the structurally related compound, all of them being applied at the same concentration as furathiocarb (i.e. 0.2 mmoles per g of clay material). The films were allowed to dry again (24 hours at 60°C). To a first set of films, 5 ml of water were added just after drying, in order to extract the carbofuran produced from the furathiocarb. A second set of films was maintained dry at room temperature (18-22°C) for 7 or 14 days before performing this extraction procedure. The percentage transformation of furathiocarb was determined by measuring the amount of carbofuran produced, using an enzymatic test based on the inhibition of acetylcholinesterase (Pussemier *et al.*, 1990). The measurements were carried out on 2 or 3 replicates and the relative standard deviation coefficients were generally close to 20%.

<u>Treated seeds.</u>

Three hundred maize seeds (c. 100 g) were introduced into an experimental seed dressing system prepared from a cylindrical drum and a commercial painting device. The treatment was performed by spraying portions of a suspension containing the seed dressing materials into the rotating cylinder and drying the seeds with hot air (60°C). The sprayed aqueous suspension contained furathiocarb (4g/L) with or without hymexazol (4g/L) as well as bentonite (80g/L) or Tylose (12g/L). The seeds were extracted and analyzed for carbofuran after 1, 7 or 14 days storage in the laboratory.

RESULTS AND DISCUSSION

<u>Furathiocarb transformation on solid films</u>

The percentages of furathiocarb converted into carbofuran after 1, 7 and 14 days of contact with the clay films are shown in Table 1. The pH values shown in the same table are those measured in the aqueous suspensions obtained when extracting the carbofuran.

In the absence of hymexazol, the rate of transformation was small (< 2.5%) and there was only a very little increase with time. This limited transformation of furathiocarb into carbofuran seemed to be associated to the acidic properties of the solid materials. From the following relationship it can be seen that after one day there was a very significant correlation with the pH :

$$\% \text{ transformation (1 day)} = 3.70 - 0.38 \text{ pH} \qquad r = -0.96 \qquad (1)$$

After 7 and 14 days, however, less significant correlations were obtained :

$$\% \text{ transformation (7 days)} = 4.13 - 0.38 \text{ pH} \qquad r = -0.90 \qquad (2)$$

$$\% \text{ transformation (14 days)} = 5.20 - 0.47 \text{ pH} \qquad r = -0.83 \qquad (3)$$

From these observations it is suggested that a limited acid-catalyzed transformation occurred at the very beginning, probably when the clay materials were still moistened and active as catalysts (Ortego *et al.*, 1991). Later, when the clay films were dry, there was very little additional production of carbofuran, and it seemed that the pH was no longer the main factor controlling this further transformation of furathiocarb.

In the presence of hymexazol, on the other hand, the transformation of furathiocarb into carbofuran was already important (up to 6.94%) after 1 day but clearly increased with time, reaching 20-30% for some of the clay materials such as bentonite, hectorite and vermiculite. For the organic adsorbent (Tylose) and for kaolin, which was the only tested clay material without swelling properties, the fraction transformed into carbofuran, whilst not negligible after 1 day, did not increase any further with time. It would seem, therefore, that the addition of hymexazol to the clay materials characterized by swelling properties may activate the surface-catalyzed transformation of furathiocarb.

TABLE 1. Percentages of furathiocarb transformation after application alone and in combination with other organic chemicals on various solid films (HYM = hymexazol; MeISO = 5-methylisoxazol; PHE = phenol; DNP = 2,4-dinitrophenol; BEN = bentonite; KAO = kaolin; VER = vermiculite; HEC = hectorite; TYL = Tylose)

Solid film	Chemical added	Interval after treatment					
		1 day		7 days		14 days	
		% transf.	pH	% transf.	pH	% transf.	pH
BEN	-	0.42	8.0	0.55	9.3	0.67	9.3
KAO	-	1.39	6.3	1.20	6.8	1.75	6.5
VER	-	0.93	7.5	1.18	8.6	2.39	7.8
HEC	-	0.10	9.6	0.18	10.2	0.46	9.7
TYL	-	1.14	6.5	1.81	6.8	2.13	6.4
BEN	HYM	4.07	6.2	15.36	6.3	23.52	6.1
KAO	HYM	3.81	5.4	2.06	5.3	2.90	5.3
VER	HYM	6.94	7.2	11.14	7.7	18.97	7.8
HEC	HYM	1.49	9.1	22.66	8.8	27.06	9.0
TYL	HYM	3.34	5.2	2.84	4.8	2.59	4.6
BEN	MeISO	0.60	9.4	0.63	9.8	0.69	9.4
KAO	MeISO	2.46	7.4	1.88	6.8	1.99	6.8
VER	MeISO	2.04	8.2	1.41	8.5	1.52	6.8
HEC	MeISO	0.58	9.3	0.14	10.0	0.40	9.0
TYL	MeISO	2.70	7.4	3.47	7.3	2.46	6.6
BEN	PHE	0.68	8.1	0.81	9.1	0.74	8.9
KAO	PHE	1.35	6.3	1.10	6.3	1.35	6.3
TYL	PHE	1.30	6.3	1.23	6.3	1.48	6.3
BEN	DNP	0.60	4.7	0.55	4.5	0.68	4.7
KAO	DNP	0.74	3.0	0.55	3.0	1.10	3.2
TYL	DNP	1.23	3.1	1.35	3.1	1.10	3.2

From the results obtained when using 5-methylisoxazol, phenol, and 2,4-dinitrophenol instead of hymexazol (Table 1), it was evident that these compounds, whilst structurally related to hymexazol, did not lead to any activation of the transformation properties of the clays. 5-Methylisoxazol is analogous to hymexazol (presence of an isoxazolic ring) but lacks the hydroxylgroup. On the other hand, both the phenol and 2,4-dinitrophenol compounds have a hydroxyl group with a pKa of 10.00 and 4.09 respectively (Albert & Serjeant, 1971). The former compound is thus less acidic than hymexazol (pKa = 5.5 ; Vanderheyden *et al.*, unpublished data) whilst the latter is more acidic. It thus seems that the properties of hymexazol are quite unique and cannot be linked to the aromatic or hydroxyl moieties alone.

<u>Transformation of furathiocarb when applied on seeds</u>

The percentages of furathiocarb transformation are presented in Table 2. It appears that, 1 day after the seed treatment, the rate of transformation of furathiocarb was very small (0.07 - 0.28% of the amount that

TABLE 2. Percentages of furathiocarb transformation after application in seed treatment with Tylose (TYL) or bentonite (BEN), with or without hymexazol (HYM).

Materials used for seed dressing	% of furathiocarb transformation after :		
	1 d	7 d	14 d
TYL	0.13	0.13	0.13
BEN	0.07	0.15	0.42
TYL + HYM	0.17	0.31	0.34
BEN + HYM	0.28	5.67	9.00

has been applied). Furathiocarb was more stable in the seed treatments without hymexazol. When both hymexazol and the clay material (bentonite) were applied simultaneously, the transformation rate was somewhat higher. These differences, however, increased afterwards during the storage of the seeds. Thus, after 14 days storage, 9% of furathiocarb had been transformed into carbofuran in the treatment combining hymexazol and bentonite, whereas no more than 0.13 - 0.42% transformation occurred with the other treatments.

CONCLUSIONS

The main results of this study can be summarized as follows :

1) Furathiocarb applied alone on clay films was quite stable. Limited transformation into carbofuran was observed after one day and seemed to be linked to the acidic properties of the supports. This transformation of furathiocarb in contact with the films did not increase very much after 7 or 14 days. It is thus suggested that this acid-catalyzed transformation occurred mainly in the liquid phase, during the time needed to evaporate the solvent in which the furathiocarb was added to the films.

2) When furathiocarb was applied together with hymexazol, the transformation was significant after 1 day and increased during the following days, especially with the films prepared from clays exhibiting swelling properties. With kaolin and Tylose, there was no increase of transformation with time. It is thus suggested that hymexazol was able to diffuse in the interlayer space of the swelling clays, thereby changing the catalytic properties of the clay surfaces, and eventually increasing the transformation of furathiocarb to carbofuran.

3) This increased transformation of furathiocarb to carbofuran in the presence
of hymexazol was also observed after performing an experimental seed treatment
on maize, as long as bentonite, a swelling clay, is added to the seed dressing
mixture. It is thus recommended to avoid the use of swelling clays when
treating seeds with a mixture of hymexazol and furathiocarb.

REFERENCES

Albert, A.; Serjeant, E.P. (1971) *The determination of ionization
constants*. Chapman & Hall, London, 115 pp.
Huijbrechts, A.W.M.; Gyssel, P.D. (1989) Dosage and release pattern of
pesticides in pelleted seeds. *Proceedings of the 52st Winter Congress
of the International Institute for Sugar Beet Research*, 131-143.
Ortego, D.J.; Kowalska, M.; Cocke, D. (1991)
Interactions of montmorillonite with organic compounds - adsorptive and
catalytic properties. *Chemosphere*, **22** (8), 769-798
Pussemier, L.; Van Elsen, Y.; Misonne, J.-F. (1990) Stabilité de
carbamates insecticides appliqués en enrobage de semences. *Parasitica*,
46 (1), 27-36.

A SMALL-SCALE LABORATORY FLUIDIZED BED SEED-COATING APPARATUS

J. S. BURRIS

Professor, Seed Science Center, Iowa State University

L. M. PRIJIC

Research Scientist, Maize Research Institute Zemun, Yugoslavia

Y. Chen

Research Scientist, George Ball Seed Company, West Chicago,IL

ABSTRACT

There is a need for a simple and efficient laboratory-sized coating machine. A fluidized bed coating apparatus developed in the Seed Science Center at Iowa State University has the capacity to coat from 100 to 500 g of seeds in 15-30 min. It is efficient, easy to operate, and suitable for coating small amounts of seed for experimental use. A fluidized-bed is formed in the coating chamber by a controlled upward flow of heated air. The coating liquid is introduced through a simple air-atomizing spray nozzle. Initial trials showed that round seed (soybean) could be coated with better success than flat seed. However, with minimum experience corn, sunflowers and other irregular shaped seeds can be successfully coated.

INTRODUCTION

Regulatory and environmental concerns combined with marketing demands have created a substantial interest in seed coating technology. Exposure of seedsmen and growers to captan and other pesticides on the seed is no longer considered acceptable. Further, production practices and techniques often result in mechanical damage, which could be corrected by coating seeds with protective materials. However, evaluation of these new materials requires greater precision than traditional seed treatments. Increasingly sophisticated equipment has been introduced to the seed industry by pharmaceutical equipment manufacturers, with experience in coating tablets, capsules, and granules with organic solvents and aqueous dispersions. Fluidized bed systems are frequently used in pharmaceutical, chemical, and food industries, and are widely used for drying of granular materials. There are many different ways to form a fluidized bed including rotary or vibrational movement, pressurized gas, impulse fluid circulation or flow, sonic vibrations, or magnetically stabilized fluidization. The vibrated fluidized bed reported by Gupta et al. (1980) refers to a shallow rectangular fluidized bed where the complete assembly is vibrated at a relatively high frequency and with a small amplitude. Kamamura and Imada (1980) used reduced pressure to dry fluidized particles. Turcaj (1983) used a fluidized bed for drying samples of wheat utilizing a laboratory apparatus. Evans et al. (1983) reported experiments with a continuous flow fluidized bed heating system capable of treating up to 500 kg/h of wheat. Bacon et al. (1988) reported on a small scale fluidized bed seed treatment device which could coat as little as 200 seed.

Following increased interest in protecting corn, soybean and other seeds by coating, and finding commercially available equipment to be too large and expensive we designed and constructed a small-scale laboratory coating apparatus to allow precise application of coating materials to approximately 500-gm of seed.

METHODS AND MATERIALS

The schematic for the coating apparatus is presented in Fig.1. All parts are mounted on a mobile table of convenient size, e.g. 75 X 60 cm X 85 cm high. A lower shelf houses the air supply, heater, and control equipment. Air is provided by a tangential fan with a capacity of 47 l/s, fan output is controlled by a variable transformer which allows nearly infinite regulation of air velocity. The 900-watt finned strip heater is mounted within a section of 7.5-cm copper tubing through which the air passes on its way to the coating chamber. The temperature is regulated by an electronic control capable of accuracy of 0.1^0C from a sensor mounted below the coating chamber. Seeds to be coated are fluidized in the coating chamber, which is 75-mm ID with a stainless-steel mesh (1 x 1 mm) screen bottom. The chamber is removable to facilitate removal of the coated seed. A spray nozzle with external mixing is supplied with the pressurized air and liquid for coating. The nozzle is constructed in our laboratory using stock 6.25-mm OD copper tubing for the outer shell and 1.5-mm OD Teflon internal liquid supply line. Coating solutions are stirred with a standard laboratory magnetic stirrer. To obtain uniform nozzle delivery of the coating solutions constant flow rates are provided with a variable speed peristaltic pump with a revolution range of 6-600 RPM.

Constant uniform movement of the seeds in the coating chamber is necessary for uniform application of coating materials. The coating material is sprayed onto the seed while it is circulating in the stream of warm air. The air pressure lifts the seed and forms the fluidized bed in the coating cylinder. Movement within the fluidized mass is created by the adjustment of the atomizing air provided by the nozzle. This pressure lifts the seeds in the middle of the chamber, and after reaching the upper level of fluidized bed, the seeds travel down the outer periphery of the chamber. The number of revolutions that the seed makes per minute (the number of exposures to the coating material) is regulated by changing the velocity of air in the chamber and the nozzle pressure. It is important to have enough seeds in the cylinder to maintain uniform movement of seeds and efficient recovery of the spray solution. If the seed volume is too small, the air will follow the path of least resistance and seed circulation in the chamber will decrease. The 75-mm chamber accommodates up to 200-250 g of corn or soybean seed.   The coating solution can be applied at various rates, depending on the nozzle capacity, physical

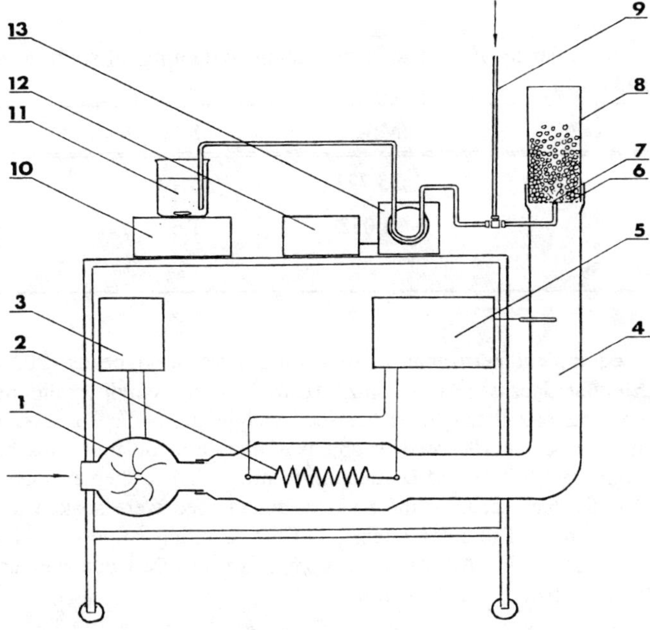

Figure 1. A fluidized bed seed-coating apparatus. 1. Fan, 2. Heater, 3. Variable transformer, 4. Pipe, 5. Temperature control, 6. Wire mesh, 7. Spray nozzle, 8. Coating chamber, 9. Regulated air supply, 10. Magnetic stirrer, 11. Coating liquid, 12. Peristaltic pump regulator 13. Peristaltic-drive pump.

properties of the coating liquid, air flow, temperature, number of revolutions of seed, and operator experience. Fast evaporation of deposited liquid in the warm air makes it possible to apply and dry the coating materials in one step.

Table 1. Coating efficiency of soybean seed as affected by seed mass and coating polymer.

Coating Polymer	Seed Mass	Coating Time	Seed Weight gm/100 Seed[1]
Sacrust	200 gm	12 min	27.273 a
Sacrust	400 gm	20 min	27.050 a
Polyvinylpyrrolidone	200 gm	15 min	26.711 b
Polyvinylpyrrolidone	400 gm	21 min	26.345 b
Control			25.833 c

[1] Values followed by the same letter are similar at the 5% level of probability.

Table 2. Analysis of variance of seed to seed coating variability of soybean as affected by seed mass.

Source	DF	MS	F	P
Treatment	1	35.721	8.40	0.0092
Replication	19	7.058	1.66	0.1391
Total	39			

To determine equipment performance, two batch sizes of soybeans (200 and 400 gm) were coated with either Sacrust (Sarea, Linz, Austria) or Polyvinylpyrrolidone (Sigma), the coating amount was calculated to be 6% seed weight gain. The process was replicated three times and the 100 seed weight was measured on four subsamples of each replicate to determine batch size and coating uniformity. To further evaluate the seed to seed variation 20 individual seeds of the Sacrust coated seed were soaked in 5 ml of distilled water for 15 min. The color intensity of the resulting solution was measured at 490 nm against a water blank. All statistical comparisons were calculated using the Statistix (Analytical Software, St. Paul MN).

RESULTS

Initial trials showed that round seed (soybean) could be coated with better success than flat seed (corn). The coat applied on soybean seed had uniform thickness, good appearance, and completely covered the seed. Corn seed tended to stick and the coating was initially uneven. After additional experience and changes in the rate of application, air pressure, solution concentrations, and temperature, however, successful coating of corn seed was achieved. Soybean seed can be coated in a relatively short time with good precision (Table 1). The time required to coat the different batch sizes varies but the resulting product is similar in overall coverage as indicated by the similarity in 100 seed weight. The seed to seed precision is not significantly effected by the amount of seed although the coating time is extended. Further the seed to seed coating level(Table 2) is similar within coating mass but significantly different between mass. The significance between seed mass levels is probably due to the large number of replications and the sensitivity of the spectrophotometric measurements. The germination and field performance of seed coated with the coater is a function of initial seed quality the polymer used and the seedbed environment. Under no circumstances did the coating procedure appear to influence seed performance. Most of the polymer systems that have been tested have not effected the germination performance of the coated seed.

ACKNOWLEDGMENTS

Journal Paper No. J-14889 of the Iowa Agriculture and Home Economics Experiment Station, Ames, Iowa. Project No. 2526.

REFERENCES

Bacon, J.R., P.A. Brocklehurst, A. Gould, R. Mahon, N.C.J. Martin and M.J. Wraith. 1988. Evaluation of a small scale fluidised bed seed treatment apparatus. Application to seeds and soil. British Crop Protection Council Mono. 39: 237-243.

Evans, D. E., G. R. Thorpe, and T. Dermott. 1983. The disinfestation of wheat in a continuous-flow fluidized bed. J. Stored Prod. Res. 19(3):125-137.

Gupta, R., P. Leung and A. S. Mujumdar. 1980. Drying of granular materials in a vibrated fluidized bed. In: Drying '80, Vol. 2, Hemisphere Publishing Corp., Washington, pp. 201-207.

Kamamura, Y., and T. Imada. 1980. Experimental study on drying in the fluidized bed under reduced pressure. In Drying '80, Vol. 1, Hemisphere Publishing Corp. Washington, pp 205-210.

Turcaj, J. 1983. Thermal drying of cereals in a fluidized bed with interruptions of the process. Sb. Vys. Sk. Chem. Technol. Praze E Potraviny E 56:55-

Session 8
Regulatory Requirements

Chairman and
Session Organiser   P J BRAIN

SEED STANDARDS IN LEGISLATION : ASSESSING SEED QUALITY AND EFFECT OF
SEED TREATMENT

J.H.B. TONKIN

Official Seed Testing Station, National Institute of Agricultural
Botany, Huntingdon Road, Cambridge, CB3 0LE

ABSTRACT

Specialist seeds legislation is put in place in most
countries with the aim of promoting the use of high quality
seed and to improve agricultural productivity. To achieve
this it must protect the interests of the breeder, the
reputable trader and the end user. Essentially there are two
types of legislation 'Truth in labelling', such as operates in
the USA, where few or no quality standards are prescribed and
'minimum standards' where a range of minimum quality standards
are set, such as is the pattern for EC countries, below which
seed may not be marketed. However legislation will only work
if there are effective means of measuring seed quality in
which everyone has confidence. The International Seed Testing
Association Rules for Seed Testing 1993 provide the standard
procedures for sampling and testing seeds, which are followed
by most countries, so that results can be readily understood
and compared. The detailed procedures for carrying out tests
and for seedling evaluation result in highly reproducible
germination results which show the capacity of a population of
seeds to produce plants under favourable field conditions.
The effects of chemical treatments, on seedling development,
if present, can be recognised in these tests. However, with
treatments currently in common use, except where serious over
treatment has occurred, acceptable germination results are
usually obtained. Classical symptoms of chemical
phytotoxicity are thickening of the seedling shoot and root
but not all chemicals coming into contact with seeds produce
the same symptoms. Correct and consistent evaluation of tests
is essential if all concerned with treating, marketing and
using seed are to have confidence that the seed will usually
perform satisfactorily and as expected.

INTRODUCTION

Because seeds are of fundamental importance to plant and food
production and therefore to the prosperity of agriculture, specialist
seeds legislation is put in place in most countries which is intended
to protect the quality of the seed in use and to improve the supply
available (Kelly, 1989). Effective legislation should promote the use
of high quality seed and thus improve agricultural productivity. To
achieve this aim it will need to protect the interests of the plant
breeder and encourage the breeding of new varieties, protect the
reputable trader in seed against unscrupulous competition, protect the
end user against being sold seed of unsatisfactory quality and protect
agriculture generally against the import in seed of pests, diseases or
weeds from other areas or countries.

Essentially there are two types of seed legislation based on 'Truth in labelling' and 'Minimum standards'. Where legislation is of the former type, a seller is required to give the buyer certain information by means of a label or otherwise, but seed of virtually any quality may be marketed. The essential point is that the declared details must be true at the time of sale.

Under 'minimum standard' legislation standards of seed quality are decreed for such attributes as cultivar purity, mechanical purity, weed content, moisture content, germination by number and sometimes for specific seedborne disease. Seed has to be tested by an officially recognised laboratory, and is approved and labelled for sale only if it complies with these standards.

In the United Kingdom a system of seed quality control is in force which is common throughout the European Economic Community. Having a standardised system allows for free movement of seed between member countries. Essentially it is a minimum standard system which lays down conditions which must be satisfied before seeds may be marketed for sowing.

However well framed, seeds legislation will only work if there are effective means of measuring seed quality in which everyone has confidence. The International Seed Testing Association (ISTA) plays an important role in world agriculture in providing the basis for uniform, and reproducible assessments of the value of seed for sowing. One of ISTA's stated objectives is 'to develop, adopt and publish standard procedures for sampling and testing seeds, and to promote uniform application of these procedures for evaluation of seed moving in international trade'. These procedures are set out in the International Rules for Seed Testing 1993 (ISTA, 1993). However, it is important to recognise that ISTA is not concerned with the actual standards seed meets, only with the accurate evaluation of the qualities of the seed being tested. Standards are a matter for individual countries to set or for traders and users to agree.

Most countries apply the ISTA Rules in their seed testing. The main exception is North America but the rules used there are in many respects the same as those of ISTA and currently there are efforts to remove the small differences which still exist.

However carefully testing is carried out the result can only be of real value if it accurately represents the true quality of the sample, and by inference the seed lot, from which the actual seed tested was taken. The production of reasonably homogeneous seed lots and the application of sampling techniques which give reliably representative samples are essential to achieving this objective.

While in this symposium, apart from the efficacy of the treatment for the purpose it is being applied, the concern is with germination and its interaction with the treatment, it should be remembered that this is only one facet of seed quality.

LEGISLATION

<u>European Economic Community and the UK</u>

As a member of the EC the UK has to conform to the EC legislation covering plant varieties and seeds, the main requirements of which are contained in seven Council Directives:

1   Common catalogue of agricultural plant varieties
2   Marketing of beet seed
3   Marketing of fodder plant seed
4   Marketing of cereal seed
5   Marketing of oil and fibre seed
6   Marketing of vegetable seed
7   Marketing of seed potatoes

Changes to and revision of Directives are achieved through discussion in a series of committees in which all member states and the EC Commission are represented. The provisions made in Directives set out the results to be achieved within a stated period but leave the implementation to national governments. In England, Scotland and Wales the seven Directives are implemented through the provisions of the 1964 Plant Varieties and Seeds Act (as amended by the European Communities Act 1972). Northern Ireland has its own legislation, the Seeds Act (Northern Ireland) 1965 as amended by the European Communities (Agriculture) Order (Northern Ireland 1972). In most respects these are the same but for example cereal seed marketed in Northern Ireland must have a greater degree of freedom from wild oats. The detailed requirements for the marketing of seeds are contained in a number of Seeds Regulations made under the Act. These are:
   The Seeds (National Lists of Varieties) Regulations 1982
   The Seeds (Registration, Licensing and Enforcement) Regulations 1985 and amendment Regulations 1987
   The Seeds (Fees) Regulations and amendment Regulations 1987 with annual amendment to fees charged.
   The Cereal Seeds Regulations 1993
   The Fodder Plant Seeds Regulations 1993
   The Oil and Fibre Plant Seeds Regulations 1993
   The Vegetable Seeds Regulations 1993
   The Beet Seeds Regulations 1993

Each member state of the Community produces lists of varieties which, together with those of other member states, make up the EC Common Catalogues of agricultural plants and vegetables. Only varieties on the Common Catalogues may be marketed in the Community.

The five specific seeds regulations cover the marketing of seed of a limited number of species which are listed in them. Even with these species there are occasions where the regulations do not apply, such as for seed intended for export to countries outside the EC, seed used for research or experiment or seed intended for further processing, treatment or cleaning prior to marketing as seeds for sowing. The regulations also permit the marketing of seeds imported from other EC countries and from countries outside the Community which have been granted 'equivalence' which comply fully with EC rules and

standards. For most kinds of seed marketed under these regulations
only certified seed may be sold. Certification provides a system of
checking and guaranteeing various aspects of seed quality during the
period seed is being multiplied through a number of generations. EC
seed certification schemes (Based on OECD schemes for varietal
certification of seed moving in International Trade) specify crop and
seed standards for varietal purity, mechanical purity and germination,
place limits on the content of disease and seeds of other crop and
weed species, and require official involvement in the verification
that seed meets the marketing standards prescribed. The main
exception is vegetable seed which is largely sold as 'standard seed',
which still has to meet marketing standards, but for which no official
examination is required. The official examinations have to show that
seed meets the prescribed standards at the time the test is carried
out if the seed is to be marketed. No further official examination is
required and thereafter it is the responsibility of the seller that at
the time of marketing the seed still meets the standards.

Traditionally in most EC countries all the official examinations
have been carried out by government organisations. However in the UK
a mixture of government organisations and licensed trained personnel
and privately owned laboratories carry out this work. This system was
in operation before our entry into the EC and other member states are
now showing interest in developing along similar lines.

Apart from setting standards for seed (see Table 1 for the
germination standard for a range of species) regulations lay down
procedures for the taking of samples and for the sealing and labelling
of packages of seed so that categories and grades of seed can be
readily identified. The regulations require that if any seeds have
been subjected to any chemical treatment this fact and the nature of
the treatment or the proprietary name of the chemical used in the
treatment shall be stated on the label or printed indelibly on the
package.

<u>Other European Countries</u>

A number of European Countries in the European Free Trade
Association are currently in the process of bringing their seeds
legislation into line with the EC and adopting the EC seed standards.
Sweden is one such country, but where its existing standards are
higher, it is considering continuing with the more stringent
requirements for certification of seed produced in its own territory.
For cereals higher standards may possibly be retained for maximum
content by number of seeds of other species and germination. While
the EC has a specific standard for loose smut (*Ustilago nuda*) in
barley no specific standards are prescribed for other seedborne
diseases on cereals like stinking smut (*Tilletia spp*), *Fusarium
nivale*, *Septoria nodorum*, *Drechslera* spp and *Bipolaris sorokinana*.
These are all currently tested for in Sweden and if present over a
specified percentage the seed must be treated before it may be
marketed. The use of routine seed health testing for assessing the
need for fungicide treatment of cereal seed has been the practice in
Sweden for more than 20 years. Germination standards are given for
some species in Table 1.

Switzerland is also bringing seed standards into line with the EC requirements. Trading in seed is currently carried out under VESKOF (Association of Swiss seed companies under official control in agricultural and vegetable seeds) rules. While for some impurities there are standards by number, standards for purity and germination are agreed between the buyer and seller. However treated seed must be marked as such and the active substance has to be declared and additional regulations (staining etc) have to be observed.

Finland is a third country looking at free movement of seed with the EC. It currently has seed standards which tend to be lower than those applied in the EC. Minimum germination levels for some species are included in Table 1.

In the longer term some East European countries will aim to make trade in seed with EC easier. Currently Poland has a complex set of standards for seed with three grades of seed with different standards for germination for each of the grades (see Table 1).

Norway currently has germination standards for home produced seed (see Table 1) but for imported seed quality is controlled by what the importer can obtain and will accept. However all seed entering Norway must be tested at their Official Seed Testing Station and the quality characteristics ascertained there must be shown on the label or seed package when the seed is sold. Norway has particularly stringent requirements for wild oats (*Avena fatua*, *A. ludoviciana* and *A. sterilis*) to try and avoid this becoming a serious weed problem. To avoid unnecessary fungicide seed treatment Norway has decided that all seed lots of cereals shall be subject to disease tests, following the Swedish practice, and for treatment to be applied only when results indicate that application is necessary.

<u>Non European Countries</u>

The USA is a major source of seed but the US Federal Seed Act (USDA, 1988) is a truth-in-labelling law prescribing very few standards that have to be met. Apart from vegetable seeds in small containers for sale to the home gardener no germination standards are prescribed in the Federal Seed Act Regulations (USDA, 1987). Even the pure live seed (purity % x germination %) requirements for imported seed listed in these regulations no longer apply leaving importers to determine the quality of seed which they are prepared to accept. The US Federal Seed Act has no requirements or standards relating to seedborne diseases. However, some States do have seedborne disease standards for seeds of certain crops. These seedborne disease standards are usually part of regulations designed to keep crop production areas free of specified diseases. The Federal Seed Act Regulations detail the labelling requirements for treated seed. The US Environmental Protection Agency (EPA) Regulations Section 153.155 require that all pesticides intended for use as seed treatments must contain an EPA approved dye. This results in all chemically treated seed sold in the USA being stained. The Federal Food, Drug and Cosmetic Act section 2.25 also has a requirement that all interstate shipments of treated cereal seed is coloured or stained.

In contrast to the USA, the Canadian Seeds Act and Regulations prescribe in detail standards for a range of quality attributes for all crop species of importance there. Seed may be marketed under a range of grades, each grade having its own set of standards, with one or more of the quality standards differing for each grade. Some germination standards are given in Table 1 showing the variation in requirement according to grade of seed. Very few seedborne diseases are covered by the Canada Seeds Act and Regulations, loose smut (*Ustilago nuda*) in barley, and ergot and/or sclerotia in cereals, sunflower, clovers, grasses and oilseed brassicas being exceptions. Where the barley has been treated with a registered product for controlling loose smut testing is not required.

Any seed treated with a pest control product must be thoroughly stained with a conspicuous colour to show that the seed has been treated and the container marked or labelled bearing the symbol and signal word presented under the Pest Control Products Act and the regulations made thereunder. These indicate the degree of risk inherent in that product and must be accompanied by a statement not to use the seed for food or feed and giving the common chemical name of the product.

The Republic of South Africa in its Agricultural Pests Act No 36 of 1983 provides for the effective control of pests that are deemed important. This legislation does specify particular pests and diseases for a very wide range of species. The Plant Improvement Act No 53 of 1976 as amended provides for the registration of seed processors, the recognition of new varieties, for a system of certification of seeds produced in South Africa and the control of import and export of seeds. The Act also covers other propagating material. In many respects the legislation is similar in content and effect to UK legislation. It requires the words "Poison - treated seed" to be conspicuous on the container or on its label if the seed has been treated with a controlled substance which is poisonous or harmful to humans or animals. A number of acts impinge on the use of chemical seed treatments. A very wide range of crop seeds are prescribed under the legislation with a range of seed quality requirements and different standards for prepacked and imported seed from other seed material (see Table 1 for germination standards). Israel also sets standards for purity, weed seed content and germination for the major crop species. In many instances germination standards are similar over a wide range of countries.

In many developed countries governments are seeking ways of withdrawing from direct involvement in seed production, quality control being recognised as being the responsibility of the seller, although the government often still sets standards below which seed may not be sold. In developing countries there is not always an established private seed trade and small scale farmers may need the greater protection provided by government quality - control systems. Even in these circumstances there is recognition of the advantages of the greater efficiency which usually results when private companies become involved in seed production. However legislation still needs to provide adequate safeguards for the end user as well as promoting the production of high quality seed whether by the public or private sector or a mixture of the two.

Table 1 Germination Standards

	EC	Sweden	Finland	Poland [+]	Israel	South Africa [*]	Canada [+]	Norway
Avena sativa (Oats)	85	90	80	94/89/83	85	80	85/75	90/88/85°
Hordeum vulgare (Barley)	85	87 or 90°	80	95/90/83	85	80	85/75	90,88,85°
Secale cereale (Rye)	85	87	80	94/89/83	75	80/70	75/65	85
Triticum aestivum (Wheat)	85	87 or 90°	80	95/90/83	85	80	85/75/70	90,88,85
Zea mays (Maize)	90	90	-	95/90/85	80	80/70	90/80/75	
Festuca pratensis (Meadow grass)	80	85	75	90/82/75	-	-	80/70	80
Festuca rubra (Red fescue)	75	85	75	90/82/75	-	-	80/70	80
Lolium multiflorum (Italian ryegrass)	75	90	75	90/80/70	80	80/50	80/70	80
Lolium perenne (Perennial ryegrass)	80	90	75	92/85/78	80	80/50	80/70	80
Phleum pratense (Timothy)	80	85	75	90/82/75	-	-	80/70	85
Medicago sativa (Lucerne)	80	75	60	85/75/65	80	80/70	85/80/70	-
Pisum sativum (Pea)	80	80	75	95/86/77	80	70/50	80/75/ 70/65	-
Trifolium pratense (Red clover)	80	75	60	86/78/70	-	70/50	85/80/75	80
Trifolium repens (White clover)	80	75	60	85/78/70	80	80/50	80/70	-
Vicia faba (Field bean)	85	80	-	93/84/75	80	80/70	85/80/ 75/65	-
Beta vulgaris (Sugar/Fodder beet)	68-80	75	80	80/65	70	70/50	75/65	-
Brassica napus (Oilseed rape)	85	80	80	95/85/75	65	(70/50)	90/80/75	85
Linum usitatissimum (Flax, Linseed)	92,85	75	80	92/84/80	80		85/70	-
Allium cepa (Onion)	70	70	70	85/80/65	75	70/50	75	80
Brassica oleracea (Cabbage)	75	80	80	90/80/70	75	70/50	80/70/60	85
Daucus carota (Carrot)	65	70	70	65/50	70	70/50	60	-
Lactuca sativa (Lettuce)	75	85	80	80/70	80	70/50	75	-
Lycopersicon lycopersicum (Tomato)	75	85	85	80/70	80	70/50	75	-
Phaseolus vulgaris (French Bean)	75	85	85	85/75	80	70/50	85/80/ 75/65	-

* South Africa Two standards: prepacked and imported and other seed

+ Germination standard varies with grade of seed

° Standard varies with winter and spring sown seed

THE INTERNATIONAL SEED TESTING ASSOCIATION (ISTA) AND ASSESSING SEED QUALITY

ISTA was formed in 1924 as a result of the recognition of the need for a universally accepted single set of seed testing rules which could be followed by all countries trading in seed so that results could be readily understood and compared. While the principles remain the same, in the intervening period many changes have been introduced to the details of the tests to improve their value for assessing suitability for sowing and to increase the uniformity of results between laboratories, both nationally and internationally. The current ISTA rules came into effect on 1 July 1993.

Assessment of any quality attribute is made on a very small amount of seed taken from a seed lot. It is vital therefore that the seed is representative if the result is to be of any value in defining the true quality of the lot. The rather time consuming procedure for drawing a sample prescribed in UK seeds legislation is not time wasted if it ensures the relevance of test results. For cereal, a minimum sample size of 1000g is prescribed for a seed lot up to a maximum size of 25,000kg. Equally as important as the drawing of the sample from the lot is the subsequent handling in the laboratory to ensure the test result accurately reflects the sample and therefore the seed lot.

<u>The Standard Germination Test</u>

The standard germination test is designed in a highly reproducible way to show the capacity of a population of seeds to produce plants in a field under favourable conditions by exposing them to conditions in the laboratory which are optimal for germination. The germination percentage is obtained by careful examination and classification of the seedlings, at a stage of development where accurate assessment of shoot and root system is possible, and the capability (normal seedlings) and inability (abnormal seedlings) for continued development into normal mature plants can by determined. In the seed testing context, therefore, germination refers to a more advanced stage of development than would normally be described by this term in physiological studies or such tests as a malting test for barley.

The test is undertaken normally on four replicates of 100 seeds. Seed is first subjected to a purity test, an essential prerequisite to the germination test, as it determines the material on which the test will be undertaken. By following internationally agreed and very precise rules, based on the principle that any crop seed which might conceivably germinate should be tested, it is ensured that equivalent material is always put forward for germination. Some degree of arbitrariness is inevitable: thus for example only broken seeds larger than half the original size are included, no attempt being made to determine the presence or absence of an embryo. The 'pure seed' from this test is sub-sampled and planted often using a vacuum planter which must be used correctly if bias, such as the preferential selection of light seed, is to be avoided (Bould and Arthur, 1978).

A basic medium is sterile sand, free from chemicals which may affect germination and standardised in particle size so that after the addition of the appropriate amount of water the right moisture/oxygen balance is obtained for germination to proceed. For cereals each replicate is planted onto the surface of a layer of moist sand 13 mm deep in a 150 mm diameter aluminium dish and after planting covered by a similar depth layer and a lid placed over the dish to restrict loss of moisture. By careful standardisation and control of the medium and other test conditions all samples and subsequent retests, in the same or different laboratories, are provided with identical growing conditions ensuring reproducibility of result. After the appropriate period if all seeds and seedlings are assessable the test is completed but if some cannot be evaluated with reliability these are left to develop further. If seedlings in the sand test are difficult to evaluate either because of the effect of seed treatment or from some other cause, a retest may be made in a proprietary seedling compost, but with the temperature and duration of the test the same as for the sand (Tonkin, 1969).

For inclusion in the germination figure reported, cereal seedlings must have a well developed root system having at least one long white, slender seminal root and an intact plumule with a well-developed green leaf, within or emerging through he coleoptile near the tip. Also included are seedlings with certain limited defects such as superficial discolouration of the shoot and, or roots, providing development of the structures is otherwise normal, or splits in the coleoptile near the tip which do not extend down more than one-third the length measured from the tip. All other seedlings are classified as abnormal and not included in the germination percentage. All remaining ungerminated seeds are examined to check whether the seed is dormant, when a retest with more extensive pretreatment may be required, or dead.

These principles apply to the testing of all species but of course details vary for substrates, temperature for optimum growth, length of test, dormancy breaking requirement and seedling evaluation. Table 2 gives information for a number of species on substrate, temperature, length of test and dormancy breaking method. Though light is not essential for any of these species to germinate it is often provided to ensure development of seedlings which can be easily evaluated. For each species the rules define, in the same way as for cereals, what seedlings can be included in the germination percentage reported.

In tests where normal germination is shown to be depressed and there are sufficient abnormal seedlings it is often possible to indicate from the types present the reason for the loss in germination. Mechanical injury, drying at too high a temperature and the application of too high levels of chemical treatment, for example, result in characteristic types of abnormal seedlings which are readily recognised by trained seed analysts. Abrol (1978) surveyed seedling abnormality in cereals and gives clear illustrations of the wide range found in tests made at the OSTS Cambridge. No similar surveys have been published for other species but the same wide range of abnormalities are found for most species where their seed is subject to the same handling and treatment procedures.

Table 2 Germination Methods Used at OSTS Cambridge

Species	Substrate	Temperature °C	First count days	Final Count (days)	Dormancy breaking Methods
Avena sativa (Oats)	S	20	5	10	Prechill
Hordeum vulgare (Barley)	S	20	4	7	Prechill : GA_3
Secale cereale (Rye)	S	20	4	7	Prechill
Triticum aestivum (Wheat)	S	20	4	8	Prechill : GA_3
Zea mays (Maize)	S	20	-	7	-
Festuca pratensis (Meadow fescue)	TP	20-30	7	14	Prechill : KNO_3
Festuca rubra (Red fescue)	TP	15-25	14	21	Prechill : KNO_3
Lolium multiflorum (Italian ryegrass)	TP	20-30	7	14	Prechill : KNO_3
Lolium perenne (Perennial ryegrass)	TP	20-30	7	14	Prechill : KNO_3
Phleum pratense (Timothy)	TP	20-30	7	10	Prechill : KNO_3
Medicago sativa (Lucerne)	BP	20	5	10	Prechill
Pisum sativum (Pea)	S	20	-	8	-
Trifolium pratense (Red clover)	BP	20	5	10	sealed polythene envelope
Trifolium repens (White clover)	BP	20	5	10	sealed polythene envelope
Vicia faba (Field bean)	S	20	-	14	-

Table 2   Continued

Beta vulgaris (Sugar/Fodder beet)	TP	20	5	14	Prewash for 2 hours at 25°C. Followed by drying for 2 hours at 25°C.
Brassica napus (Oilseed rape)	TP	20-30	5	7	Prechill
Linum usitatissimum (Flax, Linseed)	S	20	-	7	Prechill
Allium cepa (Onion)	BP	20	7	12	Prechill
Brassica oleracea (Cabbage)	TP	20-30	6	10	Prechill
Daucus carota (Carrot)	TP	20-30	7	14	-
Lactuca sativa (Lettuce)	TP	20	-	7	Prechill
Lycopersicon lycopersicum (Tomato)	BP	20-30	7	14	KNO_3
Phaseolus vulgaris (French bean)	S	20	-	9	-

Substrate	S = Sand TP = Top of paper on Copenhagen tank BP = Between paper inside polythene envelope
Temperature	A single figure indicates a constant temperature. Two figures separated by a dash indicates a 24 hour cycle with 16 hours at the lower temperature and 8 hours at the higher temperature.
First count	A deviation is permitted but seedlings must be sufficiently well developed for assessment or counting should be delayed.
Dormancy breaking	Prechill : Prechill at 5°C for 3 to 7 days GA_3 : Use GA_3 in solution at 0.05% or 0.1% KNO_3 : Use 0.2% KNO3 solution to moisten at beginning of test

The classical symptoms of chemical phytotoxicity observed in germination tests on cereals (and on other species), have been a thickening of the shoot including both the coleptile and first leaf, and or root with in extreme cases elongation being so retarded that these structures are almost spherical in shape. Not all chemicals that come into contact with seed have quite the same effect on seedling development and glyphosate, for example, sprayed onto cereal crops to control couch grass, has been observed only to affect the geotropism of the roots which start to grow upwards. Baytan and Ferrax (Tonkin, 1987) were observed to produce seedlings with marked retardation of coleptile elongation with no particularly obvious effect on root and leaf development. In this case sufficient elongation occurred if the tests were left two days longer, when seedlings could be evaluated as normal, and most treated tests had germinations above the marketing standard of 85%.

Sand (and filter paper) tests tend to accentuate the effect of chemicals present on the seed because there is no adsorption of the chemicals by the sand particles. Also no water is added after the initial wetting of the sand so increasing the likelihood of uptake of chemicals on the seeds. While accurately treated samples frequently show no phytotoxic symptoms in these tests, it has been established that it is appropriate to retest samples where phytotoxic symptoms are evident, in compost. Except where serious over treatment has occurred, even where the great majority of seedlings have been affected in the sand test, an acceptable result is usually obtained in compost.

<u>Discussion</u>

Legislation, in whatever form, sets the framework for the marketing of seed with the intended aim of supporting improved agricultural production. It must encourage the production of good quality seed but at the same time not be prohibitive in its requirements so that inadequate quantities of seed are available in the market place to satisfy growers needs. Therefore where minimum standards are prescribed these are set at levels which may be too low for specialist needs, though providing adequate safeguards against serious crop problems arising. In the EC minimum germination standards for example for sugar beet are at the highest 80% (for some classes the standard is lower) but in the UK the seed lots used for sugar beet production now show laboratory germination percentages of around 95%. Such a high level is essential to the production of the right population when sowing single seeds at spaced intervals.

With close spaced sowings such as those for cereals, where tillering can also compensate for lower establishment, such high levels may not be critical and the 85% minimum may prove satisfactory for most purposes. However in cereals many lots of seed have germinations of 95% or better. At a sowing rate of 3.0 million seeds/ha there is a very large difference (420,000) in the number of seeds being sown with the potential to produce a satisfactory plant between the best seed lots at 99% germination and those at 85%. Clearly seed with very high germination is desirable for many situations and many crops, particularly those that are wide spaced. Adjusting seed rates can compensate for low germination, but there is more risk of uneven plant distribution.

In the laboratory ideal conditions for germination are provided and all external factors such as soil fungi, insects, mice, birds and variations in soil and climatic conditions are eliminated. The grower aims for the optimum in trying to create a good seed bed and providing the best environment to support growth. The more successful he is the more closely will establishment reproduce the figure obtained in the standard germination test. However, it is not possible to achieve the ideal in the field and some depression of germination is likely. In less favourable conditions, particularly in circumstances of specific stress conditions, some seed lots in spite of above standard laboratory germination establish less satisfactorily than expected. A number of 'vigour' tests have been developed for the detection of seed lots meeting the standard for laboratory germination but showing weakness; eliminating these lots must increase the probability of obtaining satisfactory establishment when conditions are adverse. However so far such tests have not been considered sufficiently standardised to be included in ISTA rules and 'vigour' is not covered in seeds legislation. The one exception is vigour in maize using the 'cold test' which is covered in Austrian Seeds Legislation.

The correct application of seed treatments evenly applied on individual seeds should not reduce germination and may well enhance establishment in the field. Uneven application, even at the overall correct rate, may result in some reduction in germination because of phytotoxic effects on overtreated seeds, but in most seed lots this will not affect the marketability of the lot. Where legislation sets minimum standards frequently testing before treatment is permitted. This accepts that properly evaluated and correctly applied treatments rarely result in serious problems of establishment in the field. However seeds legislation frequently now stipulates identification of the treatment used and colouring of the seed so that the user is in no doubt that the seed is treated and should not be used for human or animal consumption.

For those producing new seed treatments it is important to know what standards seed has to meet, how the seed will be tested and how seedlings will be evaluated in checking the seeds suitability for marketing and sowing. The ISTA seed testing rules provide a virtually universally accepted basis for the testing. With properly trained seed analysts carrying out the tests seed can be correctly and consistently evaluated wherever the testing is carried out. All concerned with treatment, marketing and using the seed can as a result have confidence that the seed will usually perform satisfactorily and as expected.

ACKNOWLEDGEMENTS

Thanks are due to colleagues in a number of ISTA accredited seed testing stations for providing information on their legislation.

REFERENCES

Abrol, B.K. (1978) A survey of seedling abnormality in cereals. _Journal of the National Institute of Agricultural Botany 14_, 433-457.
Agriculture Canada, (1991). Office Consolidation of the Seeds Act and Seeds Regulations, Parts I and III, version SOR/91-609. _Plant_

Products Division, Canada, Ottawa 78pp + tables.

Anonymous, (1964) The Plant Varieties and Seeds Act 1964. HMSO, London.

Bould, A., Arthur, T.J. (1978) The vacuum counter as a sampling instrument. <u>Seed Science and Technology 6</u>, 495-504.

ISTA, (1993) International Rules for Seed Testing 1993. <u>Seed Science and Technology 21</u>, Supplement, Rules 1993

Kelly, A.F. (1989) <u>Seed Planning and policy for agricultural production.</u> Belhaven Press, London and New York.

MAFF, (1993) The Beet Seeds Regulations 1993. HMSO, London.

MAFF, (1993) The Cereal Seeds Regulations 1993. HSMO, London.

MAFF, (1993) The Fodder Plant Seeds Regulations 1993. HSMO, London.

MAFF, (1993) The Oil and Fibre Plant Seeds Regulations (1993). HSMO, London.

MAFF, (1993) The Vegetable Seeds Regulations 1993. HSMO, London.

Tonkin, J.H.B. (1969) Seedling evaluation: the use of soil tests. <u>Proceedings of the International Seed Testing Association 30</u>, 73-88.

Tonkin, J.H.B. (1987) Noted effects of some chemical treatments in germination tests on wheat and barley. <u>British Crop Protection Council Monograph No 39 - Application to seeds and soil</u> 113-120.

United States Department of Agriculture (1988) Federal Seed Act. US Government Printing Office.

United States Department of Agriculture (1987) Federal Seed Act Regulations Part 201-202. US Government Printing Office.

EFFICACY AND PHYSICAL/MECHANICAL DATA REQUIREMENTS FOR
APPROVAL OF SEED TREATMENTS IN THE UK

D. D. SLAWSON, M.J. GILLESPIE

MAFF, Pesticides Safety Directorate, Rothamsted, Harpenden, Herts, AL5 2SS

ABSTRACT

The scale and scope of trials and the expected level of performance required by the
UK regulatory authority to demonstrate acceptable efficacy and physical/chemical
properties of seed treatment products in support of approval/authorization, are
presented.

INTRODUCTION

In the UK, it is a legal requirement under both The Control of Pesticides Regulations
1986 (Anon., 1986a) and the EC Authorization Directive (Anon., 1991) that applicants for
approval/authorization of a plant protection product demonstrate acceptable efficacy and
physical/chemical properties. The evidence required to demonstrate acceptable efficacy and
physical/chemical properties for seed treatment products are generally less well-known than
for foliar or soil-applied plant protection products. To address this lack of information, the
official authority responsible for approval/authorization of plant protection products in the
UK, the Ministry of Agriculture, Fisheries and Food's Pesticides Safety Directorate, issued
guidance notes to applicants on the scale and scope of trials and the expected level of
performance required to demonstrate acceptable efficacy and physical/chemical properties of
seed treatment products. The guidance notes are summarized below.

EFFICACY DATA REQUIREMENTS

Effectiveness against pests/pathogens

Evidence is required (or where appropriate, a reasoned case argued) to support the
proposed claims made for each disease/pest on a product label.

Scale and scope of trials

The evidence should be provided largely by conducting field experiments under
conditions as near as possible to commonly accepted practice. Where soil type or geographic
location can affect performance, products should be tested on a range of sites/situations
relevant to the proposed area of use. This core of information can, where appropriate, be
supplemented by contrived experimental situations, such as, defined artificial plots with
introduced pests or diseases; or various pot or container trials under glass or in the open.

In the case of soil-borne pests, such as wheat bulb fly (*Delia coarctata*), sites should be
sought where pest numbers are high so that treatments under test are challenged by the

problem. The pest numbers should be determined by appropriate methods. The guidelines set out in Appendix 10 of Data Requirements for Approval under the Control of Pesticides Regulations 1986 (Anon, 1986b) should be followed where trials are to be sited in fields with natural infestations of soil- or air-borne pests, e.g. wheat bulb fly and leatherjackets (*Tipula oleracea* or *T. paludosa*) on cereals; pygmy mangold beetle (*Atomaria linearis*), millepedes and other soil pests of sugar beet; aphids, thrips; and systemic fungicides for control of air and splash-borne diseases. For example, the requirement, '*to assess the performance of the product under the conditions likely to occur there during the relevant period(s) of the year*' should be provided by distributing trials throughout the appropriate national regions.

Where activity against seed-borne diseases is to be tested, it is important to obtain seed stocks carrying moderate to high levels of disease. The proportion of seed carrying a given disease should be known and recorded. Artificially inoculated seed can be used for diseases, such as, bunt (*Tilletia caries*) on wheat. Guidance on trials techniques are given in Organisation Européenne et Mediterranéenne Pour la Protection des Plantes/European and Mediterranean Plant Protection Organization (OEPP/EPPO) guidelines on seed-borne cereal fungi (OEPP/EPPO, 1980) and seed treatments against seedling diseases (OEPP/EPPO, 1988a), the latter of which provides information on seedling diseases of larger seeded field crops, vegetable crops and beet and their damping-off diseases due to soil-borne fungi.

Numbers of trials and plot size may be limited by infected seed availability but the aim, for a major pest/disease, should be a minimum of 10 trials showing similar acceptable results. Results should be presented from tests conducted over at least two growing seasons. Preferably results should be provided from different seed batches and/or different locations. Field evidence may be supported by experiments in controlled or glasshouse conditions where it is possible to contrive challenging conditions. For pests or diseases which are extremely spasmodic in occurrence, these requirements may be reduced.

<u>Assessment criteria</u>
For seed-borne diseases on cereals, which have the potential to multiply rapidly from one generation to the next i.e. bunt (*Tilletia caries*) on wheat, covered smut (*Ustilago hordei*) on barley and oats, loose smut (*Ustilago nuda*) on wheat and barley, and leaf stripe (*Pyrenophora graminea*) on barley, evidence showing consistent control of >98%, is required to support a claim for 'control' on a product label. Demonstration of consistent control of 85-98% may be accepted for a claim of 'partial control' with the restriction that control is not sufficient for crops grown for multiplication.

For other pests/diseases, effectiveness of the test product will be judged in comparison to the untreated control and approved standards. Unless there are good reasons to accept a poorer level of control than that provided by a standard reference product, the level, duration and consistency of control of the test product must be similar to that provided by the standard. If no suitable reference product exists, the test product must be shown to give a defined benefit in terms of the level, duration and consistency of control.

<u>Safety to the crop</u>

Evidence must be provided (or where appropriate, a reasoned case argued) to support the proposed use on each crop on a product label.

<u>Scale and scope of trials</u>
Some guidance on trials techniques are given in an OEPP/EPPO guideline on phytotoxicity assessment, which includes a section on seed treatments (OEPP/EPPO, 1988b).

With the exception of germination tests, evidence should be largely provided by conducting field experiments under conditions as near as possible to commonly accepted practice. Where soil type, geographic location or sowing date can affect safety to crops, products should be tested on a range of sites/situations relevant to the proposed area of use.

With the exception of germination tests, a minimum requirement for other aspects of crop safety on a major crop e.g. winter wheat, should normally be 10 trials showing similar acceptable results. Results should be presented from tests conducted over at least two growing seasons and should include a range of common cultivars.

For more minor crops, these requirements may be reduced. For example, in the case of a seed treatment product for use on vegetable brassicas, provided satisfactory evidence of germination is provided on a range of varieties of each crop, the remaining requirements need only be satisfied for a single representative of each species e.g. evidence of safety on cabbage (*Brassica oleracea* var. *capitata*) could be extrapolated to cauliflower (*B. oleracea* var. *botrytis*) or Brussels sprouts (*B. oleracea* var. *gemmifera*).

Tests for crop safety should, where possible, be separated from tests for effectiveness on pests or diseases in order to remove an interfering variable. It is advisable to use certified seed in tests on crop safety. Seed should be free from infection by seed-borne diseases, which affect germination and emergence. Furthermore, in field trials taken to yield (essential information for crop safety), seed should be free from seed-borne diseases which adversely affect yield. N.B. It is important that all crop safety work should include the normal (N) recommended dose. For products where any adverse effects, however transitory, have been seen in trials at N dose, the margin of selectivity on the target crop must be established using twice the normal (2 N) dose. Doses less than twice the recommended dose should only be used where there are physical loading problems in applying the 2 N dose.

<u>Germination</u>
Evidence is required on the affects of seed treatment on seed germination in standard germination tests. Protocols specifying standard test procedures and assessment methods for various crop seeds are provided in the International Rules for Seed Testing (Anon., 1985). These standard tests may be supported, where necessary, with data obtained in standard soils or growing media. Normally, evidence of satisfactory germination is required from tests on a minimum of three common cultivars of each crop. Each test should consist of four replicates, each of 100 seeds.

Field emergence

Although, germination tests under controlled, artificial conditions are very challenging to a chemical seed treatment, evidence of satisfactory field emergence, which is the most important parameter for the farmer/grower, must also be demonstrated. Data may be provided using 1.0 m² plots (or larger) and records should provide comparisons with controls and standards, not only in terms of final crop stand but also speed of emergence.

Crop Vigour

Information is required on the subsequent growth and vigour of the plants. This may be done using a visual scoring system relative to untreated controls and standards. If early vigour is adversely affected by a seed treatment it is most important to provide evidence as to whether or not the treated crop recovers or gains in vigour compared with untreated and standards and the period of time over which this occurs.

Selectivity

As well as information on crop vigour, it is most important to assess the incidence and/or severity of any phytotoxic effects. It is essential that the symptoms measured and recorded are described accurately. In the absence of any observable effects, it should be clearly stated that this was the case.

Crop Yield

Evidence must be provided to show that a seed treatment does not adversely affect yield in the absence of the pest or disease. Clearly, as far as possible seed should be used which is free of seed-borne problems which affect germination, emergence and vigour of the crop. Trial design and dimensions must be suitable for yield comparisons.

Taint testing of treated crops

The requirement for taint data will depend upon the crop. Such data are normally only required for root crops e.g. carrots and radishes. However, because certain preservation processes can be sensitive to the formation of taints, discussion with the specialists, such as, Campden Food and Drink Research Association, is advised.

There are accepted procedures and numbers of taint trials required depending upon whether the crop or process is considered a major one for the food industry. Guidance on the general requirements for taint testing is given in Working Document 10/5 of Data Requirements for Approval under the Control of Pesticides Regulations, 1986 (Anon., 1986b).

Safety to following crops

Evidence or a reasoned case is required to demonstrate safety to following crops. In many instances, a case may be made for the safety to following crops based on evidence from fate and behaviour studies and results from crop screening studies.

Assessment criteria

Germination above the minimum required under the various Seeds Regulations is taken to be 'satisfactory' e.g. >85% for cereals.

For all other aspects of crop safety (emergence, vigour, selectivity, yield and quality, including, where relevant, taint), there must be no unacceptable adverse effects on treated plants or to following crops, except where the proposed label indicates appropriate limitations of use.

Biological compatibility with other seed treatments

Where the proposed label claims include recommendation for use of the product in combination with other products, whether the product is applied at the same time or before/after any other product(s), the principles for demonstration of satisfactory effectiveness and crop safety must be satisfied.

Effectiveness

In practice, evidence of effectiveness of both tank-mix / co-applied products, may often be extrapolated from evidence of satisfactory loading of active ingredients in both products.

Crop safety

Evidence of crop safety of tank-mixes or co-applications of products may often be limited to demonstration of satisfactory germination and field emergence.

Storage of treated seed

In cases where treated seed will be stored before use, the principles for demonstration of satisfactory effectiveness and crop safety after an appropriate period of storage must be satisfied. Treated seed should be stored in conditions approximating to commercial practice, e.g. in paper sacks and at ambient temperatures. The period of storage should reflect the length of time for which seed would normally be stored e.g. for most crops, 18 months should be sufficient to cover the event that sowing was missed and seed will be retained for use in the following season.

Effectiveness

In practice, evidence of effectiveness following storage, may often be extrapolated from evidence of satisfactory retention of active ingredients on stored, treated seed.

Crop safety

Evidence of crop safety following storage may often be limited to demonstration of satisfactory germination and field emergence.

PHYSICAL/MECHANICAL DATA REQUIREMENTS

Achieving and maintaining target dose

The data for a seed treatment product must cover two distinct areas. They are: (a) satisfactory retention of the chemical and physical properties of the seed treatment product in its container and (b) achieving the correct loading of seed treatment onto seed and maintaining the required loading to the point of sowing by the user. Requirements for evidence of satisfactory retention of the chemical and physical properties of the seed treatment product in its container, have already been covered in 'Guidelines for Formulation and Storage Stability Requirements for UK Registration' (Anon, 1993).

<u>Scale and scope of trials</u>

In general, at least one test should be carried out on seed from each crop (e.g. wheat, barley and oats) or crop group (e.g. brassicas) on which the product is proposed for use. Alternatively, the applicant may make a fully detailed case for the extrapolation of data from one seed type to another. Currently, there are few recognised standard methods for the tests described. Details of any method used should be submitted in full and include an assessment of the precision of the method.

Evidence must be provided to demonstrate satisfactory loading of the product on the seed when treatment is made by commercial seed treatment machinery. Evidence from small scale machines may be accepted *in lieu* of full-scale commercial machinery but because different seed treatment machines (both commercial and small scale) employ different mechanisms it is advisable to test a seed treatment using at least one commercially available seed treatment method. This is especially important should preliminary small scale testing show any suggestion of possible formulation/machinery problems.

Information must include actual results of the active ingredient(s) loading and observations on the uniformity of seed treatment distribution between seeds. Tests should be performed before and after storage of the product. Storage regimes should comply with those listed in 'Guidelines for Formulation and Storage Stability Requirements for UK Registration' (Anon., 1993).

A method for the determination of the uniformity of distribution of liquid seed treatment formulations is currently under consideration by the Collaborative International Pesticides Analytical Council (CIPAC).

Test results of studies are required to establish the retention of the seed treatment product on seed after treatment. Tests for adhesion by means of standard 'drop' tests have been used commonly (Jeffs, 1974; Maude *et al*, 1986; Suett & Maude, 1988; and Anon., 1980).

<u>Assessment criteria</u>

The suggested minimum retention of active ingredient(s) is 70% of the initial target dose.

<u>Physical compatibility with other seed treatments</u>

Evidence of satisfactory loading is required where the proposed label claims include recommendation for use of the product in mixture with other products, whether the products are applied simultaneously or sequentially.

<u>Seed drill tests</u>

Evidence must be provided to demonstrate the satisfactory flow of treated seed through the relevant seed drill mechanism(s) available commercially. Comparison should be made with untreated and standard treated seed from the same batch.

<u>Stored treated seed</u>

In cases where treated seed will be stored before use, evidence is required on the effect of storage on seed treatment retention. Alternatively, a case for satisfactory retention of active ingredient(s) may be made based on the biological effectiveness of the product.

Treated seed should be stored in conditions approximating to commercial practice e.g. in paper sacks and at ambient temperatures. The period of storage should reflect the length of time for which seed would normally be stored e.g. for most crops, 18 months should be sufficient to cover the event that sowing was missed and seed will be retained for use in the following season.

<u>Assessment criteria</u>
The suggested minimum retention of active ingredient(s) after storage of treated seed is 70% of the initial target dose.

ACKNOWLEDGEMENTS

Thanks are due to Mr Trevor Martin (Chairman of British Crop Protection Council's Seed Treatment Working Party) and to the British Agrochemical Association for helpful comments during the development of these requirements.

REFERENCES

Anon. (1985) International Rules for Seed Testing and Annexes. *Seed Science and Technology* **13**, 299-355 and 356-513.

Anon. (1986a) The Control of Pesticides Regulations 1986 (SI 1986/1510) (ISBN 0-11-067510-X).

Anon. (1886b) Data Requirements for Approval under the Control of Pesticides Regulations 1986.

Anon. (1991) Council Directive (91/414/EEC) concerning the placing of plant protection products on the market. *Official Journal of the European Community*, **L230**, 19 August 1991, pp. 1-32.

Anon. (1993) Guidelines for Formulation and Storage Stability Requirements for UK Registration. MAFF Pesticide Safety Division. February 1993.

CIPAC (1980) MT83 Seed Adhesion Test for Powders for Seed Treatment. In: *CIPAC Handbook 1A*, Ashworth, R.R. de B.; Henriet, J.; Lovett, J.S.; Martijn, A. (complilers) Harpenden, UK: Collaborative International Pesticides Analytical Council, pp. 1591-1592.

Jeffs, K. A., (1974) Tests for retention of powders. *Rothamsted Experimental Station Report for 1973*, part 1, p. 177.

Maude, R. B., Presly, A. H.,; Lovett, J. F. (1986) Demonstration of the adherence of thiram to pea seeds using a rapid method of spectrophotometric analysis. *Seed Science and Technology*, **14**, 361-369.

OEPP/EPPO. (1980) Guideline for the biological evaluation of pesticides. No. 19. Seed-borne Cereal Fungi. June 1980. PARIS. 1, rue Le Nôtre.

OEPP/EPPO. (1988a) Guideline for the biological evaluation of pesticides. No. 125. Seed Treatments against Seedling Diseases. *Bulletin OEPP/EPPO Bulletin* **18**, 743-752.

OEPP/EPPO. (1988b) Guideline for the biological evaluation of pesticides. No. 135. Phytotoxicity Assessment. *Bulletin OEPP/EPPO Bulletin*, **18**, 817-836.

Suett, D. L.; Maude, R. B. (1988) Some factors influencing the uniformity of film-coated seed treatments and their implication for biological performance. In: *Applications to Seeds and Soil*, T.J.Martin (Ed.), *BCPC Monograph No. 39*, Thornton Heath: BCPC Publications, pp. 25-32.

RESEARCH-BASED IMPROVEMENTS IN THE REGULATION OF HAZARDS TO WILDLIFE FROM PESTICIDE SEED TREATMENTS

A.D.M. HART

Central Science Laboratory, Ministry of Agriculture, Fisheries and Food, Tangley Place, Worplesdon, Guildford, Surrey, GU3 3LQ

M.A. CLOOK

Pesticides Safety Directorate, Ministry of Agriculture, Fisheries and Food, Rothamsted, Harpenden, Herts, AL5 2SS

ABSTRACT

This paper reviews the essential role of scientific research in evaluating and improving the regulation of hazards to the environment from seed treatments. Examples are given of ways in which research is improving methods of risk assessment, risk management and environmental monitoring. This helps to ensure that the regulatory system provides proper protection for both crops and the environment, without the need for unnecessary testing or undue expense.

INTRODUCTION

The formulation of pesticides as seed treatments has a number of advantages, compared to spraying or the broadcasting of pellets or granules. The chemical is concentrated on the seed which it is intended to protect, maximising its efficiency in controlling the target organism. Many of the potential risks to the user and the environment are reduced. However, the concentration of the chemical on what may be an attractive food item can present special risks to certain types of wildlife. There is a history of poisoning incidents involving grazing and seed-eating birds (Stanley & Bunyan, 1979; Greig-Smith, 1988a) which ensures that the risks to these species receive particular attention in the evaluation of seed treatments for regulatory approval. This paper reviews the contribution which research can make to improving the regulation of environmental risks, through developments in methods of risk assessment, risk management and environmental monitoring.

CURRENT REGULATORY PRACTICE

The UK regulatory system includes three components: risk assessment, risk management and environmental monitoring. The first two of these are closely interlinked, and are the responsibility of the Pesticides Safety Directorate (PSD) of the Ministry of Agriculture, Fisheries and Food (MAFF). Companies requesting approval of a pesticide submit data on its toxicity to non-target organisms including mammals, birds, aquatic organisms, non-target arthropods, honeybees, earthworms and soil micro-organisms. The data are evaluated by PSD specialists who then identify hazards to non-target organisms and assess the risks of those hazards being realised, taking into account the type of use proposed for the pesticide. Additional data will be requested if they are necessary to provide a sound basis for decision-making. Where the predicted risks are considered unacceptable, the proposed use may be modified to reduce them.

Guidance for risk assessment has recently been published by an expert panel under the auspices of the European and Mediterranean Plant Protection Organisation (EPPO) and the Council of Europe (CoE). This comprises decision-making schemes in the form of questionnaires, based on up-to-date developments in the science of risk assessment (EPPO/CoE, 1993). The EPPO/CoE schemes are compatible with the approach outlined in the final draft Uniform Principles in Annex VI of the European Community Directive 91/414/EEC. The schemes relating to terrestrial vertebrates, earthworms, honeybees, soil processes and beneficial non-target arthropods have been adopted for use by PSD, with modifications where necessary for consistency with existing PSD practice. PSD currently uses its own assessment procedure for risks to the aquatic environment, which is similar in concept to the corresponding EPPO/CoE scheme.

It is often assumed that seed treatments present a low risk to the environment compared to other pesticide applications. However, exposure cannot be ruled out, so the risk must be assessed and cannot be regarded as negligible. The risks are generally greatest for mammals and birds, and data may be required on the palatability of treated seed or seedlings. Risk to honeybees is likely to be low unless the product is both systemic and persistent, so that significant concentrations appear in nectar or aphid honeydew. Risks to the aquatic environment are also likely to be low, unless the pesticide is both mobile and persistent in soil, or unless treated seed is to be broadcast in fields adjacent to water bodies.

The third component of regulation is post-registration monitoring. In England and Wales, MAFF operates the Wildlife Incident Investigation Scheme (WIIS) to monitor the mortality of honeybees and vertebrate wildlife (Fletcher & Grave, 1992). Cases which appear to be related to pesticide use are investigated by means of field investigations, post-mortem examinations and laboratory analyses. The scheme serves three main purposes. First, it acts as a safety-net to detect mortality which was not anticipated by risk assessment prior to approval. If the actual level of risk is revealed to be unacceptable, then the Advisory Committee on Pesticides may recommend withdrawal or modification of approval. Second, the scheme provides evidence for use in prosecution, when poisonings have resulted from the abuse or misuse of a pesticide. Third, it provides data on actual risks which can be used to validate and improve risk assessment methods.

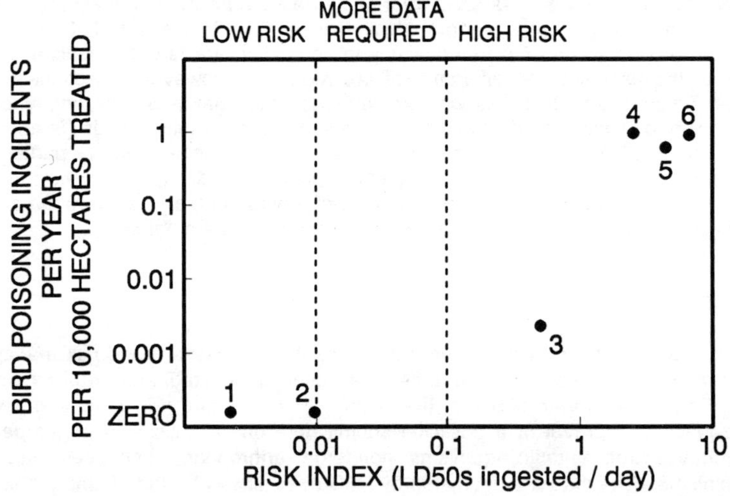

FIGURE 1. Validation of predicted risks to birds feeding on treated wheat seed, by comparison with the frequency of poisoning incidents recorded by the UK Wildlife Incident Investigation Scheme. Key to pesticides: 1, methoxyethyl mercury acetate; 2, phenyl mercury acetate; 3, lindane; 4, carbophenothion; 5, chlorfenvinphos; 6, fonofos. Dotted lines indicate the thresholds used for the initial classification of risk using the EPPO/CoE guidelines (EPPO/CoE, 1993): intermediate values usually lead to a requirement for additional data to refine the estimate of risk.

VALIDATION OF RISK ASSESSMENT METHODS

An essential and central element of research to improve regulatory procedures is the evaluation of current methods, to identify those areas in need of improvement. Risk assessment procedures should be both reliable and efficient, providing proper protection for wildlife without unnecessary testing or undue expense. Reliability can be assessed by comparing the predictions, which risk assessment produces, with data on the actual effects of pesticides in normal use. This approach has been described by Hart & Greig-Smith (1992) and has been applied to aspects of the EPPO/CoE decision-making schemes. For example, in the EPPO/CoE scheme, predicted risks to seed-eating birds from seed treatments are based initially on a risk index which estimates the number of lethal doses a bird would ingest in a day if it fed exclusively on treated seed. This risk index may be validated by comparison with data from the WIIS on the frequency of poisoning incidents involving seed treatments, expressed as a function of usage to take account of differences in the areas of land sown using the different pesticides. Results for seed treatments used on wheat show good agreement between predicted and observed risks, although the number of pesticides for which the comparison can be made is rather small (Figure 1). To complete the validation this comparison will be extended to include other crops and other types of hazard, such as that to birds grazing on seedlings.

INVESTIGATION OF BIASES IN INCIDENT MONITORING

Confidence in the regulatory system depends on the reliability of the incident monitoring scheme as well as the risk assessment procedures. The WIIS has been effective in identifying some unacceptable risks to birds and honeybees, resulting in the modification of approvals for the pesticides involved (eg. Stanley & Bunyan, 1979). However, it has been recognised for some time that only a proportion of casualties are reported to the scheme, and that the probability of reporting is likely to be lower for smaller, less conspicuous species (Greig-Smith, 1988b; Hart, 1990; Figure 2 below). MAFF has therefore funded research by the British Trust for Ornithology (BTO), in collaboration with the Central Science Laboratory, on the factors which may affect reporting rates (Baillie *et al.*, unpublished manuscript). Reporting probabilities were estimated using data on the recovery of dead or dying birds reported to the BTO ringing scheme. Probabilities for 99 species varied by over two orders of magnitude, from 0.06% for the chiffchaff (*Phylloscopus collybita*) to 16.75% for mute swan (*Cygnus olor*). Eighty-six percent of the variation in reporting rates was explained by a curvilinear relationship with body weight, confirming that small inconspicuous species are much less likely to be reported. This bias should be taken into account when interpreting data on pesticide casualties recorded by the WIIS. Consideration is now being given to how this should be done.

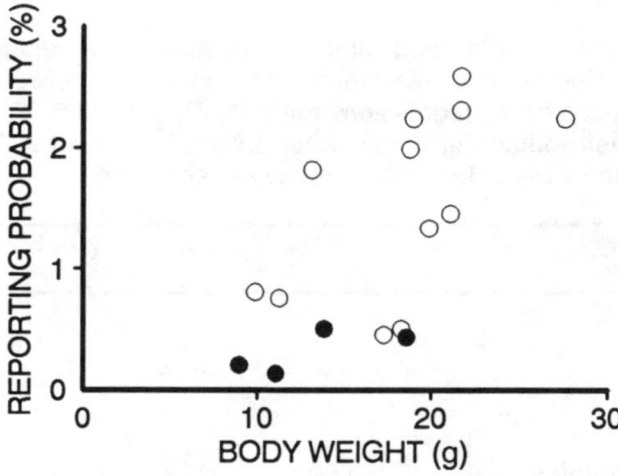

FIGURE 2. Relationship between body weight and the probability that dead birds will be reported to the BTO ringing scheme (reproduced from Hart, 1990). Each point represents one passerine species: open circles represent residents, closed circles represent summer visitors.

INVESTIGATION OF EFFECTS AT POPULATION LEVEL

The reporting rates for pesticide casualties are likely to be lower than those estimated from ringing data, except when large numbers occur in a single incident, because the estimates are based on birds marked with a leg ring bearing an address for reports. Therefore the relationship between the frequency of reported incidents and the true level of mortality in wild populations remains uncertain. This relationship needs to be understood because of the importance of the incident scheme as a safety net, and as a means of validating risk assessments. Research is therefore underway to estimate the level of pesticide-related mortality in a wild population of woodpigeons (*Columba palumbus*) in an area of Cambridgeshire where there is extensive use of fonofos, and more limited use of chlorfenvinphos, as seed treatments for winter wheat. Radio-tracking techniques are used to monitor a large sample of birds, and those which die are recovered for autopsy and residue analysis. This will provide an estimate of the contribution of these pesticides to overall mortality, which can then be compared to estimates based on data from WIIS for the same pesticides. Together with the analysis of reporting rates for ringed birds, described above, this will provide a greatly improved basis for regulatory decisions about the ecological significance of pesticide casualties.

INFLUENCE OF REPELLENCY IN THE WILD

The ability of birds to detect and avoid food contaminated with pesticides has been demonstrated in laboratory studies, and such studies are increasingly being used to argue that exposure will be limited in the wild (Grau *et al.*, 1992; Avery *et al.*, 1993). However, it has yet to be demonstrated conclusively that low palatability in the laboratory is a reliable indicator that a formulation will be avoided in the wild. Research is therefore being conducted to investigate the occurrence and significance of avoidance behaviour by woodpigeons in relation to the use of seed treatments, in conjunction with the radio-tracking study described above. This includes systematic observations of the behaviour of woodpigeons visiting fields of wheat sown with different seed treatments. Preliminary results indicate significant differences between the treatments, though differences in feeding rates are difficult to interpret (Hart *et al.*, unpublished). Most substantial are the differences in the proportion of time spent with the head withdrawn, a posture which is often adopted by resting birds but which is also symptomatic of intoxication with organophosphorus pesticides (Hart, 1993). On the first day after drilling, increased time was spent in this posture by birds on fields treated with fonofos, but not for chlorfenvinphos (Table 1). It is possible this may be related to the fact that fonofos is microencapsulated, whereas chlorfenvinphos is not. Our studies are continuing to confirm and elucidate this result, and to assess directly the relation between the repellency of seed treatments in laboratory and field. The results will provide a much improved basis for regulatory decisions in risk assessments involving evidence of repellency.

TABLE 1. Percentage of time spent in head-withdrawn posture by woodpigeons observed on fields of winter wheat sown with different seed treatments. The difference between fonofos and the other treatments is statistically significant (regression analysis, $F_{1,6}$ = 23.6, P = 0.003). There is a significant interaction between fonofos and days after drilling ($F_{2,6}$ = 10.9, P = 0.01), reflecting the fact that the difference is largely restricted to the first day after drilling.

Days after drilling	1	2	3	4
Untreated	0	-	11.9	-
Fungicide only	6.0	13.3	2.6	0
Fungicide plus chlorfenvinphos	0	7.5	-	-
Fungicide plus fonofos	32.2	11.8	6.0	-

RISK MANAGEMENT

Risk management techniques are important in enabling regulators to approve the use of products which, while offering substantial advantages to the grower, would otherwise present an unacceptable risk to wildlife. Research has a major role to play in developing new risk management techniques, and ensuring the reliability of existing ones. Current risk management for seed treatments includes general precautions imposed by the Code of Practice for the Safe Use of Pesticides on Farms and Holdings, which has statutory force in Great Britain. The Code includes a general requirement to take reasonable precautions, including preventing contamination of non-crop land. There is also a specific requirement to clear up or cover any treated seed which is spilt. In addition, the approval of particular seed treatments may include label instructions specifying that the seed should be incorporated into the soil, as opposed to being broadcast on the surface where they would present greater risks to birds and other wildlife. Some organophosphorus seed treatments, which are highly toxic to birds, are approved for sowing only in autumn. This is because in spring birds are thought more likely to concentrate their feeding on newly-sown crops, as the availability of other foods is reduced. Depth of drilling is recognised as a factor which may influence risk to birds, but label instructions for organophosphorus wheat seed treatments specify drilling depths designed to maximise germination. Research could be conducted to investigate whether significant reductions in risk could be obtained by increasing drilling depth, without causing unacceptable reductions in germination. Adding colourings or flavourings which reduce the attractiveness or palatability of treated seed may also provide a means of improving risk management, and has already been the subject of some research (eg. Greig-Smith, 1988a).

CONCLUSIONS

The examples which have been presented illustrate the importance of scientific research in evaluating and, where necessary, improving the reliability and efficiency of methods used for risk assessment, risk management and environmental monitoring. Substantial progress has been made already, through the efforts of regulators, scientists and industry, both in the UK and abroad. Continued progress will help to ensure that the regulatory system provides proper protection for both crops and the environment, without the need for unnecessary testing or undue expense.

ACKNOWLEDGEMENTS

We are grateful to P.J. Prosser, F.L. Sibson, M.R. Taylor and S. Langton for their contributions to the study on woodpigeons, to the British Ornithological Union for permission to reproduce Figure 2, and to Dr S.R. Baillie for permission to cite unpublished results. The analysis of reporting rates for dead birds was funded by a MAFF contract awarded to the British Trust for Ornithology. Other work was funded by the Pesticides Safety Directorate through contracts with the Central Science Laboratory.

REFERENCES

Avery, M.L.; Decker, D.G.; Fischer, D.L.; Stafford, T.R. (1993) Responses of captive blackbirds to a new insecticidal seed treatment. *Journal of Wildlife Management*, **57**, 652-656.

Baillie, S.R.; McCulloch, M.N.; Hart, A.D.M. (unpublished manuscript) Interspecific variation in the reporting probabilities of dead birds and its implications for the surveillance of avian mortality.

European and Mediterranean Plant Protection Organisation; Council of Europe (1993) Decision-making scheme for the environmental risk assessment of plant protection products. *EPPO Bulletin*, **23**, 1-165.

Fletcher, M.R.; Grave, R.C. (1992) Post-registration surveillance to detect wildlife problems arising from approved pesticides. *Brighton Crop Protection Conference - Pests and Diseases 1992*, **2**, 793-798.

Grau, R.; Pfluger, W.; Schmuck, R. (1992) Acceptance studies to assess the hazard of pesticides formulated as dressings, baits and granules, to birds. *Brighton Crop Protection Conference - Pests and Diseases 1992*, **2**, 787-792.

Greig-Smith, P.W. (1988a) Hazards to wildlife from pesticide seed treatments. In: *Applications to Seeds and Soil,* T.J. Martin (Ed.), *BCPC Monograph No. 39*, Thornton Heath: BCPC Publications, pp. 127-134.

Greig-Smith, P.W. (1988b) Wildlife hazards from the use, misuse and abuse of pesticides. *Aspects of Applied Biology*, **17**, 247-256.

Hart, A.D.M. (1990) The assessment of pesticide hazards to birds: the problem of variable effects. *Ibis*, **132**, 192-204.

Hart, A.D.M. (1993) Relationships between behavior and the inhibition of acetylcholinesterase in birds exposed to organophosphorus pesticides. *Environmental Toxicology and Chemistry*, **12**, 321-336.

Hart, A.D.M.; Greig-Smith, P.W. (1992) Validation of environmental risk assessment procedures for pesticides. *Brighton Crop Protection Conference - Pests and Diseases 1992*, **2**, 799-804.

Hart, A.D.M; Prosser, P.J.; Sibson, F.L.; Taylor, M.R.; Langton, S. (in prep.) Feeding ecology and behaviour of woodpigeons in relation to the use of insecticidal seed treatments.

Stanley, P.I.; Bunyan, P.J. (1979) Hazards to wintering geese and other wildlife from the use of dieldrin, chlorfenvinphos and carbophenothion as wheat seed treatments. *Proceedings of the Royal Society of London, Series B*, **205**, 31-45.

OPERATOR EXPOSURE DURING SEED TREATMENT

H. CHAMBERS

Health and Safety Executive, Magdalen House, Stanley Precinct, Bootle, Merseyside L20 3QZ

ABSTRACT

Operator exposure to pesticides and other hazardous substances may occur at various stages during the seed treatment process. The paper discusses the potential for exposure during both mobile and static seed treatment operations and the measures taken to control it.

INTRODUCTION

Operator exposure to pesticide may occur at any stage in the seed treatment cycle. That exposure, together with exposure to other hazardous substances, such as dust and micro-organisms needs to be assessed. Although the main focus of the paper is exposure to pesticides, the Control of Substances Hazardous to Health Regulations (COSHH) 1988 apply to all substances and to all stages in the procedure, from reception of the pesticide to dispatch of the treated seed and disposal of the pesticide container.

Both hazard and risk should be considered. The collection and filing of chemical data sheets is just the beginning! The actual risk of exposure to pesticide during each constituent operation must be actively assessed, the assessment acted upon and the situation regularly reviewed.

The Control of Pesticides Regulations 1986 (made under the Food and Environment Protection Act 1985) are the principal Regulations concerned with pesticides and apply in addition to COSHH. It is a duty under these regulations that users of seed treatment equipment who were born after 31.12.64 or those treating seeds as contractors have National Proficiency Test Council Certification (NPTC). Some farmers treat their own seed with or without assistance of their families or employees. Provided these operators were born after 31.12.64 or are working under the direct supervision of a certificate holder they will not require a certificate. Those involved in the sale, supply and storage for supply of seed treatment chemicals (classed as agricultural pesticides) need British Agricultural Standards Inspection Scheme (BASIS) certification. Presently approved pesticides are exempted from classification and labelling for supply under the Classification, hazard, information and packaging (CHIP) regulations.

HOW AND WHEN DOES EXPOSURE OCCUR?

The potential for exposure must be assessed during both routine and non routine operation. The routes of exposure ie inhalation (of vapour, dust and aerosol) and skin absorption should be considered at all stages. Key activities during routine operation can be identified as follows;

Reception of pesticide

Transport of pesticides on mobile treaters

Moving pesticide from store to treatment equipment

Introduction of pesticide into seed treatment equipment

Calibration

Application of pesticide to seed

Any adjustments to calibration or change over of chemical during treatment

Bagging

Sampling

Disposal of container and waste

The next step is to identify the actions of the operator during each of these activities and the ways in which exposure may result from these.

Non-routine operations include handling events such as breakdown or spillage and machinery maintenance. Unusual or small batch treatment may bring in to use otherwise little used plant which due to age and design may have inferior standards of control than other frequently used plant.

Persons carrying out maintenance or cleaning of the plant must be aware of the potential for exposure to organic dust and pesticide. This is important for those involved in changing filters, stripping plant or any other activities involving direct contact with surfaces that may have pesticide on them. Pesticide build up may occur on areas that are not normally accessed eg inside machinery housings and dust containing pesticides may be generated when these are dismantled. Better machine design should reduce the amount of pesticide and seed build up that occurs on internal surfaces. In order to maintain genetic purity of batches of seed, plant is generally cleaned out prior to a different type of grain being treated. Large tonnages of the same genetic type of grain may be treated successively but a variety of pesticides may be used. This means that when the plant is eventually cleaned out, exposure to other pesticides as well as the last one used must be considered. If cleaning and maintenance is carried out by outside contractors, it is the responsibility of the employer to ensure that the contractors are aware of the risks that they may be presented with.

FACTORS INFLUENCING EXPOSURE

Factors to be taken into account when assessing exposure include the likelihood of exposure occurring, the duration and frequency of exposure and the consequences of that exposure.   The likelihood of exposure depends on the physical form of the pesticide and the method of handling.   The consequences depend on the toxicity of active ingredient and of any solvents present and the duration of exposure.   It must also be borne in mind that seed treatment chemicals, unlike most other pesticides which are applied in a dilute form, are mainly used in their concentrated form. Climatic conditions, eg increases in temperature which can influence the rate of volatilisation of the product and also the suitability of some personal protective equipment (ppe) must also be considered.

Companies who undertake COSHH assessments centrally should be aware of regional variations in treatment regimes. For example the use of organophosphorous products against wheat bulb fly which occurs only in certain regions may require biological monitoring for those workers.

CONTROL OF EXPOSURE

After elimination or substitution and enclosure the next choice of control in the COSHH hierarchy is engineering controls (use of personal protective equipment is a last resort). More effective engineering controls especially at the design and build stage mean less reliance on ppe and given that the conditions in seed treatment are often hot and the work physically hard, anything that avoids the operators being burdened with ppe is beneficial. As old plant is phased out or modified reliance on ppe for control should reduce.

The extent to which local exhaust ventilation (lev) is used varies. On both mobile and static plant it is applied at the grain cleaning stage for dust removal and on most static plants at the pesticide application stage. Some mobile treaters have lev at the application stage. Mobile treaters have the advantage of more natural ventilation provided that this is managed properly. One of the elements of NPTC is to ensure that the mobile is positioned so that the wind takes away pesticide and dust rather than blowing it onto the operator.

It is essential that the exhaust from these controls is either collected in a suitable collection device or that particularly in the case of mobile units, where direct dispersal to atmosphere is more likely, the hose from the fan is long enough not to return dust to the work area.

<u>Setting up and calibration</u>

When introducing pesticides to the treater, exposure to pesticide can occur when removing lids and inserting bungs, tubes etc or in the case of powders when the powder is fed into the treatment equipment. Engineering controls such as 'snap on' attachments with non drip seals and valves and specially designed lids have been developed with the aim of making connection much cleaner and easier. Careful handling is also important. Treatment plants with large throughput are able to use bulk containers and multi product applicators which reduce the number of times that handling is necessary. Often, simple improvements such as having pipes short enough so that they are easy to handle and result in less splashing will reduce exposure. If containers are well sealed entry of contaminants that cause blockages will be avoided as will the need for filters that require cleaning.

However, leaks, spills etc may still occur during setting up or calibration and as it is a concentrate that is being used it is therefore essential that gloves, faceshield and suitable respiratory protection are worn.

It may be necessary to wash the plant with both solvent and water when using certain products. Care must be taken to avoid contact with the solvent as well as the rinse water.

<u>Application of pesticide to seed</u>

During the actual treatment, unless the machine has to be entered to clear blockages (in which case ppe must be used), exposure to pesticide and dust is generally prevented or controlled due to the enclosed nature of the process. However, the plant must be maintained in good condition to prevent emission from seals and joints.

<u>Bagging</u>

The ideal system is where bagging is performed by automatic means. However where such facilities are not available, exposure both by inhalation and skin absorption to pesticide at the bagging stage can be reduced by ensuring that the seal between the mouth of the bag and the chute is good. Also the use of good techniques, for example when closing the bag folding it away from the operator, can reduce exposure to any pesticide contamination that may come up off the treated seeds. It must also be borne in mind that if a final visual check of treatment is given by looking into the bag, exposure to pesticide by inhalation could occur. Bags of differing sizes may be used so that an adequate seal cannot be made, in this case suitable respiratory protection may be necessary. If the stitcher fails and improperly sealed bags are thrown onto the pallet, grain spillage results. Precautions must be taken when cleaning up because at this stage the grain is still not dry and the pesticide would easily be removed from the seed by handling.

<u>Cleaning and maintenance</u>

Since primary control methods are usually not applicable during these activities personal protective equipment (ppe) must be worn by those working on the plant (if it is still contaminated) and also by those carrying out the cleaning. Items such as respiratory protection, gloves and faceshield will probably be necessary during both routine and reactive maintenance.

CHOICE OF PERSONAL PROTECTIVE EQUIPMENT

Because ppe plays such a major role in controlling exposure it is important to ensure that it is chosen, used and maintained properly. Training in these aspects is a crucial part of an effective ppe strategy.

At present the approved pesticide label does not specify the actual type of ppe to be used except in broad generic terms. The choice depends on the pesticides in use and this should take into account both the active ingredient and the solvent in the product. The comfort and other individual requirements of the user eg beard, facial size and shape must also be considered. Equipment must also be suitable for the job, for instance gloves must not be made so unwieldy either by their size or material of construction that it is not possible to undertake the job. Gloves suitable for use when cleaning out the plant may not be suitable for use when dismantling pumps.

Advice should be sought from distributors and manufacturers of the pesticide. In many cases more detailed guidance on the specific types of ppe appropriate to that product can be found either on a non-approved section of the product label or in the safety data sheets provided to the seed treatment facility. Respiratory protection should be chosen to protect against dust and organic vapours where appropriate.

It is also essential that the ppe is used, stored and maintained properly ie away from sources of contamination when not being used. When gloves become contaminated they should be washed on the outside before removal. If this is not possible they should then be removed carefully to prevent contamination of hands with any residue that may remain on the gloves. Flock linings may be comfortable but should be avoided as these are difficult to decontaminate.

Overalls should be worn to protect clothing and skin. It is essential that overalls are kept clean because absorption of small doses may occur if there is repeated contact with contaminated clothing.

Arrangements should be made for cleaning, storage, maintenance and regular examination of respiratory protective equipment. Some disposable ppe is designed for more than one shift use and also requires examination and possibly maintenance.

Good washing facilities and eyewash are important to deal with any accidental contamination.

CONCLUSIONS

Exposure during routine operation of seed treatment equipment can be and is being controlled by engineering means backed up by the personal protective equipment. More reliance is placed on the use of personal protective equipment to control non-routine exposure. Control of exposure has to be underpinned by training (both formal and informal) in the use of all control measures and importantly the correct choice, care and maintenance of ppe.

REGULATORY CONTROLS ON FUNGICIDAL CEREAL SEED TREATMENTS -
PAST EXPERIENCE AND FUTURE PROSPECTS

H. EHLE

Biologische Bundesanstalt für Land- und Forstwirtschaft,
Abteilung für Pflanzenschutzmittel und Anwendungstechnik,
Messeweg 11 - 12, D 38104 Braunschweig

ABSTRACT

The development of non-mercurial cereal seed treatment
products and seed treaters has progressed in the Federal
Republic of Germany since the ban of mercury in plant
protection in 1982. Problems and trends in chemical
treatment of cereal seed against fungal diseases since
the change-over are described.

INTRODUCTION

Testing and authorization of plant protection products has
been mandatory in the Federal Republic of Germany since 1968.
Seed treatments are an important group of these products. The
Biologische Bundesanstalt (BBA) is responsible for the authorization of plant protection products in Germany. In the former
German Democratic Republic the situation as to seed treatment
products was different up to October 1990 and is not described
here.

Seed treatment of cereals (wheat, barley, rye and oats),
in particular to control seed-borne fungal diseases, is one of
the most effective forms of chemical plant protection. The
quantities of products required are small, thus minimizing the
environmental risk. Chemical treatment of cereal seed, especially basic and certified seed, is carried out as a routine precautionary measure. Like most European countries Germany has
strict regulations on cereal seed which make seed treatment
necessary, because certain seed-borne fungal diseases (e. g.
Ustilago and *Tilletia spp.*) cannot be controlled by fungicides
applied later to the growing crop.

The use of mercurial seed treatments has been prohibited
since 1 May 1982. The main reason for the ban was their toxicity to users. The only significant environmental hazard concerned the danger to bird species eating treated seeds. The
development of mercury-free products was boosted by the ban,
resulting in a substantial increase in the number of authorized
products. It was only after this ban came into force that
mercury-free products for cereal seed began to be used on a
large scale. But the change-over caused some problems concerning these products and their application in seed treatment
machinery. The problems and consequences are described.

PRODUCTS FOR CEREAL SEED TREATMENT

These products should be effective, not phytotoxic to cereals and not hazardous to man, wildlife (especially seed-eating birds) and the environment. In addition they should be easy to apply and adhere strongly to the seed. These prerequisites have to be fulfilled and be proved by submitting sufficient data together with the application form for authorization.

Table 1 shows the range of ingredients in authorized cereal seed treatment products and their fungicidal activity (situation 1993).

In contrast to mercury compounds many non-mercurial ingredients are systemic. Products containing such ingredients are also effective against loose smut (*Ustilago nuda*) of wheat and barley as well as foliar diseases (e. g. *Erysiphe graminis* on young plants) which cannot be controlled by mercury. But the risk of resistance to fungal diseases using mercury-free products is higher than that of mercury. Non-mercurial products often contain more than one active ingredient in order to match or exceed the fungicidal efficacy of mercury and to prevent resistance. Mercury-free ingredients have a lower vapour pressure than mercury compounds. So vapour movement of such ingredients as a factor of distribution within treated seed batches can be neglected. The amount of products needed for disease control is usually higher than that of mercury-based products (300 - 500 ml or g/dt seed instead of 200 - 300 ml or g/dt).

Cereal treatments are liquid or dry formulations. Mainly liquid formulations are used for fungicidal treatment of cereal seed in Germany. They are based on water or an organic solvent. Preference is given to waterbased products (FS) because solventbased liquids (LS) are more difficult to apply than FS-formulations due to the quick evaporation of the solvent. The adhesion of LS-formulations is so fast and their viscosity so low that no secondary distribution during mixing occurs. It is, therefore, crucial to achieve optimum distribution and most uniform loading of each seed at this first meeting between seed and product.

Powder formulations (DS) are easier to distribute from seed to seed than liquid ones. But their disadvantages can be insufficient adhesion to the seed and occurrence of dust during treatment and handling of seed. Addition of a sticking agent during application helps to overcome these adverse effects. Certain new sophisticated powder formulations for cereal seed treatment have been authorized in Germany without these disadvantages, but overall the trend is against powder treatments.

Water dispersible powders for slurry treatment (WS) are as to their physical properties (adhesion and distribution) similar to FS-formulations. However during preparation of the slurry dust may be generated. The importance of slurries for cereals is limited.

TABLE 1.   Ingredients in authorized cereal seed treatment products and their fungicidal activity.

INGREDIENTS	WHEAT							BARLEY						RYE			OATS	
	Tilletia caries, T. tritici (bunt)	*T. controversa* (dwarf bunt)	*Ustilago nuda* (loose smut)	*Fusarium nivale* (snow mould)	*F. culmorum* (seedling blight)	*Septoria nodorum* (seedling blight)	*Erysiphe graminis* (powdery mildew on young plants)	*Drechslera graminea* (leaf stripe)	*D. teres* (net blotch)	*Ustilago spp.*	*Fusarium nivale* (snow mould)	*Rhynchosporium secalis* (leaf blotch)	*Erysiphe graminis* (powdery mildew on young plants)	*Urocystis occulta* (stripe smut)	*Fusarium nivale* (snow mould)	*Erysiphe graminis* (powdery mildew on young plants)	*Ustilago avenae* (loose smut)	*Drechslera avenae* (seedling blight)
Bitertanol	x	x	x	x										x	x		x	
* Carbendazim	x		x	x											x		x	x
* Carboxin			x							x								x
Difenoconazole	x	x	x			x												
* Ethirimol													x					
* Fenfuram										x								
Fenpiclonil	x			x	x	x					x			x	x			
* Fuberidazole				x							x							
* Guazatine	x			x										x	x			
Iprodione				x				x							x			
* Imazalil								x	x									
Metsulfovax	x		x							x				x				
Nuarimol											x	x	x					
Prochloraz	x			x	x						x	x		x	x			x
Propiconazole	x			x											x			
Tebuconazole																	x	
Triadimenol	x		x	x		x					x		x	x	x	x	x	
Triaxozide								x										

* Ingredients in authorized cereal seed treatment products when mercurial treatments were banned (1 May 1982). These products are still authorized.

<u>DATA REQUIREMENTS FOR SEED TREATMENT PRODUCTS</u>

The amended German Plant Protection Act of 15 September 1986 provides the legal basis for demanding more data mainly on environmental effects of plant protection products and on the application machinery. That is why the range of this type of testing has been extended considerably in recent years.

Applicants seeking authorization for cereal seed treatment products have to

- describe the composition of the product, the content of active ingredient(s) and formulants and the physico-chemical properties of the product and its active ingredient(s) (e. g. vapour pressure, solubility in water, log Pow),
- specify the use pattern of the product (fungal pathogens, cereal species, recommended rate of application - and the minimum effective dosage, i. e., the lowest dose rate for sufficient disease control; if required the amount of water for some liquid formulations or the amount of sticking agent for powders, application technique),
- describe the mode of fungicidal action of the active ingredient(s),
- provide information on the susceptibility of the product to the development of resistance to pathogens.

In addition to this the applicants have to submit sufficient data on

<u>Efficacy:</u>
The trials are carried out by the companies themselves and on their behalf by the official plant protection stations of the federal states throughout Germany in order to obtain a balanced proportion of test reports from industry and official bodies (Ehle and Menschel, 1993).
Snow mould resistance on an economically significant extent to carbendazim and fuberidazole was recorded after seed treatment in winter wheat and winter barley after the severe winter in Northern Germany early in 1979 shortly after snow melting. Whereas mercury treated wheat and barley had little snow mould attack. Consequently, efficacy trials have been more stringent ever since and are carried out mainly with winter cereal seed infected with methyl 2-benzimidazole carbamate (MBC) resistant strains of *Fusarium nivale*.

<u>Phytotoxicity:</u>
Effects of seed treatment on germination, seed vigour and damage to seedlings are tested in the laboratory or glasshouse. If efficacy trials are conducted in the field, phytotoxicity of seed treatment products to cereal seedlings is also assessed. It is a well-known fact that some active ingredients especially triazoles (e. g. triadimenol) may affect crop emergence under unfavourable conditions during germination. Effects of seed treatment on germination and phytotoxicity are usually more pronounced in the field than in vitro. However, these effects are mostly transient.

<u>Application properties:</u>
After the ban of mercurials the application properties of
fungicidal seed treatments are tested prior to authorization.
The tests are carried out in at least one suitable seed treater
approved by the BBA. In addition to this seed samples are taken
to determine the adhesion of the product by means of an abrasion test. Moreover the quality of treatment (i. e. the ratio
of actual to target dose of a product on seed) is measured by
quantitative colourimetric single seed analysis. Measurements
of extracted dye from treated seed are related to the amount of
product on seed. They are possible because all fungicidal
cereal treatments contain a distinct dye. This range of tests
for a new product is carried out by the applicant or an official plant protection station. The BBA conducts additional laboratory tests with new products prior to authorization. Cereal
seed is treated in laboratory treaters either in the "Hege 11"
or in the "Mini-Rotostat" which can apply liquid and dry formulations equally well as commercial treaters. Both treaters are
described by Jeffs and Tuppen (1986). These further studies
concern the properties of products as to adhesion and the seed
to seed uniformity of distribution.

<u>Toxicity:</u>
Full toxicity data must be available for all compounds in
seed treatment products. The authorized non-mercurial fungicidal cereal seed treatments are classified as non-toxic. But
some furan compounds (e. g. methfuroxam) caused cancerogenic
effects on these rodents in long-term studies. Seed treatments
containing these compounds are not authorized any more for that
reason.

<u>Residues:</u>
Residue behaviour after fungicidal seed treatment is investigated in the growing crop and later on in straw and grain.
Residues exceeding maximum residue levels are usually not found
in straw and grain.

<u>Soil:</u>
Information required includes studies on adsorption /
desorption, degradation and metabolism of the active ingre-
dient(s) and on leaching of the product.

<u>Water:</u>
Hydrolysis of the active ingredient(s) is studied.

Non-target organisms such as

<u>Birds:</u>
Studies on the acute toxicity of seed treatments to Japanese quail and if necessary also to another species (e. g.
pigeons or pheasants) are required. In addition to this palatability tests are necessary for products being toxic to birds in
order to study their repellency properties on Japanese quail,
pigeons or pheasants. Products containing the ingredients
listed in Table 1 are not acutely toxic to birds except bitertanol which is moderately toxic only to pigeons.

<u>Aquatic organisms</u>:
Studies on the acute toxicity of seed treatments to fish
(mainly rainbow trout) and *Daphnia* as well as the chronic toxicity to algae are required since 1987. Like most plant protection products fungicidal seed treatments are toxic to fish.

<u>Earthworms</u>:
Studies on the acute toxicity of seed treatment products
to *Eisenia fetida* are required since 1989. Some fungicidal
active ingredients are acutely toxic to earthworms. Carbendazim
is one of them. However, up to now there is no evidence that
earthworm populations in fields might be affected by fungicidal
seed treatments.

<u>Beneficial organisms</u>:
Laboratory studies on the effects of seed treatment products on carabids and staphylinids are required since 1989.

APPLICATION TECHNIQUE AND SEED TREATMENT MACHINERY

There is a very close relationship between the product and
the application technique in seed treatment. It is important to
achieve uniform distribution of the product over the whole surface of the cereal seed. This is not easy because there may be
problems in dosing and distributing the product in the seed
treater. It is therefore necessary to consider which treater
can be used for a given product.

Mercury containing products were more or less uniform as
far as their content of active ingredients and dosages per 100
kg cereal seed were concerned. The non-mercurial products have
made higher demands on seed treatment machinery because their
physico-chemical properties vary and higher dosages are usually
needed for a sufficient disease control.

Some widespread application problems encountered during
the change-over to non-mercurials were as follows:
- inadequate flow properties, thus affecting the operation of
  the treaters and drilling of treated seed,
- some dry formulations had a poor adhesion to the seed and
  some liquid ones could not be distributed evenly,
- some solvent based products attacked plastic components of
  treaters,
- treaters frequently had to be adapted for the application of
  non-mercurials,
- cleaning of treaters had to be more frequent and thorough
  than after the application of mercurial products (Kohsiek
  and Jeffs, 1986).

Progress in the development of seed treatment machinery
has been considerable since 1982. Most of the commercial seed
treaters were tested and approved by the BBA after the ban of
mercurials (Table 2, situation 1993).

The procedure of declaration of plant protection equipment
includes seed treaters. Since 1 July 1988 such machinery can

TABLE 2. Approved and registered machines used for treating cereal seed.

name of machine (manufacturer)	method of application and mixing process	suitable for type of formulation	output in t/h
approved machines:			
* W.N.-4 (Niklas)	spinning disc and auger mixing	liquid	0.5 - 5
W.N.-7 (Niklas)	spinning disc and auger mixing	liquid and slurry	1 - 7
W.N.-12 (Niklas)	spinning disc and auger mixing	liquid and slurry	2 - 12
W.N.-20 (Niklas)	spinning disc and auger mixing	liquid and slurry	5 - 20
* BA 101-4F (Röber)	spinning disc	liquid	1 - 4
BA 101-8/F (Röber)	spinning disc	liquid	2 - 8
BA 101-12 FD (Röber)	spinning disc	liquid	2 - 12
BA 101-22 FD (Röber)	spinning disc	liquid	4 - 22
GB S 3/F-A (Goldsaat)	mist nozzle and auger mixing	liquid	0.5 - 2
Trans-Mix 45 (Amazone)	auger mixing	liquid and powder	treatment period of 30-60 s for 100 kg
registered machines:			
Macox-S (Mantis)	spinning disc	liquid	2 - 4.5
Feuchtbeizer Typ S.30 (DENIS-PRIVE)	lattice/distributor and auger mixing	liquid	0.3 - 3

* Already approved machines before mercurial treatments were banned (1 May 1982).

only be marketed in Germany "if it is designed such that when it is used correctly and in accordance with its intended purpose for the application of plant protection substances it does not produce any harmful effects on human and animal health or on the groundwater or does not have any other harmful effects, particularly on the environment, which, on the basis of the current state of the art, are avoidable."

Manufacturers, distributors or importers of new plant protection equipment have to declare to the BBA that these requirements are fulfilled and have to supply the declaration with detailed documentation and data. Registration will be granted if the requirements are met. If not, the BBA can demand the testing of the equipment concerned and can even withdraw the registration.

Furthermore according to German plant protection legislation plant protection equipment including seed treaters can be tested voluntarily. Testing is carried out by the BBA and the official plant protection offices of the federal states. It comprises complete treaters. Data on these tests allow an evaluation of the suitability of the treaters. Approval will be granted if they perform well.

Since 1987 the BBA prepares and publishes features for evaluating the maintenance of legal requirements on plant protection equipment including seed treaters.

Important features concerning seed treaters are as follows:
- They have to work reliably to ensure proper dosing and an even distribution of a product on the seed.
- They have to be manufactured in such a way that they can be filled up in a safe manner. Maximum and minimum limits of filling levels in the tanks can be determined easily and products cannot be released unintentionally.
- Full and easy to read dosing instructions have to be fixed at the treaters.
- The dosage of the product has to be adjustable at an easily accessible plaze. The adjustment of metering mechanisms has to be easily visible.
- During treatment the seed and the product have always to be in a correct ratio.
- With continuously working treaters the dosing of the product has to be interrupted automatically when the flow of seed stops and vice versa.
- At the outlet of the treater the product has to adhere to 'the seed with a tolerance of not more than ± 7 % of the mean value. The mean value must not deviate more than 10 % from the target dose.
- The applied amount of the product on at least 80 % of single seeds must not deviate more than 50 % from the mean value.
- Treaters have to be easy to clean.
- Parts of wear have to be accessible and changeable in an easy way.

LABEL INSTRUCTIONS AND PRODUCT INFORMATION

These contain details as to the application of a product to seeds. Furthermore, authorization is always connected with certain conditions which have to be shown on the label and product information: particulary special warnings and safety precautions in order to avoid damage to man, cereals and non-target organisms such as seed eating birds and fish (in case of accidental water pollution). In addition to this companies supply so-called positive lists in their instructions for use and product information to permit an effective seed treatment under practical conditions. Thus the positive lists for the equipment indicate the treatments for which it can be used. Conversely, the positive lists for the products contain information on suitable equipment with advice on adjustment, maintenance, cleaning etc. These lists are updated regularly to take account of the current situation as to seed treatment machinery and authorized products.

POST-AUTHORIZATION MONITORING OF SEED TREATMENT PRODUCTS

Because of the great diversity of factors which affect fungicidal cereal seed treatment, problems may develop in commercial use of these products. For this reason new products are monitored by the companies themselves and the official plant protection stations throughout Germany after their authorization as to factors such as disease control, phytotoxicity, yield response, resistance to pathogens and compatibility with other plant protection products (e. g. insecticidal seed treatments); application technique, quality of seed treatment and side-effects.

REFERENCES

Ehle, H.; Menschel, G. (1993) Authorization of fungicidal cereal seed treatments in Germany: official testing for efficacy, phytotoxicity and uniformity. *Pesticide Outlook*, 4, 14 - 18.

Jeffs, K. A.; Tuppen, R. J. (1986) Requirements for efficient treatment of seeds. In: *Seed Treatment*, K. A. Jeffs (ED.), *BCPC Monograph*, Thornton Heath: BCPC Publications, pp. 17 - 45.

Kohsiek, H.; Jeffs, K. A. (1986) Assessment of application methods and prevention of loss. In: *Seed Treatment*, K. A. Jeffs (ED.), *BCPC Monograph*, Thornton Heath: BCPC Publications, pp. 46 - 50.

THE NUMBER OF EXPOSED DRESSED SEEDS IN THE FIELD;
AN OUTLINE FOR FIELD RESEARCH.

W.L.M. TAMIS, M. GORREE, J. DE LEEUW, G.R. DE SNOO

Centre of Environmental Science, Leiden University P.O. Box 9518, 2300 RA Leiden,
The Netherlands

R. LUTTIK

National Institute of Public Health and Environmental Protection, P.O. Box 1, 3720 BA,
Bilthoven, The Netherlands

ABSTRACT

To evaluate hazards of treated seeds for mammals and birds, decision schemes are being
developed, in which information about the number of exposed seeds is an important
aspect. An overview has been made of factors influencing the number of exposed seeds.
The most important is the soil condition, followed by the place on the field, seeding
technique and weather and time after seeding. Also small accidents, like spills, may
offer a major exposure source of treated seeds. Information about the effect of seed
size, seeding density and depth is almost lacking. A research project into the number of
exposed seeds is outlined, focusing on seed size, seeding technique and soil type.
Preliminary research in winter wheat resulted in 18.5 exposed seeds m^{-2} in the centre
and 40 m^{-2} on the headland of the field.

INTRODUCTION

During evaluation of pesticides it is important that adverse side effects are taken into
account. In The Netherlands a step wise assessment method has been developed, to assess
the hazards of treated seeds for birds and mammals in the field (see Luttik & De Snoo,
1994 these proceedings). This hazard assessment method takes into account body weight,
daily food intake and diet of the species of concern, repellency of the pesticide and the
percentage of seeds incorporated in the soil. Luttik and De Snoo conclude that major
uncertainties exist on the extent in which pelleted seeds are taken up by birds mistaking
them as grit and about the number of exposed seeds after sowing. This article deals with
the outline of a research project into the number of exposed seeds. The different factors
that influence the number of exposed seeds are evaluated from literature and some
preliminary results are presented.

FACTORS INFLUENCING THE NUMBER OF EXPOSED SEEDS IN THE FIELD

The following factors influence the number of exposed seeds: i) seed size, seeding
density and depth ii) seeding technique and improper agricultural practice, iii) soil type
and condition (incl. weather before seeding), iv) place in the field (centre vs headland), v)
time and weather after seeding. Data are available on wheat and rapeseed: median mean
resp. 4.3 and 10.4 m^{-2} (Table 1). For only a few sources seeding densities are given and

the percentage exposed seeds ranges from c. 1.2 to 4.2% for wheat to 12% for rapeseed.

Seed size, seeding density and depth

These factors are correlated: larger seeds are mostly seeded deeper in lower densities and a lower number of exposed seeds can be expected under these conditions. The influence of seed size, seeding density and depth on the number and percentage of exposed seeds can not be derived from literature, since most research has been done with wheat (Table 1), in most cases without referring to seeding density. Only Davies (1974) mentioned a higher number of exposed wheat seeds, due to shallow drilling in order to reduce damage by insects.

Seeding technique and improper agricultural practice

The number of exposed seeds also depends on seed bed preparation, seeding implements and soil finishing. Seed bed preparation influences the soil condition and this will be dealt with later. Broadcast sowing probably causes the highest number of exposed seeds. Standard drilling probably will leave more seeds on the surface than precise drilling, because the machines for precise drilling can place the seeds more precisely at the desired place and depth, but insufficient information exists on this subject. Maze *et al.* (1991) found a higher number of exposed seeds when a press drill was used (5.5 m^{-2}) than when a hoe drill or an air seeder (with or without harrowing afterwards) was used (1.3m^2). Davies (1974) describes a decline of 60% in number of exposed seeds if harrowing took place after seeding. Also small accidents or improper management increase the number of exposed seeds. Davies (1974) and Maze *et al.* (1991) pointed out that seed spilling may be a major source of exposure of treated seeds. Davies (1974) ascribes the

Table 1. Mean number of exposed seeds m^{-2} directly after sowing (days 0 or 1) under normal conditions (fallow soil, centre or whole field numbers, soil types and machine implements combined, single seeding, soil finishing); - = data not mentioned or not relevant; * = mean of the highest reading of grains for those fields examined.

| exposed seeds m^{-2} | | | nr fields | remarks | source |
mean	min	max	x samples		
wheat					
4.3*	1.1	9.7	14 x 20	spring 1961	Murton & Vizoso, 1963
4.3*	1.1	14.0	12 x 20	autumn 1961	
4.0*	0.3	18.3	- x 20	spring 1962	
0.9	0.8	0.9	2 x 10		Jefferies *et al.*, 1973
3.0	1.0	8.0	4 x 1-2	autumn 1966	Davis, 1974
5.8	0.5	20.7	12 x 20	autumn 1972	
9.4	-	-	1 x 60		Wildlife Survey, 1975
4.0	-	-	1 x 60		Westlake *et al.*, 1980
2.3	-	-	12 x 10	experiment	Maze *et al.*, 1991
18.5	0.0	72.0	6 x 10	autumn 1991	own research, see Table 4
rapeseed					
13.9	0.0	52.0	4 x 10		Riedel *et al.*, 1992
6.9	2.0	14.4	3 x 10		Hänisch & Gemmeke, 1992

Table 2. Mean number of exposed seeds m^{-2} on the centre and headland of the field directly after sowing (days 0 or 1), see also Table 1; ratio calculated as geometric mean; * = recalculated assuming headland 5% of field area; ** = comparison single ("centre") vs double ("headland") seeding; soil type, seeding implements etc. combined.

centre field	head field	ratio H/C	number fields	source
3.6*	11.6	3.2	1	Davis, 1974
6.8	20.8	3.1	1	Wildlife Service, 1975
4.0	23.1	5.8	1	Westlake et al., 1980
2.3**	3.0	1.7	12	Maze et al., 1991
5.4	9.1	3.4	3	Hänisch & Gemmeke, 1992
18.5	40.0	2.7	6	own research, see Table 4

high number of exposed wheat seeds (4 m^{-2}) on the 8th day after sowing, as a consequence of the harrowing, rolling and reharrowing of a field to bring buried potatoes to the surface. In Table 4 the effect of an accidentally open seedpipe in field 6B is evident, in comparison with field 6A.

Place in the field (headland vs centre).

More seed often remains exposed on the headland than in the centre of the field, because the soil condition of the headland is worse, due to the turning of the machines, because seeds are spilled when the sowing equipment is taken out of the soil, and because part of the headland is double seeded when the headland is seeded crosswise. The number of exposed seeds on headlands is in general 3-4 times higher as in the centre of the fields (Table 2). In a heavy loam soil the number of exposed seeds is also high in the centre of the field, because of the overall bad field conditions (Hänisch & Gemmeke, 1992). The number of seeds after double seeding ("headland") is higher than after single seeding ("centre"), but the percentage exposed is lower if the double amount of seeds is taken into account (Maze et al., 1991).

Soil type and condition

Soil condition is the result of soil type, weather before and during seeding, seed bed preparation and the presence of previous crop debris (stubble) and influences the number of exposed seeds considerably. Hänisch & Gemmeke (1992) report a higher number of exposed rapeseeds in heavy (14.4 m^{-2}), more cloddy soils, than in light soils (2.0 m^{-2}). The highest number of exposed wheat seeds were found (Table 1) in a heavy clay soil in combination with heavy rains before seeding (Table 4). According to Davis (1974) soil condition had a greater effect than soil type. The number of exposed seeds was highest on clay after heavy rain. On a sandy loam with a better tilth the number was less, however the number of exposed seeds was even less on clay with a good tilth. The lowest number of exposed seeds occurred on a peaty soil under good weather conditions. In the Wildlife Survey (1975) 4 to 5 times more seeds remained on the cloddy part of a field compared to the good tilled part of the same field. Davis (1974) describes a negative influence of the presence of sugar beet debris on the number of exposed wheat seeds, whilst Maze et

Table 3. Percentage decrease of exposed seeds per day and number of weeks needed to reach 90% loss of exposed seeds; see also Table 1; calculations carried out with simple exponential equation; - = data not mentioned; * = recalculated assuming 10% of seeds left after stopping of the counts; ** = decrease retarded by heavy rain in first week.

Decrease % d^{-1}	nr fields	nr dates	$t_{90\%}$ (wks)	remarks	source
14.3	24	16	2.3	spring 1961	Murton & Vizoso, 1963*
16.4	27	14	2.0	autumn 1961	
38.3	-	6	0.9	spring 1962	
15.1	1	10	2.2		Jefferies *et al.*, 1973
30.8	3	3	1.0		Davis, 1974
32.2	1	5	1.0		Wildlife Survey, 1975
8.1	1	10	4.1		Westlake *et al.*, 1980**
15.8	1	5	2.1		Hänisch & Gemmeke, 1992

al. (1992) reported the contrary effect: 2.7 wheat seeds m^{-2} on summer fallow and 1.1 seeds on stubbled field.

<u>Time and weather after seeding</u>

Not only the weather before and during seeding is an important factor but also the weather *after* sowing. From the results of Westlake *et al.* (1980) an increase in seed exposure of 10-30% after a heavy rain (10 mm) can be calculated, but from 2 weeks after sowing heavy rains no longer had an effect, because the sprouted seeds were firmly anchored in the soil. Jefferies *et al.* (1973) describes a strong effect of snow and thaw, which resulted in a crumbled soil: from 0.9 exposed seeds m^{-2} also on the first day till 6.5 m^{-2} on the 15th day. Exposed seeds disappear in the course of time, by moving into the soil or by predation by mammals and birds. The median percentage decrease per day is ca. 16%, which means that within two weeks ca. 90% of the exposed seeds have disappeared (Table 3).

OUTLINE FOR FIELD RESEARCH FOCUSED ON THE NUMBER OF EXPOSED SEEDS IN THE FIELD

The most important factor influencing the number of exposed seeds is soil condition (especially soil type in combination with weather before and during seeding), followed by the place in the field and the seeding implements. From the overview it further becomes clear that little information is present about the relation between the number of exposed seeds and seed size (crops), seeding density and depth. In order to get a systematic and representative picture of the range of the number of exposed (and dressed) seeds in the field for the different crops (seed sizes) under different conditions and techniques used a research project has been started. Results of preliminary measurements are presented in Table 4; the results are already discussed in the preceding paragraph.

Table 4. Number of exposed seeds of winter wheat m^{-2} on centre and headland of field, fraction (%) of seeding density; results of preliminary field research in 1991 in the Haarlemmermeerpolder, The Netherlands; (200 kg winter wheat/ha, presumed 22 seeds/g; n = 10 squares (1 m^2) for centre field and 2 for head field; counts direct after sowing; clayey soil and rainy weather during sowing; * = accidental spilling; mean without 6B).

Field	Head field		Centre field	
	m^{-2}	%	m^{-2}	%
1	89	20.2	18	4.1
2	67	15.2	18	4.1
3	18	4.1	1	0.2
4	3	0.7	2	0.5
5	62	14.1	72	16.4
6A	0	0.0	0	0.0
6B*	95	21.6	5	1.1
mean	40	9.1	18.5	4.2

The outline of the project is as follows. In the spring of 1994 the effect of seed size (crops) and seeding technique will be investigated in a.o. the Haarlemmermeerpolder and in Flevoland in The Netherlands. Three seed size classes (small, medium, large) are combined with three groups of seeding techniques (broadcast sowing, standard drilling and precise drilling), which results in seven relevant combinations (Table 5). Within the seed size classes some major crops (area > 10000 ha) are selected. In one combination two crops are selected (bean and maize), which are seeded in a high respectively low density to investigate the relation between seeding density and fraction of exposed seeds. In the autumn of 1993 and 1994 the influence of soil type (sand, light clay, heavy clay) and soil condition, the influence of the weather after seeding, the disappearance rate (permanent quadrats on days 0, 1, 2, 5, 8 and 14) will be investigated in winter wheat. Ten counts (quadrat of 1 m^2, which lie at a regular distance from each other) will be made separately in the headland and the centre of at least 10 fields per treatment. First results of the 1993 counts will be presented during the symposium.

DISCUSSION

A lot of attention is paid to the number of exposed seeds, but also the buried seeds, sprouted seeds and seedlings (e.g. Beri *et al.*, 1969; Porter, 1977; Green, 1980) are subject to predation by a large variety of mammals and birds. The percentage decrease per day (Table 3) has been based on averaged field values and therefore are too low, because also quadrats with no or a low number of seeds are included. If only the quadrats were taken into account, in which exposed seeds are really present (e.g. Hänisch & Gemmeke, 1992), the percentage decrease per day for those quadrats would be 2-4 times as large.

Table 5. Selected crops as a result of seed sizes and seeding techniques in which the number of exposed seeds will be investigated; seed size: small < 0.015 g; 0.015 g < medium < 0.1 g, large > 0.1 g.

seed-size	broadcast sowing	standard drilling	precise drilling
small	-	spinach	sugarbeet
medium	wheat	wheat	sugarbeet (pelleted)
large	-	pea	bean & maize

ACKNOWLEDGEMENTS

We would like to thank Mark van der Wal for counting exposed seeds in the Haarlemmermeerpolder in 1991.

REFERENCES

Beri, Y.P.; Kumar, S.; Prasad, S.K. (1969) Preliminary studies with chemical seed treatment asbird deterrent for the protection of germinating peas. *Science and Culture* **35**, 473-475.

Davis, B.N.K. (1974) Levels of dieldrin in dressed wheat seed after drilling and exposure on the soil surface. *Environmental Pollution* **7**, 309-317.

Green, R.E. (1980) Food selection by sky larks and grazing damage to sugar beet seedlings. *Journal of Applied Ecology* **17**, 1613-1620.

Hänisch, D.; Gemmeke, H. (1992) Saatgutbehandlung bei Raps und Vogelgefährdung. In: *Pflanzenschutzmittel und Vogelgefährdung*, H. Gemmeke and H. Ellenberg (Eds), Berlin-Dahlem: Mitteilungen aus der Biologische Bundesanstalt für Land- und Forstwirtschaft Heft 280, pp. 123-130.

Jefferies, D.J.; Stainsby, B.; French, M.C. (1973) The ecology of small mammals in arable fields drilled with winter wheat and the increase in their dieldrin and mercury residues. *Journal of Zoology (London)* **171**, 513-539.

Luttik, R; De Snoo, G.R. (1994) Environmental hazard assessment of the use of pesticides for seed treatment, the dutch concept. *this symposium*.

Maze, R.C.; Atkins, R.P.; Mineau, P.; Collins, B.T. (1991) Measurements of surface pesticide residue in seeding operations. *Transactions of the American Society of Agricultural Engineers* **34**, 795-799.

Murton, R.K.; Vizoso, M. (1963) Dressed cereal seed as a hazard to wood-pigeons. *Annals of Applied Biology* **52**, 503-517.

Porter, R.E.R. (1977) Methiocarb protects sprouting peas from small birds. *New Zealand Journal of Experimental Agriculture* **5**, 335-338.

Riedel, B.; Wolf, K.; Litzbarski, H. (1992) Die Anwendung von inkrustiertem Rapsaatgut aus der Sicht des Vogelschutzes, In: *Pflanzenschutzmittel und Vogelgefährdung*, H. Gemmeke and H. Ellenberg (Eds), Berlin-Dahlem: Mitteilungen aus der Biologische Bundesanstalt für Land- und Forstwirtschaft Heft 280, pp. 131-137.

Westlake, G.E.; Blunden, C.A.; Brown, P.M.; Bunyan, P.J.; Martin, A.D.; Sayers, P.E., Stanley, P.I.; Tarrant, K.A. (1980) Residues and effects in mice after drilling wheat treated with chlorfenvinphos and an organomercurial fungicide. *Ecotoxicology and environmental safety* **4**, 1-16.

Wildlife Survey (1975) *Guazatine/Imazalil-1975 wildlife survey*. Internal report Murphy Chemical Limited (Archives National Institute of Public Health and Environmental Protection), 29 pp.

ENVIRONMENTAL HAZARD ASSESSMENT OF THE USE OF PESTICIDES FOR SEED TREATMENT: THE DUTCH CONCEPT

R. LUTTIK

National Institute of Public Health and Environmental Protection, P.O. Box 1, 3720 BA Bilthoven, The Netherlands

G.R. DE SNOO

Centre of Environmental Science, Leiden University, P.O. Box 9518, 2300 RA Leiden, The Netherlands

ABSTRACT

In this article a method is presented to assess the hazard/risk for birds and mammals of the use of pesticides for seed treatment. The method takes into account: body weight, daily food intake and diet of the species of concern, percentage of incorporation of the seeds into the soil and when applicable (not in this paper) repellency of the pesticide. The application rate of the active ingredient, sowing density and mean weight of 1 seed are according to the Dutch situation. The scheme has been applied for all pesticides (35 active ingredients in 1992) used for seed treatment in the Netherlands. High risk had to be assumed for 11 compounds and 2 additional combinations of two compounds for mammals and/or birds for 1 or several crops. For 5 of the compounds with high risk incidents have been reported in Great Britain or the Netherlands.

INTRODUCTION

"In about 1650 a shipload of wheat was beached by a high tide on the Somerset coast. Its cargo was salvaged, dried out with lime, and used as seed by local farmers. To their delight those crops suffered far less from stinking smut than their neighbours'. The practice of steeping seed in brine and drying it with lime, as a precaution against the disease, evolved as a result" (Anonymous, 1992).

The treatment of seeds for protection against fungi or invertebrates is a long existing phenomenon and meanwhile all kind of ingredients are used. At the end of the sixties the use of seed treatment became suspicious as a result of large bird kills, due to the use of organochlorines such as dieldrin and heptachloro-epoxide against insects and organomercurials against fungi (for instance Fuchs, 1967; Blus et al., 1984). The use of organochlorines and organomercurials was banned and the introduction of organophosphates in the beginning of the seventies resulted in a lot of new incidents but mostly with less victims. In 1992 35 active ingredients of different chemical groups were used for seed treatment in the Netherlands (Anonymous, 1991). To evaluated the hazard of these components for birds and mammals a step wise assessment method has been developed. For all active ingredients (and/or combinations) used in the Netherlands the hazard has been calculated for every crop in which the pesticide is allowed. This research is part of a larger project in which the hazards of the different application methods of pesticides for birds and mammals will be evaluated (Luttik, 1992).

In this article the stepwise hazard assessment method is presented. First the basic principles underlying the hazard assessment method are given. Afterwards the hazard assessment method itself will be explained and illustrated by an example. In the end the results of the calculations are presented and discussed.

METHOD

Basic principles

In order to carry out the hazard assessment properly, it is necessary to have information about the following subjects: acute oral toxicity, body weight (BW) and daily food intake (DFI) of bird and mammalian species, application rate of active ingredient (kg a.i./kg seeds), amount of seed (kg seed/ha), mean weight of 1 seed, efficiency of incorporation of seeds into the soil, repellency, and last but not least ecological knowledge about the bird and mammalian species used in the hazard assessment.

For the hazard assessment the lowest available LD50 has to be used. The data submitted by industry were supplemented by on-line research. Because LD50 values are conventionally expressed as mg/kg BW, it is necessary for this scheme to adjust these units to take account of the body weight of the species of concern.

The concentration of the active ingredient or a combination of several active ingredients per 1 particle (A in scheme) can be calculated when the application rate of an active ingredient and the mean weight of 1 seed is known. The number of particles per m^2 available at the surface (K in scheme) can be calculated when the amount of seeds used per hectare and the mean weight of 1 seed is known. This number can be corrected for the percentage of incorporation. The hazard assessment is carried out by using data applicable to the Dutch situation (good agricultural practice). Not much information about the percentages of incorporation of seed is available, for the time being the following percentages are used in the Netherlands (see also Tamis *et al.*, these proceedings): 99% in case of favourable sowingconditions, 95% in case of less favourable sowing conditions and 90% for the edge (turning point of sowing machines) of the field.

The PEC(seed) is the predicted environmental concentration of a compound in seed (mg a.i./kg seed) which is adjusted by the DFI for the species of concern. Preferably information about the DFI of wild species has to be used for the calculations. As a broad generalisation, it is sometimes assumed that small species (less than 100 g) eat about 30% of their body weight daily (dry weight basis), whereas larger species eat about 10% (Kenaga, 1973). More accurate predictions (used for the results in this paper) of the DFI are available from Nagy (1987), using regression equations to predict dry weight intake for an animal of a particular body weight:
Birds   Log DFI = -0.188 + 0.651 log BW (n=50, r^2 = 0.92) and
Mammals   Log DFI = -0.629 + 0.822 log BW (n=46, r^2 = 0.96).

<u>Method for hazard assessment</u>

The decision scheme for environmental hazard assessment of the use of pesticides for seed treatment is presented in Figure 1. In the first step of this scheme the amount of active ingredient in/on 1 coated or pillorized seed is compared with the LD50 of the species of concern. When the quotient is ≥ 1, high risk is assumed for the species of concern, which means that for the species 50% or more of the animals will die after consumption of 1 particle. This cut-off criterion is also adopted by the Subgroup on Terrestrial Vertebrates of the EPPO.

In step 2 it is assumed that the complete daily food intake (DFI) of the species of concern consists of the particles under consideration. When the quotient is ≤ 0.001, low risk for birds and mammals is assumed.

In step 3 one has to decide if the particles do resemble natural food or grit. Differentiation is necessary because it is assumed that natural food will be eaten until the bird or mammal is saturated, in contrary to grit consumption (not *ad libitum*).

In step 4 a trigger-value of 20 particles has been chosen for grit consumption. This trigger-value is based on research carried out by Best and Gionfriddo (1991a and 1991b). They found that the mean number of particles in the stomach of birds ranged between 0 and 70 (*Phasianus colchicus* 38 and *Passer domesticus* 69). Only for one species (*Passer domesticus*) the half-life of the grit particles is approximately known: $T\frac{1}{2} = 3$ days. When the quotient is ≤ 20 risk is assumed to be present. The "real" risk has to be assessed by comparing the characteristics of the particles and of natural grit.

In step 5 it is supposed that the risk for birds and mammals is related to the amount of available active ingredient per unit area. High risk is assumed when the quotient is ≥ 10 (this criterion is the same as the one used by the EPA, 1992), and low risk when the quotient is ≤ 0.1.

Corrections can be made for the repellency (when applicable). In step 1 one must not take repellency into consideration, because repellency is often a learning process, which can not be applicable for 1 particle. In the results presented in table 1 no correction has been carried out for repellency, because repellency studies are often not available and are difficult to translate to the environment (see Luttik, 1993 in prep.).

As example of the calculations in the scheme the case of a harvest mouse (*Micromys minutus*) of 8 gram, eating oats treated with fonofos is presented.
Basic data:
 LD50(rat) = 3.16 - 24.5 mg/kg BW (n=7); application rate of formulation = 4 ml/kg oat (250 g a.i./l); 24-33 seeds/g and 330 seeds/m^2.
Hazard assessment:
 LD50(harvest mouse) = 0.025 mg/mouse. A = 0.03-0.042 mg a.i./seed. K for 90% incorporation = 33 seeds/m^2. Outcome of criterion 1 = 0.6-0.83 ==> high risk. Outcome of criterion 5b = 40-55 ==> high risk.

FIGURE 1   Decision scheme for pesticides used for seed treatment.

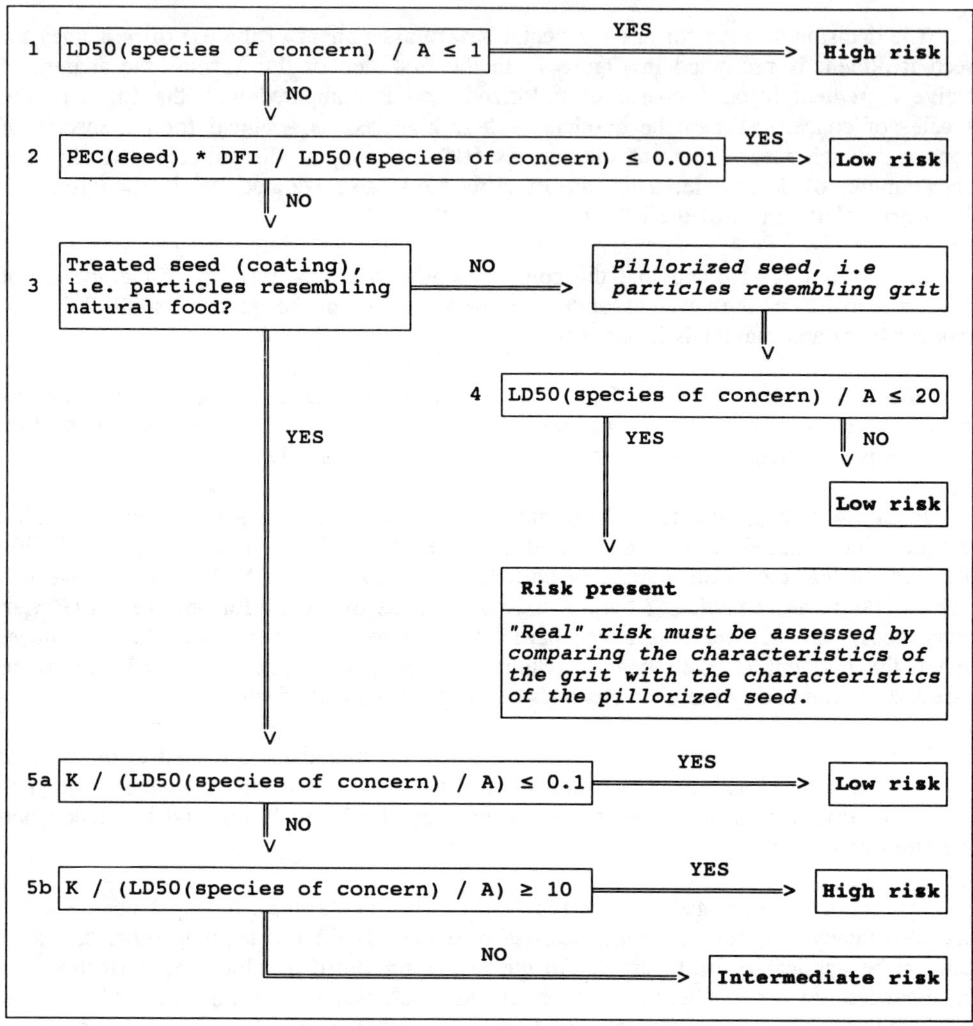

RESULTS

The scheme has been applied for all pesticides (35 active ingredients) used for seed treatment in the Netherlands. High risk had to be assumed for 11 compounds and 2 additional combinations of 2 compounds for mammals and/or birds for 1 or several crops (see Table 1). The results of all other calculations and the exact data used for the hazard assessment will be presented in Luttik & de Snoo (1993, in prep.). In Table 1 the body weight for birds and mammals at which the quotient of LD50 species of concern/A (criterion 1) is equal to 1 and at which the quotient K/(LD50 species of concern/A) is equal to 10 (criterion 5b) for 90% incorporation of the seeds into the soil. It is assumed that small birds (≤ 100 g) will not eat large seeds (maize, peas and beans) ad libitum and that the smallest seed eating mammal (6 g) will eat large seeds. For instance the outcome of criterion 1 for birds exposed to bendiocarb on maize is 430 gram. This means that for all the birds with a body weight $\leq$ 430 high risk must be assumed. For 5 of the compounds (bold in table) with "High risk" incidents have occurred in Great Britain: bendiocarb, chlorfenvinphos, fonofos and lindane (based on incident registration in the UK between 1983 and 1991) and in the Netherlands: methiocarb (written communication of the Dutch Central Veterinary Institute).

TABLE 1   Chemical/crop combinations in the Netherlands for which the outcome of the risk assessment scheme is "High risk".

Compound	Crop	Criterion 1		Criterion 5b	
		Birds	Mammals	Birds	Mammals
Antraquinone	Broad bean	11	--	--	--
Bendiocarb	Beet	34	--	--	--
	Maize	430	48	--	--
Benfuracarb	Beet	11	--	--	--
	Onion	--	--	98	28
Benomyl	Broad been	29	--	--	--
Dichlofenthion	Beans	36-107	9-12	--	--
Fonofos	Barley	--	16	14	44
	Oats	--	13	14	44
	Rye	--	9	--	27
	Wheat	--	14	15	47
Furathiocarb	Beet	375	11	--	--
	Onion	100	--	3050	92
Isofenphos	Onion	--	--	184	80
Lindane	Black radish	--	17	--	17
	Maize	--	8	--	--
	Rape-seed	--	8	--	11
	Turnip	--	6	--	6
	Mustard	--	8	--	10
Methiocarb	Beet	63	8	--	--
	Beetroot	84	11	35	8
	Green pea	376	50	338	45
	Maize	1253	167	125	17
	Pea	752	100	451	60
Thiram	Beans	10-15	7	--	--
	Pea	12	6	--	--
	Spinage	--	--	23	11
Chlorfenvinphos /thiram	Carrot	10	--	--	--
Benomyl/thiram	Green pea	10	--	--	--

DISCUSSION

In the presented hazard assessment method some gaps can be pointed out. First the exposure of birds and mammals to the amount of available seed after sowing is not clear for all crops and sowing methods from literature. Secondly the amount of grit in the stomach of several european birds species is not known. A field study to investigate both aspects is being carried out at the moment (see Tamis *et al.*, these proceedings). Further point of discussion is the use of the cut-off criteria in the scheme. As far as possible they are based on literature and conventions of international organizations. But for example the criterion of 0.001% in step 2 is maybe too strict.

From the results it is concluded that of the 35 a.i.s used for seed treatment in the Netherlands, 11 a.i.s and 2 combinations result in a high risk for wild living birds and mammals in certain crops. In some cases high risks were calculated for all bird species and for many mammals. 10 of the 11 a.i.s result in a high risk when a bird or mammal eats one seed only! Some of these pesticides are used in large scale crops such as beet and maize. The indications from the reported incidents in the UK and the Netherlands show that the outcome of the hazard assessment is realistic.

REFERENCES

Anonymous, 1991. *Gewasbeschermingsgids 1991 [Dutch Crop Protection Guide 1991].* Ed. W.C.A. van Geel, IKC-AT/PD. Wageningen, pp. 606.

Anonymous, 1992. Post-mercury seed treatment choices. *Farmers Weekly* **116** (6) Fungicides Supplement, 8-9.

Best, L.B.; Gionfriddo, J.P. (1991a) Integrity of five granular insecticide carriers in house sparrow gizzards. *Environmental Toxicology and Chemistry* **10**, 1487-1492.

Best, L.B.; Gionfriddo, J.P. (1991b) Characterization of grit use by cornfield birds. *Wilson Bulletin* **103**, 68-82.

Blus, L.J., C.J. Henny, D.J. Lenhart and T.E. Kaiser (1994). Effects of heptachlor- and lindane-treated seed on Canada Geese. *J. Wildlife Management* **48** (4): 1097-1111.

EPA (1992) *Comparative analysis of acute avian risk from granular pesticides.* Office of pesticides Programs U.S. Environmental Protection Agency, Washington, D.C., pp. 71.

Fuchs, P. (1967). Death of birds caused by application of seed dressings in the Netherlands. *Mededelingen van de Faculteit Landbouwetenschappen Rijksuniversiteit Gent* **32**, (3/4): 855-859.

Kenaga, E.E. (1973) Factors to be considered in the evaluation of toxicity of pesticides to birds in their environment. *Environmental Quality and Safety.* Academic Press, New York, II, pp. 166-181.

Luttik, R. (1992). *Environmental hazard/risk assessment of pesticides used in agriculture for birds and mammals. The Dutch concept. Part 1 Introduction and synopsis of the decision scheme.* RIVM Report 679101006, pp. 30.

Luttik, R. (1993, in prep). *Environmental hazard/risk assessment of pesticides used in agriculture for birds and mammals. The Dutch concept. Part 3 Avian food avoidance behaviour.* RIVM Report 679101012. December 1993.

Luttik, R. & G.R. de Snoo (1993, in prep). *Environmental hazard/risk assessment of pesticides used in agriculture for birds and mammals. The Dutch concept. Part 2 Exposure by pesticides used for seed treatment.* RIVM Report 679101007. December 1993.

Nagy, K.A. (1987) Field metabolic rate and food requirement scaling in mammals and birds. *Ecological Monographs* **57** (2), 111-128.